Springer Proceedings in Mathematics & Statistics

Volume 312

Springer Proceedings in Mathematics & Statistics

This book series features volumes composed of selected contributions from workshops and conferences in all areas of current research in mathematics and statistics, including operation research and optimization. In addition to an overall evaluation of the interest, scientific quality, and timeliness of each proposal at the hands of the publisher, individual contributions are all refereed to the high quality standards of leading journals in the field. Thus, this series provides the research community with well-edited, authoritative reports on developments in the most exciting areas of mathematical and statistical research today.

More information about this series at http://www.springer.com/series/10533

Martin Bohner · Stefan Siegmund ·
Roman Šimon Hilscher · Petr Stehlík
Editors

Difference Equations and Discrete Dynamical Systems with Applications

24th ICDEA, Dresden, Germany, May 21–25, 2018

Springer

Editors
Martin Bohner
Department of Mathematics and Statistics
Missouri University of Science
and Technology
Rolla, MO, USA

Roman Šimon Hilscher
Department of Mathematics
and Statistics, Faculty of Science
Masaryk University
Brno, Czech Republic

Stefan Siegmund
Institute for Analysis, CfD
Dresden University of Technology
Dresden, Germany

Petr Stehlík
Department of Mathematics
University of West Bohemia
Pilsen, Czech Republic

ISSN 2194-1009 ISSN 2194-1017 (electronic)
Springer Proceedings in Mathematics & Statistics
ISBN 978-3-030-35501-2 ISBN 978-3-030-35502-9 (eBook)
https://doi.org/10.1007/978-3-030-35502-9

Mathematics Subject Classification (2010): 34NXX, 37-XX, 39-XX

This Springer imprint is published by the registered company Springer Nature Switzerland AG
The registered company address is: Gewerbestrasse 11, 6330 Cham, Switzerland

Conference Committees

Local Organizing Committee

Stefan Siegmund, Technical University Dresden, Germany (Chair)
Christine Heinke, Technical University Dresden, Germany
Norbert Koksch, Technical University Dresden, Germany
Sebastian Mildner, Technical University Dresden, Germany
Petr Stehlík, University of West Bohemia, Czech Republic
Sascha Trostorff, Technical University Dresden, Germany
Hannes Weiße, Technical University Dresden, Germany

Scientific Committee

Zuzana Došlá, Masaryk University, Czech Republic (Chair)
Martin Bohner, Missouri University of Science and Technology, USA
Saber Elaydi, Trinity University, USA
Małgorzata Guzowska, University of Szczecin, Poland
Christian Pötzsche, University of Klagenfurt, Austria
Adina L. Sasu, West University of Timişoara, Romania
Stefan Siegmund, Technical University Dresden, Germany
Hal Smith, Arizona State University, USA
Petr Stehlík, University of West Bohemia, Czech Republic

Preface to Proceedings of ICDEA 2018

These proceedings of the 24th International Conference on Difference Equations and Applications cover the theory and applications of difference equations and discrete dynamical systems. The conference was held at the Institute of Analysis at the Technical University Dresden (Saxony, Germany), under the auspices of the International Society of Difference Equations (ISDE) May 21–25, 2018. Its purpose was to bring together renowned researchers working actively in the respective fields, to discuss the latest developments, and to promote international cooperation on the theory and applications of difference equations. The main topics in ICDEA 2018 were difference equations, discrete dynamical systems, discrete biomedical models, and discrete models in natural science and social sciences. More than 100 participants attended the 84 talks of the conference, including 10 plenary talks given by invited speakers.

These proceedings contain 16 articles written by participants of ICDEA 2018. Each manuscript underwent a rigorous refereeing process to ensure scientific quality. Three of the articles are prepared by the plenary speakers Elena Braverman, Bernd Krauskopf, and Erik Van Vleck. This book will appeal to researchers and scientists working in the fields of difference equations, discrete dynamical systems, and their applications. We would like to take this opportunity to give our special thanks to all the participants for their active contributions to the success of ICDEA 2018. Our gratitude and appreciation go to the organizers for their efforts that made possible the success of the conference, the members of the scientific committee who ensured the high standards of the conference scientific activities, the administration of the Institute of Analysis of Faculty of Mathematics at the Technical University Dresden for providing its facilities and resources to the conference participants, and last but not least the sponsors "Deutsche Forschungsgemeinschaft (DFG)" and "Gesellschaft von Freunden und Förderern der TU Dresden e.V." for their financial support.

Rolla, USA	Martin Bohner
Dresden, Germany	Stefan Siegmund
Brno, Czech Republic	Roman Šimon Hilscher
Pilsen, Czech Republic	Petr Stehlík
June 2019	

Contents

Papers by Plenary Speakers

Linearized Oscillation Theory for a Nonlinear Nonautonomous Difference Equation

Elena Braverman and Başak Karpuz

Abstract We review some theorems and mistakes in linearized oscillation results for difference equations with variable coefficients and constant delays, as well as develop linearized oscillation theory when delays are also variable. Main statements are applied to discrete models of population dynamics. In particular, oscillation of generalized Pielou, Ricker and Lasota–Wazewska equations is considered.

Keywords Difference equations · Variable delays · Linearized oscillation · Pielou equation · Ricker model · Lasota–Wazewska equation

1 Introduction

Linearized theory usually relates properties of nonlinear equations to their linearized versions. Results connecting oscillation properties of a nonlinear delay difference equation

$$x(n+1) - x(n) + \sum_{k=1}^{m} r_k(n) f_k[x(n-\tau_k)] = 0 \quad \text{for } n \geq n_0, \tag{1}$$

The authors were partially supported by NSERC, grant RGPIN-2015-05976.
Elena Braverman acknowledges the support of both ICDEA-2018 in Dresden and NSERC, grant RGPIN-2015-05976.

E. Braverman (✉)
Department of Mathematics and Statistics, University of Calgary, 2500 University Dr. NW, Calgary, AB T2N 1N4, Canada
e-mail: maelena@ucalgary.ca

B. Karpuz
Department of Mathematics, Dokuz Eylül University, Buca, 35160 İzmir, Turkey
e-mail: bkarpuz@gmail.com

M. Bohner et al. (eds.), *Difference Equations and Discrete Dynamical Systems with Applications*, Springer Proceedings in Mathematics & Statistics 312,
https://doi.org/10.1007/978-3-030-35502-9_1

where $\{r_k(n)\} \subset \mathbb{R}_0^+ := [0, \infty)$, $\tau_k \in \mathbb{N}_0 := \{0, 1, \ldots\}$, $k = 1, 2, \ldots, m$, with the linear equation

$$x(n+1) - x(n) + \sum_{k=1}^{m} r_k(n)x(n-\tau_k) = 0 \quad \text{for } n \geq n_0 \tag{2}$$

go back to 1990ies [14]. Here in (1), $uf_k(u) > 0$ for $u \in \mathbb{R}\backslash\{0\}$ and either $\liminf_{u\to 0} \frac{f_k(u)}{u} = 1$ or $|f_k(u)| \geq |u|$ in some neighborhood of the origin for any $k = 1, 2, \ldots, m$.

In [14, Theorem 6], under natural assumptions, Yan and Qian stated that oscillation of (2) implies oscillation of (1). However, in [10, p. 478], Tang and Yu disproved the statement by Yan and Qian by constructing a counterexample in the case of a single constant delay to show that (1) may have a nonoscillatory solution while every solution of (2) is oscillatory. This happens in the critical case when the variable coefficient is close to the boundary of the oscillation domain, see [11] for more details.

It is mentioned in [9] that to explore linearized oscillation of the autonomous equation

$$x(n+1) - x(n) + \sum_{k=1}^{m} p_k f_k[x(n-\tau_k)] = 0 \quad \text{for } n \geq n_0, \tag{3}$$

it is sufficient to imply limitations on nonlinear functions in a small neighbourhood of zero.

Proposition 1 ([9, p. 570]) *Assume that $p_k \in \mathbb{R}^+ := (0, \infty)$, $f_k \in C(\mathbb{R}, \mathbb{R})$ satisfies $uf_k(u) > 0$ for all $u \in (-\delta, \delta)\backslash\{0\}$, where $\delta \in \mathbb{R}^+$, and $\lim_{u\to\infty} \frac{f_k(u)}{u} = 1$ for $k = 1, 2, \ldots, m$. Every solution of (3) is oscillatory if and only if every solution of the autonomous equation*

$$x(n+1) - x(n) + \sum_{k=1}^{m} p_k x(n-\tau_k) = 0 \quad \textit{for } n \geq n_0 \tag{4}$$

is oscillatory.

However, a slight modification of [9, p. 574] disproves this statement.

Example 1 Let $\tau \in \mathbb{N}$ and $p \in [1, \infty)$. Consider the nonlinear equation

$$x(n+1) - x(n) + p\sin\big(x(n-\tau)\big) = 0 \quad \text{for } n \geq 0. \tag{5}$$

As any eventually positive solution of the linearized equation

$$x(n+1) - x(n) + px(n-\tau) = 0 \quad \text{for } n \geq 0$$

should be monotone decreasing for $n \geq n_0$, we get $x(n_0+1) < 0$. Similarly, there are no eventually negative solutions. However, (5) has an infinite number of constant nonoscillatory solutions $\{x(n)\} = \{j\pi\}$, $j = \pm 1, \pm 2, \ldots$, both positive and negative.

Reducing conditions to a small neighbourhood of zero as in Proposition 1 is allowed only when all nonoscillatory solutions $\{x(n)\}$ tend to zero as $n \to \infty$.

The purpose of the present paper is to establish connections between oscillation properties of the nonlinear equation with variable coefficients and delays

$$x(n+1) - x(n) + \sum_{k=1}^{m} r_k(n) f_k\big[x\big(h_k(n)\big)\big] = 0 \quad \text{for } n \geq n_0 \tag{6}$$

and the linear equation

$$x(n+1) - x(n) + \sum_{k=1}^{m} r_k(n) x\big(h_k(n)\big) = 0 \quad \text{for } n \geq n_0, \tag{7}$$

as well as to apply the obtained results to some models of population dynamics models.

We consider (6) under some the following assumptions:

(A1) For $k = 1, 2, \ldots, m$, $\{r_k(n)\} \subset \mathbb{R}_0^+$.
(A2) For $k = 1, 2, \ldots, m$, $\{h_k(n)\} \subset \mathbb{Z}$, $h_k(n) \leq n$, $n \geq n_0$ and $\lim\limits_{n\to\infty} h_k(n) = \infty$.
(A3) For $k = 1, 2, \ldots, m$, $f_k \in \mathrm{C}(\mathbb{R}, \mathbb{R})$ satisfies $u f_k(u) > 0$ for all $u \in \mathbb{R}\backslash\{0\}$.
(A4) $\lim\limits_{u\to 0} \dfrac{f_k(u)}{u} = 1$ for $k = 1, 2, \ldots, m$.
(A5) There exists $\delta \in \mathbb{R}^+$ such that either

$$0 \leq f_k(u) \leq u \quad \text{for all } u \in [0, \delta] \text{ and } k = 1, 2, \ldots, m, \tag{8}$$

or

$$0 \geq f_k(u) \geq u \quad \text{for all } u \in [-\delta, 0] \text{ and } k = 1, 2, \ldots, m. \tag{9}$$

For all results concerning (6), we assume that (A1) and (A2) hold. Define

$$n_{-1} := \min_k \min\{h_k(n) : n \geq n_0\},$$

which exists and is finite by (A2). By a solution of (6), we mean a sequence $\{x(n)\}_{n=n_{-1}}^{\infty}$ for which

$$x(n+1) = x(n) - \sum_{k=1}^{m} r_k(n) f_k\big[x\big(h_k(n)\big)\big] \quad \text{for } n = n_0, n_0+1, \ldots .$$

It is well known that (6) has a unique solution satisfying the initial condition

$$x(n) = \varphi_{n-n_{-1}} \quad \text{for } n = n_{-1}, n_{-1}+1, \dots, n_0,$$

where $\varphi_0, \varphi_1, \dots, \varphi_{n_0-n_{-1}}$ are prescribed real numbers.

A solution $\{x(n)\}$ of (6) is said to be *oscillatory* if $x(n)$ are neither eventually positive nor eventually negative. Equation (6) is *oscillatory* if all its solutions are oscillatory. Otherwise, (6) is called *nonoscillatory*.

After linearization, we have to apply oscillation results for linear equation (7), so we present below some of them.

Proposition 2 ([7, Corollary 7.1.1]) *Assume that $p_k \in \mathbb{R}$ and $\tau_k \in \mathbb{Z}$ for $k = 1, 2, \dots, m$. Linear autonomous equation* (4) *is oscillatory if and only if the characteristic equation*

$$\lambda - 1 + \sum_{k=0}^{m} p_k \lambda^{-\tau_k} = 0 \tag{10}$$

has no positive roots.

Proposition 3 ([12, Theorem 1]) *Let (A1) and (A2) hold and*

$$\liminf_{n\to\infty} \inf_{\lambda\in(0,1)} \left\{ \sum_{k=1}^{m} \frac{r_k(n)}{\lambda(1-\lambda)^{n-h_k(n)}} \right\} > 1.$$

Then, (7) *is oscillatory.*

Proposition 4 ([15, Corollary 3]) *Assume that (A1) and (A2) are satisfied and there exist $\lambda_0 \in (0, 1)$ and $n_1 \geq n_0$ such that*

$$\sum_{k=1}^{m} \frac{r_k(n)}{\lambda_0(1-\lambda_0)^{n-h_k(n)}} \leq 1 \quad \text{for } n \geq n_1.$$

Then, (7) *is nonoscillatory.*

Denoting

$$\{\mu(n)\} = \left\{ \frac{u(n)}{1-u(n)} \right\}$$

in [3, Theorem 2.1], we get the following.

Proposition 5 ([15, Corollary 3]) *If (A1) and (A2) hold, the following statement are equivalent.*

(i) Equation (7) *is nonoscillatory.*
(ii) There exists $n_0 \in \mathbb{Z}$ such that the inequality

$$x(n+1) - x(n) + \sum_{k=1}^{m} r_k(n) x\big(h_k(n)\big) \leq 0$$

has an eventually positive solution for $n \geq n_0$ *and/or*

$$x(n+1) - x(n) + \sum_{k=1}^{m} r_k(n) x\big(h_k(n)\big) \geq 0$$

has an eventually negative solution for $n \geq n_0$.

(iii) *There exist* $n_0 \in \mathbb{Z}$ *and a sequence* $\{\mu(n)\} \subset \mathbb{R}_0^+$ *such that*

$$\mu(n) \geq \sum_{k=1}^{m} r_k(n) \prod_{j=h_k(n)}^{n} [1 + \mu(j)] \quad \text{for } n \geq n_1, \tag{11}$$

where $n_1 \geq n_0$ *satisfies* $h_k(n) \geq n_0$ *for all* $n \geq n_1$ *and* $k = 1, 2, \ldots, m$.

Remark 1 Proposition 4 follows from Proposition 5 with $\{\mu(n)\} \equiv \left\{\dfrac{\lambda_0}{1 - \lambda_0}\right\}$.

Next, we quote an oscillation result for nonlinear equation (6).

Proposition 6 ([7, Theorem 7.4.1]) *Assume that (A4) and (A5) hold,* $p_k \in \mathbb{R}^+$ *and* $\tau_k \in \mathbb{N}_0$ *for* $k = 1, 2, \ldots, m$. *The autonomous nonlinear equation*

$$x(n+1) - x(n) + \sum_{k=1}^{m} p_k f_k[x(n - \tau_k)] = 0 \quad \text{for } n \geq n_0 \tag{12}$$

is oscillatory if and only if the autonomous linear equation

$$x(n+1) - x(n) + \sum_{k=1}^{m} p_k x(n - \tau_k) = 0 \quad \text{for } n \geq n_0$$

is oscillatory.

Proposition 7 *Assume (A4) and (A5). The autonomous nonlinear equation* (12) *is oscillatory if and only if the characteristic equation* (10) *has no positive roots.*

By analyzing the characteristic equation

$$\lambda - (1 - p) + q\lambda^{-\tau} = 0,$$

we get the following result.

Proposition 8 *Assume* $f, g \in \mathrm{C}(\mathbb{R}, \mathbb{R})$ *satisfy (A4) and (A5),* $p \in [0, 1)$, $q \in \mathbb{R}_0^+$ *and* $\tau \in \mathbb{N}_0$. *The autonomous nonlinear equation*

$$x(n+1) - x(n) + pf[x(n)] + qg[x(n - \tau)] = 0 \quad \text{for } n \geq n_0$$

is oscillatory if and only if

$$q(\tau+1)^{\tau+1} > (1-p)^{\tau+1}\tau^{\tau}.$$

Further, Sect. 2 contains known and some new auxiliary statements that will be required to prove main results in Sect. 3. In Sect. 4, we apply linearization theorems to equations of mathematical biology generalizing Pielou's equation, Ricker's model and Lasota–Wazewska equation. Some final comments are presented in Sect. 5.

2 Auxiliary Results

We will assume without further mentioning that (A1)–(A3) hold. Let us start with the statement that under (A1)–(A3), unlike Example 1, there is a nonoscillatory solution which does not tend to zero if and only if the series of the sum of coefficients converges.

Theorem 1 *The following statements are equivalent.*

(i) $\sum_{j}^{\infty}\sum_{k=1}^{m} r_k(j) < \infty.$

(ii) Equation (6) *has a nonoscillatory solution* $\{x(n)\}$ *such that* $\lim_{n\to\infty} x(n) \neq 0.$

Proof (i) $\implies$ (ii): Pick $L>0$ and denote $M := \max_{k} \max_{L\le x\le 2L} \{f_k(x)\}$. Since the series $\sum_{j}^{\infty}\sum_{k=1}^{m} r_k(j)$ converges, we can find $n_1 \ge n_0$ such that

$$\sum_{j=n}^{\infty}\sum_{k=1}^{m} r_k(j) \le \frac{L}{M} \quad \text{for } n \ge n_1.$$

By (A2), there is $n_2 \ge n_1$ such that $h_k(n) \ge n_1$ for $n \ge n_2$. Define $x_0(n) :\equiv 1$ for $n \ge n_1$ and $\{x_\ell(n)\}$ for $\ell \in \mathbb{N}$ as

$$x_\ell(n) := \begin{cases} L + \sum_{j=n}^{\infty}\sum_{k=1}^{m} r_k(j) f_k\big[x_{\ell-1}\big(h_k(j)\big)\big], & n \ge n_2, \\ 2L, & n_1 \le n \le n_2. \end{cases}$$

Inductive arguments yield that $2L \ge x_\ell(n) \ge x_{\ell+1}(n) \ge L$ for $n \ge n_1$ and $\ell \in \mathbb{N}$. Define $\{x(n)\}$ by $x(n) := \lim_{\ell\to\infty} x_\ell(n)$ for $n \ge n_1$.

Then, we see that $\{x(n)\}$ is a positive solution of (6) satisfying $\lim_{n\to\infty} x(n) = L > 0$.

(ii) $\Longrightarrow$ (i): We may suppose without loss of generality that $\{x(n)\}$ is an eventually positive solution of (6) such that $\lim_{n\to\infty} x(n) \neq 0$. By (A1)—(A3) and (6), $\{x(n)\}$ is eventually nonincreasing. Then, $\lim_{n\to\infty} x(n) =: L > 0$. We can find $n_1 \geq n_0$ such that $x(n) > 0$ and $\frac{L}{2} \leq x(h_k(n)) \leq \frac{3L}{2}$ for all $n \geq n_1$. Set $m := \min_k \min_{|L-x|\leq\frac{L}{2}} \{f_k(x)\}$, then $m > 0$ by (A3). Summing (6) from n_1 to $(n-1)$, we get

$$0 = x(n) - x(n_1) + \sum_{j=n_1}^{n-1}\sum_{k=1}^{m} r_k(j) f_k\big[x\big(h_k(n)\big)\big] \quad \text{for all } n \geq n_1,$$

which yields

$$\sum_{j=n_1}^{n-1}\sum_{k=1}^{m} r_k(j) \leq \frac{x(n_1)}{m} \quad \text{for all } n \geq n_1.$$

This proves (i) provided that $\{x(n)\}$ is eventually positive. The case of $\{x(n)\}$ being eventually negative is similar and thus is omitted.

Let us illustrate that in Theorem 1 the limit assumption on variable delays in (A2) is necessary, as well as continuity and the sign condition on f in (A3), with two examples.

Example 2 The equation

$$x(n+1) - x(n) + x(-1) = 0 \quad \text{for } n \geq 0$$

with the initial conditions $x(-1) = -1$ and $x(0) = 0$ has an eventually positive solution $\{x(n)\} = \{n\}$, which does not tend to zero, because the delay obviously does not satisfy $\lim_{n\to\infty} h(n) = \infty$ of (A2).

Example 3 For the equation

$$x(n+1) - x(n) + 2f[x(n-1)] = 0 \quad \text{for } n \geq 0, \tag{13}$$

where f is either

$$f_1(u) := \begin{cases} u, & u \leq 1 \\ \frac{1}{8}(u-1), & u > 1 \end{cases} \quad \text{or} \quad f_2(u) := \begin{cases} \frac{1}{2} - \left|\frac{1}{2} - u\right|, & u \leq 1 \\ \frac{1}{8}(u-1), & u \geq 1, \end{cases}$$

all the conditions but one (continuity for f_1 or sign condition for f_2) in (A3) are satisfied. However, all solutions of its linearized counterpart are obviously oscillatory, while $\{x(n)\} = \{1 + 2^{-n}\}$ is a positive solution of (13) with the initial conditions $x(-1) = 3$ and $x(0) = 2$ since

$$x(n+1) = 1 + 2^{-(n+1)} = 1 + 2^{-n} - \frac{1}{2}2^{-n} = x(n) - 2f_1[x(n-1)] \quad \text{for } n = 0, 1, \ldots .$$

With $f = f_2$, it is easy to verify that $\{x(n)\} = \{1\}$ is also a solution of (13) with the initial conditions $x(-1) = 1$ and $x(0) = 1$.

Now, consider the condition

$$\sum_{j}^{\infty} \sum_{k=1}^{m} r_k(j) = \infty. \tag{14}$$

Corollary 1 *Every nonoscillatory solution $\{x(n)\}$ of* (6) *tends to zero as $n \to \infty$ if and only if* (14) *holds.*

Remark 2 If (6) is oscillatory, then (14) holds.

Lemma 1 *Assume that every nonoscillatory solution $\{x(n)\}$ of* (7) *satisfies $\lim_{n\to\infty} x(n) = 0$. Then, there exists a solution $\{\mu(n)\}$ of inequality* (11) *satisfying*

$$\lim_{n\to\infty} \left(\prod_{j}^{n-1} [1 + \mu(j)] \right)^{-1} = 0.$$

Proof Without loss of generality, let $\{x(n)\}$ be an eventually positive solution of (7) such that $x(n) \to 0$ as $n \to \infty$. Then, we can find $n_1 \geq n_0$ such that $x(n) > 0$ and $x(h_k(n)) > 0$ for all $n \geq n_1$ and $k = 1, 2, \ldots, m$. From (7), $x(n+1) \leq x(n)$ for all $n \geq n_1$. Now, define $\mu(n) := \dfrac{x(n)}{x(n+1)} - 1 \geq 0$ for $n \geq n_1$, then

$$x(n) = x(n_1) \left(\prod_{j=n_1}^{n-1} [1 + \mu(j)] \right)^{-1} \quad \text{for } n \geq n_1. \tag{15}$$

Substituting (15) into (7), we get

$$\mu(n) = \sum_{k=1}^{m} r_k(n) \prod_{j=h_k(n)}^{n} [1 + \mu(j)] \quad \text{for } n \geq n_2,$$

where $n_2 \geq n_1$ is such that $h_k(n) \geq n_1$ for all $n \geq n_1$ and $k = 1, 2, \ldots, m$, i.e., $\{\mu(n)\}$ satisfies (11) with equality. Further, we have

$$\lim_{n\to\infty} \left(\prod_{j=n_1}^{n-1} [1 + \mu(j)] \right)^{-1} = \lim_{n\to\infty} \frac{x(n)}{x(n_1)} = 0.$$

This completes the proof.

3 Main Results

Our first main result states that oscillation of the perturbed linear equation implies oscillation of the nonlinear equation.

Theorem 2 *Assume that (A4) holds, and there exists* $\theta \in (0, 1)$ *such that the linear equation*

$$x(n+1) - x(n) + \theta \sum_{k=1}^{m} r_k(n) x\big(h_k(n)\big) = 0 \quad \textit{for } n \geq n_0 \tag{16}$$

is oscillatory. Then, (6) *is oscillatory.*

Proof Assume the contrary that $\{x(n)\}$ is a nonoscillatory solution of (6). First, suppose that $\{x(n)\}$ is eventually positive. Then, we can find $n_1 \geq n_0$ such that $x(n) > 0$ and $x(h_k(n)) > 0$ for all $n \geq n_1$ and $k = 1, 2, \ldots, m$. From (6), $x(n+1) \leq x(n)$ for all $n \geq n_1$. By Corollary 1, we see that $\lim_{n\to\infty} x(n) = 0$. Thus, we can find $n_2 \geq n_1$ such that

$$f_k\big[x\big(h_k(n)\big)\big] \geq \theta x\big(h_k(n)\big) \quad \text{for all } n \geq n_2 \text{ and } k = 1, 2, \ldots, m.$$

From (6), we obtain the inequality

$$x(n+1) - x(n) + \theta \sum_{k=1}^{m} r_k(n) x\big(h_k(n)\big) \leq 0 \quad \text{for all } n \geq n_2.$$

By Proposition 5, Eq. (16) also has a nonoscillatory solution. The case where $\{x(n)\}$ is eventually negative is similar, which concludes the proof.

Next, we show that oscillation of a nonlinear equation implies oscillation of its linearized counterpart.

Theorem 3 *Assume that (A5) holds and* (6) *is oscillatory. Then,* (7) *is also oscillatory.*

Proof Assume the contrary, let $\{x(n)\}$ be a nonoscillatory solution of (7). By Corollary 1 and Remark 2, we have $\lim_{n\to\infty} x(n) = 0$. By Proposition 5, there exists a positive sequence $\{\mu_0(n)\}$ such that

$$\mu_0(n) \geq \sum_{k=1}^{m} r_k(n) \prod_{j=h_k(n)}^{n} [1 + \mu_0(j)] \quad \text{for all } n \geq n_1,$$

where $n_1 \geq n_0$. Thus, by Lemma 1, we have

$$\lim_{n\to\infty} \left(\prod_{j=n_1}^{n-1} [1 + \mu(j)] \right)^{-1} = 0.$$

First, suppose that there exists $\delta \in \mathbb{R}^+$ satisfying (A5) with (8). We can find $n_2 \geq n_1$ such that $\left(\prod_{j=n_1}^{n-1}[1+\mu(j)]\right)^{-1} \leq \delta$ for all $n \geq n_2$. By (A2), we can find $n_3 \geq n_2$ such that $h_k(n) \geq n_2$ for all $n \geq n_3$ and $k = 1, 2, \ldots, m$. Note that $\left(\prod_{j=n_1}^{h_k(n)-1}[1+\mu(j)]\right)^{-1}$ $\leq \delta$ for all $n \geq n_3$ and $k = 1, 2, \ldots, m$. Define $\{\mu_\ell(n)\}$ by

$$\mu_\ell(n) = \sum_{k=1}^{m} r_k(n) f_k\left[\left(\prod_{j=n_1}^{h_k(n)-1}[1+\mu_{\ell-1}(j)]\right)^{-1}\right]\prod_{j=n_1}^{n}[1+\mu_{\ell-1}(j)]$$

for any $n \geq n_3$ and $\ell \in \mathbb{N}$. Clearly, $\mu_0(n) \geq \mu_1(n) \geq \cdots \geq \mu_\ell(n) \geq \mu_{\ell+1}(n) > 0$ for $n \geq n_3$ and $\ell \in \mathbb{N}$. Let $\mu(n) := \lim_{\ell\to\infty} \mu_\ell(n)$ for $n \geq n_3$, and define

$$y(n) := \left(\prod_{j=n_3}^{n-1}[1+\mu(j)]\right)^{-1} \quad \text{for } n \geq n_3.$$

Then $\{y(n)\}$ is an eventually positive solution of (6). If (A5) holds with (9), we can proceed similarly and show that (6) has an eventually negative solution, which completes the proof.

4 Applications

Discrete population models are usually constructed assuming that per capita production rate g is density-dependent $x(n+1) - x(n) = x(n)g[x(n)]$. However, this rate may depend on population size at one of the previous stages $x(n+1) - x(n) = x(n)g[x(h(n))]$. To account for reference population sizes at different moments in the past, either additive

$$x(n+1) - x(n) = x(n)\sum_{k=1}^{m} r_k(n) g_k\big[x\big(h_k(n)\big)\big] \quad \text{for } n \geq n_0$$

or multiplicative

$$x(n+1) - x(n) = x(n)\prod_{k=1}^{m} r_k(n) g_k\big[x\big(h_k(n)\big)\big] \quad \text{for } n \geq n_0$$

extensions can be considered.

4.1 Pielou's Equation with Several Arguments

First, consider the following Pielou's difference equation with variable delays

$$N(n+1) = N(n)\prod_{k=1}^{m}\left[\frac{\alpha_k}{1+\beta_k N\big(h_k(n)\big)}\right]^{p_k(n)} \quad \text{for } n \geq n_0, \tag{17}$$

where $\{h_k(n)\}$ satisfies (A2), $\alpha_k \in (1, \infty)$, $\beta_k \in \mathbb{R}^+$ and $\{p_k(n)\} \subset \mathbb{R}_0^+$ for $k = 1, 2, \ldots, m$ (see [8, p. 22]). One can show that if $N(n) \geq 0$ for $n < n_0$ and $N(n_0) > 0$, Eq. (17) has a unique positive solution.

In the case of a single delay term, (17) includes the so-called logistic equation

$$N(n+1) - N(n) = \gamma N(n+1)\left(1 - \frac{N\big(h(n)\big)}{K}\right) \quad \text{for } n \geq n_0,$$

where $\{h(n)\}$ satisfies (A2), $\gamma \in (0, 1)$ and $K \in \mathbb{R}^+$.

Let us suppose that there exists $K \in \mathbb{R}^+$ such that

$$\alpha_k - 1 = K\beta_k \quad \text{for } k = 1, 2, \ldots, m. \tag{18}$$

If we let

$$x(n) := \ln\left[\frac{N(n)}{K}\right] \quad \text{for } n \geq n_0, \tag{19}$$

then (17) takes the form

$$x(n+1) - x(n) + \sum_{k=1}^{m} p_k(n)\ln\Big[1 + \gamma_k\big(\mathrm{e}^{x(h_k(n))} - 1\big)\Big] = 0 \quad \text{for } n \geq n_0, \tag{20}$$

where $\gamma_k := 1 - \frac{1}{\alpha_k} \in (0, 1)$ for $k = 1, 2, \ldots, m$.

We therefore showed the equivalence between oscillation of all solutions of nonlinear equation (17) about K and oscillation of nonlinear equation (20) about zero.

Note that for $k = 1, 2, \ldots, m$, the function $f_k(u) := \frac{1}{\gamma_k}\ln[1 + \gamma_k(\mathrm{e}^u - 1)]$ for $u \in \mathbb{R}$ satisfies $0 \geq f_k(u) \geq u$ for $x \leq 0$, i.e., for $k = 1, 2, \ldots, m$, f_k fulfills (A5) with (9) and any $\delta \in \mathbb{R}^+$. In view of our discussion in Sect. 3, we associate (17) with the linear equation

$$x(n+1) - x(n) + \sum_{k=1}^{m} \gamma_k p_k(n) x\big(h_k(n)\big) = 0 \quad \text{for } n \geq n_0. \tag{21}$$

Thus, we obtain some explicit oscillation and nonoscillation tests for (17).

Proposition 9 *Assume that* $\{h_k(n)\}$ *satisfies (A2),* $\alpha_k \in (1, \infty)$, $\beta_k \in \mathbb{R}^+$, $\{p_k(n)\} \subset \mathbb{R}_0^+$ *for* $k = 1, 2, \ldots, m$. *Assume further that there exists* $K \in \mathbb{R}^+$ *such that* (18) *holds.*

(i) If there exists $\theta \in (0, 1)$ *such that the linear equation*

$$x(n+1) - x(n) + \theta \sum_{k=1}^{m} \gamma_k p_k(n) x\big(h_k(n)\big) = 0 \quad \text{for } n \geq n_0 \tag{22}$$

is oscillatory then (17) *is oscillatory about* K.

(ii) If (21) *is nonoscillatory then* (17) *is nonoscillatory about* K.

Corollary 2 *Assume that* $\{h_k(n)\}$ *satisfies (A2),* $\alpha_k \in (1, \infty)$, $\beta_k \in \mathbb{R}^+$, $\{p_k(n)\} \subset \mathbb{R}_0^+$ *for* $k = 1, 2, \ldots, m$. *Assume further that there exists* $K \in \mathbb{R}^+$ *such that* (18) *holds.*

(i) If

$$\liminf_{n\to\infty} \inf_{\lambda\in(0,1)} \left\{ \sum_{k=1}^{m} \frac{\gamma_k p_k(n)}{\lambda(1-\lambda)^{n-h_k(n)}} \right\} > 1 \tag{23}$$

then (17) *is oscillatory about* K.

(ii) If there exist $\lambda_0 \in (0, 1)$ *and* $n_1 \geq n_0$ *such that*

$$\sum_{k=1}^{m} \frac{\gamma_k p_k(n)}{\lambda_0(1-\lambda_0)^{n-h_k(n)}} \leq 1 \quad \text{for } n \geq n_1 \tag{24}$$

then (17) *is nonoscillatory about* K.

Proof (i) From (23), there exists $\theta \in (0, 1)$ such that

$$\liminf_{n\to\infty} \inf_{\lambda\in(0,1)} \left\{ \theta \sum_{k=1}^{m} \frac{\gamma_k p_k(n)}{\lambda(1-\lambda)^{n-h_k(n)}} \right\} > 1. \tag{25}$$

Due to Proposition 3, (25) implies that (22) is oscillatory. An application of Proposition 9 completes the proof.

(ii) The proof follows from Propositions 4 and 9.

The following result for autonomous equations follows from Proposition 7.

Proposition 10 *Assume that* $\tau_k \in \mathbb{N}_0$, $\alpha_k \in (1, \infty)$, $\beta_k, p_k \in \mathbb{R}^+$, *and there exists* $K \in \mathbb{R}^+$ *such that* (18) *holds. The equation*

$$N(n+1) = N(n) \prod_{k=1}^{m} \left[\frac{\alpha_k}{1 + \beta_k N(n-\tau_k)} \right]^{p_k} \quad \text{for } n \geq n_0$$

is oscillatory about K if and only if the characteristic equation

$$\lambda - 1 + \sum_{k=1}^{m} \gamma_k p_k \lambda^{-\tau_k} = 0, \tag{26}$$

where $\gamma_k := 1 - \frac{1}{\alpha_k}$ *for* $k = 1, 2, \ldots, m$, *has no positive roots.*

4.2 Generalized Ricker Model with Variable Arguments

Next, consider Ricker's stock and recruitment model with variable delays

$$N(n+1) = N(n) \exp\left\{\sum_{k=1}^{m} p_k(n)\left(1 - \left[\frac{N\big(h_k(n)\big)}{K}\right]^{\gamma_k}\right)\right\} \quad \text{for } n \geq n_0, \tag{27}$$

where all $\{h_k(n)\}$ satisfy (A2), $\{p_k(n)\} \subset \mathbb{R}_0^+$, $\gamma_k \in \mathbb{R}^+$ and $K \in \mathbb{R}^+$ (see [1, p. 91]). Substitution (19) transforms (27) into

$$x(n+1) - x(n) + \sum_{k=1}^{m} p_k(n)\big[\mathrm{e}^{\gamma_k x(h_k(n))} - 1\big] = 0 \quad \text{for } n \geq n_0. \tag{28}$$

This implies the equivalence of oscillation of nonlinear equation (27) about K to oscillation of (28) about zero.

Note that for $k = 1, 2, \ldots, m$, the function $f_k(u) := \dfrac{1}{\gamma_k}(\mathrm{e}^{\gamma_k u} - 1)$ for $u \in \mathbb{R}$ satisfies $0 \geq f_k(u) \geq u$ for $u \leq 0$, i.e., for $k = 1, 2, \ldots, m$, f_k fulfills (A5) with (9) and any $\delta \in \mathbb{R}^+$. We associate linear equation (21) with (27), see Sect. 3. Since (27) is associated with the same equation as Pielou's equation (17), we can give the following results without a proof.

Proposition 11 *Assume that* $K \in \mathbb{R}^+$, $\{h_k(n)\}$ *satisfies (A2),* $\{p_k(n)\} \subset \mathbb{R}_0^+$ *and* $\gamma_k \in \mathbb{R}^+$ *for* $k = 1, 2, \ldots, m$.

(i) If there exists $\theta \in (0, 1)$ *such that* (22) *is oscillatory then* (27) *is oscillatory about* K.
(ii) If (21) *is nonoscillatory then* (27) *is nonoscillatory about* K.

Corollary 3 *Assume that* $K \in \mathbb{R}^+$, $\{h_k(n)\}$ *satisfies (A2),* $\{p_k(n)\} \subset \mathbb{R}_0^+$ *and* $\gamma_k \in \mathbb{R}^+$ *for* $k = 1, 2, \ldots, m$.

(i) If (23) *holds then* (27) *is oscillatory about* K.
(ii) If there exists $\lambda_0 \in (0, 1)$ *such that* (24) *holds then* (27) *is nonoscillatory about* K.

Proposition 12 *Assume that* $K \in \mathbb{R}^+$, $\tau_k \in \mathbb{N}_0$ *and* $p_k, \gamma_k \in \mathbb{R}^+$ *for* $k = 1, 2, \ldots, m$. *The equation*

$$N(n+1) = N(n)\exp\left\{\sum_{k=1}^{m} p_k\left(1 - \left[\frac{N(n-\tau_k)}{K}\right]^{\gamma_k}\right)\right\} \quad \textit{for } n \geq n_0$$

is oscillatory about K *if and only if the characteristic equation* (26) *has no positive roots.*

4.3 Lasota–Wazewska Equation

Finally, consider the discrete retarded Lasota–Wazewska equation for the survival of red-blood cells (see [13])

$$N(n+1) - N(n) = -p(n)N(n) + q(n)\mathrm{e}^{-\gamma N(h(n))} \quad \text{for } n \geq n_0, \tag{29}$$

where $\{h(n)\}$ satisfies (A2), $\{p(n)\} \subset [0, 1)$ describes probability of cell death at each step, $\{q(n)\} \subset \mathbb{R}_0^+$ and $\gamma \in \mathbb{R}^+$ are production parameters such that $p(n) = Kq(n)$ for some $K \in \mathbb{R}^+$ and $n = n_0, n_0 + 1, \ldots$. We will suppose that $\{p(n)\}$ or $\{q(n)\}$ does not vanish eventually. Then, there exists a unique number $N^* \in \mathbb{R}^+$ such that

$$KN^* = \mathrm{e}^{-\gamma N^*},$$

which is called the equilibrium of (29). By applying the change of variables

$$x(n) := \gamma\big[N(n) - N^*\big] \quad \text{for } n \geq n_0,$$

we transform (29) into another nonlinear equation

$$x(n+1) - x(n) + p(n)x(n) + \gamma N^* p(n)\Big[1 - \mathrm{e}^{-x(h(n))}\Big] = 0 \quad \text{for } n \geq n_0. \tag{30}$$

Denote $r_1(n) := p(n)$, $f_1(u) := u$, $h_1(n) := n$, $r_2(n) := \gamma N^* p(n)$, $f_2(u) := 1 - \mathrm{e}^{-u}$, $h_2(n) := h(n)$. Obviously f_1 and f_2 satisfy (A4) and (A5).

Therefore oscillation of nonlinear equation (29) about N^* is equivalent to oscillation of (30) about zero.

Proposition 13 *Let* $\{h(n)\} \subset \mathbb{Z}$, $h(n) \leq n$ *for all* $n \geq n_0$, $\lim_{n\to\infty} h(n) = \infty$, $\{p(n)\} \in [0, 1)$, $\{q(n)\} \subset \mathbb{R}_0^+$, $\gamma \in \mathbb{R}^+$ *and there exist* $K \in \mathbb{R}^+$ *such that* $p(n) = Kq(n)$ *for* $n = n_0, n_0 + 1, \ldots$.

(i) If there exists $\theta \in (0, 1)$ *such that the linear equation*

$$x(n+1) - x(n) + \theta p(n)x(n) + \theta\gamma N^* p(n)x\big(h(n)\big) = 0 \quad \textit{for } n \geq n_0 \tag{31}$$

is oscillatory then (29) *is oscillatory about* N^*.

(ii) If the linear equation

$$x(n+1) - x(n) + p(n)x(n) + \gamma N^* p(n) x\big(h(n)\big) = 0 \quad \textit{for } n \geq n_0$$

is nonoscillatory, Eq. (29) is nonoscillatory about N^*.

Corollary 4 *Let $\{h(n)\} \subset \mathbb{Z}$, $h(n) \leq n$ for all $n \geq n_0$, $\lim_{n\to\infty} h(n) = \infty$, $\{p(n)\} \in [0, 1)$, $\{q(n)\} \subset \mathbb{R}_0^+$ and $\gamma \in \mathbb{R}^+$. Assume further that there exists $K \in \mathbb{R}^+$ such that $p(n) = Kq(n)$ for $n = n_0, n_0 + 1, \ldots$.*

(i) If

$$\liminf_{n\to\infty} \inf_{\lambda\in(0,1)} \left\{ \frac{\gamma N^* p(n)}{\lambda(1-\lambda)^{n-h(n)} \prod_{j=h(n)}^{n} \big(1 - p(j)\big)} \right\} > 1, \tag{32}$$

then (29) is oscillatory about N^*.

(ii) If there exists $\lambda_0 \in (0, 1)$ such that

$$\frac{\gamma N^* p(n)}{\lambda_0(1-\lambda_0)^{n-h(n)} \prod_{j=h(n)}^{n} \big(1 - p(j)\big)} \leq 1 \quad \textit{for all large } n,$$

then (29) is nonoscillatory about N^*.

Proof (i) From (32), there exists $\theta \in (0, 1)$ such that

$$\liminf_{n\to\infty} \inf_{\lambda\in(0,1)} \left\{ \frac{\theta\gamma N^* p(n)}{\lambda(1-\lambda)^{n-h(n)} \prod_{j=h(n)}^{n} \big(1 - \theta p(j)\big)} \right\} > 1. \tag{33}$$

Note that (31) transforms into

$$y(n+1) - y(n) + \frac{\theta\gamma N^* p(n)}{\prod_{j=h(n)}^{n} \big(1 - \theta p(j)\big)} y\big(h(n)\big) = 0 \quad \text{for } n \geq n_0 \tag{34}$$

by the sign-preserving substitution

$$y(n) := \frac{x(n)}{\prod_{j=n_0}^{n-1} \big(1 - \theta p(j)\big)} \quad \text{for } n \geq n_0.$$

Due to Proposition 3, (33) yields that (34) (and hence (31)) is oscillatory. Therefore, an application of Proposition 13 completes the proof.

(ii) The proof follows from Propositions 4 and 13.

Theorem 4 ([5, Theorem 1]) *Assume that $p \in [0, 1)$, $q \in \mathbb{R}_0^+$, $\gamma \in \mathbb{R}^+$ and $\tau \in \mathbb{N}_0$. The equation*

$$N(n+1) - N(n) = -pN(n) + q\mathrm{e}^{-\gamma N(n-\tau)} \quad \textit{for } n \geq n_0 \tag{35}$$

is oscillatory about N^* *if and only if*

$$p\gamma N^*(\tau+1)^{\tau+1} > (1-p)^{\tau+1}\tau^{\tau}.$$

Proof For (35), linearized Eq. (30) has the form

$$x(n+1) - x(n) + px(n) + p\gamma N^*\Big[1 - \mathrm{e}^{-x(n-\tau)}\Big] = 0 \quad \text{for } n \geq n_0$$

for which Proposition 8 applies, which concludes the proof.

5 Final Comments

In the present paper, we have reviewed some known results and mistakes connected to linearized oscillation of difference equations. Sufficient linearization results are obtained for equations with variable coefficients and delays. They are illustrated with examples and applications to discrete delay models of population dynamics. Let us note that Proposition 12 solves [6, Problems 1–3 of Exercise 7.3]. Theorem 4, obtained here as an illustration of the main linearization method, is the main result of [5].

It is well known that the properties of difference equation with constant and variable delays and variable coefficients are usually essentially different when delays are unbounded. It would be interesting to consider linearization in the case of pantograph-type difference equations. In particular, it is possible to explore models studied in the present paper:

$$N(n+1) = N(n)\prod_{k=1}^{m}\left[\frac{\alpha_k}{1+\beta_k N(\lfloor \frac{n}{\tau_k}\rfloor)}\right]^{\frac{p_k}{n}} \quad \text{for } n \geq n_0,$$

$$N(n+1) = N(n)\exp\left\{\sum_{k=1}^{m}\frac{p_k}{n}\left(1-\left[\frac{N(\lfloor \frac{n}{\tau_k}\rfloor)}{K}\right]^{\gamma_k}\right)\right\} \quad \text{for } n \geq n_0,$$

$$N(n+1) = N(n)\exp\left\{\sum_{k=1}^{m}\frac{p_k}{n}\left(1-\left[\frac{N(\lfloor \frac{n}{\tau_k}\rfloor)}{K}\right]^{\gamma_k}\right)\right\} \quad \text{for } n \geq n_0,$$

where $n_0 \in \mathbb{N}$ and $\lfloor\cdot\rfloor$ is the floor function, i.e., $\lfloor u\rfloor$ is the greatest integer not exceeding $u \in \mathbb{R}$. We expect that the result on monotonicity of oscillation properties on time scales [4] can be applied to connect pantograph differential and difference equations.

In addition, careful treatment of the critical case known for differential equations [2] will also be interesting for difference equations. As mentioned in the introduction, it is the area of parameters where the discrepancy between the properties of linearized and original equations is observed.

References

1. Allen, L.J.S.: An Introduction to Mathematical Biology. Pearson Prentice Hall, Upper Saddle River (2006)
2. Baštinec, J., Berezansky, L., Diblík, J., Šmarda, Z.: On the critical case in oscillation for differential equations with a single delay and with several delays. Abstr. Appl. Anal. Article ID 417869 (2010)
3. Berezansky, L., Braverman, E.: On existence of positive solutions for linear difference equations with several delays. Adv. Dyn. Syst. Appl. **1**, 29–47 (2006)
4. Braverman, E., Karpuz, B.: On monotonicity of nonoscillation properties of dynamic equations in time scales. Z. Anal. Anwend. **31**, 203–216 (2012)
5. Chen, M.P., Yu, J.S.: Oscillation and global attractivity in the discrete Lasota–Wazewska model. Soochow J. Math. **25**, 1–9 (1999)
6. Elaydi, S.: An Introduction to Difference Equations. Undergraduate Texts in Mathematics, 3rd edn. Springer, New York (2005)
7. Győri, I., Ladas, G.: Oscillation Theory of Delay Differential Equations with Applications. Oxford University Press, New York (1991)
8. Pielou, E.C.: An Introduction to Mathematical Ecology. Wiley, New York (1969)
9. Tang, S., Xiao, Y., Chen, J.F.: Linearized oscillations in nonlinear delay difference equations. Acta Math. Sin. (Engl. Ser.) **15**, 569–574 (1999)
10. Tang, X.H., Yu, J.S.: Oscillation of nonlinear delay difference equations. J. Math. Anal. Appl. **249**, 476–490 (2000)
11. Tang, X.H., Yu, J.S.: Oscillations of delay difference equations in a critical state. Appl. Math. Lett. **13**, 9–15 (2000)
12. Wang, Z.C., Zhang, R.Y.: Nonexistence of eventually positive solutions of a difference inequality with multiple and variable delays and coefficients. Comput. Math. Appl. **40**, 705–712 (2000)
13. Ważewska-Czyżewska, M., Lasota, A.: Mathematical problems of the dynamics of a system of red blood cells. Mat. Stos. **3**(6), 23–40 (1976)
14. Yan, J.R., Qian, C.X.: Oscillation and comparison results for delay difference equations. J. Math. Anal. Appl. **165**, 346–360 (1992)
15. Zhou, Y.: Oscillation and nonoscillation for difference equations with variable delays. Appl. Math. Lett. **16**, 1083–1088 (2003)

Generalized Mandelbrot and Julia Sets in a Family of Planar Angle-Doubling Maps

Stefanie Hittmeyer, Bernd Krauskopf and Hinke M. Osinga

Abstract We study a planar noninvertible map that acts as a nonanalytic generalization of the complex quadratic family. It maintains the property of angle doubling, but is no longer an analytic map on the complex plane. Rather, the role of the critical point is now played by a critical circle. We generalize the notion of Julia sets to this new setting and show how these invariant sets interact with stable and unstable sets of saddle fixed and periodic points. We employ state-of-the-art numerical techniques to find and characterize new types of Julia sets, which are associated with the behavior of points on the critical circle under iteration. In parameter space this is encoded by the (generalized) Mandelbrot set.

Keywords Noninvertible planar map · Angle doubling · Nonanalytic perturbation · Mandelbrot set · Julia set · Stable and unstable sets · Critical set

1 Introduction

Our object of study is the map

$$f_\lambda(z) = (1 - \lambda + \lambda|z|^a)\left(\frac{z}{|z|}\right)^2 + c, \tag{1}$$

for $z \in \mathbb{C}\backslash\{0\}$, $\lambda \in [0, 1]$, and two additional parameters $a, c \in \mathbb{C}$. It was introduced in [6] for a near 1 and fixed $c = 1$ as a map reduction of a wild Lorenz-like vector field in $\mathbb{R}^n$ for $n \geq 5$; see also the geometric studies of (1) in [20–22, 31].

S. Hittmeyer · B. Krauskopf (✉) · H. M. Osinga
Department of Mathematics, The University of Auckland,
Private Bag 92019, Auckland 1142, New Zealand
e-mail: b.krauskopf@auckland.ac.nz

S. Hittmeyer
e-mail: stefanie.hittmeyer@auckland.ac.nz

H. M. Osinga
e-mail: h.m.osinga@auckland.ac.nz

M. Bohner et al. (eds.), *Difference Equations and Discrete Dynamical Systems with Applications*, Springer Proceedings in Mathematics & Statistics 312,
https://doi.org/10.1007/978-3-030-35502-9_2

Here, on the other hand, we fix $a = 2$ from now on because we are concerned with (1) as a one-parameter family in λ that contains the well-known complex quadratic family

$$f_1(z) = z^2 + c \tag{2}$$

for $z \in \mathbb{C}$ as a special case for $\lambda = 1$. The quadratic map (2) is actually complex analytic on $\mathbb{C}$; geometrically, it is the composition of angle doubling around the *critical point* 0 with translation by the *critical value* c.

A main object in the study of the dynamics of the quadratic map (2) on the complex plane is the *Julia set*. The Julia set can equivalently be defined as the boundary of the basin of attraction of infinity, the closure of the set of repelling periodic points, or the set of non-normal points. The complex quadratic map features a *fundamental dichotomy* which states that the Julia set of (2) is connected if, and only if, the orbit of the critical value c is bounded; if the orbit of c goes to infinity, the Julia set is a Cantor set, that is, it is totally disconnected [9]. This dichotomy is encoded in the parameter space of $c \in \mathbb{C}$ by the well-known *Mandelbrot set*, which we refer to as $\mathscr{M}_1$. It is defined as the set of parameter values c for which the corresponding Julia set of the quadratic family (2) is connected and, hence, constitutes the bifurcation diagram of (2) in the complex c-plane. See already Fig. 1 for an approximation of the Mandelbrot set $\mathscr{M}_1$ and, for example, [9, 13, 29] as an entry point to the extensive literature on the complex quadratic map.

We proceed by generalizing the notions of Julia set and Mandelbrot set to the family of maps (1) for any $\lambda \in [0, 1]$. A key property of this family is that (1) maps the punctured plane $\mathbb{C} \setminus \{0\}$ to the outside of the closed disk $\overline{\mathbb{D}}_{1-\lambda}(c)$ with radius $1 - \lambda$ around c, namely, also by angle doubling followed by translation by c. Hence, for $\lambda < 1$ the map (1) generalizes or 'unfolds' the quadratic family (2) in such a way that the basic mapping property of angle doubling is preserved, while complex analyticity on $\mathbb{C}$ is obviously lost. Points in $\overline{\mathbb{D}}_{1-\lambda}(c)$ have no preimage, while points in $\mathbb{C} \setminus \overline{\mathbb{D}}_{1-\lambda}(c)$ have two preimages. More specifically, the critical value c of (2) is replaced by the *critical circle* $J_1 := \partial\overline{\mathbb{D}}_{1-\lambda}(c)$ of (1), which can be thought of as the multi-valued image of the critical point $J_0 := 0$. When λ is decreased from 1 the radius $1 - \lambda$ of J_1 increases from 0.

We now define, for any $\lambda \in (0, 1]$ and $c \in \mathbb{C}$, the *(generalized) Julia set* $\mathscr{Y}$ as

$$\mathscr{Y} := \partial\mathscr{B}(\infty), \tag{3}$$

where $\mathscr{B}(\infty)$ is the basin of attraction of infinity, consisting of all points in $\mathbb{C}$ that escape to infinity under iteration of f_λ for the given value of c. The basin $\mathscr{B}(\infty)$ is well defined for any $\lambda \in (0, 1]$, but it retracts to infinity as λ goes to 0. Hence, for $\lambda = 0$, infinity is no longer attracting; therefore, we define the Julia set to be $\mathscr{Y} := \{\infty\}$ for $\lambda = 0$. We remark that we refer to the Julia set as $\mathscr{Y}$ (and not J as is common in the literature) to avoid confusion with the critical sets J_0 and J_1.

Clearly, definition (3) of the Julia set $\mathscr{Y}$ of (1) encompasses and, hence, generalizes the Julia set of the quadratic family (2). However, there is a second possibility of generalizing the notion of Julia set, namely by considering the repelling periodic points of (1). We consider the *set of preperiodic repelling points*

$$\mathscr{P} := \{z \in \mathbb{C}\backslash\{0\} \mid \exists\, m \geq 0 \text{ such that } f_\lambda^m(z) \text{ is a repelling periodic point}\}.$$

Here we refer to a period-k point z_0 as repelling when both eigenvalues of the linearization of f_λ^k at z_0 have modulus larger than 1. We choose to use the larger set $\mathscr{P}$ of preperiodic repelling points instead of just the set of repelling periodic points because it may also contain points in the disk bounded by the critical circle J_1 that do not have a preimage and, hence, cannot be periodic. In particular, we have $\mathscr{Y} = \overline{\mathscr{P}}$ for $\lambda = 1$. Therefore the closure $\overline{\mathscr{P}}$ of the set of preperiodic repelling points is also a generalization of the notion of Julia set. However, as we will see in Sect. 4.2, one can encounter situations where $\overline{\mathscr{P}}$ is a true subset of $\mathscr{Y}$. We will compute approximations of both $\mathscr{Y}$ and $\mathscr{P}$ to illustrate the properties of different types of Julia sets of the map (1); in particular, we will utilize the properties of $\mathscr{P}$ when classifying different types of Julia sets $\mathscr{Y}$ in Sect. 4.2.

An important additional ingredient of the dynamics one finds in the map (1) for $\lambda < 1$ are saddle fixed points and saddle periodic orbits and their stable and unstable sets. The latter are generalizations of stable and unstable manifolds in the context of noninvertible maps; they are introduced more formally in Sect. 2.2. In [20] we showed that there are four types of tangency bifurcations where the stable and unstable sets interact with the forward and backward critical sets, given by the images and preimages of the critical circle J_1 and the critical point J_0, respectively; these tangency bifurcation are presented in Sect. 2.3. Moreover, as was discussed in [22], stable and unstable sets interact with the (generalized) Julia set, and this may lead to types of Julia sets that are not found for the quadratic family. The emphasis in this earlier work was on how the Julia set for a given fixed $c \in \mathbb{R}$ changes and interacts with other emerging invariant objects for decreasing $\lambda < 1$; we will review some of the results in Sect. 4.1. Note, in particular, that saddle fixed and periodic points and their (un)stable sets can only exist in nonanalytic maps, that is, they do not exist for $\lambda = 1$.

Here, we take a complimentary point of view and consider the generalization of the Mandelbrot set of the map (1) for all $\lambda \in [0, 1]$. In fact, it is sufficient to consider only the λ-range $[0.5, 1]$; this is due to a non-obvious symmetry, which we show via a rescaling of the map (1) in Sect. 3.3. It is a key realization that there is now a trichotomy, rather than a dichotomy, when it comes to the fate of the critical circle J_1 under iteration: (1) J_1 may lie entirely in the basin $\mathscr{B}(\infty)$, that is, all points on J_1 escape to infinity under iteration; (2) all points on J_1 may remain bounded under iteration; and (3) the ambivalent case where some points on J_1 escape to infinity and some points on J_1 remain bounded under iteration. As we will show, the third and new possibility needs to be taken into consideration when studying the generalized Mandelbrot set of (1) and associated Julia sets.

We now define the *(generalized) Mandelbrot set*

$$\mathscr{M} := \{(c, \lambda) \in \mathbb{C} \times [0, 1] \mid c \in \mathscr{M}_\lambda\}$$

as the λ-union of the Mandelbrot sets

$$\mathcal{M}_\lambda := \{c \in \mathbb{C} \mid \exists z \in J_1 \text{ so that } f_\lambda^k(z) \text{ is bounded } \forall k \in \mathbb{N}\}.$$

This is an appropriate generalization of the Mandelbrot set $\mathcal{M}_1$ of the quadratic family, because we observe that the Julia set $\mathcal{Y}$ is connected whenever $c \in \mathcal{M}_\lambda$, while $\mathcal{Y}$ is a Cantor set whenever $c \notin \mathcal{M}_\lambda$.

In light of the trichotomy for $\lambda < 1$, we also define the *core* of the Mandelbrot set

$$\mathcal{M}^o := \{(c, \lambda) \in \mathbb{C} \times [0, 1] \mid c \in \mathcal{M}_\lambda^o\},$$

where

$$\mathcal{M}_\lambda^o := \{c \in \mathbb{C} \mid f_\lambda^k(z) \text{ is bounded } \forall z \in J_1 \text{ and } k \in \mathbb{N}\},$$

as well as the respective *ambivalence regions*

$$\widetilde{\mathcal{M}} := \mathcal{M} \setminus \mathcal{M}^o \quad \text{and} \quad \widetilde{\mathcal{M}}_\lambda := \mathcal{M}_\lambda \setminus \mathcal{M}_\lambda^o.$$

The latter correspond to the new case of parameter values for which some points on J_1 remain bounded under iteration and some do not; note that we define $\widetilde{\mathcal{M}}_1 := \emptyset$.

As we will see in Sect. 3, for $\lambda < 1$, the map (1) admits parameter regimes where several attractors co-exist. This is possible because different points on J_1 can go to different attractors under iteration. Therefore, the hyperbolic components of the Mandelbrot set $\mathcal{M}_1$ that correspond to the existence of a unique attractor of a certain period generally become components of $\mathcal{M}_\lambda$ for $\lambda < 1$ that correspond to the existence of several, possibly co-existing, periodic and even chaotic attractors.

In this paper, we consider how the Mandelbrot set $\mathcal{M}_\lambda \subset \mathbb{C}$ of the map (1) changes with λ. Closely associated is the complimentary question of how the Julia set changes across the boundary of $\mathcal{M}_\lambda$, that is, when $\lambda \in (0, 1)$ remains fixed and c is changed. This means, in particular, that the parameter c may take any value in $\mathbb{C}$. This work follows on from our previous work in [22], where we considered the case of fixed real c and found, for $\lambda < 1$, four new types of Julia sets, called *critically connected Cantor sets*, *Cantor bouquets*, *Cantor tangles*, and *Cantor cheeses*, which cannot occur in the quadratic map; see already Sect. 4.1. Here, we find yet more types of Julia sets, which we call *Cantor shrub* and *Cantor beetle*. Regarding the properties of the sets $\mathcal{M}^o$ and $\widetilde{\mathcal{M}}$, we find that there are the following three cases for $c \in \mathbb{C}$: in the complement of $\mathcal{M}$ the Julia set $\mathcal{Y}$ is a Cantor set; in the core $\mathcal{M}^o$ the Julia set $\mathcal{Y}$ is a connected union of Jordan curves; and in the ambivalence region $\widetilde{\mathcal{M}}$ we find the additional types of Julia sets mentioned above, which are effectively all derived from Cantor sets by generating different types of connectedness. We will show how these new types of Julia sets arise in the transition between different components of $\mathcal{M}^o$. Our results are obtained by careful numerical investigations with state-of-the-art tools and, mathematically speaking, have the status of conjectures. More specifically, we combine the computation of Julia sets, both as the boundary of $\mathcal{B}(\infty)$ via an escape algorithm and as the set of preperiodic points, with the computations of (un)stable sets in phase space. Moreover, we determine the Mandelbrot set $\mathcal{M}_\lambda$ by determining the fate under iteration of a large number of test points on the critical circle J_1, in combination with the direct computation of loci of different types of bifurcations, which include saddle-node, period-doubling and Neimarck–Sacker bifurcations, as well as the loci of the different tangency bifurcations.

Other analytic and singular perturbations of the complex quadratic family (2) have been considered by different authors; see also the literature review in [22]. Quite closely related is the work in [8, 10], which considers the case that the exponent of the quadratic term is close to 2; this corresponds in the map (1) to fixed $\lambda = 1$ and varying a near $a = 2$. For $a \neq 2$, the resulting map is nonanalytic and no longer quadratic, but still has a critical point $J_0 = 0$. In particular, this map is a map on $\mathbb{C} \setminus \{0\}$ and the forward critical orbit consists of the unique orbit of the critical value c. These authors define the *filled Julia set* as the set of points with bounded orbits and the (generalized) Julia set as the set of points that have no neighborhood in which the iterates of the map form an equicontinuous family. The latter notion is a generalization of the definition of the Julia set as the set of non-normal points. We do not consider it here, because for the map (1), the set of non-normal points contains all saddle points and their stable sets even if they do not lie in the boundary $\partial\mathscr{B}(\infty)$; moreover, the attracting set may consist of several or even chaotic attractors, to which convergence is not equicontinuous. Furthermore, the authors of [8, 10] study the bifurcation diagram of their perturbed map in the c-plane. To this end, they define a generalized version of the Mandelbrot as the connectedness locus of the filled Julia set. They also define the set of parameter values c such that the filled Julia set is not a Cantor set. There are parameter values $a < 2$ such that the two sets differ, and the difference set, where the filled Julia set is disconnected but not totally disconnected, could be interpreted as an "ambivalence region". However, this appears to be a relatively small set, while we find that the ambivalence region $\widetilde{\mathscr{M}}$ for the map (1) considered here is very large; this is due to the fact that one needs to consider the forward orbit of an entire critical circle, rather than of a single critical point.

The paper is structured as follows. In Sect. 2, we introduce the necessary background and notation. In particular, we discuss the definitions, properties and bifurcations of the critical and (un)stable sets in Sects. 2.1–2.3. In Sect. 3, we study the structure of the generalized Mandelbrot set $\mathscr{M}$ by discussing the Mandelbrot set $\mathscr{M}_1$ of the quadratic family (2) in Sect. 3.1 and the transition of the Mandelbrot set $\mathscr{M}_\lambda$ in the c-plane for decreasing λ from $\lambda = 1$ to $\lambda = 0.75$ in Sect. 3.2. We introduce a rescaling of the phase and parameter spaces of the map (1) in Sect. 3.3 and present the bifurcation diagram in the rescaled $(\mathrm{Re}(c), \lambda)$-plane, that is, along the period-doubling route to chaos in the quadratic family (2). Section 4 focuses on the Julia sets. In Sect. 4.1, we review the types of Julia sets that can occur in (1) for $c \in \mathbb{R}$, and in Sect. 4.2, we study the transition for three sequences of parameter values through the ambivalence region $\widetilde{\mathscr{M}_\lambda}$ of $\mathscr{M}_\lambda$ in the transition between the period-one and the period-two component for fixed $\lambda = 0.98$. As part of these transitions, we identify and characterize new types of Julia sets. In the concluding Sect. 5, we formulate our results regarding the occurrence of the different types of Julia sets and point out open questions and directions for future research.

2 Background

We now introduce the necessary definitions and properties of the forward and backward critical sets in Sect. 2.1 and the stable and unstable sets in Sect. 2.2. We then recall in Sect. 2.3 the four types of tangency bifurcations from [20] between these different invariant sets.

2.1 The Critical Set

To generalize the notions of critical point and critical value to the map (1) with $\lambda < 1$, we use notions from the theory of noninvertible maps; see [20] and, for example, [2, 30] as entry points to the relevant literature. The origin is the only point where the Jacobian of f_λ (as a real function on $\mathbb{R}^2$) is not defined and, hence, it forms the critical set J_0; for simplicity we still refer to J_0 as the *critical point*. The image J_1 of J_0 under f_λ does not exist for $\lambda < 1$, but it makes perfect sense to define it as $J_1 := \partial\overline{\mathbb{D}}_{1-\lambda}(c)$ [20]. Therefore, one can think of J_1 as the multivalued image of J_0 and we refer to it as the *critical circle*. Note that this circle collapses down to the critical value for $\lambda = 1$.

The points in the closed disk $\overline{\mathbb{D}}_{1-\lambda}(c)$ have no preimages and every point $z \in \mathbb{C} \setminus \overline{\mathbb{D}}_{1-\lambda}(c)$ has two preimages $f_0^{-1}(z)$ and $f_1^{-1}(z)$ given by

$$f_{0,1}^{-1}(z) = \pm\sqrt{\left(\frac{|z-c|-1+\lambda}{\lambda}\right)\frac{z-c}{|z-c|}}.$$

The kth preimage $f^{-k}(z)$ of z is the set of preimages

$$f_{s_k\cdots s_1}^{-k}(z) := f_{s_k}^{-1} \circ \cdots \circ f_{s_1}^{-1}(z),$$

for $(s_l)_{1\leq l\leq k} \in \{0, 1\}^k$ and consists of up to 2^k points.

For any $\lambda \in [0, 1]$, the dynamics of f_λ is organized by the *backward critical set*

$$\mathcal{J}^- := \bigcup_{k\geq 0} J_{-k},$$

and by the *forward critical set*

$$\mathcal{J}^+ := \bigcup_{k\geq 1} J_k,$$

where $J_k := f^k(J_0)$ with the convention that $f(J_0) = J_1$. The backward critical set $\mathcal{J}^-$ consists of points, which map to J_0 under a finite number of iterates. On the other hand, the forward critical set $\mathcal{J}^+$ consists of closed curves (except for $\lambda = 1$ when it is the forward orbit of the critical value c).

2.2 Stable and Unstable Sets

The map (1) is (complex) analytic for $\lambda = 1$ and nonanalytic for $\lambda < 1$. More specifically, for $\lambda \in (0, 1)$, the noninvertible map f_λ may have saddle fixed points and saddle periodic points. These have one-dimensional stable and unstable sets, which are generalizations of stable and unstable manifolds of diffeomorphisms. More precisely, for any hyperbolic saddle fixed point p of (1) there is a neighborhood V of p, such that V contains the *local stable manifold*

$$W^s_{\text{loc}}(p) := \{z \in V \mid f^k(z) \in V \text{ for all } k \geq 0\},$$

which has the same dimension as and is tangent to the stable eigenspace of p [32]. The *stable set* $W^s(p)$ is given by all preimages of $W^s_{\text{loc}}(p)$, that is,

$$W^s(p) := \bigcup_{k \geq 0} f^{-k}(W^s_{\text{loc}}(p)).$$

Similarly, there is a neighborhood V of p that contains the *local unstable manifold* $W^u_{\text{loc}}(p)$, defined as the local stable manifold with respect to the local inverse f^{-1}_{loc} of f that satisfies $f^{-1}_{\text{loc}}(p) = p$; in other words,

$$W^u_{\text{loc}}(p) := \{z \in V \mid (f^{-1}_{\text{loc}})^k(z) \in V \text{ for all } k \geq 0\},$$

which has the same dimension as and is tangent to the unstable eigenspace of p. The *unstable set* $W^u(p)$ is given by all images of $W^u_{\text{loc}}(p)$, that is,

$$W^u(p) := \bigcup_{k \geq 0} f^k(W^u_{\text{loc}}(p)).$$

The (un)stable set of a saddle periodic point of period k is defined as the (un)stable set of this point as a fixed point of kth iterate f^k, and the (un)stable set of the corresponding saddle periodic orbit is the union of the (un)stable sets of the periodic points on this orbit. By definition, stable and unstable sets are backward and forward invariant. More precisely, stable sets are invariant under the map f and both inverses f_0^{-1} and f_1^{-1}. It follows that a stable set $W^s(p)$ consists of infinitely many branches and, in particular, is not an immersed manifold [30]. We call the branch that contains the local manifold $W^s_{\text{loc}}(p)$ the primary branch. The unstable set $W^u(p)$, on the other hand, is invariant under f but not under both inverses. It can be an immersed manifold. However, note that $f(z) = f(-z)$ for the map (1), so that an occurrence when both z and $-z$ lie on $W^u(p)$ leads to points of self-intersection at $f(z)$ and its images; in this case $W^u(p)$ is not an immersed manifold either.

2.3 Tangency Bifurcations

The critical set and (un)stable sets of saddle (periodic) points may interact with each other in four types of tangency bifurcations of codimension one; see [20] for more details.

(H) The *homoclinic tangency* is a tangency between the stable and unstable sets $W^s(p)$ and $W^u(p)$; this leads to the creation of two homoclinic orbits that approach the saddle p under forward iteration and under a sequence of backward iterations. This bifurcation is the direct equivalent of the homoclinic tangency for diffeomorphisms;

(F) At the *forward critical tangency* the forward critical set $\mathcal{J}^+$ is tangent to the stable set $W^s(p)$; this leads to a branching of $W^s(p)$ at the points in the backward critical set $\mathcal{J}^-$;

(B) At the *backward critical tangency* a segment of the unstable set $W^u(p)$ moves over a sequence of points in $\mathcal{J}^-$; this leads to the formation of loops in $W^u(p)$ around the closed curves in $\mathcal{J}^+$;

(FB) At the *forward-backward critical tangency* a segment of a curve in $\mathcal{J}^+$ moves over a sequence of points in $\mathcal{J}^-$; this leads to the formation of loops in the images of this curve segment in $\mathcal{J}^+$ around other curves in $\mathcal{J}^+$.

These four tangency bifurcations were identified in [20] as generating a bifurcation structure that constitutes a geometric mechanism for the transition to so-called wild chaos in the map (1) for the parameter a near $a = 1$ [20]. We will encounter them here as well as part of the bifurcation diagram for $a = 2$.

3 The Generalized Mandelbrot Set $\mathcal{M}$

In this section we investigate the Mandelbrot set $\mathcal{M}$ as the bifurcation diagram of the map (1) in the three-dimensional (c, λ)-space for $c \in \mathbb{C}$ and $\lambda \in [0, 1]$. More specifically, we compute the Mandelbrot sets $\mathcal{M}_\lambda$ in the two-dimensional c-plane for $\lambda = 1$ and five fixed values of $\lambda < 1$, and we compute the corresponding bifurcation set in a rescaled $(\mathrm{Re}(\tilde{c}), \lambda)$-plane for fixed $\mathrm{Im}(\tilde{c}) = 0$.

3.1 The Mandelbrot Set $\mathcal{M}_1$ of the Quadratic Map

Figure 1 shows the Mandelbrot set $\mathcal{M}_1$ of (2) in terms of its *hyperbolic components* in $[-1.8, 0.45] \times [-1.05\,i, 1.05\,i]$. The parameter point c lies in a hyperbolic component of $\mathcal{M}_1$ if and only if c is in the basin of an attracting (hyperbolic) periodic orbit of a given period; in particular, there can be at most one attracting periodic orbit for any c. In this case, the Julia set bounds the basin of this attractor and is given

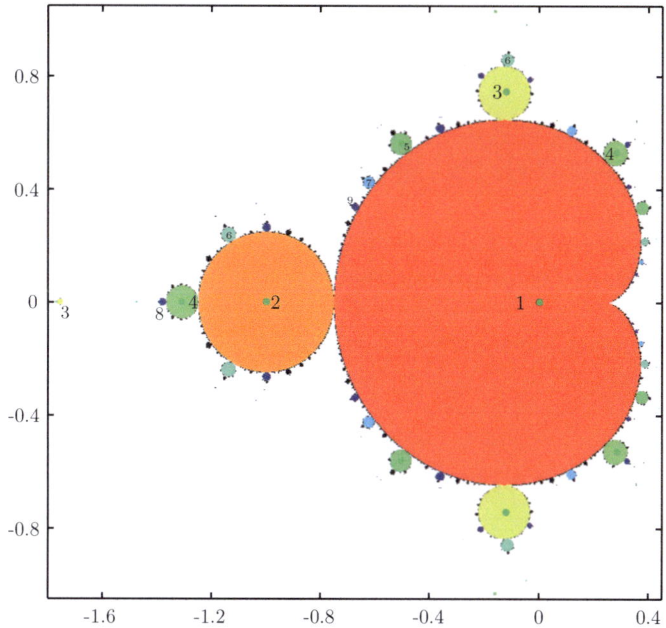

Fig. 1 The Mandelbrot set of (2) in the region $[-1.8, 0.45] \times [-1.05\,i, 1.05\,i]$. The colored domains are hyperbolic components where f_1 has an attracting periodic orbit of period 1 (red), 2 (orange), 3 (yellow), 4 (light green), 5 (dark green), 6 (cyan), 7 (light blue), 8 (dark blue) and 9 (purple); for black points the orbit of c is also bounded, while for white points $c \in \mathscr{B}(\infty)$. Also shown are the centers (green dots) of some hyperbolic components up to period 5

by a countable union of Jordan curves. The Mandelbrot set $\mathscr{M}_1$ is connected and its interior consists of hyperbolic components. To find $\mathscr{M}_1$ numerically, we use an escape-algorithm approach of checking for each point of a sufficiently fine grid in the c-plane (we use 4000×4000 points) whether the orbit of the critical point c remains bounded, that is, whether it escapes the disk with radius 2 after 4000 iterations. In the white regions the orbit of c is unbounded, the Julia set is a Cantor set and this value of c is, hence, not in $\mathscr{M}_1$. For bounded orbits we test convergence to a periodic attractor with a period up to 9; this allows us to identify and color the hyperbolic components up to this period. Black points remain bounded and the associated orbit of c converges to an attractor of period at least 10 or corresponds to other bounded dynamics.

The main cardioid (red) in Fig. 1 corresponds to parameter values c for which f_1 has an attracting fixed point. The critical point $z = c = 0$ is a super-attracting fixed point at $c = 0$ (green dot), and this point is also referred to as the center of the main cardioid. The hyperbolic components that correspond to attracting periodic orbits of periods 2 up to 9 are also highlighted in color as indicated; they are referred to as bulbs when their boundary is smooth and as cardioids when they have a cusp.

They also have center points (green dots) that correspond to super-attracting periodic points of the corresponding periods.

3.2 The Mandelbrot Set $\mathcal{M}_\lambda$ for Decreasing λ

The Mandelbrot set $\mathcal{M}_\lambda$ of the map (1) for $\lambda < 1$ is shown in Fig. 2 for $\lambda = 0.98$. It and subsequent such figures were computed with a generalization and refinement of the escape algorithm. For each point of a 3000×3000 grid in a region of interest in the parameter plane, we iterate a fixed number of equally distributed sample points on J_1; we used 25 points throughout, which we found to be a good balance between computational expense and accuracy. Regions that correspond to parameter points for which none of the sample points stay bounded are white; this can be checked

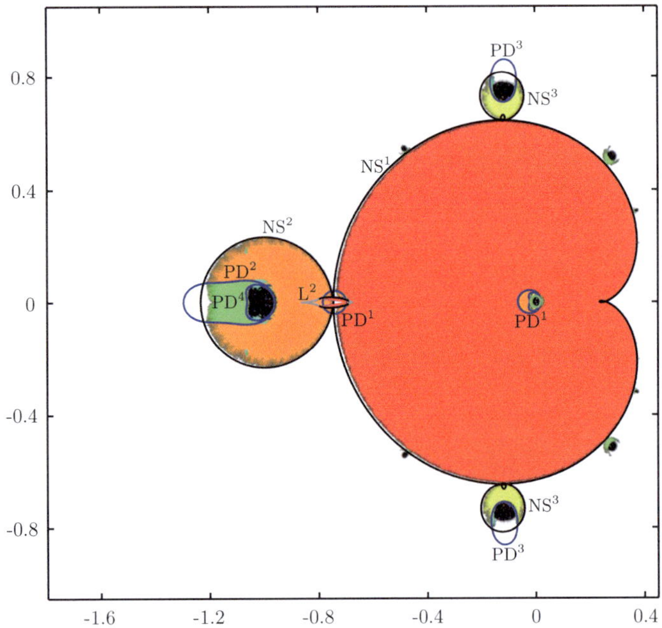

Fig. 2 The Mandelbrot set $\mathcal{M}_{0.98}$ for $\lambda = 0.98$ in the region $[-1.8, 0.45] \times [-1.05\,i, 1.05\,i]$; color corresponds to the periods of the respective periodic attractors as in Fig. 1; here full color means that all test points on J_1 converge to the periodic attractor, while gray shading indicates that some points on J_1 escape to infinity instead; black indicates convergence to a periodic attractor of period larger than 9, to an attracting invariant circle or to a chaotic attractor; white indicates that all sample orbits on J_1 are unbounded. Also shown are the curves NS^1, NS^2 and NS^3 (black) of Neimarck–Sacker bifurcations, PD^1, PD^2, PD^3 and PD^4 (blue) of period-doubling bifurcations, L^2 (gray) of saddle-node bifurcations, H (magenta; near L^2) of first homoclinic tangencies, B (purple; hidden by H) of first backward critical tangencies and the circle (green; near the origin) of last forward-backward critical tangencies

efficiently by determining that the modulus of the respective iterate is large enough (generally, we use a bound of 3000). On the other hand, if all sample points of J_1 stay bounded, then we conclude that the respective parameter point lies in $\mathscr{M}_\lambda$. We then check for each sample point whether it converges to an attracting periodic orbit of some given period (up to period 9); if all sample points converge to a k-periodic orbit with $k \leq 9$ then the parameter point is colored accordingly (where we use the color coding from Fig. 1). Otherwise, the parameter point is colored black. Hence, colored and black points lie in the core $\mathscr{M}_\lambda^o$. For $\lambda < 1$, however, the points on J_1 may be converging under iteration to an attractor that is much more complicated than an attracting periodic point; such points are also colored black. Finally, when some sample points are detected as staying bounded and some escape to infinity, then we conclude that the corresponding parameter point lies in the ambivalence region $\widetilde{\mathscr{M}_\lambda}$. We still run the detection of convergence to attracting k-periodic points for the bounded sample points, and we give the parameter point the respective color with an additional gray scale that indicates how many of the sample points diverge; here, full color corresponds to no sample points diverging, while mostly gray corresponds to all but one sample point diverging. Parameter points with some bounded sample points that do not appear to converge to a k-periodic point with $k \leq 9$ are colored similarly in shades of gray, again depending on how many of the sample points diverge; here, lighter gray corresponds to more sample points diverging. Hence, any grayish region in Fig. 4 corresponds to the ambivalent case of some bounded and some unbounded orbits in $\mathscr{J}^+$ and, therefore, belongs to the ambivalence region $\widetilde{\mathscr{M}_\lambda}$.

Figure 2 shows the Mandelbrot set $\mathscr{M}_{0.98}$ together with several bifurcation curves; compare with Fig. 1. The curves NS^1, NS^2 and NS^3 (black) are curves of Neimarck–Sacker bifurcations, PD^1, PD^2, PD^3 and PD^4 (blue) are curves of period-doubling bifurcations, and L^2 (gray) are curves of saddle-node bifurcations of period-one, -two, -three or -four points. The curves H (magenta) and B (purple) correspond to the first homoclinic and first backward critical tangencies between the sets $W^s(p)$ and $W^u(p)$, and $W^u(p)$ and $\mathscr{J}^-$, respectively; these curves cannot be distinguished in Fig. 2, but lie near L^2. The green circle around the origin is the locus of last forward-backward critical tangency, namely the values of c with $|c| = 0.02 = 1 - \lambda$ for which the critical point J_0 lies on the critical circle J_1. These curves were computed by continuation using MatContM [15, 18, 19] with the methods outlined in [20], which are adaptations of the boundary value setup for following homoclinic or heteroclinic tangencies from [7].

When comparing the sets $\mathscr{M}_{0.98}$ in Fig. 2 and $\mathscr{M}_1$ in Fig. 1, we note that many regions corresponding to the existence of higher periodic attractors of $\mathscr{M}_1$ have disappeared already; moreover, the components of $\mathscr{M}_{0.98}$ are no longer connected to each other. In particular, the red shaded region of $\mathscr{M}_{0.98}$ (formerly the main cardioid of $\mathscr{M}_1$) and the orange shaded region to its left (formerly the period-two bulb) no longer connect at a point of period-doubling bifurcation. Rather, there is a complicated transition from a unique attracting fixed point to a unique attracting period-two point; we will discuss this region of the c-plane for fixed $\lambda = 0.98$ and the corresponding transitions in the phase plane in more detail in Sect. 4.2. Moreover, areas of existence of higher periodic or other attractors have appeared in $\mathscr{M}_{0.98}$ near the centers of the

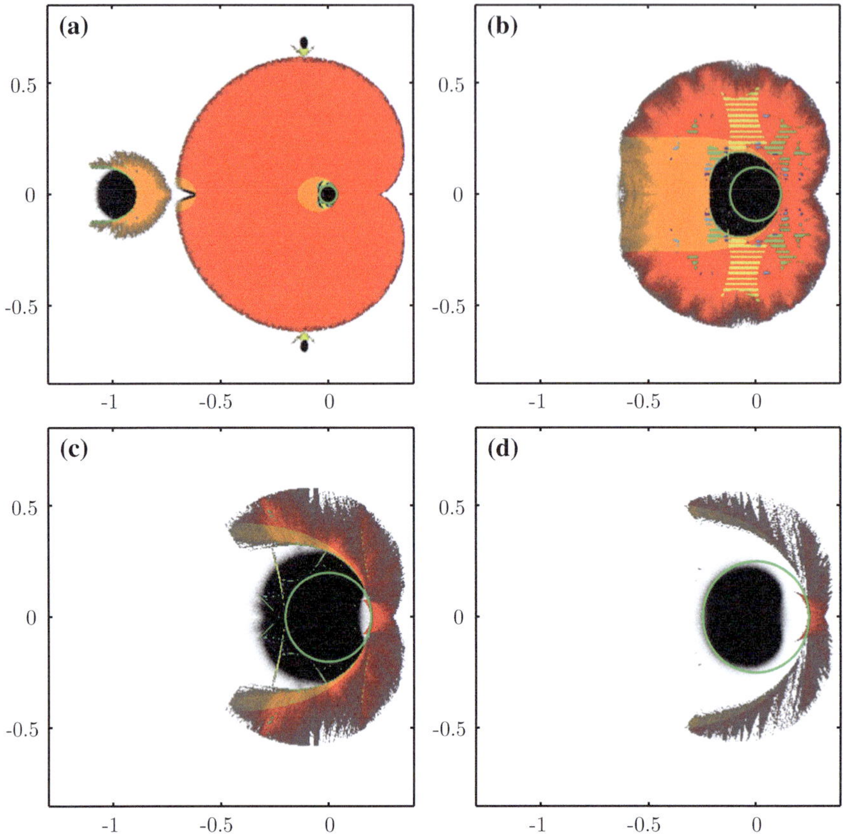

Fig. 3 The Mandelbrot set $\mathscr{M}_\lambda \subset \mathbb{C}$ in the region $[-1.3, 0.4] \times [-0.85\,i, 0.85\,i]$ for $\lambda = 0.96$ (**a**), $\lambda = 0.88$ (**b**), $\lambda = 0.8$ (**c**), and $\lambda = 0.75$ (**d**); compare with Fig. 2

remaining components. In particular, the center $c = 0$ of the Mandelbrot set $\mathscr{M}_1$ in Fig. 1 has opened up to the circle with $|c| = 1 - \lambda$ and regions corresponding to attracting periodic points of periods 2, 3, 4, 6 and 8 and to higher-periodic or other attractors have appeared nearby.

Figure 3 shows the Mandelbrot set $\mathscr{M}_\lambda$ of the map (1) in the c-plane for $\lambda = 0.96$ in panel (a), $\lambda = 0.88$ in panel (b), $\lambda = 0.8$ in panel (c) and $\lambda = 0.75$ in panel (d). As in Figs. 1 and 2, the colored areas in Fig. 3 correspond to values of c that admit a periodic attractor of a given period up to 9, with gray indicating how many test points on J_1 are unbounded. As λ is decreased further from $\lambda = 0.98$, more regions of higher-period attractors and of other attractors disappear. Already for $\mathscr{M}_{0.96}$, in Fig. 3a, there appear to be only regions left of period-one, -two and -three attractors. Moreover, the orange-shaded regions with period-two attractors have increased in the component that used to be the main cardioid of M_1, including near its center. At the same time, the gaps between former hyperbolic components have increased and there is considerably more gray shading, meaning that the ambivalence region $\widetilde{\mathcal{M}}$ is increasing in size. For $\lambda = 0.88$, all former bulbs of the Mandelbrot set $\mathcal{M}_1$ have disappeared; see Fig. 3b. Notice that effectively only the main cardioid remains, and only a small part of it represents a unique attracting fixed point. We find (striped) regions of multistability with higher-period attractors; here some of the sample points on J_1 converge to one attractor, and others to a different attractor. We even found values of c at which sample points on J_1 go to more than two different attractors; however, for simplicity, we show in Fig. 3 only the colors of the two lowest periods, or black in case of an attractor of period larger than 9 or other attractor. Note further that the ambivalence region $\widetilde{\mathcal{M}}_{0.88}$, indicated by gray shading, has increased considerably and appears quite frayed. For $\lambda = 0.8$ and $\lambda = 0.75$ as in Fig. 3c and d, respectively, the ambivalence region $\widetilde{\mathcal{M}}_\lambda$ as well as the black regions near the origin corresponding to attractors of periods larger than 9 or other bounded attractors appear to make up the entire Mandelbrot set $\mathcal{M}_\lambda$.

3.3 Rescaling and Bifurcation Set for $c \in \mathbb{R}$

In [22], we presented the bifurcation diagram of the map (1) in the $(\mathrm{Re}(c), \lambda)$-plane for $\mathrm{Im}(c) = 0$ and found evidence that the sequence of bifurcations reverse as $\lambda = 0.5$ is crossed. This motivates us to make the underlying symmetry explicit by introducing a rescaling of parameter and phase space, and to present the rescaled bifurcation diagram.

Consider the family of maps

$$\begin{aligned} F_\lambda &: \mathbb{C}\backslash\{0\} \to \mathbb{C}, \\ &w \mapsto (\lambda(1-\lambda) + |w|^2)\left(\frac{w}{|w|}\right)^2 + \tilde{c}\,(\lambda(1-\lambda)+1), \end{aligned} \tag{4}$$

with parameters $\lambda \in [0, 1]$ and $\tilde{c} \in \mathbb{C}$. The map F_λ is conjugate to f_λ for $\lambda \in (0, 1]$ via the coordinate change

$$w = \lambda z,$$

subject to the parameter rescaling

$$\tilde{c} := c/(1 - \lambda + \lambda^{-1}).$$

The rescaled map F_λ has the parameter symmetry about $\lambda = 0.5$ given by $F_\lambda = F_{1-\lambda}$. Due to the conjugacy, for all $\lambda \in (0, 1)$, the dynamics of F_λ for a given $\tilde{c}$ is conjugate to the dynamics of f_λ for the corresponding c. Moreover, $F_0 = F_1$ emerges as a well-defined, rescaled limit of f_λ for $\lambda \to 0$; note that, in contrast, f_0 collapses the entire punctured plane onto the unit circle, followed by angle doubling plus a translation

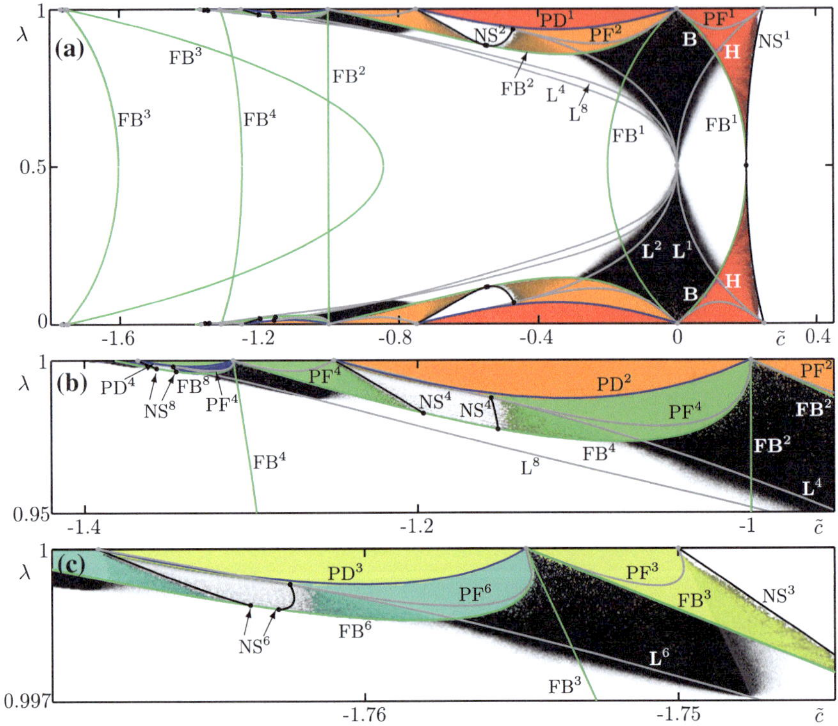

Fig. 4 The bifurcation set of (4) in the $(\mathrm{Re}(\tilde{c}), \lambda)$-plane in the regions $[-1.8, 0.45] \times [0, 1]$ (**a**), $[-1.42, -0.95] \times [0.95, 1]$ (**b**) and $[-1.77, -1.745] \times [0.997, 1]$ (**c**). Shown are the Mandelbrot set $\mathscr{M}$ (colored, gray and black regions) and curves L^1, L^2, L^4, L^6 and L^8 (gray) of saddle-node bifurcations, PF^1, PF^2, PF^3, PF^4 and PF^6 (gray) of pitchfork bifurcations, PD^1, PD^2, PD^3 and PD^4 (blue) of period-doubling bifurcations, NS^1, NS^2, NS^3, NS^4, NS^6 and NS^8 (black) of Neimarck–Sacker bifurcations, H (magenta) of first homoclinic tangencies, B (purple) of first backward critical tangencies, and FB^1, FB^2, FB^3, FB^4, FB^6 and FB^8 (green) of forward-backward critical tangencies

by c [22]. We conclude that f_λ is conjugate to $f_{1-\lambda}$ for all $\lambda \in [0, 1]$, and this means that we can restrict our attention to $\lambda \in [\frac{1}{2}, 1]$.

Figure 4 shows the bifurcation diagram of f_λ (or of F_λ) plotted in the rescaled $(\mathrm{Re}(\tilde{c}), \lambda)$-plane. Shown are the intersection of the Mandelbrot set $\mathscr{M}$ with this parameter plane, together with numerous bifurcation curves. Panel (a) clearly brings out the symmetry in λ about $\frac{1}{2}$ by presenting the full λ-range from 0 to 1. Panels (b) and (c) are enlargements near the period-2 and period-3 regions of the period-doubling sequence to chaos for $\lambda = 1$ and real c, respectively.

Figure 4a shows that the Mandelbrot set changes dramatically with decreasing $\lambda < 1$, and it practically disappears for $\lambda = 0.5$ (and then reappears for $\lambda < 0.5$). Most prominent are the regions with an attracting fixed point (red) and a two-periodic point (orange). Notice also the large black region where the critical circle converges to some other attractors. Panels (b) and (c) of Fig. 4 are enlargements near the period-two and period-four bulbs, and near the period-three cardioid of $\mathscr{M}_1$, respectively. The boundary between white and color/black is marked by the emergence of gray, which indicates the ambivalence region $\widetilde{\mathscr{M}}$. Some of the transitions appear to be well aligned with certain bifurcation curves. In particular, the attracing fixed point that exists in the main cadioid of $\mathscr{M}_1$ bifurcates at a pitchfork bifurcation curve PF^1 (gray) for $\tilde{c} > 0$ and a period-doubling curve PD^1 (blue) for $\tilde{c} < 0$, respectively. Furthermore, the overall bifurcation structure that emerges from the centers is very similar for other bulbs and cardioids, with corresponding bifurcations of periodic orbits of higher period; we refer to [22] for further details.

Overall, we conclude from Fig. 4 that not only the boundary ∂M_1 of the Mandelbrot set $\mathscr{M}_1$ but also the centers of the hyperbolic components in $\mathscr{M}_1$ give rise to new dynamics as soon as λ is decreased from 1; this clearly agrees with the images of $\mathscr{M}_\lambda$ in Figs. 2 and 3. Note that the gray-shaded ambivalence region $\widetilde{\mathscr{M}}_\lambda$ seems to emerge only from the boundary $\partial\mathscr{M}_1$ and, hence, we conclude that $\partial\mathscr{M}_1$ is the limit of $\widetilde{\mathscr{M}}_\lambda$ as $\lambda \to 1$. On the other hand, for $\lambda = 0.5$, the Mandelbrot set $\mathscr{M}_\lambda \cap \mathbb{R}$ shrinks down to the points $\tilde{c} = 0$ and $\tilde{c} = 0.2$, which correspond to $c = 0$ and $c = 0.5$, respectively.

For simplicity, from now on we will continue to present the Mandelbrot sets $\mathscr{M}_\lambda$ in the original c-plane, and Julia sets and other invariant objects in the original z-plane of f_λ, respectively.

4 Generalized Julia Sets

We now study how the phase portrait of the map (1) changes when parameters are changed away from the case $\lambda = 1$ of the complex quadratic family. A phase portrait in this context includes the Julia set as the boundary of the basin $\mathscr{B}(\infty)$, as well as stable and unstable sets of saddle points. The key is to understand the interplay between these different types of sets, and how the Julia set changes in the process. As the starting point we present first some new types of Julia sets that we found in our previous work in [22] when decreasing λ from 1 for fixed $c \in \mathbb{R}$. We then focus in Sect. 4.2 on the transition in $\mathscr{M}_{0.98}$ from a period-one to a period-two attractor as c is changed.

4.1 Julia Sets for Fixed $c \in \mathbb{R}$

In [22], we studied the map (1) with $a = 2$ for the special case $\mathrm{Im}(c) = 0$ when the Julia set is symmetric with respect to complex conjugation (reflection on the $\mathrm{Re}(z)$-axis). When λ is decreased from 1, we found that some new types of Julia sets arise in the process, and we present them in Fig. 5 as our starting point.

Each panel of Fig. 5 and subsequent similar figures shows a phase portrait consisting of the basin of attraction of infinity $\mathscr{B}(\infty)$ (shaded gray), the Julia set $\mathscr{J}$ (black), the backward critical set $\mathscr{J}^-$ (green dots), the critical circle J_1 (green circle), as

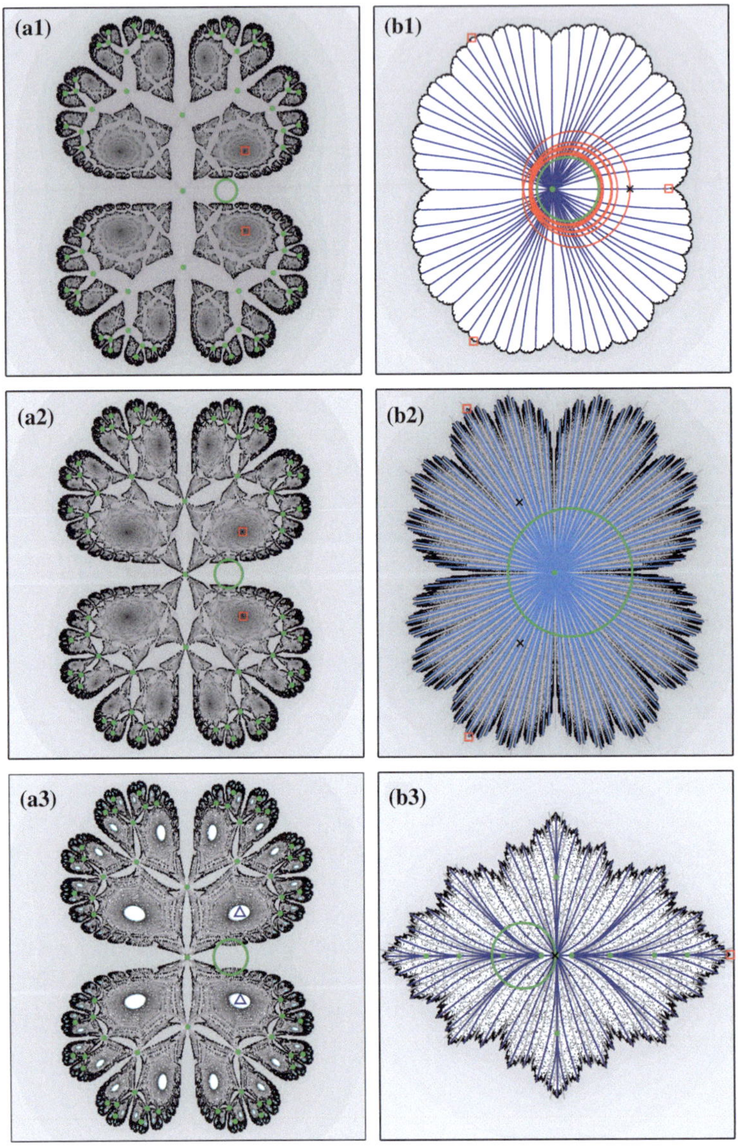

Fig. 5 Phase portraits of the map (1) for fixed $a = 2$, showing the Julia set $\mathscr{Y}$ and stable and unstable sets of selected saddle objects. In panel **a1** for $c = 0.28$ and $\lambda = 0.93$ the Julia set $\mathscr{Y}$ is a Cantor set; in **a2** for $c = 0.28$ and $\lambda = 0.91$ it is a critically connected Cantor set; in **a3** for $c = 0.28$ and $\lambda = 0.89$ it is a Cantor cheese; in **b1** for $c = 0.1$ and $\lambda = 0.8$ it is a Jordan curve; in **b2** for $c = 0.1$ and $\lambda = 0.6$ it is a Cantor bouquet; and in **b3** for $c = -0.25$ and $\lambda = 0.76$ it is a Cantor tangle

well as selected saddle points and their stable sets (blue) and/or unstable sets (red). We compute the Julia set $\mathscr{J}$ in two complimentary ways, namely, as the boundary of the basin $\mathscr{B}(\infty)$ (gray) of infinity and as the closure of the preperiodic repelling points $\mathscr{P}$ (black). More precisely, we color the points on the plane that escape to infinity with a gray scale that corresponds to the number of iterates these points need to go beyond a certain escape circle. Shown in black are about 2000 periodic and preperiodic repelling points in $\mathscr{P}$. For this set of parameter values the set $\mathscr{P}$ seems to lie dense in $\mathscr{J}$. However, later we will encounter examples in which this is not the case; see already Fig. 7c–f. The stable and unstable sets are computed with the methods outlined in [20], which are based on the method proposed in [25] and implemented in the DsTool environment [5, 17, 26].

Column (a) of Fig. 5 shows the transition as λ is decreased for the case when the Julia set $\mathscr{J}$ for $\lambda = 1$ is a Cantor set; specifically, we consider $c = 0.28$. For $\lambda < 1$ sufficiently close to 1, the set $\mathscr{J}$ is still a Cantor set; namely, as long as the critical circle J_1 does not interact with the Julia set, that is, $J_1 \subset \mathscr{B}(\infty)$; see Fig. 5a1. This means that the backward critical set $\mathscr{J}^-$ is not contained in the Julia set $\mathscr{J}$. Moreover, $\mathscr{B}(\infty)$ is simply connected and there is no bounded attractor. When λ is decreased sufficiently from 1, the critical circle J_1 does interact with the Julia set, meaning that some points on J_1 remain bounded under iteration while some points on J_1 still escape to infinity; see Fig. 5a2 for $\lambda = 0.91$. The backward critical set $\mathscr{J}^-$ is now contained in the Julia set $\mathscr{J}$, which is connected as a result, while $\mathscr{B}(\infty)$ is multiply connected. We refer to this type of Julia set as a *critically connected Cantor set* and it is characterized by being connected at a countable dense set of point and containing Jordan curves that bound the bounded subsets of $\mathscr{B}(\infty)$. Two symmetric repelling fixed points $q^{\pm}$ are contained in $\mathscr{J}$ and are indicated by red squares in panels (a1) and (a2) of Fig. 5. When λ is decreased further, the points $q^{\pm}$ undergo a Neimarck–Sacker bifurcation, where they become attractors that are surrounded by repelling smooth invariant circles (sufficiently close to the bifurcation and under the assumption that there are no strong resonances); see Fig. 5a3 for $\lambda = 0.89$. The invariant circles and their preimages (cyan curves) effectively create infinitely many holes by bounding the multiply connected basins of attractions $\mathscr{B}(q^{\pm})$. We conjecture that the boundary of $\mathscr{B}(q^{\pm})$ is dense in $\mathscr{J}$ and we refer to this type of Julia set as a *Cantor cheese* [22], which one can think of as a critically connected Cantor set but now containing a dense set of Jordan curves bounding the basins of the bounded attractors.

Column (b) of Fig. 5 shows the transition as λ is decreased for the case when the Julia set $\mathscr{J}$ for $\lambda = 1$ is a Jordan curve that bounds the simply connected basin of an attracting fixed point; initially, we consider $c = 0.1$, which lies inside the main cardioid of the Mandelbrot set. For $\lambda < 1$ sufficiently close to 1, the set $\mathscr{J}$ is still a Jordan curve. However, the nature of the attracting set with its basin bounded by $\mathscr{J}$ may change dramatically. For $\lambda = 0.8$, as in Fig. 5b1, we find a saddle point p with an unstable set $W^u(p)$ (red curve) that accumulates on a chaotic attractor. Its stable set $W^s(p)$ (blue curves) consists of the interval from 0 to a repelling fixed point r (red square) on the real line, and of all its preimages; also shown is a period-two repellor (red squares) consisting of the symmetric points $s^{\pm} \in \mathscr{J}$. Note that the critical point

J_0 lies inside the disk bounded by critical circle J_1; as a result, all branches of $W^s(p)$ connect at $J_0 = 0$. When λ is decreased, the set Julia $\mathscr{J}$ interacts with the saddle point and its stable and unstable sets, which happens at a saddle-node bifurcation where p and r meet. In Fig. 5b2 for $\lambda = 0.6$, the fixed points p and r, as well as the sets $W^s(p)$ and $W^u(p)$ have disappeared. This means that there is no longer a bounded attractor. In particular, points on the positive real axis lie now in $\mathscr{B}(\infty)$, which implies that there are points on the critical circle J_1 that escape to infinity under iteration. On the other hand, there are two period-two saddle points $p^{\pm}$ (black crosses) such that the stable sets $W^s(p^{\pm})$ (light-blue curves) consist of the two arcs that connect $J_0 = 0$ with the points $s^{\pm} \in \mathscr{J}$, as well as all their preimages. Since J_1 intersects $W^s(p^{\pm})$, it has points that remain bounded under iteration. We conclude that the Julia set $\mathscr{J}$ now forms a *Cantor bouquet*, which is an infinite union of arcs, connected in one point, the so-called *explosion point*, such that the end points of the arcs are dense. Note that Cantor bouquet Julia sets have been found in the study of the complex exponential family $z \mapsto \lambda \exp(z)$ for $\lambda < e^{-1}$; see [1, 11, 14, 24, 28], where the explosion point is at infinity. In contrast, the Cantor bouquet of (1) is the closure of the stable sets $W^s(p^{\pm})$ and the explosion point is $J_0 = 0$, which for this choice of parameters lies inside the disk bounded by the critical circle J_1.

Figure 5b3 is for $c = -0.25$ and $\lambda = 0.76$, where the critical point J_0 lies outside the critical circle J_1 and, therefore, has preimages in $\mathscr{I}^-$. In particular, there is still no finite attractor, but $\mathscr{I}^-$ is now dense in $\mathscr{J}$. Since every point of $\mathscr{I}^-$ is a connection point, we conjecture that the Julia set $\mathscr{J}$ is a *Cantor tangle* [22], which is given by the closure of the stable set $W^s(p)$ of a saddle fixed point p (black cross). A Cantor tangle is an infinite union of arcs, connected in a dense set of points such that the end points of the arcs are dense. Hence, a Cantor tangle can be thought of as a Cantor bouquet, but with a dense set of points where arcs connect instead of a single explosion point.

We remark that in previous work we referred to the critically connected Cantor set shown in Fig. 5a2 also as a Cantor tangle, because these sets have in common that they are connected at the dense subset $\mathscr{I}^-$ and that there is no bounded attractor. We now distinguish between them, because we wish to emphasize an important difference. Namely, in case of the Cantor tangle the basin $\mathscr{B}(\infty)$ is simply connected and the set of points on J_1 that remain bounded under iteration is dense in J_1; whereas in case of the critically connected Cantor set the basin $\mathscr{B}(\infty)$ is multiply connected and the critical circle J_1 contains segments that escape to infinity.

Overall, we found in [22] the following trichotomy for $c \in \mathbb{R}$ in the map (1): If all orbits in $\mathscr{I}^+$ are unbounded then $\mathscr{J}$ is a Cantor set; if all orbits in $\mathscr{I}^+$ are bounded then $\mathscr{J}$ is connected and a countable union of Jordan curves; in the ambivalent case of some bounded and some unbounded orbits in $\mathscr{I}^+$, the Julia set is connected but may be much more complicated; specifically, we found in [22] new types of Julia sets, namely, the cases of critically connected Cantor set, Cantor cheese, Cantor bouquet and Cantor tangle. We now proceed to consider other transitions through parameter space, and these will reveal even more types of interactions between Julia sets and other invariant sets.

4.2 From the Period-One to the Period-Two Region of $\mathscr{M}_{0.98}$

In order to understand better the dynamics of the map (1) relating to the ambivalence region $\widetilde{\mathscr{M}}$, we now study phase portraits of the map (1) when the parameter c is moved through $\widetilde{\mathscr{M}}$. More specifically, we discuss three parameter paths for $\lambda = 0.98$ related to the transition from a period-one attractor and a period-two attractor. Figure 6 shows two successive enlargements of $\mathscr{M}_{0.98}$ from Fig. 2 near the boundaries of the period-one and period-two regions (corresponding to the main cardioid and the period-two bulb for $\lambda = 1$). The first transition we consider is for $\mathrm{Im}(c) = 0$ and decreasing $\mathrm{Re}(c)$, and it is indicated in Fig. 6a by gray dots labeled (7a) to (7h), which correspond to the panels of phase portraits in Fig. 7(a to h). Similarly, the second transition for $\mathrm{Im}(c) = 0.008$ and decreasing $\mathrm{Re}(c)$ is indicated in Fig. 6a by gray dots labeled (8a) to (8l), which correspond to Fig. 8(a to l). Finally, we consider the transition between these two cases by decreasing $\mathrm{Im}(c)$ for fixed $\mathrm{Re}(c) = -0.707$; the gray dots labeled

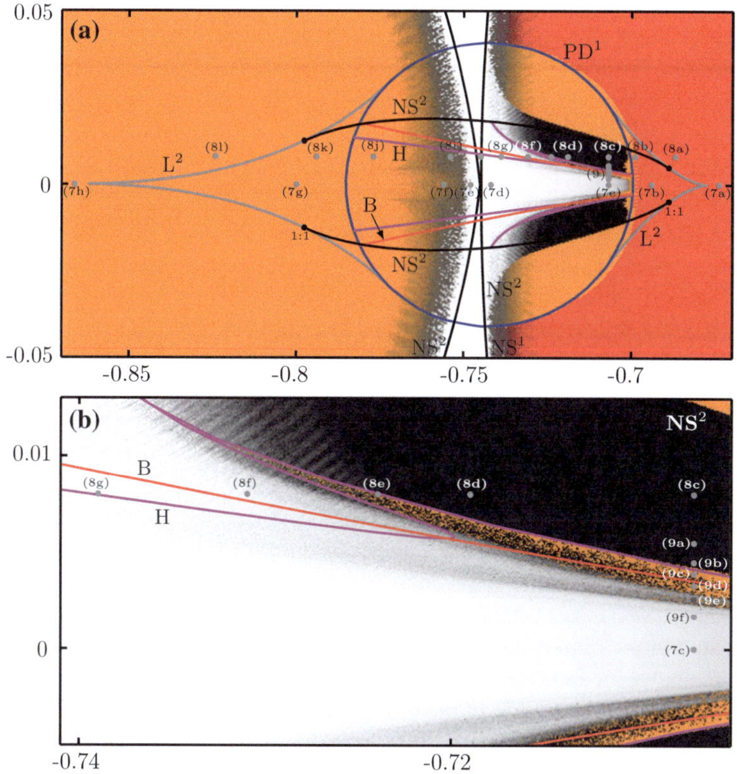

Fig. 6 Enlargements of $\mathscr{M}_{0.98}$ from Fig. 2 for $c \in [-0.87, -0.67] \times [-0.05\,i, 0.05\,i]$ (**a**) and $c \in [-0.741, -0.705] \times [-0.005\,i, 0.013\,i]$ (**b**). The points labeled (7a)–(7h), (8a)–(8l), and (9a)–(9f) correspond to the parameter values used for the corresponding panels of Figs. 7, 8 and 9, respectively

(9a) to (9f) in Fig. 6b correspond to Fig. 9(a to f). We remark that we focus our analysis on qualitative changes and new types of generalized Julia sets by presenting phase portraits at these distinct parameter values throughout the three transitions; hence, we show only the main bifurcation curves in Fig. 6. The phase portraits we present show the Julia set $\mathscr{Y}$ (black), the basin of attraction of infinity (shaded gray), the backward critical set $\mathscr{J}^-$ (green dots), the forward critical set $\mathscr{J}^+$ (green curves), as well as certain periodic orbits (blue triangles when attracting, black crosses when saddles, red squares when repelling) with their stable sets (dark-blue and light-blue curves) and unstable sets (red and magenta curves) as appropriate.

4.2.1 Transition for $\mathrm{Im}(c) = 0$

Figure 7 shows the transition through the ambivalence region $\widetilde{\mathscr{M}}_{0.98}$ between the period-one and period-two regions of $\mathscr{M}_{0.98}$ for the symmetric case of $c \in \mathbb{R}$, that is, $\mathrm{Im}(c) = 0$. Panel (a) is for $\mathrm{Re}(c) = -0.674$ inside the period-one region and shows the case of a unique, bounded fixed point attractor p (blue triangle). All points of J_1 converge to p and its basin is bounded by the Julia set $\mathscr{Y}$, which is a Jordan curve. Also shown is a repelling fixed point r and a period-two cycle $s^{\pm}$ (red squares), which are part of $\mathscr{Y}$. Note that this dynamics and, in particular, the topology of the Julia set $\mathscr{Y}$ are qualitatively the same as for parameter values from the main cardioid of the Mandelbrot set $\mathscr{M}_1$; the only difference is that the orbit of c is replaced by a sequence of small closed curves in $\mathscr{J}^+$.

When $\mathrm{Re}(c)$ is decreased the two conjugate points s^+ and s^- undergo a pitchfork bifurcation when the curve L^2 (gray) in Fig. 6a is crossed exactly through the cusp (since c is real). Just after this bifurcation, as in Fig. 7b for $\mathrm{Re}(c) = -0.694$, the attracting fixed point p is still the only bounded attractor and all orbits in $\mathscr{J}^+$ still converge to p. However, the points $s^{\pm}$ are now saddles and the primary branches of their stable sets $W^s(s^{\pm})$ extend to period-two repellors nearby, which are part of $\mathscr{Y}$; hence, the entire stable sets $W^s(s^{\pm})$ (light-blue curves) consist of all preimages of these branches. The unstable sets $W^u(s^{\pm})$ (magenta curves) each have a branch that converges to the attractor p and a branch that extends to infinity. In particular, this means that the sets $W^s(s^{\pm})$ are part of the boundary of $\mathscr{B}(\infty)$ and, hence, $W^s(s^{\pm}) \subset \mathscr{Y}$. In fact, the evidence suggests that the union of $W^s(s^{\pm})$ is dense in the Julia set $\mathscr{Y}$, which is still a Jordan curve; moreover, the set of preperiodic points $\mathscr{P}$ is no longer dense in $\mathscr{Y}$, because $\mathscr{P}$ does not accumulate on $W^s(s^{\pm})$.

The next bifurcation is the period-doubling bifurcation PD^1 (blue) in Fig. 6a. As Figure 7c shows for $\mathrm{Re}(c) = -0.707$, at PD^1 the attracting fixed point p bifurcates with the period-two cycle $s^{\pm}$; in the process p becomes a saddle and the points $s^{\pm}$ disappear. The stable set $W^s(p)$ (blue curves) consists of the segment on the real line that is bounded by the Julia set $\mathscr{Y}$ and all its preimages, such that it forms a quad-tree with branches that meet at the backward critical orbit $\mathscr{J}^-$, which is bounded. On the other hand, the unstable set $W^u(p)$ (red) extends to infinity in both directions. Hence, $W^s(p) \subset \mathscr{Y}$. The basin $\mathscr{B}(\infty)$ is simply connected and contains J_1, except for the two intersection points $J_1 \cap \mathbb{R}$ with the real line; this has been checked with

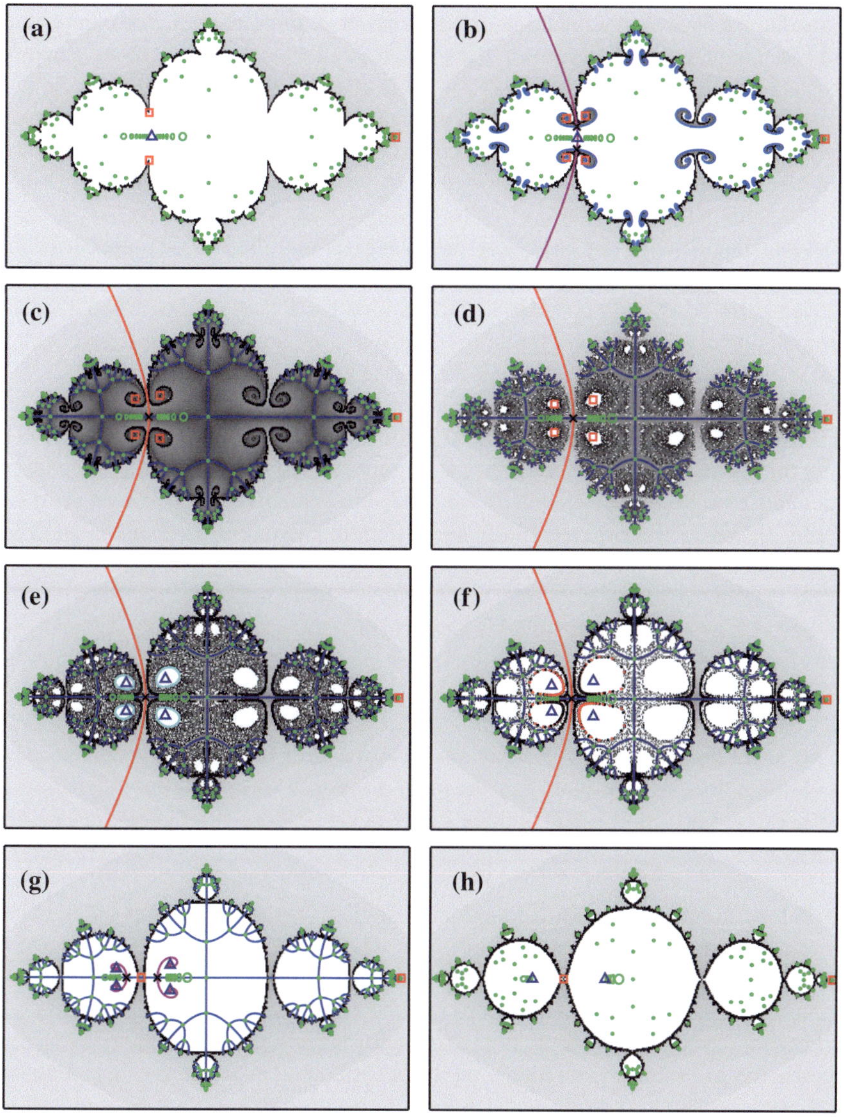

Fig. 7 Phase portraits for the parameter points labeled (7a)–(7h) in Fig. 6, namely, for $\lambda = 0.98$ and $\mathrm{Im}(c) = 0$, and for $\mathrm{Re}(c) = -0.674$ in (**a**), $\mathrm{Re}(c) = -0.694$ in (**b**), $\mathrm{Re}(c) = -0.707$ in (**c**), $\mathrm{Re}(c) = -0.7451$ in (**d**), $\mathrm{Re}(c) = -0.748$ in (**e**), $\mathrm{Re}(c) = -0.756$ in (**f**), $\mathrm{Re}(c) = -0.8$ in (**g**), and $\mathrm{Re}(c) = -0.866$ in (**h**)

separate numerical simulation of many more sample points (not shown). Therefore, we conclude that these parameter values correspond to the ambivalent case of some bounded and some unbounded orbits in $\mathscr{J}^+$. However, since there is no attractor, the Julia set $\mathscr{Y}$ bounds a set with empty interior, even though it contains the set $W^s(p)$.

Note that the backward critical set $\mathcal{J}^-$ and the set $\mathcal{P}$ of preperiodic repelling points are both contained in $\mathcal{Y}$ but are not dense in the Julia set. Overall, we observe that $\mathcal{Y}$ is the closure of $W^s(p)$. Because of the quad-tree structure of $W^s(p)$ we conclude that $\mathcal{Y}$ in Fig. 7c is a *dendrite*, that is, a locally connected, compact and connected set that does not contain any Jordan curves. The closure of the end-points of the branches of the tree is the closure of the set $\mathcal{P}$, which is a Cantor set. In other words, $\mathcal{Y}$ is the union of a Cantor set (given by the closure of $\mathcal{P}$) and the stable set $W^s(p)$; compare Fig. 7c with Fig. 5a1. Note that dendrites as Julia sets also appear in the quadratic family (2) for c in the boundary of the Mandelbrot set $\mathcal{M}_1$ when the point c is preperiodic [12]. In contrast to the situation for $\lambda = 1$, the quad-tree dendrite presented here is structurally stable because p and $W^s(p)$ persist under sufficiently small variations of the parameters of the map (1).

When $\mathrm{Re}(c)$ is decreased further, the forward critical set $\mathcal{J}^+$ interacts with the Julia set $\mathcal{Y}$ for the first time. Figure 7d is the phase portrait for $\mathrm{Re}(c) = -0.7451$, after this bifurcation. Numerical simulation (not shown) suggests that $\mathcal{J}^+$ contains unbounded orbits. More specifically, the set J_1 still contains arcs, on either side of the points $J_1 \cap \mathbb{R}$, that escape to infinity under iteration. On the other hand, in addition to the two bounded orbits of $J_1 \cap \mathbb{R}$, the Julia set $\mathcal{J}^+$ now contains other bounded orbits that appeared due to its interaction with $\mathcal{Y}$. In particular, while $W^s(p)$ is still branched at $\mathcal{J}^-$, the closure of $\mathcal{P}$ is now critically connected along $\mathcal{J}^-$. Hence, $\mathcal{Y}$ is no longer a dendrite and $\mathcal{B}(\infty)$ is no longer simply connected. Rather, the Julia set $\mathcal{Y}$ is the union of a critically connected Cantor set, still given by the closure of $\mathcal{P}$, and the stable set $W^s(p)$; compare Fig. 7d with Fig. 5a2. We refer to this new type of set as a *Cantor shrub* and we think of this set as a dendrite that is also connected at all the points in $\mathcal{J}^-$. Note that the main difference from a Cantor tangle is that the backward critical set $\mathcal{J}^-$ and the set $\mathcal{P}$ of preperiodic repelling points are not dense in $\mathcal{Y}$.

Just after the Neimarck–Sacker bifurcation NS^2 (black) in Fig. 6a, both repelling period-two orbits become attractors (blue triangles) that are surrounded by repelling invariant circles (cyan curves); see Fig. 7e, where $\mathrm{Re}(c) = -0.748$. Hence, these invariant circles and all of their preimages lie in the Julia set. Otherwise, the structure remains unchanged. By this we means that the Julia $\mathcal{Y}$ still contains $\mathcal{J}^-$ and does not accumulate on $W^s(p)$. Since now a countably infinite subset is replaced by Jordan curves, the Julia set $\mathcal{Y}$ is no longer a Cantor shrub. Our evidence strongly suggests that these Jordan curves are not dense in $\mathcal{Y}$ because they do not accumulate on $W^s(p) \subset \mathcal{Y}$; on the other hand, these Jordan curves appear to be dense in $\mathcal{P}$ and, hence, its closure. In other words, the Julia set $\mathcal{Y}$ is the union of a Cantor cheese, given still by the closure of $\mathcal{P}$, and the stable set $W^s(p)$; compare Fig. 7e with Fig. 5a3. We refer to this type of Julia set as a *Cantor beetle* and we think of it as a Cantor shrub where a set of preperiodic repelling points is replaced by preimages of repelling invariant circles.

When $\mathrm{Re}(c)$ is varied, the dynamics on the invariant circles may be quasiperiodic or phase-locked. Figure 7f shows an example with $\mathrm{Re}(c) = -0.756$, where the dynamics is phase-locked; more precisely, there are two period-70 repellors (red dots) and two period-70 saddles (black dots) on the invariant curves, which are given by the closures of the corresponding primary branches of the stable sets of these saddles, respectively (not shown). Otherwise, the structure of the phase portrait is unchanged from that shown in panel (e), and $\mathscr{Y}$ is still a Cantor beetle. We remark that the invariant curves and, hence, the Cantor beetle are structurally stable irrespective of whether the dynamics is locked or not (provided strong resonances are avoided) [3, 27, 33]. On the other hand, it is possible that the invariant curves lose their smoothness (technically, their normal hyperbolicity) in different scenarios associated with overlapping resonance tongues [4, 23]; what this means for the structure of the Julia set is not considered in detail here.

When $\text{Re}(c)$ is decreased further, the unstable set $W^u(p)$ interacts with $\mathscr{Y}$ and the invariant circles; for details, we refer to a similar (unsymmetric) situation in Fig. 8j that will be discussed in Sect. 4.2.2. After the period-doubling bifurcation PD^1 (blue) in Fig. 6a is crossed a second time, the saddle fixed point p has become a repellor (red square) and is, hence, in the Julia set $\mathscr{Y}$; at the same time, a period-two saddle $q = \{q_1, q_2\}$ (black crosses) has appeared on the real line near p; see Fig. 7g, where $\text{Re}(c) = -0.8$. Here, $W^u(q)$ (magenta curves) converges to the two period-two attractors (blue triangles). The stable set $W^s(q)$ (light-blue curves) separates the interior of the set bounded by $\mathscr{Y}$ into the two basins of the period-two attractors. In particular, the Julia set $\mathscr{Y}$ now consists of a union of Jordan curves, which are connected at p and its preimages in $\mathscr{Y}$. The orbits in $\mathscr{J}^+ \cap \mathbb{R}$ go to the saddle q and the points in $\mathscr{J}^+ \backslash \mathbb{R}$ go to one of the attractors. Therefore, all orbits in $\mathscr{J}^+$ are now bounded and the parameter value $\text{Re}(c) = -0.8$ does not lie in the ambivalence region $\widetilde{\mathscr{M}}_{0.98}$ but in the period-two region of the core $\mathscr{M}^o_{0.98}$. The Julia set $\mathscr{Y}$ bounds the closure of the basins of the two period-two attractors and is a connected union of Jordan curves. Finally, the saddle-node bifurcation L^2 (gray) in Fig. 6a is crossed again, at the second cusp (corresponding to a pitchfork bifurcation), where the two symmetric period-two attractors disappear and the period-two saddle q becomes a period-two attractor. As Fig. 7h shows for $\text{Re}(c) = -0.866$, all points on J_1 now converge to this unique attractor. Indeed, the Julia set $\mathscr{Y}$ is still a connected union of Jordan curves, and the phase portrait is qualitatively as for parameter values from the period-two bulb of the Mandelbrot set $\mathscr{M}_1$, except that the orbit of c is replaced by a sequence of small closed curves.

4.2.2 Transition for $\text{Im}(c) = 0.008$

We now consider the case that $c \in \mathbb{C}$ has a nonzero imaginary part, which we fix at $\text{Im}(c) = 0.008$, and focus again on the transition through the ambivalence region $\widetilde{\mathscr{M}}_{0.98}$ between the period-one and period-two regions of $\mathscr{M}_{0.98}$. The corresponding parameter points we chose for this transition are labeled (8a)–(8l) in Fig. 6; notice that the parameter path for decreasing $\text{Re}(c)$ and fixed $\text{Im}(c) = 0.008$ now also crosses the curves NS^2 (black) of Neimarck-Sacker bifurcations, traverses a region with more complicated attractors and then crosses curves B (purple) and H (magenta) of backward critical and homoclinic tangencies, respectively. As a result, this transition is much more complicated with a lot of fine detail.

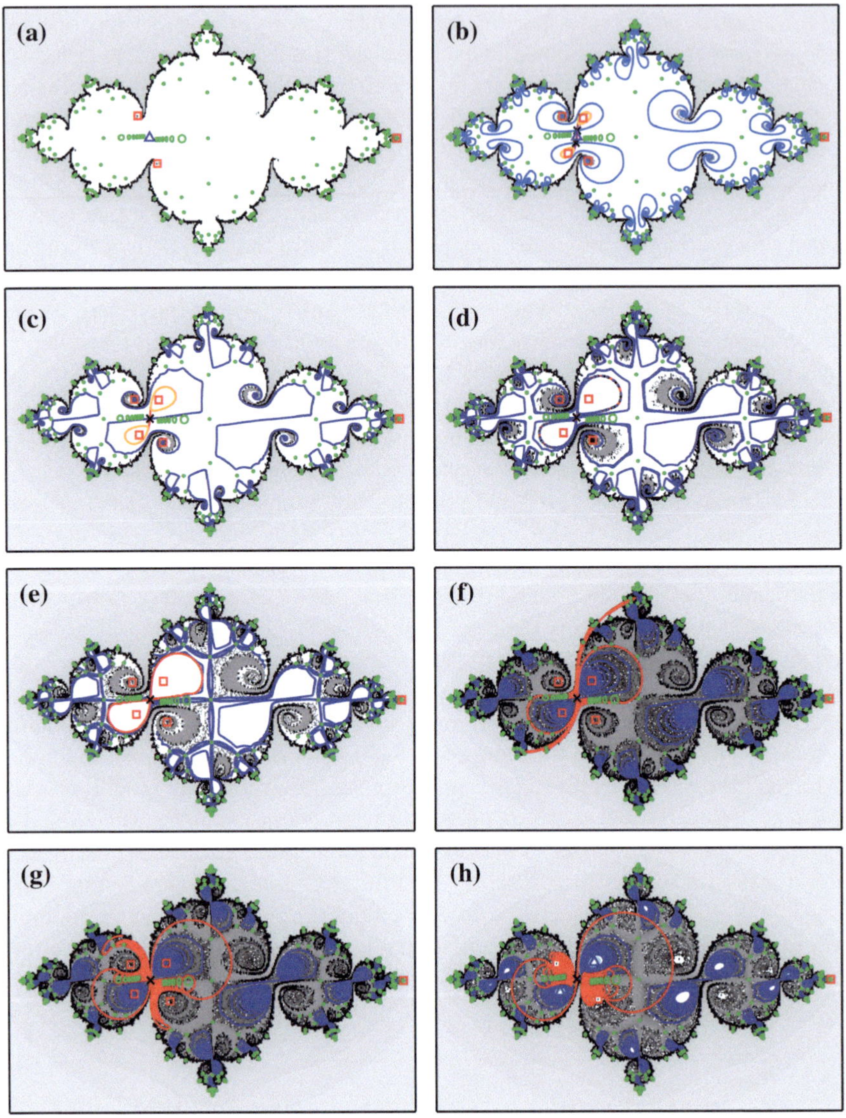

Fig. 8 Phase portraits for the parameter points labeled (8a)–(8l) in Fig. 6, namely, for $\lambda = 0.98$ and $\mathrm{Im}(c) = 0.008$, and for $\mathrm{Re}(c) = -0.687$ in (**a**), $\mathrm{Re}(c) = -0.699$ in (**b**), $\mathrm{Re}(c) = -0.707$ in (**c**), $\mathrm{Re}(c) = -0.719$ in (**d**), $\mathrm{Re}(c) = -0.724$ in (**e**), $\mathrm{Re}(c) = -0.731$ in (**f**), $\mathrm{Re}(c) = -0.739$ in (**g**), and $\mathrm{Re}(c) = -0.745$ in (**h**), for $\mathrm{Re}(c) = -0.754$ in (**i**), $\mathrm{Re}(c) = -0.777$ in (**j**), $\mathrm{Re}(c) = -0.794$ in (**k**), and $\mathrm{Re}(c) = -0.824$ in (**l**)

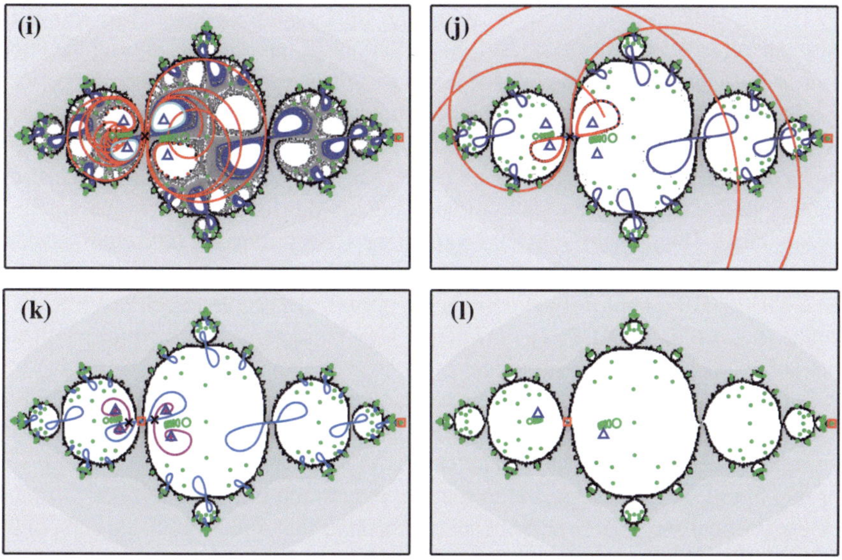

Fig. 8 (continued)

Figure 8 shows a series of phase portraits that represent the main gist of this transition. In panel (a) the phase portrait is for $\text{Re}(c) = -0.687$ from the period-one region of the core $\mathscr{M}^o_{0.98}$. Therefore, it is qualitatively as in Fig. 7a, with a unique attractor p that admits a basin bounded by the Jordan curve $\mathscr{Y}$. However, this and all phase portraits of Fig. 8 are no longer symmetric with respect to complex conjugation since $\text{Im}(c) \neq 0$; note that taking $\text{Im}(c) = -0.008$ leads to the mirror images of the respective phase portraits.

Figure 8b shows the phase portrait for $\text{Re}(c) = -0.699$, after the saddle-node bifurcation L^2 (gray) and the Neimarck–Sacker bifurcation NS^2 (black) in Fig. 6; note that NS^2 ends at a 1 : 1-resonance point on L^2. At L^2 a period-two saddle (black crosses) and a period-two attractor are created; when NS^2 is crossed, the attractor becomes a repellor (red squares) and it is surrounded by attracting invariant circles $\mathscr{C}^{\pm}$ (orange curves) that map to each other, that is, they are also two-periodic. As a result, the attracting fixed point p is no longer the only finite attractor, because it now coexists with the attracting period-two invariant circles. The two basins are bounded by the stable set of the period-two saddle, which we again refer to as $s^{\pm}$ (even though the two points are now not related by complex conjugation). Note that both primary branches of $W^s(s^{\pm})$ (light-blue curves) end up at repelling periodic points that are part of the Julia set $\mathscr{Y}$. One each of the branches of $W^u(s^{\pm})$ (magenta curves) ends at p while the other accumulates on $\mathscr{C}^{\pm}$. Still, J_1 lies in the basin $\mathscr{B}(p)$ of p and $\mathscr{Y}$ is a Jordan curve, but it no longer bounds a unique basin; moreover, because of the existence of the additional repelling period-two orbit, $\mathscr{Y}$ is now a strict subset of $\mathscr{P}$.

After the period-doubling bifurcation PD[1] (blue) is crossed, which is shown in Fig. 8c for $\mathrm{Re}(c) = -0.707$, the period-two saddle $s^{\pm}$ has disappeared and the fixed point p is no longer attracting but now a saddle (black cross). As a result, the period-two invariant circles $\mathscr{C}^{\pm}$ form the only attractor and $\mathscr{J}^{+}$ accumulates on them. The unstable set $W^{u}(p)$ (red curve) accumulates on $\mathscr{C}^{\pm}$, while the stable set $W^{s}(p)$ (blue curves) bounds the basins $\mathscr{B}(\mathscr{C}^{\pm})$ of $\mathscr{C}^{\pm}$; its two primary branches (emanating from p) end at the repelling period-two points on the Julia set $\mathscr{Y}$, which is still a Jordan curve. The dynamics on the invariant circles $\mathscr{C}^{\pm}$ may again be quasiperiodic or phase-locked. Figure 8d shows an example with $\mathrm{Re}(c) = -0719$ for which there are a period-318 attractor (blue dots) and a period-318 saddle (black dots) on $\mathscr{C}^{\pm}$. Notice that the stable set $W^{s}(p)$ is already very close to $\mathscr{J}^{+}$, which means that a forward critical tangency is imminent.

In Fig. 8e, for $\mathrm{Re}(c) = -0.724$, the forward critical tangency has taken place and the primary branch of $W^{s}(p)$ now intersects $\mathscr{J}^{+}$, which also means that $W^{s}(p)$ now consists of curve segments that connect at $\mathscr{J}^{-}$. Moreover, the parameter point for this figure seems to lie just after the first interaction of the Julia set $\mathscr{Y}$ with the forward critical set $\mathscr{J}^{+}$. Now there are points on J_1 that remain bounded and points that escape to infinity, meaning that $c \in \widetilde{\mathscr{M}}_{0.98}$. Furthermore, J_0 and, hence, all points in $\mathscr{J}^{+}$ are now in $\mathscr{Y}$. This means that the Julia set $\mathscr{Y}$ is critically connected. Notice further, that $\mathscr{Y}$ accumulates on $W^{s}(p)$ from one side. Another feature of Fig. 8e is that this phase portrait is very close to a first homoclinic tangency of $W^{s}(p)$ and $W^{u}(p)$, which marks the disappearance of the invariant circles $\mathscr{C}^{\pm}$ (in a complicated scenario that involves the loss of normal hyperbolicity of $\mathscr{C}^{\pm}$). Numerical evidence (not shown) indicates that $W^{u}(p)$ remains bounded so that its closure contains an attractor. In particular, the unstable set $W^{u}(p)$ does not appear to intersect the Julia set $\mathscr{Y}$; therefore, $\mathscr{Y}$ still bounds a bounded basin of attraction (white region) and appears to be a (critically connected) union of Jordan curves.

Figure 8f shows the situation for $\mathrm{Re}(c) = -0.731$, after a last homoclinic tangency between $W^{s}(p)$ and $W^{u}(p)$. Notice that the primary branch of $W^{s}(p)$ now converges to the period-two repelling cycle (red squares). In particular, $W^{s}(p)$ no longer intersects $\mathscr{J}^{+}$ and, hence, is no longer connected at $\mathscr{J}^{-}$. Notice further that the unstable set $W^{u}(p)$ now intersects the Julia set $\mathscr{Y}$; therefore, there are points on $W^{u}(p)$ that remain bounded and segments of $W^{u}(p)$ that do not. We conclude that the interaction of $W^{u}(p)$ with $\mathscr{Y}$ is a boundary crisis, where the attractor in the closure of $W^{u}(p)$ and its basin disappear at some parameter value between panels (e) and (f). Indeed, numerical simulation (not shown) suggests that there is no finite attractor for $\mathrm{Re}(c) = -0.731$ shown in Fig. 8f. In particular, the basin $\mathscr{B}(\infty)$ is dense and the Julia set $\mathscr{Y}$ is the closure of the stable set $W^{s}(p)$. Now that the branch points $\mathscr{J}^{-}$ are contained in $\mathscr{Y}$, the circle J_1 seems to contain segments in $\mathscr{B}(\infty)$, and $\mathscr{B}(\infty)$ seems to be simply connected, we conclude that $\mathscr{Y}$ is a critically connected Cantor set; compare with Fig. 5b3.

In the phase portrait of Fig. 8g, for $\mathrm{Re}(c) = -0.739$, the first backward critical tangency B (purple) has been crossed. This means that the unstable set $W^{u}(p)$ has crossed over J_0 and now forms loops around the circles in $\mathscr{J}^{+}$; yet $\mathscr{Y}$ is still a critically connected Cantor set. Notice the intersections between $W^{u}(p)$ and backward images of the primary branch of $W^s(p)$, meaning that some subsequences of points on $W^u(p)$ converge to the period-two repellor. Moreover, this phase portrait is very close to yet another homoclinic tangency of the primary branch of $W^s(p)$ and $W^u(p)$. After this homoclinic tangency, the two invariant sets intersect, as is illustrated in Fig. 8h for $\mathrm{Re}(c) = -0.745$. Moreover, the Neimarck–Sacker bifurcation NS^2 (black) has been crossed as well. Therefore, one of the repelling period-two orbits is now a unique period-two attractor (blue triangles), which is surrounded by repelling invariant circles (cyan curves) that form its basin boundary together with all their preimages. We conclude that $\mathscr{Y}$ is now a Cantor cheese; compare with Fig. 5f.

Subsequently, the Neimarck–Sacker bifurcation NS^2 (black) is crossed for the second time and the second period-two repellor becomes a period-two attractor, surrounded by repelling invariant circles $\mathscr{D}^{\pm}$; see Fig. 8i, where $\mathrm{Re}(c) = -0.754$. For the chosen parameter point, the dynamics on $\mathscr{D}^{\pm}$ is phase-locked of period 56, with a corresponding repellor (red dots) and saddle (black dots) periodic points. Note that the previous invariant circles $\mathscr{C}^{\pm}$ still exist and have dynamics that is quasiperiodic (or of very high period). Numerical simulation (not shown) suggests that $\mathscr{J}^+$ still contains both bounded and unbounded orbits. We conclude that $\mathscr{Y}$ is still a Cantor cheese, but now bounds the basins of the two different period-two attractors. When $\mathrm{Re}(c)$ is decreased to $\mathrm{Re}(c) = -0.777$, shown in Fig. 8j, the invariant circle $\mathscr{C}^{\pm}$ still exists and the dynamics on it is locked with a period-274 repellor (red dots) and a period-274 saddle (black dots). The repelling invariant circles $\mathscr{D}^{\pm}$, on the other hand, have disappeared and $\mathscr{J}^+$ now lies in the basin of attraction of the corresponding period-two attractor. The set $\mathscr{P}$ appears to accumulate on $W^s(p)$ and the two primary branches of $W^s(p)$ accumulate on the repelling invariant circles $\mathscr{C}^{\pm}$. Overall, we conclude that c now again lies in the core $\mathscr{M}^o_{0.98}$ and that $\mathscr{Y}$ is a countable union of Jordan curves, which is connected at p and its preimages. Notice that $W^u(p)$ still intersects $\mathscr{Y}$, but it has much longer segments in $\mathscr{B}(\infty)$.

As $\mathrm{Re}(c)$ is decreased further, the period-doubling bifurcation PD^1 (blue) is crossed and the point p becomes a repelling periodic point again (red square). At the same time, a period-two saddle is created. We remark that crossing PD^1 is preceded by $W^u(p)$ no longer intersecting $\mathscr{Y}$, from when on $W^u(p) \subset \mathscr{B}(\infty)$. The stable and unstable sets of the period-two saddle then encounter first and last homoclinic tangencies, which result in the disappearance of the invariant circles $\mathscr{C}^{\pm}$. As Fig. 8k shows for $\mathrm{Re}(c) = -0.794$, the stable sets (light-blue curves) now bound the basin of the two different period-two attractors, at which the respective branches of the unstable sets (magenta curves) end. Finally, $\mathrm{Re}(c)$ is decreased beyond the saddle-node bifurcation L^2 (gray), where a period-two saddle and a period-two attractor disappear. Figure 8l is for $\mathrm{Re}(c) = -0.824$ and shows that there is now a unique period-two attractor left and its basin is bounded by $\mathscr{Y}$, which remains a countable union of Jordan curves. Hence, c is now in the period-two region of the core $\mathscr{M}^o_{0.98}$, and the phase portrait is qualitatively as that in Fig. 7h.

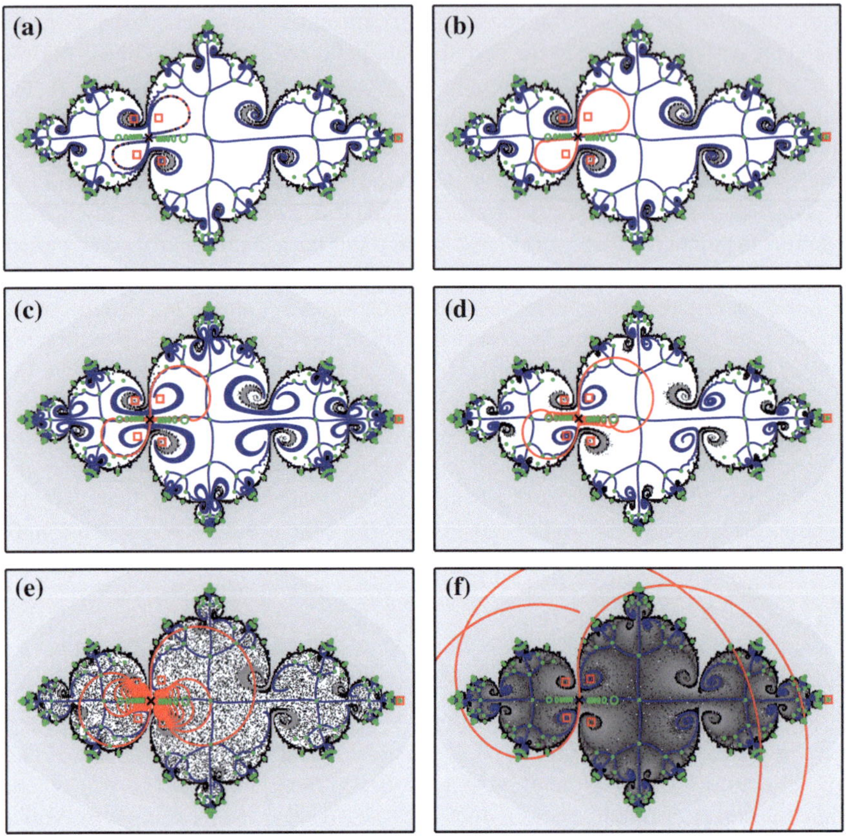

Fig. 9 Phase portraits for the parameter points labeled (9a)–(9f) in Fig. 6, namely, for $\lambda = 0.98$ and $\text{Re}(c) = -0.707$, and for $\text{Im}(c) = 0.0055$ in (**a**), $\text{Im}(c) = 0.0045$ in (**b**), $\text{Im}(c) = 0.0039$ in (**c**), $\text{Im}(c) = 0.0033$ in (**d**), $\text{Im}(c) = 0.0026$ in (**e**), and $\text{Im}(c) = 0.0017$ in (**f**)

4.2.3 Transition for $\text{Re}(c) = -0.707$

An important feature of the transition for $\text{Im}(c) = 0.008$ is that the parameter crosses the region shown in Fig. 6b with dynamics on tori, resonance tongues and many homoclinic, forward and backward tangency bifurcations (only very few of which are shown in Fig. 6). In order to illustrate how the additional complexity arises, we present in Fig. 9 six phase portraits of the map (1) for fixed $\text{Re}(c) = -0.707$ at the parameter points labeled (9a)–(9f) in Fig. 6; they illustrate the transition from Fig. 8c to Fig. 7c.

Initially, for c in the black region past the Neimarck–Sacker bifurcation NS^2, the phase portrait is very similar to that in Fig. 8c. Figure 9a is for $\text{Im}(c) = 0.0055$ and shows that the fixed point p is a saddle and the period-two orbits are repellors surrounded by attracting invariant circles $\mathscr{C}^{\pm}$. The dynamics on them is phase-locked with period-230 attracting points (blue dots) and period-230 saddle points (blue dots). Note that the stable set $W^s(p)$ intersects J_1, which means that a forward tangency has taken place; as a result, $W^s(p)$ is a quad-tree that is branched at the points of $\mathscr{J}^-$. Nevertheless, $W^s(p)$ still forms the boundary between the basins of the two invariant circles $\mathscr{C}^\pm$. The two branches of the unstable set $W^u(p)$ accumulate on $\mathscr{C}^\pm$. The Julia set $\mathscr{Y}$ is a Jordan curve. The phase portrait for $\mathrm{Im}(c) = 0.0045$ in Fig. 9b is qualitatively the same, except that the dynamics on $\mathscr{C}^\pm$ is quasiperiodic or of very high period. Note that $W^s(p)$ and $W^u(p)$ are already close to each other, which means that the first homoclinic tangency H is imminent.

After the first homoclinic tangency, as is shown in Fig. 9c for $\mathrm{Im}(c) = 0.0039$, the invariant sets $W^s(p)$ and $W^u(p)$ intersect and form a homoclinic tangle. Numerical simulation indicates that the closure of $W^u(p)$ is a 'very thin' chaotic attractor and that there is no other attractor. Notice also that $W^u(p)$ is now very close to J_0, meaning that the first backward critical tangency B is imminent.

After the first backward critical tangency, as shown in Fig. 9d for $\mathrm{Im}(c) = 0.0033$, the unstable set $W^u(p)$ has moved over J_0 and a sequence of its preimages in $\mathscr{J}^-$ and, therefore, forms loops around the circles in $\mathscr{J}^+$. We observe that the closure of $W^u(p)$ is still a chaotic attractor.

When $\mathrm{Im}(c)$ is decreased further, the chaotic attractor appears to reach the boundary of its basin, which is the Julia set $\mathscr{Y}$. At this point, $\mathscr{Y}$ ceases to be a Jordan curve. As Fig. 9e illustrates for $\mathrm{Im}(c) = 0.0026$, $\mathscr{Y}$ now also contains $W^s(p)$. Moreover, we conjecture that the disappearance in this type of boundary crisis results in the creation of a hyperbolic set Λ (or chaotic saddle or invariant Cantor set of a horseshoe); this is reminiscent of, for example, the transition from turbulence to preturbulence in the Lorenz system [16]. The stable set of Λ will, therefore, also be in $\mathscr{Y}$ and, since there are branch points on $W^s(p)$, we conjecture that the Julia set is a 'very thick' Cantor shrub (it has a large Hausdorff dimension) in this case. As $\mathrm{Im}(c)$ is decreased, we conjecture that the set Λ undergoes infinitely many bifurcations where periodic and homoclinic orbits disappear. As a result, the Cantor shrub becomes 'thinner' (it has a smaller Hausdorff dimension), as in Fig. 9f for $\mathrm{Im}(c) = 0.0017$. Finally, Λ disappears and only p and $W^s(p)$ are left, in which case the Julia set $\mathscr{Y}$ is a dendrite as in Fig. 7c.

5 Conclusions

In this paper we considered the family of angle-doubling maps, given by map (1) with fixed $a = 2$, as a perturbation of the complex quadratic map (2), which one finds for $\lambda = 1$. As soon as $\lambda < 1$, the map (1) is no longer a complex analytic map on $\mathbb{C}$. Rather, it is a map on $\mathbb{C} \setminus \{0\}$, which maps to the outside of a circle J_1 of radius $1 - \lambda$ around the former critical value c in a two-to-one fashion. We showed that the dynamics can still be understood by considering the forward critical orbit of the critical circle J_1, as well as the backward orbit of $J_0 = 0$, which we still refer to as the critical point. The fact that J_1 is a one-dimensional set now gives rise to the fundamental trichotomy that all points on J_1 are unbounded, all points of J_1 are bounded, or some points of J_1 are unbounded and some others are bounded under iteration of the map.

This trichotomy is encoded in the generalized Mandelbrot set $\mathcal{M}$, which we define as the set of parameter values with at least one bounded orbit on J_1, with its core $\mathcal{M}^o$ for which all points of J_1 are bounded. The third possibility of the trichotomy is associated with the ambivalence region $\widetilde{\mathcal{M}} = \mathcal{M} \setminus \mathcal{M}^o$, which gives rise to a lot of extra possibilities for the structure of the generalized Julia set $\mathcal{Y}$. We define $\mathcal{Y}$ as the boundary of the basin of attraction $\mathcal{B}(\infty)$ of infinity, because $\mathcal{Y}$ is, in general, not the closure of the preperiodic points when $\lambda < 1$. We studied $\mathcal{M}$ as the λ-union of the Mandelbrot sets $\mathcal{M}_\lambda$ in the complex c-plane and found that $\mathcal{M}_\lambda$ changes very rapidly with λ. More specifically, periodic components disappear, and growing regions with new types of dynamics arise near the boundary of the standard Mandelbrot set $\mathcal{M}_1$, as well as near the centers of its hyperbolic components. We pointed out a hidden symmetry, made visible via a rescaling, which means that it is sufficient to consider only $\lambda \in [0.5, 1]$; this was illustrated with the bifurcation diagram in the $(\mathrm{Re}(c), \lambda)$-plane.

As we showed, key roles in this overall complex picture are played by additional invariant objects that may exist as soon as $\lambda < 1$, namely, stable and unstable sets of saddle periodic orbits and invariant curves with quasiperiodic or locked dynamics. In particular, these objects can interact with J_0 and J_1 and the set of preperiodic points in different types of bifurcations. The result is a complicated interplay between invariant objects associated with complex dynamics (the basin of infinity and the closure of the set of preperiodic points), invariant objects and bifurcations associated with diffeomorphisms (invariant curves, resonance tongues, homoclinic tangencies and chaotic attractors) and invariant objects and bifurcations associated with noninvertible maps (the critical circle, stable and unstable sets and associated tangency bifurcations).

To make this point, we considered how the phase portrait changes in the transition through the ambivalence region from the period-one region to the period-two region of $\mathcal{M}_{0.98}$. More precisely, for fixed $\lambda = 0.98$, we presented the changes to the Julia set, the attractors, the forward and backward critical sets, and stable and unstable sets of saddle points as $\mathrm{Re}(c)$ decreases for the two cases that $\mathrm{Im}(c) = 0$ (when the map and the phase portrait are symmetric under complex conjugation) and $\mathrm{Im}(c) = 0.008$ (when they are not). Moreover, we also presented a transition between these two cases by varying $\mathrm{Im}(c)$ for a fixed value of $\mathrm{Re}(c)$. These transitions reveal an intriguing and complicated bifurcation structure that involves, in particular, Neimarck–Sacker bifurcations and resonance tongues, as well as infinite sequences of global bifurcations, namely tangency bifurcations of different kinds. Our numerical evidence supports the natural conjecture that quite similar transitions arise when passing through any two adjoining hyperbolic components (with a likely difference between cases involving strong resonances and those that do not).

We focused here on the implications for the structure of the Julia set $\mathcal{Y}$ in these transitions. This allowed us to identify the following basic and structurally stable types of Julia sets of the map (1), most of which appear in (open regions of) the ambivalence region $\widetilde{\mathcal{M}}$ of the generalized Mandelbrot set.

Result 1 (Trichotomy and types of Julia sets for the map (1))

(A) *In the complement of* $\mathscr{M}$, *all orbits in* $\mathscr{J}^+$ *are unbounded, there is no bounded attractor, and* $\mathscr{Y}$ is a Cantor set; *see, for example, Fig. 5a1.*

(B) *In the core* $\mathscr{M}^o$, *all orbits in* $\mathscr{J}^+$ *are bounded, there is set of bounded attractors, and* $\mathscr{Y}$ is a connected union of Jordan curves *bounding the union of their basins; note that if there is more than one attractor then the stable set of one or more saddle objects act as a boundary between their basins. Examples are Figs. 5b1, 7a, g, and 8a–c, k–l.*

(C) *In the ambivalent region* $\widetilde{\mathscr{M}}$ *there is at least one bounded and one unbounded orbit in* $\mathscr{J}^+$, *and we find the following structurally stable subcases:*

(i) *There is no bounded attractor, the sets* J_1 *and* $\mathscr{Y}$ *intersect in a set that is not dense in* J_1, *and there are no isolated points of* J_1 *that remain bounded; hence,* $J_1 \setminus \mathscr{Y}$ *consists of segments of* J_1 *that go to infinity under iteration. Then* $\mathscr{Y}$ is a critically connected Cantor set, *with connection points given by the set* $\mathscr{J}^-$, *which is dense in* $\mathscr{Y}$; *hence,* $\mathscr{Y} \setminus \mathscr{J}^-$ *is not connected. Moreover,* $\mathscr{P}$ *is dense in* $\mathscr{Y}$; *see Fig. 5a2.*

(ii) *The sets* J_1 *and* $\mathscr{Y}$ *intersect in a set that is not dense in* J_1, *there are no isolated points of* J_1 *that remain bounded, and there is at least one bounded hyperbolic attractor; hence,* $J_1 \setminus \mathscr{Y}$ *consists of segments of* J_1 *that go to infinity or to the bounded hyperbolic attractor(s) under iteration. Then* $\mathscr{Y}$ is a Cantor cheese, *with connection points given by the set* $\mathscr{J}^-$, *which is dense in* $\mathscr{Y}$; *hence,* $\mathscr{Y} \setminus \mathscr{J}^-$ *is not connected and contains a dense set of Jordan curves bounding the basin(s). Moreover,* $\mathscr{P}$ *is dense in* $\mathscr{Y}$; *see Fig. 5a3.*

(iii) *There is no bounded attractor, the sets* J_1 *and* $\mathscr{Y}$ *intersect in a dense set in* J_1, *and the critical point* J_0 *lies in the disk bounded by* J_1. *Then* $\mathscr{Y}$ is a Cantor bouquet, *with explosion point* J_0; *that is,* $\mathscr{Y}$ *is a connected union of arcs that are connected at* J_0, *such that their other end points, which include* $\mathscr{P}$, *are dense in* $\mathscr{Y}$; *see Fig. 5b2.*

(iv) *There is no bounded attractor, the sets* J_1 *and* $\mathscr{Y}$ *intersect in a dense set in* J_1, *and the critical point* J_0 *does not lie in the disk bounded by* J_1. *Then* $\mathscr{Y}$ is a Cantor tangle, *consisting of arcs with* $\mathscr{J}^-$ *as a dense set of explosion points; see Fig. 5b3.*

(v) *There is no bounded attractor, yet* $\mathscr{J}^+$ *intersects the stable set of a saddle point* p *so that it has exactly two bounded orbits. Then* $\mathscr{Y}$ is a dendrite, *given as the disjoint union of* $W^s(p)$ *and the closure of the set* $\mathscr{P}$ *of repelling preperiodic points, which is a Cantor set. Note that the sets* $\mathscr{P}$ *and* $\mathscr{J}^-$ *are contained in, but are not dense in* $\mathscr{Y}$; *see Fig. 7c.*

(vi) *There is no bounded attractor, yet* $\mathscr{J}^+$ *intersects the stable set of a saddle point* p *as well as* $\mathscr{Y}$. *Then* $\mathscr{Y}$ is a Cantor shrub, *given as the disjoint union of* $W^s(p)$ *and the closure of the set* $\mathscr{P}$, *which is a critically connected Cantor set. Note that the sets* $\mathscr{P}$ *and* $\mathscr{J}^-$ *are contained in, but are not dense in* $\mathscr{Y}$; *see Fig. 7d.*

(vii) *The forward critical set* $\mathscr{J}^+$ *intersects the stable set of a saddle point* p *as well as* $\mathscr{Y}$, *and there is at least one bounded hyperbolic attractor. Then* $\mathscr{Y}$

is a Cantor beetle, *given as the disjoint union of* $W^s(p)$ *and the closure of the set* $\mathscr{P}$, *which is a Cantor cheese. Note again that the sets* $\mathscr{P}$ *and* $\mathscr{J}^-$ *are contained in, but are not dense in* $\mathscr{Y}$; *see Fig. 7e.*

We remark that these findings have been obtained by careful numerical studies with state-of-the-art, effective and accurate methods for the bifurcation analysis of planar maps. These include the computation of basins and their boundaries, of large numbers of preperiodic points, of the forward and backward critical sets, and of stable and unstable sets; furthermore, we determined underlying bifurcations, including different types of tangency bifurcations, and continued them in suitable parameter planes. Therefore, we are confident that our observations are correct regarding the generalized Mandelbrot set and associated different types of main Julia sets as listed above.

What we presented here has the status of a numerically derived conjecture, which we hope will stimulate future research. To put Result 1 into perspective, cases (A) and (B) are direct generalizations of what the Julia set looks like away from the boundary of the standard Mandelbrot set $\mathscr{M}_1$, that is, in its complement and for c from a hyperbolic component. A slight difference here is that there may be more than one attractor (since J_1 is now a circle, rather than a critical value). Since the Julia set $\mathscr{Y}$ remains bounded away from the critical orbit $\mathscr{J}^+$, we believe that it may be possible to construct a proof by considering a remainder of the complex structure in an open region around $\mathscr{Y}$, possibly by using quasiconformal arguments. Proving any of the cases under (C) will likely require quite different methods, due to the interaction between $\mathscr{Y}$ and J_1, which is somewhat reminiscent of the case that the critical value c lies on the boundary of $\mathscr{M}_1$. Finally, we remark that the new cases (C)(v)–(vii) are closely related to (A), (C)(i) and (C)(ii), the difference being the existence of a saddle point with a stable set in the form of a quad-tree.

Result 1 constitutes a high-level summary of prominent basic types of Julia sets in the map (1), without a claim of being exhaustive. In light of the many subtleties arising from overlapping regions of resonance tongues and cascades of tangency bifurcations, there may well be further types. In particular, we expect that saddle periodic orbits of higher periods and other invariant objects of saddle type may well need to be studied in much more detail. An example of this is the creation of a hyperbolic set in a boundary crisis, for which we found its stable set to be part of the Julia set. We conjectured that this type of Julia set is a very thick Cantor shrub, which then disappears. We suspect that this happens via the untangling of the underlying horshoe in cascades of homoclinic bifurcations. The further study of this structure and its subsequent bifurcations will be an interesting topic of future research. Another open direction for further investigation is to determine the exact nature of the dynamics that arises for $\lambda < 1$ near the centers of the hyperbolic components of $\mathscr{M}_1$; we suspect again that there is some underlying overall bifurcation structure, irrespective of the period of the superattracting periodic point (certainly when the period is at least 5).

Acknowledgements The research of BK and HMO was supported by grant 16-UOA-286 from the Marsden Fund managed by the Royal Society Te Apārangi.

References

1. Aarts, J.M., Oversteegen, L.G.: The geometry of Julia sets. Trans. Am. Math. Soc. **338**, 897–918 (1993)
2. Abraham, R.H., Gardini, L., Mira, C.: Chaos in Discrete Dynamical Systems: A Visual Introduction in 2 Dimensions. Springer, New York (1997)
3. Arnol′d, V.I.: Catastrophe Theory, 3rd ed. Springer, Berlin (1992). Translated from the Russian by G. S. Wassermann, Based on a translation by R. K. Thomas
4. Aronson, D.G., Chory, M.A., Hall, G.R., McGehee, R.P.: Bifurcations from an invariant circle for two-parameter families of maps of the plane: a computer-assisted study. Comm. Math. Phys. **83**, 303–354 (1982)
5. Back, A., Guckenheimer, J., Myers, M., Wicklin, F., Worfolk, P.: DsTool: Computer assisted exploration of dynamical systems. Notices Am. Math. Soc. **39**, 303–309 (1992)
6. Bamón, R., Kiwi, J., Rivera-Letelier, J.: Wild Lorenz-like attractors, 0508045 (2006). http://arxiv.org/abs/math/0508045
7. Beyn, W.-J., Kleinkauf, J.-M.: The numerical computation of homoclinic orbits for maps. SIAM J. Numer. Anal. **34**, 1207–1236 (1997)
8. Bielefeld, B., Sutherland, S., Tangerman, F., Veerman, J.J.P.: Dynamics of certain nonconformal degree-two maps of the plane. Experiment. Math. **2**, 281–300 (1993)
9. Blanchard, P.: Complex analytic dynamics on the Riemann sphere. Bull. Am. Math. Soc. **11**, 85–141 (1984)
10. Bruin, H., van Noort, M.: Nonconformal perturbations of $z \mapsto z^2 + c$: the 1:3 resonance. Nonlinearity **17**, 765–789 (2004)
11. Bula, W.D., Oversteegen, L.G.: A characterization of smooth Cantor bouquets. Proc. Am. Math. Soc. **108**, 529–534 (1990)
12. Carleson, L., Gamelin, T.W.: Complex Dynamics. Springer, New York (1993)
13. Devaney, R.L.: An Introduction to Chaotic Dynamical Systems. Westview Press, Boulder (2003)
14. Devaney, R.L., Krych, M.: Dynamics of $\exp(z)$. Ergodic Theory Dynam. Syst. **4**, 35–52 (1984)
15. Dhooge, A., Govaerts, W., Kuznetsov, Yu.A.: MatCont: a Matlab package for numerical bifurcation analysis of ODEs. ACM Trans. Math. Soft. **29**, 141–164 (2003). http://sourceforge.net/projects/matcont
16. Doedel, E.J., Krauskopf, B., Osinga, H.M.: Global invariant manifolds in the transition to preturbulence in the Lorenz system. Indag. Math. (N.S.) **22**, 222–240 (2011)
17. England, J.P., Krauskopf, B., Osinga, H.M.: Computing one-dimensional stable manifolds and stable sets of planar maps without the inverse. SIAM J. Appl. Dyn. Syst. **3**, 161–190 (2004)
18. Ghaziani, R.K., Govaerts, W., Kuznetsov, Yu.A., Meijer, H.G.E.: Numerical continuation of connecting orbits of maps in Matlab. J. Differ. Equ. Appl. **15**, 849–875 (2009)
19. Govaerts, W., Kuznetsov, Yu.A., Khoshsiar Ghaziani, R., Meijer, H.G.E.: Cl_MatContM: a toolbox for continuation and bifurcation of cycles of maps (2008). Software. http://sourceforge.net/projects/matcont
20. Hittmeyer, S., Krauskopf, B., Osinga, H.M.: Interacting global invariant sets in a planar map model of wild chaos. SIAM J. Appl. Dyn. Syst. **12**, 1280–1329 (2013)
21. Hittmeyer, S., Krauskopf, B., Osinga, H.M.: From wild lorenz-like to wild rovella-like dynamics. Dynam. Syst. **30**, 525–542 (2015)
22. Hittmeyer, S., Krauskopf, B., Osinga, H.M.: Interactions of the Julia set with critical and (un)stable sets in an angle-doubling map on $\mathbb{C}\backslash\{0\}$. Int. J. Bifurcat. Chaos **25**, 1530013 (2015)
23. Kim, S.-H., MacKay, R.S., Guckenheimer, J.: Resonance regions for families of torus maps. Nonlinearity **2**, 391–404 (1989)
24. Krauskopf, B., Kriete, H.: Approximating transcendental Julia sets. In: Kriete, H. (ed.) Progress in Holomorphic Dynamics. Pitman Research Notes in Mathematics Series, vol. 387, pp. 116–138. Longman, Harlow (1998)
25. Krauskopf, B., Osinga, H.M.: Growing 1D and quasi-2D unstable manifolds of maps. J. Comput. Phys. **146**, 404–419 (1998)

26. Krauskopf, B., Osinga, H.M.: Investigating torus bifurcations in the forced Van der Pol oscillator. In: Numerical Methods for Bifurcation Problems and Large-Scale Dynamical Systems, Minneapolis, 1997. The IMA Volumes in Mathematics and its Applications, vol. 119, pp. 199–208. Springer, New York (2000)
27. Kuznetsov, Yu.A.: Elements of Applied Bifurcation Theory. Applied Mathematical Sciences, vol. 112, 3rd edn. Springer, New York (2004)
28. Mayer, J.C.: An explosion point for the set of endpoints of the Julia set of $\lambda \exp(z)$. Ergodic Theory Dynam. Syst. **10**, 177–183 (1990)
29. Milnor, J.: Dynamics in One Complex Variable. Annals of Mathematics Studies, vol. 160, 3rd edn. Princeton University Press, Princeton (2006)
30. Mira, C., Gardini, L., Barugola, A., Cathala, J.-C.: Chaotic Dynamics in Two-Dimensional Noninvertible Maps. World Scientific Series on Nonlinear Science Series A Monographs and Treatises, vol. 20. World Scientific, Singapore (1996)
31. Osinga, H.M., Krauskopf, B., Hittmeyer, S.: Chaos and wild chaos in Lorenz-type systems. In: AlSharawi, Z., Cushing, J., Elaydi, S. (eds.) Theory and Applications of Difference Equations and Discrete Dynamical Systems, pp. 75–98. Springer, Berlin (2014)
32. Palis, J., de Melo, W.: Geometric Theory of Dynamical Systems. Springer, New York (1982)
33. Palis, J., Takens, F.: Hyperbolicity and Sensitive Chaotic Dynamics at Homoclinic Bifurcations. Cambridge Studies in Advanced Mathematics, vol. 35. Cambridge University Press, Cambridge (1993)

Traveling Waves and Pattern Formation for Spatially Discrete Bistable Reaction-Diffusion Equations

Hermen Jan Hupkes, Leonardo Morelli, Willem M. Schouten-Straatman and Erik S. Van Vleck

Abstract We survey some recent results on traveling waves and pattern formation in spatially discrete bistable reaction-diffusion equations. We start by recalling several classic results concerning the existence, uniqueness and stability of travelling wave solutions to the discrete Nagumo equation with nearest-neighbour interactions, together with the Fredholm theory behind some of the proofs. We subsequently discuss extensions involving wave connections between periodic equilibria, long-range interactions and planar lattices. We show how some of the results can be extended to the two-component discrete FitzHugh–Nagumo equation, which can be analyzed using singular perturbation theory. We conclude by studying the behaviour of the Nagumo equation when discretization schemes are used that involve both space and time, or that are non-uniform but adaptive in space.

Keywords Bistable lattice differential equations · Traveling waves · Pattern formation

1 Introduction

The purpose of this article is to survey recent results concerning the pattern forming properties of spatially discrete reaction diffusion equations. Our guiding example will be the Nagumo lattice differential equation (LDE)

H. J. Hupkes · L. Morelli · W. M. Schouten-Straatman
Mathematisch Instituut, Universiteit Leiden, P.O. Box 9512, 2300 Leiden, RA, The Netherlands
e-mail: hhupkes@math.leidenuniv.nl

L. Morelli
e-mail: l.morelli@math.leidenuniv.nl

W. M. Schouten-Straatman
e-mail: w.m.schouten@math.leidenuniv.nl

E. S. Van Vleck (✉)
Department of Mathematics, University of Kansas, Lawrence, KS 66045, USA
e-mail: erikvv@ku.edu

M. Bohner et al. (eds.), *Difference Equations and Discrete Dynamical Systems with Applications*, Springer Proceedings in Mathematics & Statistics 312,
https://doi.org/10.1007/978-3-030-35502-9_3

$$\dot{u}_j(t) = d\big[u_{j-1}(t) + u_{j+1}(t) - 2u_j(t)\big] + g\big(u_j(t); a\big) \tag{1}$$

posed on $j \in \mathbb{Z}$, with the bistable cubic nonlinearity

$$g(u; a) = u(1-u)(u-a), \qquad a \in (0, 1). \tag{2}$$

However, we also consider variants of this LDE that involve planar lattices, inhomogeneous diffusion coefficients, long-range interactions and additional components. Furthermore, we discuss the consequences of replacing the time-derivative in (1) by appropriate discretizations, which leads to difference equations in both space and time.

The recurring theme throughout this paper is that the discrete nature of the model (1) allows it to exhibit a much richer class of behaviour than its continuous counterpart, the Nagumo PDE

$$u_t = u_{xx} + g\big(u; a\big). \tag{3}$$

Indeed, we will explore the profound effects that different discretization choices for the Laplacian can have on the dynamical behaviour of the underlying systems. In fact, important differences already appear at the level of the equilibrium solutions.

Bistable systems The pair of stable equilibria for the nonlinearity g can be used to represent material phases or biological species that compete for dominance in a spatial domain [5]. The high frequency oscillations caused by this bistability are damped by the diffusion term in (3), which leads to interesting pattern forming behaviour. In particular, solutions generally develop interface layers that separate the spatial domain into regions governed by the two stable phases [3, 112]. The evolution of these interfaces can be seen as a desingularized version of the mean curvature flow [2, 66].

The PDE (3) has served as a prototype system for the understanding of many basic concepts at the heart of dynamical systems theory. An important role is reserved for travelling wave solutions, which can be written as

$$u(x, t) = \Phi(x + ct), \qquad \Phi(-\infty) = 0, \qquad \Phi(+\infty) = 1 \tag{4}$$

and hence provide a mechanism by which one of the two stable phases can invade the spatial domain. Such pairs (c, Φ) must satisfy the travelling wave ODE

$$c\Phi' = \Phi'' + g\big(\Phi; a\big), \tag{5}$$

which admits the explicit solutions

$$\Phi(\xi) = \frac{1}{2} + \frac{1}{2}\tanh\big[\tfrac{1}{4}\sqrt{2}\xi\big], \qquad c(a) = \frac{1}{2}\sqrt{2}(1 - 2a). \tag{6}$$

These special solutions play a key role in the dynamics of the full PDE (3). Indeed, the classical work by Fife and McLeod [72] shows that these waves are stable under a broad range of perturbations that need not be small.

These pairs (c, Φ) can also be interpreted as solutions to the planar Nagumo PDE

$$u_t = u_{xx} + u_{yy} + g\big(u; a\big) \tag{7}$$

by writing

$$u(x, y, t) = \Phi(x \cos \zeta + y \sin \zeta + ct). \tag{8}$$

This is a direct consequence of the isotropy of the Laplacian, which causes ζ to drop out from the resulting wave equation.

These planar waves are also stable [17, 111], but the underlying analysis is much more subtle as deformations of the wave interface decay at an algebraic rate; see Sect. 4. The seminal work by Weinberger [6] uses these planar waves as building blocks to study the speed at which sufficiently large but *compact* sets of the favourable species spread throughout the domain.

Spatially discrete systems For many physical phenomena such as crystal growth in materials [25], the formation of fractures in elastic bodies [146] and the motion of dislocations [32] and domain walls [52] through crystals, the discreteness and the topology of the underlying spatial domain both have a major impact on the dynamical behaviour. For example, the spreading of auxin through plant leaves depends crucially on the active PIN-induced transport through the cell-membranes [129]. Peierls–Nabarro barriers typically prevent small defects from spreading through discrete media, but can be eliminated by carefully tuning system parameters [46]. Recent experiments show that even light waves can be trapped inside well-designed photonic lattices [131, 156].

Motivated by these considerations, LDEs have been used to describe a wide range of propagation phenomena in spatially discrete systems. Examples include the propagation of electrical signals through transmission lines [63] and myelinated nerve fibres [115], the formation of compression waves through Hertzian chains of elastic beads [144], the occurrence of denaturation bubbles in strands of DNA [47] and the self-trapping of wave packets in coupled optical waveguides [121].

Bistable discrete reaction-diffusion systems such as (1) have been used to describe action potentials in myelinated nerve fibers [115], phase transitions in Ising models [12] and predator-prey interactions in patchy environments [145]. They were even implemented on circuit boards in order to develop fast pattern recognition algorithms in image processing [41, 42].

The non-local diffusion in (1) introduces a natural length scale into applications and breaks the continuous translational symmetry of $\mathbb{R}$. The coupling with the bistable nonlinearity leads to a wide range of interesting behaviour that we explore in this paper. On the other hand, (1) still admits a comparison principle, which in some cases allows global effects to be captured rigorously. For these reasons, (1) has served as a prototype system to explore the effects of non-locality and spatial discreteness.

Spatially discrete travelling waves Substituting the travelling wave ansatz

$$u_j(t) = \Phi(j + ct) \tag{9}$$

into the LDE (1), one arrives at the system

$$c\Phi'(\xi) = d\big[\Phi(\xi - 1) + \Phi(\xi + 1) - 2\Phi(\xi)\big] + g\big(\Phi(\xi); a\big). \tag{10}$$

Our main interest here is in connections between the two stable equilibria of g, which leads to the boundary conditions

$$\Phi(-\infty) = 0, \qquad \Phi(+\infty) = 1. \tag{11}$$

No explicit solutions are known to exist for this wave equation. However, this changes if the cubic g is replaced by suitable specially constructed nonlinearities [1] or if the nonlinear term in the LDE (1) is allowed to involve the values of u at multiple lattice sites [9, 53, 76, 116, 148]. For example, the LDE

$$\dot{u}_j = u_{j-1} + u_{j+1} - 2u_j + \frac{1}{2}u_j(u_{j+1} + u_{j-1} - 2a)(1 - u_j) \tag{12}$$

admits the explicit solutions

$$u_j(t) = \frac{1}{2} + \frac{1}{2}\tanh\left(\operatorname{arcsinh}\left(\frac{1}{4}\sqrt{2}\right)(j + ct)\right), \qquad c(a) = \frac{(1 - 2a)}{4\operatorname{arcsinh}(\frac{1}{4}\sqrt{2})}. \tag{13}$$

A useful procedure that has been widely used to gain insight into the behaviour of system such as (10) is to replace the cubic g by the discontinuous piecewise linear 'caricature'

$$g_{\mathrm{pl}}(u; a) = \begin{cases} u & \text{for } u < a, \\ u - 1 & \text{for } u \geq a. \end{cases} \tag{14}$$

This allows linear techniques such as the Fourier transform to be applied. The resulting expressions for Φ involve explicit Fourier sums that allow interesting conclusions to be extracted. These results have often helped to guide the exploration of systems with smooth nonlinearities. In Sect. 3.3 we discuss an analysis of this type for a version of (1) that includes next-to-nearest-neighbour couplings. Earlier results of this type can be found in [27, 56, 57, 62, 67, 91, 151–153].

Returning to (10), we refer the reader to Sect. 2.2 for an overview of the different types of techniques that can be used to obtain the existence of travelling waves. For now, we simply note that the comparison principle can be used [35, 126] to show that c is uniquely determined by the detuning parameter a. However, the character of (10) depends crucially on whether c vanishes or not. Indeed, the continuous translational symmetry of $\mathbb{R}$ is broken by the transition $\mathbb{R} \to \mathbb{Z}$, causing the wavespeed c to act in (10) as a singular parameter. It is therefore natural to discuss the two cases $c = 0$ and $c \neq 0$ separately.

Pinned waves For $c = 0$ one can restrict the spatial variable in (10) to the integers and look for solutions $\Phi : \mathbb{Z} \to \mathbb{R}$. In fact, the travelling wave system (10) reduces to the difference equation

$$(r_{j+1}, s_{j+1}) = \big(s_j, 2s_j - r_j - d^{-1}g(s_j; a)\big) \tag{15}$$

upon writing $\big(\Phi_j, \Phi_{j+1}\big) = (r_j, s_j)$. The fixed points of this system are $(0, 0)$, (a, a) and $(1, 1)$. In particular, the boundary conditions (11) mean that for any $j \in \mathbb{Z}$, the pair (Φ_j, Φ_{j+1}) lies in the intersection of the stable manifold $\mathcal{W}^s(1, 1)$ and the unstable manifold $\mathcal{W}^u(0, 0)$ associated to (15).

The symmetry of the cubic at $a = \frac{1}{2}$ allows us to apply a result by Qin and Xiao [134, Theorem 2.7] and conclude that (10)–(11) admits (at least) two solutions

$$\Phi^{(s)} : \mathbb{Z} \to \mathbb{R}, \qquad \Phi^{(u)} : \mathbb{Z} \to \mathbb{R}. \tag{16}$$

These solutions are referred to as site-centered and bond-centered, since they are point symmetric around $(0, \frac{1}{2})$ respectively $(\frac{1}{2}, \frac{1}{2})$. In particular, the intersection $\mathcal{W}^s(1, 1) \cap \mathcal{W}^u(0, 0)$ is non-empty and contains (at least) the two families $(\Phi_j^{(s)}, \Phi_{j+1}^{(s)})$ and $(\Phi_j^{(u)}, \Phi_{j+1}^{(u)})$.

If these intersections are transverse, both solutions $\Phi^{(s)}$ and $\Phi^{(b)}$ will persist as pinned waves for $a \approx \frac{1}{2}$. It is then natural to expect that these branches coalesce and terminate in a saddle–node bifurcation at $a = a_{\pm}$; see panels (i) and (ii) in Fig. 1. In particular, this would mean that $c(a) = 0$ for a in the nontrivial interval $[a_-, a_+]$.

This phenomenon is referred to as propagation failure and has been observed among a wide class of discrete systems. In fact, Hoffman and Mallet-Paret show [88] that it is 'generic' in a suitable sense by establishing a Melnikov condition on the nonlinearity in (1) that is sufficient to guarantee its occurrence. For the cubic nonlinearity that we consider here, early results by Keener [114] guarantee that (1) admits propagation failure when $d > 0$ is sufficiently small. For general $d > 0$ the problem is still open, but the results in [18, 71] suggest that the width of the pinning interval $[a_-, a_+]$ decreases exponentially with $d > 0$. Further results confirming this phenomenon can be found in [1, 27, 61, 65, 67, 127].

Let us emphasize that these issues depend very subtly on the nonlinearity. For example, the explicit solutions (13) clearly indicate that (12) does not suffer from propagation failure. In this case, the two solutions $\Phi^{(s)}$ and $\Phi^{(b)}$ are part of a continuous one-parameter family of standing waves; see panel (iii) in Fig. 1. In such settings the intuition developed above fails and more subtle bifurcation scenarios can arise [98]. Similar observations can already be found in early work by Elmer [54]. Indeed, after replacing g by a piecewise linear zigzag nonlinearity, propagation failure can be excluded for a countable set of diffusion coefficients d.

Returning to (1), we illustrate the consequences of this behaviour on the pairs (c, Φ) in Fig. 2. These numerical results confirm that wave profiles lose their smoothness as the pinning region is approached, while the derivative $\partial_a c$ blows up. This latter fact appears to be related to the discreteness of the LDE (1) rather than its non-locality.

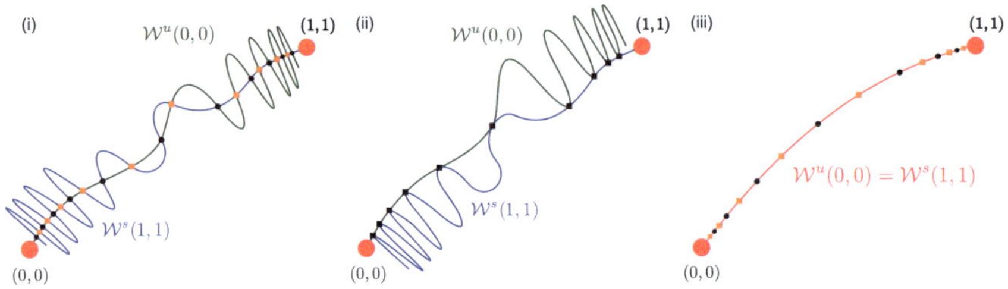

Fig. 1 The stable and unstable manifolds $\mathcal{W}^s(1,1)$ and $\mathcal{W}^u(0,0)$ associated to the discrete system (10) with $c = 0$ can intersect in several ways. Left: the generic situation expected at $a = \frac{1}{2}$. Middle: the saddle-node bifurcation at $a = a_{\pm}$. Right: the degenerate situation at $a = \frac{1}{2}$ corresponding to the multi-site discretization (13)

Indeed, pinning phenomena have also been investigated in nonlocal problems featuring smooth convolution kernels [4, 140]. In certain cases the behaviour of c near the pinning boundary can be described by power laws with various exponents.

Functional differential equations When $c \neq 0$, the travelling wave equation (10) is referred to as a functional differential equation of mixed type (MFDE), since it involves delayed terms and advanced terms simultaneously. Such equations are generally not well-posed as initial value problems. This can be seen by looking at the linear equation [137]

$$v'(\xi) = v(\xi + 1) + v(\xi - 1), \qquad v(\theta) = 1 \text{ for } -1 \leq \theta \leq 1, \tag{17}$$

which only has discontinuous solutions.

When studying such ill-posed infinite-dimensional problems, exponential dichotomies, Fredholm theory and dimension reduction techniques become the methods of choice. Initiated by the pioneering work of Rustichini [137, 138], research in this area features spectral flow results to compute the Fredholm index for operators with finite range shifts [125], infinite-range shifts [68] or neutral terms [119], various state-space decompositions based on exponential dichotomies [82, 93, 108, 128], techniques to construct local [70, 106, 107] and global [99] center manifolds and extensions of geometric singular perturbation theory [100].

Several of these results are discussed in Sects. 2.1 and 5.1. We emphasize that this theory is useful not only for establishing the existence of travelling wave solutions, but also for the analysis of their stability.

Planar travelling waves The two-dimensional analogue of (1) is given by the planar Nagumo LDE

$$\dot{u}_{ij} = d[u_{i+1,j} + u_{i-1,j} + u_{i,j+1} + u_{i,j-1} - 4u_{ij}] + g\big(u_{ij}; a\big), \tag{18}$$

where we take $(i, j) \in \mathbb{Z}^2$. Planar travelling wave solutions propagating at an angle ζ relative to the horizontal can be written in the form

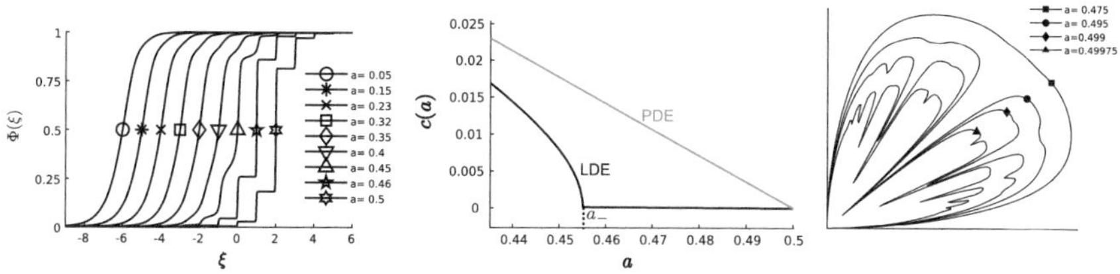

Fig. 2 Numerical simulations for the Nagumo LDEs (1) and (18) with $d = 0.1$. Left: the behaviour of the waveprofiles as a is increased towards $a = 0.5$. The zero-speed profiles for $a \in \{0.46, 0.5\}$ are discontinuous step functions. Center: the behaviour of $c(a)$ near the critical value a_- where pinning sets in. Right: Polar plots of the speed c_ζ as a function of the propagation angle ζ for different values of a. The curves have been rescaled by a factor r_a for comparison purposes, so that every curve can be written as $c_\zeta r_a\big(\cos(\zeta), \sin(\zeta)\big)$ for $0 \le \zeta \le \pi/2$

$$u_{ij}(t) = \Phi(i \cos\zeta + j \sin\zeta + ct), \tag{19}$$

where we again impose the boundary condition (11). Substitution into the LDE (18) now yields the travelling wave MFDE

$$\begin{aligned} c\Phi'(\xi) = d\big[&\Phi(\xi + \cos\zeta) + \Phi(\xi + \sin\zeta) + \Phi(\xi - \cos\zeta) + \Phi(\xi - \sin\zeta) - 4\Phi(\xi)\big] \\ &+ g\big(\Phi(\xi); a\big). \end{aligned} \tag{20}$$

The broken rotational invariance in the $\mathbb{R}^2 \to \mathbb{Z}^2$ transition is manifested by the explicit presence of the propagation direction in (20).

In particular, the speed $c = c_\zeta$ of planar waves typically depends on the angle ζ. This dependence can be very intricate when $a \approx \frac{1}{2}$; see panel (iii) in Fig. 2 and [27, 58, 105]. In fact, planar waves can fail to propagate in certain directions that resonate with the lattice, whilst travelling freely in other directions [27, 88, 127]. Several of the resulting peculiarities will be discussed in detail in Sect. 4.

Equilibrium patterns Besides the pinned waves discussed above, we have so far only considered equilibrium solutions to (1) that are spatially homogeneous. However, the discrete nature of the problem allows this LDE to have a much richer set of equilibria than the corresponding PDE (3). For example, in Sect. 3.1 we discuss solutions to the difference equation (15) that admit the periodicity $(r_{j+n}, s_{j+n}) = (r_j, s_j)$. Such solutions exist for small $d > 0$ and correspond with equilibria for (1) that are spatially n-periodic. In fact, one can show [114] that (15) admits a horseshoe for small $d > 0$, which implies the presence of chaos in the set of equilibria for (1).

It is not hard to see that n-periodic equilibria for (1) can be seen as equilibria for the Nagumo equation posed on the cyclic graph C_n. In [96] exact expressions are given for the number of equivalence classes of such equilibria, after factoring out the graph symmetries. It is natural to ask which other finite graphs can arise in this fashion by studying regular lattices. A brief discussion can be found in [130, Sect. 1], where various colouring schemes are described for hexagonal lattices that lead to the crown graph $K_{3,3}$, the utility graph S_4^0 and the complete graph K_4. In general however, full classifications are still unavailable. Results that can be used to count the number of equilibria on several types of graphs can be found in [149].

For the planar LDE (18), equilibrium patterns that are 2-periodic in both horizontal and vertical directions again lead to the graphs C_2 or C_4, allowing us to reuse the one-dimensional results discussed in Sect. 3.1. However, one can also construct patterns that are 3×2-periodic which lead to the ladder graph CL_3, or use five colours to arrive at the complete graph K_5. In general, the study of planar equilibrium patterns is still in its infancy. An illuminating discussion can be found in [40, 124], where the periodicity and chaos present in the set of such patterns is explored using simplified nonlinearities.

Outline This paper is organized as follows. In Sect. 2 we outline the basic machinery that has been developed for MFDEs and discuss a range of approaches that can be used to establish the existence, uniqueness and stability of the travelling waves (9). We move on in Sect. 3 to apply this theory to study the pattern forming properties of several extensions to (1) that are all posed on the one-dimensional lattice $\mathbb{Z}$.

We subsequently consider the planar lattice $\mathbb{Z}^2$ in Sect. 4. A large part of this section is concerned with the stability of the planar waves (19) under various types of perturbations, which includes the removal of lattice points. However, we also use these waves as building blocks to analyze more general types of solutions such as corners and expanding interfaces.

In Sect. 5 we return to the one-dimensional lattice $\mathbb{Z}$, but now consider discrete FitzHugh–Nagumo equations. These are two-component LDEs that arise by coupling a second variable to (1). Various types of singular perturbations are considered that allow us to reduce the complexity of the underlying problem by invoking familiar results for the limiting subsystems.

We conclude in Sect. 6 by discussing discretization schemes for the Nagumo PDE (3) that are relevant in the context of numerical analysis. In particular, we construct fully-discrete travelling waves that solve (1) after discretizing the only remaining derivative with a family of schemes referred to as backward differentiation formula (BDF) methods. In addition, we explore the effects of using an adaptive spatial grid to solve (3), instead of the standard spatially uniform grid that leads to (1).

2 Background: Tools and Techniques

We here review the basics of some of the crucial techniques that have been developed to analyze travelling waves in spatially discrete settings. For simplicity and explicitness, we present all the main ideas in the context of the spatially homogeneous Nagumo LDE

$$\dot{u}_j(t) = d\big[u_{j-1}(t) + u_{j+1}(t) - 2u_j(t)\big] + g\big(u_j(t); a\big) \tag{21}$$

with $d > 0$ and the cubic nonlinearity $g(u; a) = u(1 - u)(u - a)$. However, as we will discuss in detail below, many of the tools can be extended to a diverse range of more complicated systems. In this case the travelling wave MFDE is given by

$$c\Phi'(\xi) = d\big[\Phi(\xi - 1) + \Phi(\xi + 1) - 2\Phi(\xi)\big] + g\big(\Phi(\xi); a\big) \tag{22}$$

and we look for solutions that satisfy the limits

$$\Phi(-\infty) = 0, \qquad \Phi(+\infty) = 1. \tag{23}$$

The linearization around any solution to this MFDE can be described by the linear operator

$$[L_{\mathrm{tw}}]v = -cv'(\xi) + d\big[v(\xi + 1) + v(\xi - 1) - 2v(\xi)\big] + g'(\Phi(\xi); a)v(\xi). \tag{24}$$

We start in Sect. 2.1 by discussing several key aspects of the theory that has been developed for operators such as $\mathcal{L}_{\mathrm{tw}}$ that are associated to linear MFDEs. We proceed in Sect. 2.2 by exploring several different approaches that can be used to obtain the existence and uniqueness of solutions to (22) and to understand their parameter-dependence. In Sect. 2.3 we subsequently investigate the nonlinear stability of these waves under the dynamics of (21). Finally, in Sect. 2.4 we discuss numerical techniques that can be used to visualize the behaviour of (21) and compute approximations for the speed and shape of the waves.

2.1 *Linear Theory*

An important feature of the linear operator $\mathcal{L}_{\mathrm{tw}}$ is that it reduces to a constant coefficient operator in the limits $\xi \to \pm\infty$. In particular, looking for solutions of the form $e^{z\xi}$ for these limiting systems, we readily see that the spatial eigenvalues z are given by the roots of the characteristic functions

$$\Delta_{\pm}(z) = -cz + d\big[e^{z} + e^{-z} - 2\big] + g'\big(\Phi(\pm\infty); a\big). \tag{25}$$

For $0 < a < 1$, the inequalities $g'(0; a) < 0$ and $g'(1; a) < 0$ imply that both

$$\Delta_{\pm}(i\nu) \neq 0, \ \text{ for all } \nu \in \mathbb{R}. \tag{26}$$

On account of this fact, the linear operator $\mathcal{L}_{\mathrm{tw}}$ is said to be asymptotically hyperbolic. Several powerful results are available for such operators.

Fredholm properties For $c \neq 0$, the results in [125, 137] show that $\mathcal{L}_{\mathrm{tw}}$ is a Fredholm operator from $W^{1,p}$ into L^p for all $1 \leq p \leq \infty$. In particular, we have the inequalities

$$\dim \mathrm{Ker}(\mathcal{L}_{\mathrm{tw}}) < \infty, \qquad \mathrm{codim}\,\mathrm{Range}(\mathcal{L}_{\mathrm{tw}}) < \infty. \tag{27}$$

In addition, upon writing

$$\begin{aligned} \Delta_\mu(z) &= -cz + d\big[e^z + e^{-z} - 2\big] + \mu g'\big(\Phi(-\infty); a\big) + (1-\mu) g'\big(\Phi(+\infty); a\big) \\ &= \mu \Delta_-(z) + (1-\mu)\Delta_+(z), \end{aligned} \tag{28}$$

we see that $\Delta_\mu(i\nu) \neq 0$ for all $\nu \in \mathbb{R}$ and $\mu \in [0, 1]$. In particular, one can construct a homotopy between the limiting systems at $\pm\infty$ during which no spatial eigenvalues cross the imaginary axis. The spectral flow result [125, Theorem C] for the Fredholm index of $\mathcal{L}_{\mathrm{tw}}$ then implies that

$$\begin{aligned} 0 = \mathrm{ind}(\mathcal{L}_{\mathrm{tw}}) &= \dim \mathrm{Ker}(\mathcal{L}_{\mathrm{tw}}) - \mathrm{codim}\,\mathrm{Range}(\mathcal{L}_{\mathrm{tw}}) \\ &= \dim \mathrm{Ker}(\mathcal{L}_{\mathrm{tw}}) - \dim \mathrm{Ker}(\mathcal{L}_{\mathrm{tw}}^{\mathrm{adj}}), \end{aligned} \tag{29}$$

in which the adjoint linear operator $\mathcal{L}_{\mathrm{tw}}^{\mathrm{adj}} : W^{1,p} \to L^p$ is given by

$$\mathcal{L}_{\mathrm{tw}}^{\mathrm{adj}} w = cw' + d\big[w(\xi+1) + w(\xi-1) - 2w(\xi)\big] + g'(\Phi; a) w(\xi). \tag{30}$$

This terminology is motivated by the fact that

$$\langle \mathcal{L}_{\mathrm{tw}} v, w \rangle_{L^2} = \langle v, \mathcal{L}_{\mathrm{tw}}^{\mathrm{adj}} w \rangle_{L^2} \tag{31}$$

holds for any pair of smooth functions v and w that decay sufficiently fast.

In [68] this Fredholm theory was extended to nonlocal linear operators that involve infinite-range interactions and/or smooth convolution kernels. Such operators arise frequently when studying neural field models [22, 69, 70]; see also Sect. 5.3.

Kernel of $\mathcal{L}_{\mathrm{tw}}$ In order to further exploit the identity (29), it is essential to gain a thorough understanding of the kernel of $\mathcal{L}_{\mathrm{tw}}$. The translational invariance of (22) immediately implies that $\mathcal{L}_{\mathrm{tw}} \Phi' = 0$, but the specific structure of this wave equation must be used to rule out the presence of additional linearly independent kernel elements.

Such an analysis is performed in [126], where one establishes

$$\mathrm{Ker}(\mathcal{L}_{\mathrm{tw}}) = \mathrm{span}\{\Phi'\}, \qquad \mathrm{Ker}(\mathcal{L}_{\mathrm{tw}}^{\mathrm{adj}}) = \mathrm{span}\{\Psi\}. \tag{32}$$

In addition, both Φ' and Ψ are found to be strictly positive exponentially decaying functions, which means that we can impose the normalization

$$\langle \Psi, \Phi' \rangle_{L^2} = 1. \tag{33}$$

The key point in the analysis above is to obtain precise asymptotics for the behaviour of Φ' as $\xi \to \pm\infty$. In particular, the spatial eigenvalue system $\Delta_+(z) = 0$ has a single real negative root $z = -\beta_+$, which using contour integration [125, Proposition 7.2] allows one to show that

$$\Phi'(\xi) = C_+ e^{-\beta_+ \xi} + O(e^{-(\beta_+ + \epsilon)\xi}), \qquad \xi \to \infty \tag{34}$$

for some $\epsilon > 0$.

Using the comparison principle it is easy to show that $\Phi' \geq 0$ and hence $C_+ \geq 0$, but the delicate task is to sharpen this to $C_+ > 0$. Assuming to the contrary that $C_+ = 0$, it is natural to expect the asymptotic behaviour of Φ to be governed by the roots of $\Delta_+(z) = 0$ with $\Re z < -\beta_+$. This possibility can easily be excluded by the comparison principle, since the non-zero imaginary parts of these roots all lead to forbidden oscillatory behaviour.

However, in this setting one also needs to exclude the case that Φ' decays at a rate that is faster than any exponential. This requires subtle estimates on (22) involving differential inequalities. In a sense, these computations can be seen as a non-autonomous counterpart to the completeness theory discussed in [51, 78]. This theory can be used to decide whether a given set of eigenfunctions for a delay differential operator can be used as a basis for the underlying statespace.

By developing a graph-theoretic framework to keep track of the interactions [103], it is possible to generalize this completeness result to multi-component versions of (21). In this way, the characterization (32) can also be obtained for multi-component systems such as those encountered in Sect. 3.

The resolvent of $\mathcal{L}_{\text{tw}}$ By applying the general Fredholm theory in [125] to the identifications (32), we can identify the range of $\mathcal{L}_{\text{tw}}$ as the set

$$\text{Range}(\mathcal{L}_{\text{tw}}) = \{f \in L^p : \langle \Psi, f \rangle_{L^2} = 0\}. \tag{35}$$

The normalization condition (33) hence implies that $\Phi' \notin \text{Range}(\mathcal{L}_{\text{tw}})$.

This fact can be reformulated as the statement that the algebraic multiplicity of the zero eigenvalue of $\mathcal{L}_{\text{tw}}$ is equal to one. Indeed, a Lyapunov–Schmidt decomposition can be used [143, Lemma 6.10] to obtain the meromorphic expansion

$$(\lambda - \mathcal{L}_{\text{tw}})^{-1} f = \lambda^{-1} \Phi' \langle \Psi, f \rangle_{L^2} + \mathcal{B}(\lambda) f, \tag{36}$$

in which the map

$$\lambda \mapsto \mathcal{B}(\lambda) \in \mathcal{L}(L^p; W^{1,p}) \tag{37}$$

is analytic in a neighbourhood of $\lambda = 0$. As we will see in Sect. 2.3, such expansions are very powerful tools when analyzing the stability of travelling waves. In the continuous setting, Zumbrun and Howard [161] pioneered the use of such expansions to obtain pointwise bounds on the Green's functions associated to a wide range of linear and nonlinear stability problems.

Exponential dichotomies While the theory above focussed on the properties of $\mathcal{L}_{\text{tw}}$ acting on functions that are defined on the full line, it is often very advantageous to consider half-lines and compact intervals. In particular, we consider the homogeneous equation

$$cv'(\xi) = d\big[v(\xi - 1) + v(\xi + 1) - 2v(\xi)\big] + g'(\Phi(\xi); a)v(\xi) \tag{38}$$

and look for solutions in the space of bounded continuous functions

$$BC(\mathcal{I}; \mathbb{R}) = \{v \in C(\mathcal{I}; \mathbb{R}) : \|v\|_\infty := \sup_{\xi \in \mathcal{I}} |v(\xi)| < \infty\} \tag{39}$$

defined on intervals $\mathcal{I} \subset \mathbb{R}$.

In order to consider the half-lines $\mathbb{R}_-$ and $\mathbb{R}_+$, we now introduce the solution spaces

$$\begin{aligned} \mathfrak{P} &= \big\{\mathfrak{p} \in BC((-\infty, 1]; \mathbb{R}) : \mathfrak{p} \text{ satisfies (38) for all } \xi \leq 0\big\}, \\ \mathfrak{Q} &= \big\{\mathfrak{q} \in BC([-1, \infty); \mathbb{R}) : \mathfrak{q} \text{ satisfies (38) for all } \xi \geq 0\big\} \end{aligned} \tag{40}$$

and look at their initial segments

$$\begin{aligned} P &= \big\{p \in C([-1, 1]; \mathbb{R}) : p = \mathfrak{p}_{|[-1,1]} \text{ for some } \mathfrak{p} \in \mathfrak{P}\big\}, \\ Q &= \big\{q \in C([-1, 1]; \mathbb{R}) : q = \mathfrak{q}_{|[-1,1]} \text{ for some } \mathfrak{q} \in \mathfrak{Q}\big\}. \end{aligned} \tag{41}$$

Functions that are in both P and Q can be extended to solutions on the entire real line, which means that

$$P \cap Q = \Phi'_{|[-1,1]}. \tag{42}$$

In addition, the expansions in [125, Proposition 7.2] together with the hyperbolicity condition (26) imply that functions in $\mathfrak{P}$ and $\mathfrak{Q}$ decay exponentially as $\xi \to -\infty$ respectively $\xi \to +\infty$.

The results in [128] show that the sum $S = P + Q$ forms a set of codimension one. More precisely, we have

$$S = \{\phi \in C([-1, 1]; \mathbb{R}) : \langle \Psi, \phi \rangle_{\text{Hale}} = 0\}, \tag{43}$$

in which the so-called Hale inner product is given by

$$\langle \psi, \phi \rangle_{\text{Hale}} = c\psi(0)\phi(0) + \int_{-1}^{0} \psi(\vartheta + 1)\phi(\vartheta)\, d\vartheta - \int_{0}^{1} \psi(\vartheta - 1)\phi(\vartheta)\, d\vartheta. \tag{44}$$

Such a splitting of the space S is referred to as an exponential dichotomy on the full line. Related results in a Hilbert space setting were obtained independently by Härterich, Sandstede and Scheel [82].

It is also worthwhile to consider exponential dichotomies on half-lines, in which case one introduces for $L \geq 0$ the solution spaces

$$\begin{aligned} \mathfrak{P}_L &= \big\{\mathfrak{p} \in BC([-L - 1, 1]; \mathbb{R}) : \mathfrak{p} \text{ satisfies (38) for all } -L \leq \xi \leq 0\big\}, \\ \mathfrak{Q}_L &= \big\{\mathfrak{q} \in BC([-1, L + 1); \mathbb{R}) : \mathfrak{q} \text{ satisfies (38) for all } 0 \leq \xi \leq L\big\}, \end{aligned} \tag{45}$$

together with their initial segments

$$\begin{aligned} P_L &= \{p \in C([-1, 1]; \mathbb{R}) : p = \mathfrak{p}_{|[-1,1]} \text{ for some } \mathfrak{p} \in \mathfrak{P}_L\}, \\ Q_L &= \{q \in C([-1, 1]; \mathbb{R}) : q = \mathfrak{q}_{|[-1,1]} \text{ for some } \mathfrak{q} \in \mathfrak{Q}_L\}. \end{aligned} \tag{46}$$

In this case, the results in [108] show that the full space $C([-1, 1]; \mathbb{R})$ can be decomposed as

$$C([-1, 1]; \mathbb{R}) = P_L \oplus Q = P \oplus Q_L \tag{47}$$

for any $L \geq 0$.

In [108] we show how the characterization (35) can be exploited to construct solutions to equations of the form $\mathcal{L}_{\text{tw}} v = f$ for functions f that are only defined on half-lines. These solutions are by no means unique. Indeed, the power of the exponential splittings above is that they correspond precisely to the freedom that we have to solve such problems. This allows a general technique referred to as Lin's method [123] to be lifted from the finite dimensional world of ODEs to the infinite dimensional setting of MFDEs.

For example, one can vary the parameters c and a in (22) and construct so-called quasi-solutions, which are two-valued on $[-1, 1]$ with a difference that can be explicitly quantified by the Hale inner product. By closing the gap one can obtain information on the $c(a)$ relation. Alternatively, these quasi-solutions can be combined and used as building blocks to construct travelling waves to more complicated problems; see Sect. 5.1.

2.2 *Existence of Waves*

By now it is well-known that for each $a \in [0, 1]$ there is a unique $c = c(a)$ for which the wave MFDE (22)–(23) admits a solution. When $c(a) \neq 0$, the waveprofile Φ is unique up to translations and the pair (c, Φ) depends smoothly on a.

In this subsection we discuss several different methods that can be used to establish these properties. Each of these methods has its own strengths and weaknesses, but together they form a powerful toolbox that can be applied in a wide range of different settings.

Qualitative features As a preparation, we first discuss three important properties of the $c(a)$ relationship that can be deduced directly from (22) and the linear theory discussed in Sect. 2.1. We remark that all the arguments here can be found in [126].

First of all, we claim that in the boundary cases $a \in \{0, 1\}$ one must have $c \neq 0$. Indeed, assuming to the contrary that $c = 0$, one can integrate (22) and use the identity

$$\int_{-\infty}^{\infty} \big(\Phi(\xi) - \Phi(\xi - 1)\big)\, d\xi = \int_{-\infty}^{\infty} \big(\Phi(\xi + 1) - \Phi(\xi)\big)\, d\xi \tag{48}$$

to conclude that

$$0 = \int_{-\infty}^{\infty} g(\Phi(\xi); a)\, d\xi. \tag{49}$$

Since we have the inequalities

$$g(u; 1) < 0 < g(u; 0) \tag{50}$$

for all $u \in (0, 1)$, this implies that Φ must be a constant function, contradicting (23).

The second claim is that a solution (c, Φ) to (22)–(23) that has $c \neq 0$ extends smoothly to a branch of solutions for nearby parameters (a, d). To see this, one can rephrase (22) in the form

$$\mathcal{G}(c, \Phi, a, d) = 0 \tag{51}$$

and compute the derivative

$$D_{c,\Phi}\mathcal{G}(c, \Phi, a, d)[\tilde{c}, v] = -\tilde{c}\Phi' + \mathcal{L}_{\mathrm{tw}} v. \tag{52}$$

This operator has full range on account of the characterization (35), which allows the implicit function theorem to be applied.[1]

Under the assumption $c \neq 0$, this smoothness allows us to differentiate (22) with respect to a, which yields

$$[\partial_a c]\Phi' = \mathcal{L}_{\mathrm{tw}}[\partial_a \Phi] + \partial_a g(\Phi; a). \tag{53}$$

Applying (35) and recalling the normalization (33), we hence obtain the useful monotonicity condition

$$\partial_a c = \langle \Psi, \partial_a g(\Phi; a)\rangle = -\langle \Psi, \Phi(1 - \Phi)\rangle < 0. \tag{54}$$

Global homotopy The main idea in [126] is to establish the existence of travelling waves by constructing a global homotopy between families of bistable MFDEs. In particular, the collection of MFDEs (22) formed by taking $a \in [0, 1]$ is connected to the family

$$c\phi'(\xi) = \gamma^{-1}\big(\phi(\xi - 1) - \phi(\xi)\big) + 4\gamma g\big(\phi(\xi); a\big)\big(1 + \gamma - 2\gamma\phi(\xi)\big)^{-1} \tag{55}$$

with $\gamma = \tanh 1$. For $a = \frac{1}{2}$ and $c = -1$, one can explicitly verify that this latter MFDE has the solution

$$\phi(\xi) = \frac{1}{2}(1 + \tanh \xi). \tag{56}$$

The monotonicity property (54) also holds in this case, which allows us to conclude that (55) has solutions with $c < 0$ for all $a \in [\frac{1}{2}, 1]$.

By using a similar continuation argument as above, these solutions can be continued to the original family (22). The key issue is that the inequality $c \neq 0$ must be main-

[1] Actually, in order to ensure that the boundary conditions (23) are satisfied one needs to consider perturbations $v \in W^{1,p}$ for $1 \leq p < \infty$ while taking $\Phi \in W^{1,\infty}$.

tained throughout this homotopy. This can be done uniformly for all $a \in [1-\epsilon, 1]$ by applying coercivity arguments similar to those described above.

By exploiting the comparison principle, one can subsequently show that there is a unique wavespeed $c = c(a)$ for which (22)–(23) admits solutions. If $c \neq 0$, then the waveprofile Φ is also unique up to translations. These arguments can be readily generalized to general scalar equations of bistable type, but the use of the specific reference system (55) prevents an easy generalization to the multi-component case.

Spectral convergence In [11] the authors develop a perturbative technique to construct solutions to (22)–(23) near the continuum regime. In particular, they pick $d = h^{-2} \gg 1$ and spatially rescale (22) to arrive at the MFDE

$$c\Phi'(\xi) = \frac{1}{h^2}\big[\Phi(\xi - h) + \Phi(\xi + h) - 2\Phi(\xi)\big] + g\big(\Phi(\xi); a\big). \tag{57}$$

The main goal is to construct solutions that can be written as

$$\Phi = \Phi_0 + v, \qquad c = c_0 + \tilde{c} \tag{58}$$

for small perturbations $(\tilde{c}, v)$, in which the pair (c_0, Φ_0) satisfies the travelling wave ODE

$$c_0\Phi_0' = \Phi_0'' + g(\Phi_0; a) \tag{59}$$

associated to the Nagumo PDE (3).

The crucial linear operator to understand in this respect is

$$[\mathcal{L}_h v](\xi) = -c_0 v'(\xi) + \frac{1}{h^2}[v(\xi - h) + v(\xi + h) - 2v(\xi)] + g'(\Phi_0(\xi); a)v(\xi), \tag{60}$$

which arises upon linearizing the MFDE (57) around the PDE wave (c_0, Φ_0). This operator is a singularly perturbed version of its PDE counterpart

$$[\mathcal{L}_0 v](\xi) = -c_0 v'(\xi) + v''(\xi) + g'(\Phi_0(\xi); a)v(\xi). \tag{61}$$

The main contribution in [11] is that Fredholm properties of $\mathcal{L}_0$ are transferred to $\mathcal{L}_h$. In particular, the authors fix a constant $\delta > 0$ and use the invertibility of $\mathcal{L}_0 - \delta$ to show that also $\mathcal{L}_h - \delta$ is invertible for small $h > 0$. To achieve this, they consider bounded weakly-converging sequences $\{v_j\} \subset H^1$ and $\{w_j\} \subset L^2$ with $(\mathcal{L}_h - \delta)v_j = w_j$ and set out to find a lower bound for w_j that is uniform in δ and h.

The strategy is to show that such a lower bound can be found even if w_j is restricted to a large but compact interval K. Indeed, one can then extract a subsequence of $\{v_j\}$ that converges strongly in $L^2(K)$ and use properties of the limiting operator $\mathcal{L}_0$ to obtain the desired bounds.

Special care must therefore be taken to rule out the limitless transfer of energy into oscillatory or tail modes, which are not visible in this strong limit. Spectral properties of the (discrete) Laplacian together with the bistable structure of the nonlinearity g provide the control on $\{v_j\}$ that is necessary for this.

The power of this approach is that it does not use the comparison principle and is not limited to finite range interactions. In particular, the results in [11] also hold for general LDEs of the form

$$\dot{u}_j = \frac{1}{h^2} \sum_{k>0} \alpha_k \big[u_{j-k} + u_{j+k} - 2u_j \big] + g\big(u_j; a\big), \tag{62}$$

provided that the coefficients $\{\alpha_k\}$ satisfy the second-moment constraints

$$\sum_{k>0} \alpha_k k^2 = 1, \qquad \sum_{k>0} |\alpha_k| k^2 < \infty, \tag{63}$$

together with the spectral bounds

$$\sum_{k>0} \alpha_k \big(1 - \cos(kz)\big) \geq 0 \text{ for all } z \in [0, 2\pi]. \tag{64}$$

Together, these properties ensure that several key properties of the Laplacian carry over to the discretization (62).

Recently, we generalized this approach to multi-component systems such as the FitzHugh–Nagumo LDE [142, 143]; see Sects. 5.2 and 5.3. In addition, we considered situations where the discrete Laplacian includes a (localized) spatial dependence or where the Nagumo PDE is fully discretized; see Sect. 6.1.

Monotonic iteration The approach taken by Chen, Guo and Wu in [35] is to recast (22) as an integral equation that can be interpreted as a fixed point problem. In particular, for any $\nu > 0$ and $c \neq 0$ the authors introduce the linear operator

$$\begin{aligned} [T_c \Phi](\xi) = \textstyle\int_{-\infty}^{0} e^{\nu t} \Big[&\nu \Phi(\xi - ct) + g\big(\Phi(\xi - ct); a\big) \\ &+ d\big[\Phi(\xi - ct + 1) + \Phi(\xi - ct - 1) - 2\Phi(\xi - ct)\big]\Big] \, dt. \end{aligned} \tag{65}$$

Using $\exp[-\nu\xi/c]$ as an integrating factor, the travelling wave MFDE (22) can be integrated to yield

$$\Phi = T_c \Phi. \tag{66}$$

It turns out to be worthwhile to attack (66) indirectly by first restricting the problem to finite intervals $[-m, m]$. To this end, let us introduce the truncation operator P_m that acts as

$$[P_m \Phi](\xi) = \begin{cases} 0 & \text{if } \xi < -m, \\ \Phi(\xi) & \text{if } -m \leq \xi \leq m, \\ 1 & \text{if } \xi > m. \end{cases} \tag{67}$$

By choosing $\nu \gg 1$ to be sufficiently large, one can ensure that the integrand in (65) increases monotonically with respect to the arguments $\Phi(\cdot)$ and $\Phi(\cdot \pm 1)$. In particular, if $\Phi_A \leq \Phi_B$ holds in a pointwise fashion, then also $P_m T_c \Phi_A \leq P_m T_c \Phi_B$.

In order to exploit this, the authors develop a monotonic iteration scheme by writing

$$\Phi_0^- \equiv 0, \qquad \Phi_0^+ \equiv 1 \tag{68}$$

and subsequently

$$\Phi_j^- = P_m T_c \Phi_{j-1}^-, \qquad \Phi_j^+ = P_m T_c \Phi_{j-1}^+. \tag{69}$$

Indeed, the ordering properties above imply that

$$0 \leq \Phi_{j-1}^- \leq \Phi_j^- \leq \Phi_j^+ \leq \Phi_{j+1}^+ \leq 1, \tag{70}$$

which allows us to take the $j \to \infty$ limit and obtain two limiting functions that we denote by $\Phi_{(m,c)}^-$ and $\Phi_{(m,c)}^+$.

Naturally, one now wishes to take the limit $m \to \infty$ in order to obtain a solution to (22). The hard part in this procedure is to show that the limiting waveprofile is not constant and indeed satisfies the boundary conditions (23).

For finite m, the truncation P_m helps in this respect as one can readily see that

$$0 < \Phi_{(m,c)}^-(\xi) \leq \Phi_{(m,c)}^+(\xi) < 1, \qquad \xi \in (-m, m). \tag{71}$$

When $c \neq 0$, one can actually show that $\Phi_{(m,c)}^- = \Phi_{(m,c)}^+$. In addition, whenever $c_1 < c_2$ we have

$$\Phi_{(m,c_1)}^+(\xi) < \Phi_{(m,c_2)}^-(\xi), \qquad \xi \in (-m, m). \tag{72}$$

Suppose now that $\liminf_{m\to\infty} \Phi_{(m,0)}^+ \leq \frac{1}{2}$. This allows us to define a sequence $c_l > 0$, $m_l > l$ for which

$$\Phi_{(m_l,c_l)}^\pm(0) = \frac{1}{2} + l^{-1}. \tag{73}$$

By taking the limit $m \to \infty$ we obtain a non-trivial solution to the travelling wave problem (22)–(23). A similar procedure works when $\limsup_{m\to\infty} \Phi_{(m,0)}^- \geq \frac{1}{2}$, but more delicate arguments are needed if both these conditions fail.

This procedure can be readily generalized to multi-component versions of (21). In fact, the results in [35] are formulated for (21) with a diffusion coefficient $d = d_j$ that depends periodically on j. Besides the existence of travelling wave solutions, the authors also establish the uniqueness and stability of these solutions by exploiting the comparison principle.

Topological methods The techniques developed in [7, 8] aim to exploit the variational structure present in the wave MFDE (22). Indeed, upon picking a primitive G that has $G'(\cdot; a) = g(\cdot; a)$, the authors introduce the energy functional

$$\mathcal{H}_A(\Phi)(\xi) = \frac{d}{2}\big(\Phi(\xi - 1) + \Phi(\xi + 1) - 2\Phi(\xi)\big)\Phi(\xi) + G\big(\Phi(\xi); a\big), \quad (74)$$

together with a boundary term

$$\mathcal{H}_B(\Phi)(\xi) = \frac{1}{2}\int_{\xi}^{\xi+1} \big(\Phi(\xi - 1)\Phi'(\xi) - \Phi'(\xi - 1)\Phi(\xi)\big)\, d\xi. \quad (75)$$

A short computation shows that any solution to (22) must also satisfy the identity

$$\frac{d}{d\xi}\big[\mathcal{H}_A(\Phi) + \mathcal{H}_B(\Phi)\big](\xi) = c\Phi'(\xi)^2, \quad (76)$$

which means that $\mathcal{H}_A + \mathcal{H}_B$ can be used as a spatial Lyapunov function.

Exploiting this observation, one can construct a global Conley-Floer homology that encodes information concerning the travelling wave connections between the three equilibria of g. Such connections satisfy (22) but not necessarily (23). This homology is invariant under a large class of perturbations of g. In particular, this allows the authors to reduce the problem to the much simpler case where $g(u) = u^3$, which only admits a single equilibrium. The Conley-Floer homology can be explicitly computed in this case, which subsequently allows one to formulate a forcing theorem to conclude that the travelling wave MFDE admits a solution for each $c \neq 0$. However, this theorem (at present) provides no information concerning the possible limiting values of the waveprofile Φ.

This technique can be generalized to multi-component versions of (21) as long as the nonlinearity has the structure of a gradient. However, there is no need for the discretization of the Laplacian to have only positive off-diagonal coefficients. In addition, it is also possible to consider models that have infinite range interactions or that involve smooth non-local convolution kernels.

Spatial regularization The main feature of this indirect approach is that it replaces (21) by the spatially smoothened system

$$u_t = \gamma u_{xx} + d\big[u(x - 1) + u(x + 1) - 2u(x)\big] + g(u; a). \quad (77)$$

The first step is to show that this regularized system with $\gamma > 0$ admits travelling wave solutions. By taking the limit $\gamma \downarrow 0$ and considering an appropriate subsequence, one can subsequently recover a solution to the original wave problem (22)–(23).

The travelling waves for (77) can be obtained by studying the time-evolution of the initial condition

$$u(x, 0) = \frac{1}{2}\big(1 + \tanh(x)\big). \quad (78)$$

In particular, one can show that the corresponding solution $u(x, t)$ converges to a travelling wave for $t \to \infty$. To accomplish this, one first proves that $u(\cdot, t)$ is strictly increasing for all $t > 0$ and that is satisfies the limits $u(-\infty, t) = 0$ and $u(+\infty, t) = 1$. This allows us to define quantities $\xi_-(t)$ and $\xi_+(t)$ by demanding

$$u(\xi_-(t), t) = \delta, \qquad u(\xi_+(t), t) = 1 - \delta. \tag{79}$$

The key technical step is to obtain time uniform bounds

$$\xi_+(t) - \xi_-(t) \leq m(\delta), \tag{80}$$

which can be seen as an a-priori steepness result for the interface.

One can subsequently use the comparison principle to build a sequence of shifts $\{\xi_j\}$ and times $\{t_j\} \to \infty$ so that we have the convergence

$$u(\cdot - \xi_j, t_j) \to \Phi \tag{81}$$

in the space of bounded continuous functions. The uniform bounds (80) allow us to conclude that the limiting function satisfies $\Phi' > 0$ and has the limits (23). After some manipulations this limiting function turns out to be the desired waveprofile. These arguments were first introduced by Chen [34] in the context of general nonlocal PDEs. In [103] we generalized this procedure to spatially regularized multi-component versions of (21).

In order to construct a converging subsequence for the singular limit $\gamma \downarrow 0$, it suffices to obtain a uniform bound on the wavespeed c for small γ. The key issue is to show that the limits (23) are maintained throughout this limiting procedure. This can be achieved by establishing that the difference $\xi_+(t) - \xi_-(t)$ also remains bounded as $\gamma \downarrow 0$. Arguments of this type can be found in [103, 105].

2.3 *Stability of Waves*

Here we discuss several different methods that can be used to establish the nonlinear stability of the travelling waves (22)–(23) in the case where $c \neq 0$. In particular, one can show that there exists $\delta > 0$ so that any solution to (21) with

$$\sup_{j \in \mathbb{Z}} \left| u_j(0) - \Phi(j) \right| < \delta \tag{82}$$

must satisfy the limit

$$\sup_{j \in \mathbb{Z}} \left| u_j(t) - \Phi(j + ct + \vartheta) \right| \to 0 \tag{83}$$

as $t \to \infty$ for some asymptotic phaseshift ϑ. In addition, the rate of convergence is exponentially fast.

Wave squeezing This method builds upon the classic approach developed by Fife and McLeod for the Nagumo PDE (3). In particular, one sets out to exploit the comparison principle by constructing sub- and super-solutions of the form

$$u_j^-(t) = \Phi\big(j + ct - Z(t)\big) - z(t), \qquad u_j^+(t) = \Phi\big(j + ct + Z(t)\big) + z(t), \quad (84)$$

with z decreasing from z_0 to 0 and with Z increasing from 0 to Z_∞. These functions hence transform initial global additive perturbation of size z_0 into phaseshifts of size Z_∞.

To better understand the crucial relationship between the asymptotic phaseshift Z_∞ and the additive perturbation z, we note that the super-solution residual

$$\mathcal{J}_j^+ = \dot{u}_j^+ - d\big[u_{j-1}^+ + u_{j+1}^+ - 2u_j^+\big] - g(u_j; a) \quad (85)$$

is given by

$$\mathcal{J}_j^+(t) = \dot{Z}(t)\Phi'\big(\xi_j(t)\big) + \dot{z}(t) + g\Big(\Phi\big(\xi_j(t)\big); a\Big) - g\Big(\Phi\big(\xi_j(t)\big) + z(t); a\Big), \quad (86)$$

with $\xi_j(t) = j + ct + Z(t)$. Close to the interface, the term $g(\Phi) - g(\Phi + z) \sim -g'(\Phi)z$ is negative and must be dominated by the positive term $\dot{Z}\Phi'$. This requires that $\dot{Z}$ dominates z and $\dot{z}$. On the other hand, close to the spatial limits $\Phi \to 0$ and $\Phi \to 1$ we have $g'(\Phi; a) < 0$, so this regime requires z to dominate $\dot{z}$. These observations allow us to define $z(t)$ as a slowly decaying exponential, which gives the relation

$$Z_\infty \sim \int_0^\infty z(t)\, dt \sim z_0 \quad (87)$$

between the asymptotic phase shift and the size of the initial perturbation.

In [35, Sect. 5.4] the authors provide an elaborate argument to show how these explicit sub- and super-solutions can be used to trap a broad class of perturbations from Φ and steer them towards a phase-shifted version of this wave. The key ingredient is a careful mechanism to track the position of the perturbed wave by means of a phase condition. A variant of this method was used in [87] to establish the existence of an entire solution to (21) in the presence of obstacles; see Sect. 4.6.

Temporal Green's function In the remainder of this subsection we refrain from using the comparison principle, which widens the class of LDEs that can be considered. In this case one can only expect to handle perturbations from Φ that are small, since the linear behaviour of (21) can then be used to control the nonlinear effects.

Looking for solutions to (21) of the form

$$u_j(t) = \Phi(j + ct) + v_j(t), \quad (88)$$

we find that v must satisfy the time-dependent LDE

$$\dot{v}_j(t) = d\big[v_{j-1}(t) + v_{j+1}(t) - 2v_j(t)\big] + g'\big(\Phi(j + ct); a\big)v_j(t) + \mathcal{N}_j\big(t, v\big), \quad (89)$$

in which

$$\mathcal{N}_j(t, v) = g\big(\Phi(j + ct) + v_j(t); a\big) - g\big(\Phi(j + ct); a\big) - g'\big(\Phi(j + ct); a\big)v_j(t). \tag{90}$$

This nonlinearity behaves as $\big|\mathcal{N}_j(t, v)\big| = O(\|v(t)\|_\infty^2)$ for $v \to 0$, uniformly for $t \in \mathbb{R}$.

This allows us to focus on the linear part

$$\dot{w}_j(t) = d\big[w_{j-1}(t) + w_{j+1}(t) - 2w_j(t)\big] + g'\big(\Phi(j + ct); a\big)w_j(t), \tag{91}$$

which for each pair $(j_0, t_0) \in \mathbb{Z} \times \mathbb{R}$ has a unique solution $W^{t_0 j_0}$ that satisfies the initial condition

$$W_j^{t_0 j_0}(t_0) = \delta_{j j_0}. \tag{92}$$

These special solutions can be used to define a Green's function $\mathcal{G}(t, t_0) \in \mathcal{L}(\ell^\infty)$ that acts as the convolution

$$\mathcal{G}(t, t_0)y = \sum_{j_0 \in \mathbb{Z}} W_j^{t_0 j_0}(t)y_{j_0}. \tag{93}$$

Indeed, for any pair $t > t_0$ solutions to (91) can be found by writing

$$w(t) = \mathcal{G}(t, t_0)w(t_0). \tag{94}$$

In general, it is very hard to analyze non-autonomous evolution equations. In this case we are aided by the special structure of wave solutions. Indeed, the system (89) remains invariant under the replacements

$$(j, t) \mapsto \big(j - 1, t + c^{-1}\big). \tag{95}$$

In particular, upon introducing the right-shift operator $\mathcal{S} \in \mathcal{L}(\ell^\infty)$ that acts as $(\mathcal{S}U)_j = U_{j-1}$, we have the shift-periodicity

$$\mathcal{G}(t + c^{-1}, t_0 + c^{-1}) = \mathcal{S}^{-1}\mathcal{G}(t, t_0)\mathcal{S}. \tag{96}$$

In a sense, this compactifies the set of time differences $t - t_0$ that need to be considered.

Shift-periodic Floquet theory A first approach to exploit the identity (96) was developed by Chow, Mallet-Paret and Shen in [39]. In particular, they introduce the operator

$$\mathcal{R} = \mathcal{S}\mathcal{G}(c^{-1}, 0), \tag{97}$$

which can be seen as a shift-periodic version of the monodromy map that is usually considered for periodic systems. By checking that $v_j(t) = \Phi'(j + ct)$ is a solution to (91), one readily sees that $\mathcal{R}\Phi' = \Phi'$. In particular, we have

$$1 \in \sigma(\mathcal{R}) = \{\zeta \in \mathbb{C} \mid \mathcal{R} - \zeta : \ell^\infty \to \ell^\infty \text{ is not invertible}\}. \tag{98}$$

If this unit eigenvalue is simple and the remainder of the spectrum satisfies the inclusion

$$\sigma(\mathcal{R}) \setminus \{1\} \subset \{\zeta : |\zeta| < 1\}, \tag{99}$$

the general results in [39] show that the travelling wave is nonlinearly stable. This was achieved by constructing a local ℓ^∞-coordinate system around the travelling wave and analyzing the monodromy map in these new coordinates. The arguments used to construct this coordinate system are very technical and it is therefore not straightforward to generalize this approach to spectral scenarios that are more complicated than (99).

For general problems the spectrum of $\mathcal{R}$ is hard to analyze in a direct fashion. However, for our specific system (21) one can use the comparison principle to show that (99) indeed holds.

Spatial Green's functions A second approach towards analyzing the temporal Green's function $\mathcal{G}$ was pioneered by Benzoni-Gavage and her coworkers [14, 15]. The main idea is to sacrifice the discreteness of the spatial variable by passing to a reference frame that moves along with the wave. In particular, one replaces the non-autonomous linear system (91) by the autonomous non-local PDE

$$\begin{aligned} \partial_t w(t,\xi) &= -c\partial_\xi w(t,\xi) + d\big[w(t,\xi-1) + w(t,\xi+1) - 2w(t,\xi)\big] \\ &\quad + g'\big(\Phi(\xi); a\big)w(t,\xi). \\ &= \Big[\mathcal{L}_{\mathrm{tw}} w(t,\cdot)\Big](\xi). \end{aligned} \tag{100}$$

In order to solve this system by Laplace transform techniques, it is very useful to understand the pointwise behaviour of the resolvents of $\mathcal{L}_{\mathrm{tw}}$. This is encoded in the spatial Green's functions G_λ, which solve the resolvent problem

$$(\mathcal{L}_{\mathrm{tw}} - \lambda)G_\lambda(\cdot, \xi_0) = \delta(\cdot - \xi_0) \tag{101}$$

in the sense of distributions.

The link between the temporal and spatial Green's functions is provided by the representation formula

$$\mathcal{G}_{jj_0}(t, t_0) = -\frac{1}{2\pi i}\int_{\gamma - i\pi c}^{\gamma + i\pi c} e^{\lambda(t-t_0)} G_\lambda(j + ct, j_0 + ct_0)\, d\lambda, \tag{102}$$

which was established in [15, Theorem 4.2]. A-priori one must take $\gamma \gg 1$, but in order to obtain useful decay rates on $\mathcal{G}$ one needs to shift the contour as far to the left as possible.

In order to achieve this in the context of (21), one can exploit the meromorphic splitting (36) to write

$$G_\lambda(\xi, \xi_0) = -\lambda^{-1}\Phi'(\xi)\Psi(\xi_0) + \widetilde{G}_\lambda(\xi, \xi_0), \tag{103}$$

with a remainder term $\widetilde{G}_\lambda(\xi, \xi_0)$ that is analytic for small $|\lambda|$. This allows us to shift the contour in (102) to the left of the imaginary axis, picking up a simple pole in the process. For small $\beta > 0$ this yields the pointwise bound

$$\mathcal{G}_{jj_0}(t, t_0) = \Phi'(j + ct)\Psi(j_0 + ct_0) + O\big(e^{-\beta(t-t_0)}e^{-\beta|j+ct-j_0-ct_0|}\big). \tag{104}$$

With this bound in hand, one can proceed to construct stable manifolds for the family of travelling wave solutions $u_j(t) = \Phi(j + ct + \vartheta)$ to (21) obtained by varying the phase $\vartheta \in \mathbb{R}$. Together these manifolds span the entire ℓ^p-neighbourhood of the travelling wave solution, which leads to the nonlinear stability result.

This procedure is explained in detail in [101]. Generalizations to problems with infinite-range interactions can be found in [143]. On the other hand, problems with complicated spectral scenarios involving curves of essential spectrum that touch the origin were analyzed in [13].

2.4 Numerical Aspects

Since the travelling waves (22)–(23) are stable, they can in principle be observed numerically by simulating the dynamics of the LDE (21) with an initial condition that resembles the wave profile. In practice, the choice

$$u_j(0) = \frac{1}{2}\big[1 + \tanh(j)\big] \tag{105}$$

often suffices for the Nagumo (21), but other systems can require a more subtle construction. Naturally, one must truncate the problem to a finite domain $j \in \{-L, \ldots, L\}$ and enforce appropriate boundary conditions. The Dirichlet conditions $u_{-L} = 0$ and $u_L = 1$ give reliable results, but more refined choices also fit the leading order coefficients in the asymptotic expansions (34).

Naturally, simulations of (21) only allow travelling waves to be tracked for the limited amount of time that their 'core' is contained in the domain under consideration. In principle, this can be improved by using so-called freezing methods [20], which introduce an extra phase variable to keep track of the position of wave-like solutions. In a sense, this is the numerical equivalent of passing to a reference frame that moves along with the wave.

An alternative - more direct - approach is to look for numerical solutions of the travelling wave system (22)–(23). The early work by Chi, Bell and Hassard [38] already contains computations of this nature. This numerical work was continued by Elmer and Van Vleck, who have performed extensive calculations on MFDEs in [1, 55–58].

To our knowledge, the most recent and versatile tools in this area are collocation solvers that are able to solve MFDEs on finite intervals [1, 105]. In particular, these codes solve n-dimensional problems of the form

$$\tau(\xi)\phi'(\xi) = f\Big(\phi(\xi), \phi\big(\xi + \sigma_1(\xi, \phi(\xi))\big), \ldots, \phi\big(\xi + \sigma_N(\xi, \phi(\xi))\big)\Big), \tag{106}$$

for given functions $f : \mathbb{R}^{n(N+1)} \to \mathbb{R}^n$, $\tau : \mathbb{R} \to \mathbb{R}^n$ and shifts $\sigma_i : \mathbb{R}^{1+n} \to \mathbb{R}$. This is achieved by subdividing the interval under consideration into M subintervals $[t_i, t_{i+1}]$ and representing $\phi(\xi)$ on each subinterval in terms of a standard Runge-Kutta monomial basis. The representation is required to be continuous at the boundary points between subintervals and in addition to satisfy (106) at the Md collocation points $t_i + (t_{i+1} - t_i)c_j$, for $i = 1 \ldots M$ and $j = 1 \ldots d$, where $0 < c_1 < \ldots < c_d < 1$ are the Gaussian collocation points of degree d for some $d \geq 3$. Using Newton iterations such a piecewise polynomial ϕ can be found, provided that a sufficiently close initial estimate is available.

Various types of boundary conditions can be used to close the system (106). Since the shifts σ_i may depend on the spatial variable ξ as well as on the function value $\phi(\xi)$ itself, this allows us to compute periodic solutions to MFDEs even when the period is unknown [92].

The occurrence of propagation failure presents serious difficulties for this numerical scheme when solving (22)–(23), since solutions may lose their smoothness in the singular limit $c \to 0$. This difficulty can be overcome by introducing a term $-\gamma\Phi''$ to the left hand side of (22) and using numerical continuation techniques to take the positive constant γ as small as possible. In [103, 105] this approach is analyzed from a theoretical viewpoint. In particular, convergence results are provided to show that this approximation still allows us to uncover the behaviour that occurs at $\gamma = 0$.

3 Periodic Patterns

The interplay between different interactions on different length scales in a spatially discrete system often leads to the formation of periodic patterns. Examples include twinning microstructures in shape memory alloys [21], domain-wall microstructures in dielectric crystals [150] and oscillations in neural networks [64]. The dynamics of these patterns is both interesting and relatively unexplored.

Early results in this area were obtained in the context of material science, where spatially discrete equations have long been considered. The model of Hillert [84] is a one-dimensional model to describe phase separation in binary solids that allows for periodic patterns with no restriction on the amplitude of oscillations. The three-dimensional counterpart in [43] allows for periodic patterns with small amplitude. Subsequent work of Cahn and Novick-Cohen [28] developed a quasi-continuum approach to the model by deriving formal limiting PDEs that are able to capture the oscillatory behaviour; see also [141].

In [26] several spatially periodic structures were investigated for spatially discrete Allen–Cahn and Cahn–Hilliard equations with 'negative diffusion' and with both first and second neighbor couplings. It was noted in [27] that two-periodic patterns could be smoothened out by rewriting the scalar LDE as a vector LDE in order to split the odd and even numbered lattice sites. In [152] a model for martensitic phase transformations was derived that resulted in a bistable system with repelling first neighbor interactions.

In this section we focus on four recent projects related to this theme. The first of these [95, 97] reconsiders the Nagumo LDE (21) but now looks at the existence of spatially periodic equilibria and the possibility of forming connections between such patterns by multi-component travelling waves. In Sect. 3.2 we consider repelling interactions by taking $d < 0$ in (21). We show how the system can be reformulated as a vector equation and discuss results from [24] concerning the existence and stability of bistable and monostable traveling waves for a large range of parameter values. Co-existence of traveling solutions and propagation failure are investigated as well. Finally, in Sect. 3.3 we add next-to-nearest neighbour interactions to (21). Several results from [153, 155] are summarized to show how the competition between the two types of interactions affects the patterns that can be formed.

3.1 Multichromatic Waves

We are here again interested in the Nagumo LDE

$$\dot{u}_j(t) = d\big[u_{j-1} + u_{j+1} - 2u_j(t)\big] + g\big(u_j; a\big), \tag{107}$$

with $d > 0$. The discrete nature of this problem allows for a much richer class of equilibrium solutions than its spatially continuous counterpart. For example, one can search for n-periodic equilibria by writing

$$u_i = \mathrm{u}_{\mathrm{mod}(i,n)} \tag{108}$$

for some $\mathrm{u} \in \mathbb{R}^n$.

Upon introducing the nonlinear mapping $G(\cdot; a, d) : \mathbb{R}^n \to \mathbb{R}^n$ that acts as[2]

$$\big[G(\mathrm{u}; a, d)\big]_i = d(\mathrm{u}_{i-1} - 2\mathrm{u}_i + \mathrm{u}_{i+1}) + g\big(\mathrm{u}_i; a\big), \tag{109}$$

such equilibria must satisfy $G(\mathrm{u}; a, d) = 0$. Since the system decouples at $d = 0$, we have $G(\mathrm{w}_a; a, 0) = 0$ for any $\mathrm{w}_a \in \{0, a, 1\}^n$. Using the implicit function theorem, these 3^n simple roots can be locally continued for $d > 0$ until they collide with another branch.

---

[2]Here we use modulo arithmetic on i.

To formalize this procedure, we say that a solution to $G(\mathrm{u}; a, d) = 0$ is an equilibrium of type $\mathrm{w} \in \{\mathtt{o}, \mathtt{a}, \mathtt{1}\}^n$ if there exists a path through $[0, 1]^n \times (0, 1) \times [0, \infty)$ of simple roots that connects $(\mathrm{u}; a, d)$ to $(\mathrm{w}_a; a, 0)$. Here w_a is obtained from w by replacing $\{\mathtt{o}, \mathtt{a}, \mathtt{1}\}$ by $\{0, a, 1\}$. We use this symbolic distinction to emphasize the fact that the *type* w remains fixed, while the actual root u varies as the parameters (a, d) are changed. Indeed, we can now define the parameter set

$$\Omega_{\mathrm{w}} = \{(a, d) : \text{the system } G(\cdot\,; a, d) = 0 \text{ admits an equilibrium of type w}\}; \tag{110}$$

see Fig. 3 for several examples.

We will be specially interested in the cases $\mathrm{w} \in \{\mathtt{o}, \mathtt{1}\}^n$, since equilibria of this type are stable solutions to (107). Indeed, the monotonic iteration theory in [35] that was described in Sect. 2.2 can then be used to establish the existence of wave solutions that connect ordered pairs of such stable equilibria.

For $n = 2$, the upper boundary of $\Omega_{\mathtt{o1}}$ can be written as a graph $d = d_{\mathtt{o1}}(a)$. Writing $\mathrm{u} = \mathrm{u}_{\mathtt{o1}}(a, d)$ for the corresponding equilibrium (108), the wave solution that connects the homogeneous equilibrium $u_j \equiv 0$ to this two-periodic pattern can be written as

$$u_j(t) = \begin{cases} \Phi_e(j - ct) & j \text{ is even,} \\ \Phi_o(j - ct) & j \text{ is odd.} \end{cases} \tag{111}$$

Here we write $c = c_{\mathtt{o}\to\mathtt{o1}}(a, d)$ and impose the limits

$$\lim_{\xi\to-\infty} \big(\Phi_e(\xi), \Phi_o(\xi)\big) = (0, 0), \qquad \lim_{\xi\to+\infty} \big(\Phi_e(\xi), \Phi_o(\xi)\big) = \mathrm{u}_{\mathtt{o1}}(a, d). \tag{112}$$

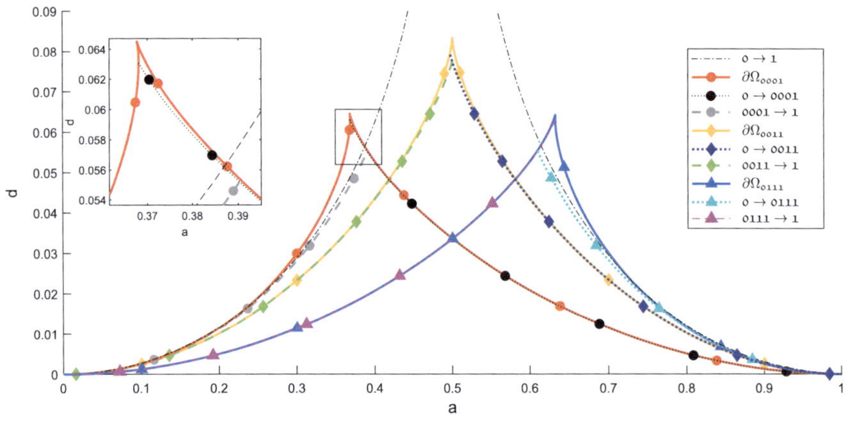

Fig. 3 The solid lines denote the upper boundary of the regions Ω_{w} where roots of type $\mathrm{w} \in \{\mathtt{0001}, \mathtt{0011}, \mathtt{0111}\}$ are defined. The dashed lines denote the boundaries above which the indicated wave connections have non-zero speed. Notice that regions exist in which the $\mathtt{o} \to \mathtt{0001}$, $\mathtt{0001} \to \mathtt{1}$ and $\mathtt{o} \to \mathtt{1}$ waves all travel

It is an interesting and delicate problem to decide whether these so-called bichromatic waves can travel or whether they are pinned. For small $d > 0$ the latter is the case, which allows us to consider the bichromatic pinning boundary

$$d_{0\to 01}(a) = \max\{0 \le d \le d_{01}(a) : c_{0\to 01}(a, d) = 0\}. \tag{113}$$

The main result in [95] is that there exists $a_- \in (0, \frac{1}{2})$ so that

$$d_{0\to 01}(a) < d_{01}(a) \tag{114}$$

holds for all $a \in (a_-, 1)$, which forces waves to travel for intermediate values of d. This can be seen as a generalization of the coercivity results for the standard monochromatic waves discussed in Sect. 2.2.

The cases $n = 3$ and $n = 4$ were analyzed numerically in [97]. In these cases there exist (small) intervals of a in which the upper boundaries of the sets Ω_w can no longer be written as graphs $d = d_w(a)$; see the inset of the cusp in Fig. 3. This is the reason that we must allow a to vary along the defining paths through the regions Ω_w. As a consequence, for fixed a the number of n-periodic equilibria does not decrease monotononically as the diffusion d is increased.

Upon writing $d_{0\to 1}(a)$ for the pinning boundary corresponding to the standard monochromatic waves (22) and generalizing the notation (113), the numerical results indicate that the ordering

$$d_{01}(a) < d_{0\to 0111}(a) < d_{0\to 1} < d_{0111}(a) \tag{115}$$

holds for all $a \in (0.63, 1)$. In particular, while bichromatic waves can only exist for parameters (a, d) where monochromatic waves are pinned, travelling quadrichromatic waves can co-exist with travelling monochromatic waves; see Fig. 3. This allows several types of intricate collision processes to take place. For example, a travelling $0 \to 0111$ wave can collide with a counter-propagating $0111 \to 1$ connection to form a travelling monochromatic $0 \to 1$ wave; see Fig. 4.

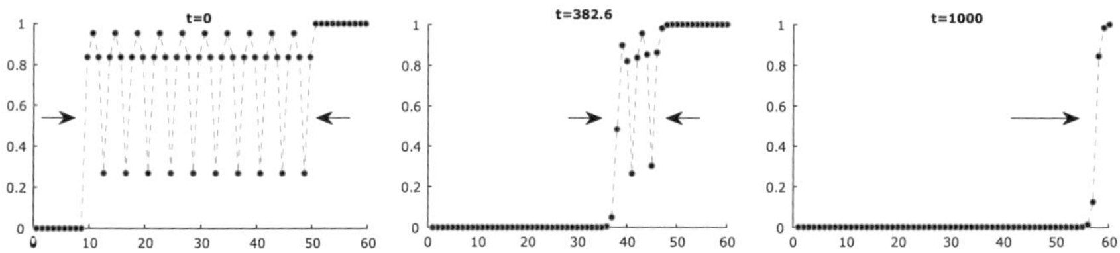

Fig. 4 This sequence of snapshots from a simulation of the Nagumo LDE (21) with $a = 0.63$ and $d = 0.0625$ features a collision between a wave of type $0 \to 0111$ and a counterpropagating wave of type $0111 \to 1$. The right part of the intermediate buffer zone is first pulled towards one by the incoming wave on the right, but then gets pulled towards zero by the final travelling monochromatic wave

3.2 Negative Diffusion

In [24], the existence and stability of traveling waves solutions with 'negative diffusion' is investigated. In particular, let us consider the LDE

$$\dot{u}_j(t) = -d\big[u_{j-1} + u_{j+1} - 2u_j(t)\big] + g\big(u_j; a\big), \tag{116}$$

again with $d > 0$. This system has been used in [152] to describe martensitic phase transitions in a chain of particles that features interactions between first and second neighbours. The latter interactions are modelled by standard springs, but the former are governed by a non-convex biquadratic potential representing two elastic phases. After a suitable rescaling, (116) can be seen as the overdamped limit of the resulting system.

Instead of flattening variations, the diffusion term in (116) now promotes oscillatory behaviour. Unlike its PDE counterpart, this system is still well-posed on account of the fact that the discrete Laplacian is simply a bounded linear operator. In particular, the cubic g can keep the oscillations under control.

Of course, the comparison principle no longer holds for (116). It can be recovered by introducing the new variables $v_j = (-1)^j u_j$, which satisfy the spatially periodic system

$$\dot{v}_j = d\big[v_{j-1} + v_{j+1} + 2v_j\big] + h_j\big(v_j; a\big) \tag{117}$$

with alternating nonlinearities

$$h_j(v; a) = \begin{cases} g(v; a) & \text{for even } j, \\ -g(-v; a) & \text{for odd } j. \end{cases} \tag{118}$$

In principle, this equation fits into the general spatially periodic framework of [35]. The challenge is to identify and classify the equilibrium solutions to (117). To this end, we look for stationary solutions of the form

$$v_j = \begin{cases} v_e & \text{for even } j, \\ v_o & \text{for odd } j, \end{cases} \tag{119}$$

which must hence satisfy the coupled system

$$2d(v_o + v_e) = -g(v_e; a), \qquad 2d(v_o + v_e) = g(-v_o; a). \tag{120}$$

Eliminating v_o from these equations, we must find the roots $G(v_e; a, d) = 0$ of the ninth-degree polynomial

$$G(v_e; a, d) = g(v_e; a) + g\big(v_e + \frac{g(v_e; a)}{2d}\big), \tag{121}$$

which can be factored as

$$G(v_e; a, d) = g(v_e; a)\tilde{G}(v_e; a, d) \tag{122}$$

for some sixth-degree polynomial $\tilde{G}(\cdot; a, d)$. This polynomial can have between zero and six real roots. Taking into account the three equilibria $(0, 0)$, $(a, -a)$ and $(1, -1)$ induced by the factor g, there are hence up to 81 potential connecting orbits between two-periodic stationary patterns.

In [24], we give a detailed description of the roots of G and their dependence on the parameters (a, d). This allows us us to study the existence, uniqueness and stability of traveling wavefront solutions to the antidiffusion Nagumo LDE (116). In particular, we uncover the different regions in the (a, d)-plane where bistable and monostable dynamics occur.

We analyze the propagation failure phenomenon in the bistable region and provided expressions for minimum wavespeeds in the monostable region. Our computations show that there are values for the parameters (a, d) where bistable and monostable connections co-exist.

3.3 Competing Interactions

In order to explore the effects of interactions with different length scales, we extend (21) by including a diffusion term that involves next-to-nearest neighbours. In particular, we introduce the discrete diffusions

$$(\Delta_1 u)_j := u_{j-1} + u_{j+1} - 2u_j, \qquad (\Delta_2 u)_j := u_{j-2} + u_{j+2} - 2u_j \tag{123}$$

and consider the system

$$\dot{u}_j = d_1(\Delta_1 u)_j + d_2(\Delta_2 u)_j + g\big(u_j; a\big) \tag{124}$$

on the one-dimensional lattice $j \in \mathbb{Z}$. Motivated by our earlier work [152], we focus much of our attention on the case where at least one of the interactions is repulsive (d_1 and/or d_2 are negative). Note that this is certainly not a necessary condition for pattern formation, as we have seen in Sect. 3.1.

The results in [153, 155] consider spatially p-periodic equilibrium solutions to (124) and the potential for connecting them via travelling waves. The focus is on $p \in \{1, 2, 4\}$, but larger periods, longer interaction lengths and higher dimensional lattices can also be considered using similar techniques.

In order to gain insight, we started in [153] by replacing the cubic g with its piecewise linear caricature g_{pl} defined in (14). As in Sect. 3.2, the first step was to find the periodic equilibria and identify the range of parameters where these solutions and their homogeneous counterparts are stable. We then took advantage of the piecewise linear form of the nonlinearity g_{pl} to apply Fourier transform techniques and constructed explicit traveling wave solutions along the lines of the approach in [27]. We conducted a detailed analysis of the solution in the $p = 2$ case and consid-

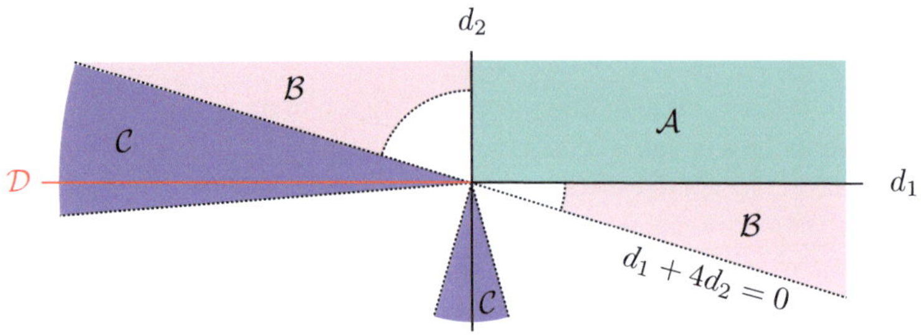

Fig. 5 Summary of results on existence of traveling wave solutions with spatially periodic patterns in (d_1, d_2) parameter space. The results by Chen et al. [36] are directly applicable to region $\mathcal{A}$. When $|(d_1, d_2)| \gg 1$, the techniques by Bates et al. [11] can be applied in regions $\mathcal{A}$ and $\mathcal{B}$. In region $\mathcal{C}$ the computations in [155] can be used, which for $d_2 < 0$ require either $|d_1/d_2| \ll 1$ or $|d_2/d_1| \ll 1$. The line $\mathcal{D}$ was investigated in [24]

ered some examples in the $p = 3$ and $p = 4$ cases. In particular, we investigated the dependence of the detuning parameter a on the wavespeed c and the parameters d_1 and d_2.

In [155] we returned to (124) with the cubic nonlinearity and used techniques similar to those in [11] to establish the existence of traveling wave solutions in the presence of competing interactions. When $d_2 > 0$ and $d_1 < 0$, the variable transformation $v_j = (-1)^j u_j$ can again be used to recover the comparison principle and apply the framework of [35]. However, when $d_2 < 0$ one must resort to perturbative techniques. Figure 5 summarizes the parameter regions where the various approaches can be applied.

4 Nagumo Equations on Planar Lattices

In this section we focus on bistable equations posed on two-dimensional lattices. To set the stage, let us consider the Nagumo LDE

$$\dot{u}_{ij} = u_{i+1,j} + u_{i-1,j} + u_{i,j+1} + u_{i,j-1} - 4u_{ij} + g(u_{ij}; a) \tag{125}$$

with $(i, j) \in \mathbb{Z}^2$. As in the continuous setting [6], we will see that planar travelling waves (see Fig. 6) can be used as a skeleton to help uncover the global dynamics of (125). Such waves can be written in the form

$$u_{ij}(t) = \Phi_\zeta\big(i \cos\zeta + j \sin\zeta + c_\zeta t\big), \qquad \Phi_\zeta(-\infty) = 0, \qquad \Phi_\zeta(+\infty) = 1, \tag{126}$$

in which ζ denotes the angle of propagation. Substituting this expression into (125) leads to the ζ-dependent MFDE

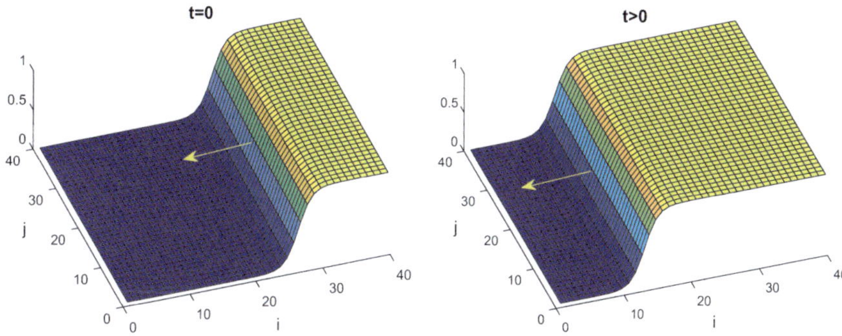

Fig. 6 Schematic representation of a planar wave (126) travelling in the horizontal direction $\zeta = 0$

$$c_\zeta \Phi_\zeta'(\xi) = \Phi_\zeta(\xi + \cos\zeta) + \Phi_\zeta(\xi + \sin\zeta) + \Phi_\zeta(\xi - \cos\zeta) + \Phi_\zeta(\xi - \sin\zeta) \\ -4\Phi_\zeta(\xi) + g\big(\Phi_\zeta(\xi); a\big). \tag{127}$$

The global homotopy argument discussed in Sect. 2.2 can also be used to construct solutions to this system [126]. In particular, after fixing $a \in (0, 1)$, the wavespeed c_ζ depends uniquely on ζ and the waveprofiles Φ_ζ are unique up to translation provided that $c_\zeta \neq 0$.

It turns out to be a much more subtle question to decide upon the stability of these planar waves. Early work in this direction for four-dimensional nonlocal problems can be found in [10], but in two dimensions the relevant decay rates associated to a brute-force linearization of (125) are simply too slow. One encounters similar problems when analyzing the Nagumo PDE, but two main approaches have been developed in recent years to overcome them [17, 111]. Both these approaches build upon the realization that transverse deformations of the wave interface are governed by a heat equation that can be analyzed separately from the remainder of the perturbation.

In order to discuss these issues, it is convenient to introduce the new variables

$$n = i\cos\zeta + j\sin\zeta, \qquad l = j\cos\zeta - i\sin\zeta, \tag{128}$$

which are parallel respectively tranverse to the direct of propagation of the wave. We will write $\Lambda_\zeta \subset \mathbb{R}^2$ for the collection of all pairs (n, l) obtained by applying (128) to the original lattice $(i, j) \in \mathbb{Z}^2$. This set is dense when $\tan\zeta \notin \mathbb{Q}$. On the other hand, when $\tan\zeta$ is rational or infinite it is possible to find a $\delta > 0$ so that we have the inclusion

$$\Lambda_\zeta \subset \delta\mathbb{Z}^2. \tag{129}$$

In this case, it is often convenient to extend the problem (by repetition) onto the entire grid $(n, l) \in \delta\mathbb{Z}^2$.

For any ζ, the coordinates (128) transform the LDE (125) into

$$\dot{u}_{nl} = u_{n+\cos\zeta,l-\sin\zeta} + u_{n+\sin\zeta,l+\cos\zeta} + u_{n-\cos\zeta,l+\sin\zeta} + u_{n-\sin\zeta,l-\cos\zeta} \\ -4u_{nl} + g\big(u_{nl}; a\big), \tag{130}$$

which now admits the planar wave solutions

$$u_{nl}(t) = \Phi_\zeta\big(n + c_\zeta t\big). \tag{131}$$

The refined splitting alluded to above is now given by

$$u_{nl}(t) = \Phi_\zeta\big(n + c_\zeta t + \theta_l(t)\big) + v_{nl}(t). \tag{132}$$

In Sect. 4.1 we discuss the linearized equations for the pair (θ, v) and show how to extend the one-dimensional Green's functions from Sect. 2.3. We use this in Sect. 4.2 to generalize [111] and establish the local nonlinear stability of the planar waves (131) using bootstrapping arguments. On the other hand, in Sect. 4.3 we obtain the global stability of these waves by building on the arguments in [17], which are based on the comparison principle.

Moving on to solutions that are more general than travelling waves, we discuss the expansion of compactly supported initial conditions in Sect. 4.4. The limiting shapes often resemble polygons that can be predicted from the structure of the $\zeta \mapsto c_\zeta$ function. This leads naturally to the study of travelling corners in Sect. 4.5.

The last part of this section is focused on versions of (125) with spatially-varying diffusion coefficients, i.e.,

$$\dot{u}_{ij} = d_{ij}\big[u_{i+1,j} + u_{i-1,j} + u_{i,j+1} + u_{i,j-1} - 4u_{ij}\big] + g(u_{ij}; a). \tag{133}$$

In Sect. 4.6 we allow the diffusion coefficient to be modified on a single lattice site, which can be interpreted as an obstacle. We discuss the results in [87] that show that planar waves can pass around such an obstacle and regain their shapes.

We conclude in Sect. 4.7 by allowing d_{ij} to vary periodically or become negative, analogous to the one-dimensional setup in Sect. 3. In particular, we describe results from [103] on the existence of travelling checkerboard patterns.

4.1 Linear Theory

A key ingredient towards analyzing the behaviour of (125) in the neighbourhood of the planar waves (126) is to linearize the LDE around these waves. The anistropy of the lattice plays an important role here, as the resulting structure of the linear equations depends heavily on the direction of propagation ζ.

Lattice directions Performing this linearization for the horizontal direction $\zeta = 0$, we arrive at the system

$$\begin{aligned}\dot{v}_{ij}(t) = v_{i+1,j}(t) + v_{i,j+1}(t) + v_{i-1,j}(t) + v_{i,j-1}(t) - 4v_{ij}(t) \\ +g'\big(\Phi_0(i + c_0 t); a\big)v_{ij}(t),\end{aligned} \tag{134}$$

which as in the one-dimensional case discussed in Sect. 2.3 is temporally non-autonomous. The j-direction however is unaffected by this issue, which allows us to take a discrete Fourier transform in this direction. This readily leads to the decoupled family of systems

$$\begin{aligned}\tfrac{d}{dt}[\hat{v}_\omega]_i(t) = [\hat{v}_\omega]_{i+1}(t) + e^{i\omega}[\hat{v}_\omega]_i(t) + [\hat{v}_\omega]_{i-1}(t) + e^{-i\omega}[\hat{v}_\omega]_i(t) - 4[\hat{v}_\omega]_i(t) \\ +g'(\Phi(i + ct); a)[\hat{v}_\omega]_i(t)\end{aligned} \tag{135}$$

for $\omega \in [-\pi, \pi]$.

Using the procedure outlined in Sect. 2.3, it is possible to construct Green's functions for each of these spatially one-dimensional systems. The relevant linear operators can be found by searching for solutions of the form

$$[\hat{v}_\omega]_i(t) = e^{\lambda t} w_\omega(i + ct), \tag{136}$$

which leads to the eigenvalue problem

$$\mathcal{L}_{i\omega}^{(\zeta=0)} w_\omega = \lambda w_\omega \tag{137}$$

for the linear operator

$$\begin{aligned}[\mathcal{L}_z^{(0)} p](\xi) = -cp'(\xi) + 2\cosh(z)p(\xi) + p(\xi + 1) + p(\xi - 1) - 4p(\xi) \\ +g'\big(\Phi(\xi); a\big)p(\xi).\end{aligned} \tag{138}$$

Upon writing $\lambda_z = 2(\cosh(z) - 1)$, we have

$$\mathcal{L}_z^{(0)}\Phi' = \lambda_z \Phi'. \tag{139}$$

To find the evolution of perturbations of the form

$$v_{ij}(t) = \theta_j(t)\Phi'(i + ct) \tag{140}$$

under (134), it now suffices to solve the discrete heat equation

$$\dot{\theta}_j = \theta_{j+1} + \theta_{j-1} - 2\theta_j. \tag{141}$$

Indeed, taking the discrete Fourier transform of this system yields $\frac{d}{dt}\hat{\theta}_\omega = \lambda_{i\omega}\hat{\theta}_\omega$. These perturbations are important because they correspond at the linear level with transverse deformations of the planar wave interface. We remark that the situation here closely resembles that encountered for the Nagumo PDE in [111].

General directions For general angles ζ, the relevant linear system can be written as

$$\begin{aligned}\dot{v}_{nl}(t) = {} & v_{n+\cos\zeta,l-\sin\zeta}(t) + v_{n+\sin\zeta,l+\cos\zeta}(t) \\ & + v_{n-\cos\zeta,l+\sin\zeta}(t) + v_{n-\sin\zeta,l-\cos\zeta}(t) - 4v_{nl}(t) \\ & + g'\big(\Phi_\zeta(n + c_\zeta t); a\big)v_{nl}(t).\end{aligned} \tag{142}$$

The corresponding operator is given by

$$\begin{aligned}[\mathcal{L}_z^{(\zeta)} v](\xi) = {} & -c_\zeta v'(\xi) + e^{-z\sin\zeta} v(\xi + \cos\zeta) + e^{z\cos\zeta} v(\xi + \sin\zeta) \\ & + e^{z\sin\zeta} v(\xi - \cos\zeta) + e^{-z\cos\zeta} v(\xi - \sin\zeta) - 4v(\xi) \\ & + g'\big(\Phi_\zeta(\xi); a\big)v(\xi).\end{aligned} \tag{143}$$

In particular, the dependence on z is now non-trivial. This means that one can no longer hope to find an exact solution similar to (140).

On the other hand, for small values of $|z|$ it is possible to find ζ-dependent branches $z \mapsto (\lambda(z), \phi_z)$ of the eigenvalue problem

$$\lambda(z)\phi_z = \mathcal{L}_z^{(\zeta)}\phi_z. \tag{144}$$

Upon introducing the quantities

$$\alpha_\nu = \langle \Psi, \Phi'(\cdot + \nu)\rangle_{L^2}, \qquad \beta_\nu = \langle \Psi, [\partial_z \phi_z]_{z=0}(\cdot + \nu)\rangle_{L^2}, \tag{145}$$

it is not hard to compute [94]

$$\begin{aligned}\lambda'(0) &= \cos\zeta\big[\alpha_{\sin\zeta} - \alpha_{-\sin\zeta}\big] + \sin\zeta\big[\alpha_{-\cos\zeta} - \alpha_{\cos\zeta}\big], \\ \lambda''(0) &= \sin^2\zeta\big[\alpha_{\cos\zeta} + \alpha_{-\cos\zeta}\big] + \cos^2\zeta\big[\alpha_{\sin\zeta} + \alpha_{-\sin\zeta}\big] \\ & \quad + 2\cos\zeta\big[\beta_{\sin\zeta} - \beta_{-\sin\zeta}\big] + 2\sin\zeta\big[\beta_{-\cos\zeta} - \beta_{\cos\zeta}\big].\end{aligned} \tag{146}$$

For $\zeta \in \frac{1}{4}\pi\mathbb{Z}$ we have $\lambda'(0) = 0$ and $\lambda''(0) > 0$. We have numerical evidence to show that the second inequality persists for all angles ζ, but the first identity is typically violated. In fact, making the dependence on ζ explicit, we have

$$\lambda'_\zeta(0) = \partial_\zeta c_\zeta \tag{147}$$

and this quantity is typically referred to as the group velocity. However, we do remark here that the function $\zeta \mapsto c_\zeta$ can behave rather wildly in the critical regime where $a \approx \frac{1}{2}$, allowing the group velocity to vanish at specific values for a even if $\zeta \notin \frac{\pi}{4}\mathbb{Z}$.

Since the group velocity is a real number, we have

$$\Re\, \lambda_{i\omega} \sim -\lambda''(0)\omega^2 \tag{148}$$

for $\omega \approx 0$, which means that we can still expect linear decay estimates similar to those of the discrete heat equation.

Let us now assume that $\tan\zeta$ is rational and recall the inclusion (129). Extending our solution to the entire grid $(n, l) \in \delta\mathbb{Z}^2$ and imposing the decomposition

$$v_{nl}(t) = \theta_l(t)\Phi'(n+ct) + w_{nl}(t), \tag{149}$$

we normalize w by demanding

$$\sum_{n\in\delta\mathbb{Z}} \Phi'(n+ct)w_{nl}(t) = 0 \tag{150}$$

for all $t \geq 0$ and $l \in \delta\mathbb{Z}$. Formally writing the coupled system for (v, θ) as

$$\big(\dot{\theta}, \dot{v}\big) = B_\zeta[\theta, v], \tag{151}$$

we can combine the Green's function techniques from Sect. 2.3 with the Fourier decomposition above to obtain a full two-dimensional Green's function $\mathcal{G}_\zeta$ for (151); see [86, Sect. 4]. The decay rates for θ correspond to those for the discrete heat equation, while v decays at rates that are $t^{-1/2}$ faster. In the diagonal direction $\zeta = \frac{\pi}{4}$ this factor increases to t^{-1}, while in the horizontal direction $\zeta = 0$ the v component even decays exponentially rather than algebraically.

4.2 Local Stability

Our first approach [86] to establish the stability of planar waves only considers small perturbations from the wave and hence avoids using the comparison principle. In particular, this technique can be applied to a wide range of problems, as long as several standard spectral conditions can be verified. It is based on a surprisingly recent result due to Kapitula [111], who used semigroup methods and fixed-point arguments to obtain the corresponding result for the Nagumo PDE.

In particular, we pick an angle ζ for which $c_\zeta \neq 0$ and for which $\tan\zeta$ is rational. Based on the intuition obtained in Sect. 4.1, we consider the evolution under the LDE (125) of patterns of the form

$$u_{ij}(t) = \Phi\big(n + ct + \theta_l(t)\big) + v_{nl}(t), \tag{152}$$

with (n, l) given by (128). We assume that the initial perturbation from the travelling wave is small in the sense that

$$\sum_{l\in\delta\mathbb{Z}} |\theta_l(0)| + \sup_{n\in\delta\mathbb{Z}} \sum_{l\in\delta\mathbb{Z}} |v_{nl}(0)| < \delta. \tag{153}$$

In order to fix the coordinate system in such a way that the variables $\theta_l(t)$ represent genuine phaseshifts, we impose the normalization condition

$$\sum_{n\in\delta\mathbb{Z}} \Psi(n+ct)v_{nl}(t) = 0. \tag{154}$$

After some algebra, the resulting system for (θ, v) can be written in the abstract form

$$\big(\dot{v}, \dot{\theta}\big)(t) = B_\zeta(t)\big[v, \theta\big] + \big(\mathcal{N}_\zeta^{(v)}(t; v, \theta), \mathcal{N}_\zeta^{(\theta)}(t; v, \theta)\big), \tag{155}$$

in which the linear operator $B_\zeta(t)$ was introduced in (151). Using the Green's function for this latter system, one can use Duhamel's principle to recast (155) into the mild form

$$\begin{aligned} \big(v, \theta\big)(t) = {} & \mathcal{G}_\zeta(t, 0)[v(0), \theta(0)] \\ & + \int_0^t \mathcal{G}_\zeta(t, s)\big[\mathcal{N}_\zeta^{(v)}\big(s; v(s), \theta(s)\big), \mathcal{N}_\zeta^{(\theta)}\big(s; v(s), \theta(s)\big)\big]\, ds. \end{aligned} \tag{156}$$

Based on the linear theory in Sect. 4.1, the choice for ℓ^1-based norms in the transverse direction implies that θ can be expected to decay as $t^{-1/4}$ in ℓ^2. However, the decay rate for v depends on the angle ζ.

We now explore the effect of the nonlinear terms $\mathcal{N}_\zeta$, which also depend crucially on ζ. As a preparation, we introduce the first differences

$$\theta_l^\diamond = \big(\theta_{l+\sin\zeta} - \theta_l, \theta_{l-\cos\zeta} - \theta_l, \theta_{l-\sin\zeta} - \theta_l, \theta_{l+\cos\zeta} - \theta_l\big), \tag{157}$$

together with the second differences

$$\theta^{\diamond\diamond} = \big(\theta^\diamond\big)^\diamond. \tag{158}$$

The linear theory predicts the decay rates

$$\big\|\theta^\diamond\big\|_{\ell^2} \sim t^{-3/4}, \qquad \big\|\theta^{\diamond\diamond}\big\|_{\ell^2} \sim t^{-5/4} \tag{159}$$

for these expressions.

Horizontal and diagonal directions For $\zeta \in \frac{\pi}{4}\mathbb{Z}$ the nonlinear term for v satisfies the pointwise bounds

$$\mathcal{N}_\zeta^{(v)}(t; v, \theta) = O(|v|^2) + O(|v\theta|) + O\big(\big|\theta^\diamond\big|^2\big) + O\big(\big|\theta\theta^{\diamond\diamond}\big|\big), \tag{160}$$

with similar behaviour for $\mathcal{N}_\zeta^{(\theta)}$. In particular, the standard Cauchy–Schwarz bound $\|ab\|_{\ell^1} \leq \|a\|_{\ell^2}\, \|b\|_{\ell^2}$ together with the linear behaviour

$$\sup_{n \in \delta\mathbb{Z}} \sum_{l \in \delta\mathbb{Z}} |v_{nl}(t)|^2 \sim t^{-5/4} \tag{161}$$

show that in the norm (153) we can expect $(\mathcal{N}^{(v)}, \mathcal{N}^{(\theta)})$ to decay as $t^{-3/2}$. This is very similar to the corresponding bounds obtained in [111] for the PDE setting. In particular, setting up a variation-of-constants argument using the Green's function $\mathcal{G}_\zeta$ defined in Sect. 4.1 allows us to conclude that the expected linear decay bounds carry over to the nonlinear setting.

General directions For general rational directions ζ the geometry of the lattice affects both the linear and nonlinear terms in our problem. Indeed, we have seen in Sect. 4.1 that the expected linear decay in v slows down to

$$\sup_{n \in \delta\mathbb{Z}} \sum_{l \in \delta\mathbb{Z}} |v_{nl}(t)|^2 \sim t^{-3/4}. \tag{162}$$

In addition, the lack of symmetry weakens the pointwise bounds for $\mathcal{N}_\zeta^{(v)}$ to

$$\mathcal{N}_\zeta(t; v, \theta) = O(|v|^2) + O(|v\theta|) + O\big(\left|\theta^\diamond\right|^2\big) + O\big(\left|\theta\theta^\diamond\right|\big). \tag{163}$$

In particular, the terms of order $|v\theta|$ and $|\theta\theta^\diamond|$ now cause complications, since they can only be expected to decay as t^{-1} in ℓ^1. Indeed, the crude estimate

$$\int_0^t (1+t-s)^{-1/4}(1+s)^{-1} ds \sim \log(1+t)(1+t)^{-1/4} \tag{164}$$

shows that the $t^{-1/4}$ decay for $\|\theta\|_{\ell^2}$ cannot be recovered from a simple application of the Duhamel formula.

In order to refine this estimate, it is essential to consider the explicit structure of the problematic terms. Consider for example the simple, yet crucial, identity

$$\theta_l(\theta_{l+1} - \theta_l) = \frac{1}{2}[\theta_{l+1}^2 - \theta_l^2 - (\theta_{l+1} - \theta_l)^2]. \tag{165}$$

The square of the first difference can be expected to behave as $t^{-3/2}$, which does not pose a problem. The difference of the squares though only decays like t^{-1}. However, using a summation by parts procedure the difference operator can be transferred to the Green's function. The bad estimate (164) can now be replaced by the good estimate

$$\int_0^t (1+t-s)^{-3/4}(1+s)^{-1/2} ds \sim (1+t)^{-1/4}. \tag{166}$$

A careful analysis of the troublesome nonlinear terms shows that they can all be treated by tricks of this kind, allowing us to close the nonlinear stability argument. We remark that the identity (165) can be regarded as a discrete form of $uu_x = \left(\frac{1}{2}u^2\right)_x$, which plays a crucial role when studying the stability of travelling waves for PDEs with a conservation law structure [160].

4.3 Global Stability

The approach in [87] leverages the comparison principle to show that the planar waves (126) are stable under a much larger class of initial perturbations. This generalizes the techniques developed in the landmark paper [17], where the authors constructed explicit super- and sub-solutions to the Nagumo PDE (3) that trap perturbations that can be arbitrarily large (but localized).

These sub-solutions can be seen as two-dimensional refinements of (84). Indeed, they take the form

$$u^-(x, y, t) = \Phi\big(x + ct - \theta(y, t) - Z(t)\big) - z(t), \tag{167}$$

with global functions

$$z(t) = \epsilon e^{-\eta t}, \qquad Z(t) = \int_0^t z(s)\, ds \tag{168}$$

for small $\epsilon > 0$ and $\eta > 0$ that turn small additive perturbations into a phaseshift. The function θ is used to control transverse perturbations and is given by

$$\theta(y, t) = \beta(t + 1)^{-2\gamma^{-1}} e^{-y^2/\left(4\gamma(t+1)\right)}. \tag{169}$$

Here $\gamma = \gamma(\beta) \gg \beta$ and $\beta \gg 1$ can be chosen to be arbitrarily large, allowing localized perturbations of any size to be controlled. In particular, the function θ can be seen as a modified heat-kernel where the diffusion is sped up by a factor γ and the decay rate at the center $y = 0$ is slowed down to $(t + 1)^{-2\gamma^{-1}}$.

In [16] this idea was generalized to the LDE (125) for waves travelling in the horizontal direction $\zeta = 0$. This was achieved by replacing the Gaussian in (169) by the corresponding kernel for the discrete heat equation (141). This kernel can be expressed in the form of modified Bessel functions of the first kind.

This procedure relies on a specific factorization similar to (140) that fails to hold for general directions ζ. In order to fully account for all the slowly decaying resonances spawned by the anisotropy of the lattice, the sub-solution (167) needs to be adjusted significantly. In particular, for suitably chosen functions

$$P^{\diamond} : \mathbb{R} \to \mathbb{R}^{1\times 4}, \qquad P^{\diamond\diamond} : \mathbb{R} \to \mathbb{R}^{1\times 16}, \qquad Q^{\diamond\diamond} : \mathbb{R} \to \mathbb{R}^{4\times 4}, \tag{170}$$

it is possible [87] to use

$$\begin{aligned} u_{ij}^-(t) = {} & \Phi\big(n + ct - \theta_l(t) - Z(t)\big) + P^{\diamond}\big(n + ct - \theta_l(t) - Z(t)\big)\theta_l^{\diamond}(t) \\ & + P^{\diamond\diamond}\big(n + ct - \theta_l(t) - Z(t)\big)\theta_l^{\diamond\diamond}(t) \\ & + \theta_l^{\diamond}(t)^T Q^{\diamond\diamond}\big(n + ct - \theta_l(t) - Z(t)\big)\theta_l^{\diamond}(t) - z(t). \end{aligned} \tag{171}$$

Here we interpret the first and second difference $\theta_l^{\diamond}$ and $\theta_l^{\diamond\diamond}$ defined in (157)–(158) as vectors in $\mathbb{R}^{4\times 1}$ and $\mathbb{R}^{16\times 1}$ respectively. In addition, the decay of the function z needs to be slowed down to $z(t) \sim t^{-3/2}$ for $t \gg 1$ and the function θ needs to be modified to

$$\theta_l(t) = \beta t^{-\frac{1}{4}\gamma^{-1}} \sqrt{\frac{2\pi}{\lambda''(0)}} \exp\Big[- \frac{(l + \lambda'(0)t)^2}{2\lambda''(0)\gamma t} \Big]. \tag{172}$$

Notice that the group velocity (146) appears here as an extra convective term.

One can use normal form techniques to find the specific form of the functions (170). Indeed, their purpose is to neutralize any terms in the sub-solution residual that decay at rates that cannot be integrated uniformly in γ. Since we are artificially slowing down the decay of θ, this means that we need to pay special consideration to terms of order $\theta^{\diamond\diamond}$ and $(\theta^{\diamond})^2$, which could be neglected in Sect. 4.2.

4.4 Spreading Phenomena

Let us here assume that $c_\zeta > 0$ for all angles ζ and consider the compactly supported initial condition

$$u_{ij}(0) = \begin{cases} 1 & \text{for } i^2 + j^2 \leq R^2, \\ 0 & \text{for } i^2 + j^2 > R^2 \end{cases} \tag{173}$$

for some sufficiently large $R > 0$. In the PDE setting, a classic result [6, Theorem 5.3] obtained by Weinberger states that such regions containing the energetically favourable state will spread throughout the space with a radial speed equal to the speed of the planar travelling waves. The proof of this result uses radially expanding sub- and super-solutions that can be constructed by glueing together planar travelling waves travelling in different directions.

In [87] a weak version of this expansion result was established for the LDE (125) with (173). However, the underlying sub- and super-solutions expand at the speeds $\min_{0 \leq \zeta \leq 2\pi} c_\zeta$ and $\max_{0 \leq \zeta \leq 2\pi} c_\zeta$ respectively, which generally differ from each other. This still leaves a considerable hole in our knowledge of the expansion process.

The numerical results in [154] provide strong evidence that the limiting shape formed by the evolution of (173) can be found by applying the Wulff construction [136] to the polar plot of the $\zeta \mapsto c_\zeta$ relation. In particular, for each $\zeta \in [0, 2\pi]$ one considers the line

$$\ell_\zeta = c_\zeta(\cos\zeta, \sin\zeta) + \mathbb{R}(\sin\zeta, -\cos\zeta), \tag{174}$$

which is tangent to the circle of radius c_ζ at the angle ζ. The set that is enclosed by the collection of these lines is referred to as the Wulff shape; see Fig. 7. For a large subset of parameters a this limiting shape resembles a polygon.

Let us mention here that the Wulff shape plays an important role in the field of crystallography, where it is used to predict the limiting shape of growing crystals [158]. In such applications it is usually constructed from a polar plot of a free energy functional.

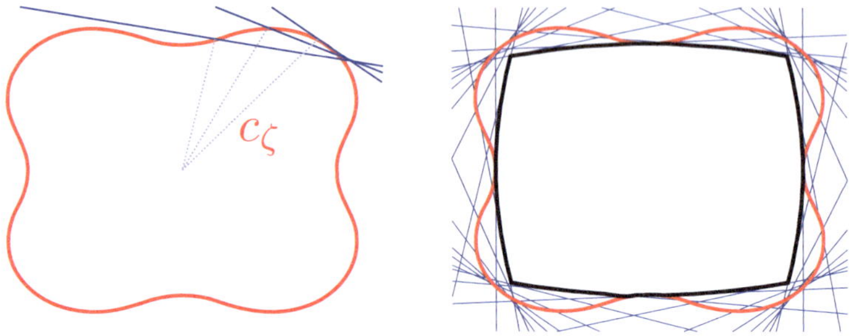

Fig. 7 The left panel depicts three of the perpendicular lines ℓ_ζ that play a role in the Wulff construction. The boundary of the Wulff shape is displayed in the right panel and resembles a slightly rounded polygon

4.5 Travelling Corners

The main motivation behind [94] is to take a step towards understanding the polygonal expansion process described in Sect. 4.4 by looking at the evolution of a single corner. Indeed, when the expanding blob is sufficiently large, it would seem to be very reasonable to assume that the corners of the polygon behave in an almost independent fashion.

A natural starting point is to look at so-called travelling corner solutions, which arise from the regular planar travelling waves by 'bending' the level sets of the waves. More specifically, the level sets are transformed from straight lines | perpendicular to the direction of propagation into shapes resembling a > or < sign; see Fig. 8.

For technical reasons the results in [94] only apply to propagation directions for which the group velocity vanishes. Recalling (147), this means that

$$\lambda_\zeta'(0) = \partial_\zeta c_\zeta = 0. \tag{175}$$

Although the group velocity can always be transformed away in the PDE setting by a simple change of variables, the discreteness prevents us from doing this for LDEs. In particular, (175) is a genuine restriction. On the other hand, an important role in the Wulff construction is reserved for the directions where the wavespeed is minimal and these are all obviously covered.

The admissable bending angles are determined by the directional dispersion

$$d_\zeta(\varphi) = \frac{c_{\zeta+\varphi}}{\cos(\varphi)\cos\zeta}, \tag{176}$$

which measures the speed at which the level-sets of the planar wave travelling in the direction $\zeta + \varphi$ move along the vector $(\cos\zeta, \sin\zeta)$. In view of (175) we have $d_\zeta(0) = c_\zeta$ and $d_\zeta'(0) = 0$.

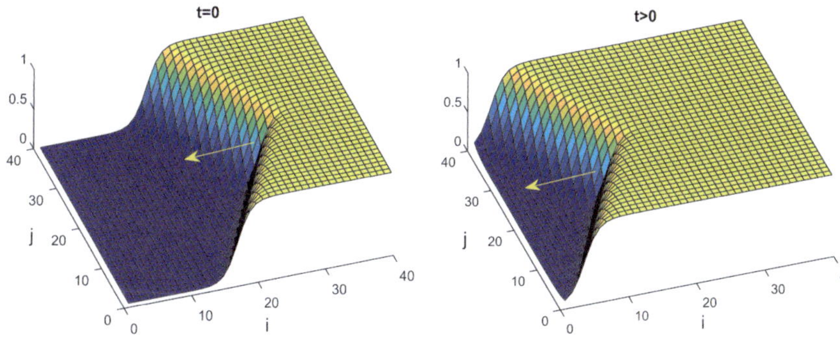

Fig. 8 Schematic representation of a corner travelling in the horizontal direction

Assuming that $d''(0) > 0$, we hence see that for all $\tilde{c} > c_\zeta$ sufficiently close to c_ζ, there exist two small angles $\varphi_- < 0 < \varphi_+$ for which

$$d_\zeta(\varphi_\pm) = \tilde{c} \tag{177}$$

holds. The main result in [94] now states that there exists a sequence of phaseshifts $\{\theta_l\}_{l \in \delta\mathbb{Z}}$ that satisfies the limits

$$\delta^{-1}(\theta_{l+\delta} - \theta_l) \to \tan(\varphi_\pm), \qquad l \to \pm\infty, \tag{178}$$

together with a sequence of perturbations $\{v_l\}_{l \in \delta\mathbb{Z}} \subset H^1$ so that the function

$$u_{ij}(t) = \Phi(n + \tilde{c}t + \theta_l) + v_l(n + \tilde{c}t) \tag{179}$$

is a solution to (125). In particular, the level sets are bent away from the vector $(\cos\zeta, \sin\zeta)$ to form a $>$ shape. When $d''(0) < 0$ this orientation is reversed and one must take $\tilde{c} < c_\zeta$.

The main idea behind the proof of this result was inspired by the series of papers written by Scheel and Haragus [79–81] for the planar Nagumo PDE, where they used a two-dimensional global center manifold to capture the corner solutions. The key feature of this approach is that the phaseshifts θ can be factored out from the analysis due to the translational symmetry of the problem. However, the loss of regularity in the discrete case causes severe complications and forces one to analyze the full coupled system for v and θ along the lines of [99].

In any case, the delicate behaviour of the c_ζ map for the Nagumo LDE (125) leads to a rich class of behaviour. For example, the second derivative $d''(0)$ can have both signs, while in the spatially homogeneous case one always has $d''(0) > 0$. In addition, it is possible to find so-called bichromatic corners, which connect spatially homogeneous equilibria to checkerboard patterns.

4.6 Obstacles

In many situations it is desirable to understand the impact that localized spatial impurities have on the dynamical behaviour of systems. In the context of (133), this can be modelled by choosing diffusion coefficients that are homogeneous everywhere except at (0, 0), i.e., $d_{ij} = d + \beta\delta_{i0}\delta_{j0}$ with $d > 0$ and $\beta > -d$. Alternatively, one can consider the system on the punctured grid $\mathbb{Z}^2 \setminus \{(0,0)\}$, replacing the nearest-neighbour discrete Laplacian by the graph Laplacian. In particular, the equation for $u_{1,0}$ is modified to

$$\dot{u}_{1,0} = u_{2,0} + u_{1,1} + u_{1,-1} - 3u_{1,0} + g\big(u_{1,0}; a\big) \tag{180}$$

with similar modifications for $u_{0,\pm1}$ and $u_{-1,0}$.

In both cases, one can study the existence of *entire solutions* that resemble a planar travelling wave as $t \to \pm\infty$ but have large transients in between caused by the impurity. In [87] we show how the subsolutions (171) can be adapted to control these large transients. In essence, we allow the global term $z(t)$ to grow instead of decay during the time that the steep part of the waveprofile is in the vicinity of the obstacle. In particular, the results show that planar waves can pass around large compact obstacles and still eventually recover their shape.

Our techniques are a variant of those developed fairly recently in [17] for the Nagumo PDE. However, we do need to require that $c_\zeta \neq 0$ for all $\zeta \in [0, 2\pi]$ in order to utilize the spreading result discussed in Sect. 4.4. It is an interesting open problem to determine the behaviour of the large transients when certain directions are pinned.

4.7 Travelling Checkerboards

In [103], we focussed on (133) with coefficients that are allowed to vary periodically, e.g. $d_{ij} = d_o > 0$ whenever $i + j$ is odd and $d_{ij} = d_e > 0$ whenever $i + j$ is even. Following the approach in Sect. 3.1, we look for solutions that can be written in the form of travelling checkerboards

$$u_{ij}(t) = \begin{cases} \Phi_e(i\cos\zeta + j\sin\zeta + ct), & i + j \text{ is even}, \\ \Phi_o(i\cos\zeta + j\sin\zeta + ct), & i + j \text{ is odd}, \end{cases} \tag{181}$$

in which we impose the spatially homogeneous limiting values

$$\lim_{\xi\to-\infty} \big(\Phi_e(\xi), \Phi_o(\xi)\big) = (0, 0), \qquad \lim_{\xi\to+\infty} \big(\Phi_e(\xi), \Phi_o(\xi)\big) = (1, 1). \tag{182}$$

These waves must satisfy the coupled system

$$\begin{aligned}c\Phi_e'(\xi) &= d_e\big[\Phi_o(\xi+\cos\zeta)+\Phi_o(\xi+\sin\zeta)\\&\qquad +\Phi_o(\xi-\cos\zeta)+\Phi_o(\xi-\sin\zeta)-4\Phi_e(\xi)\big]+g\big(\Phi_e(\xi);a\big),\\c\Phi_o'(\xi) &= d_o\big[\Phi_e(\xi+\cos\zeta)+\Phi_e(\xi+\sin\zeta)\\&\qquad +\Phi_e(\xi-\cos\zeta)+\Phi_e(\xi-\sin\zeta)-4\Phi_o(\xi)\big]+g\big(\Phi_o(\xi);a\big).\end{aligned} \tag{183}$$

In this multi-component case it is no longer clear how the global homotopy argument described in Sect. 2.2 can be extended. In addition, the potential irrationality of the shifts prevents a direct application of the results from [35]. In [103] we generalized the ideas in [34] to the present multi-component setting, which allowed us to develop a spatial regularization method to obtain the existence and uniqueness of travelling waves. We also extended the Fredholm theory for the linearized operator $\mathcal{L}_{\mathrm{tw}}$, which for $c \neq 0$ allows us to conclude that these checkerboards are nonlinearly stable and depend smoothly on the parameters (a, ζ, d_o, d_e).

These techniques can also be applied to (133) with homogeneous negative diffusion coefficients $d_{ij} = d < 0$. As before, one can introduce new variables $v_{ij} = (-1)^{i+j}u_{ij}$ to transform the problem into a two-component system that admits a comparison principle. Under certain restrictions on the nonlinearity, we were able to again obtain travelling checkerboards. However, in this case the limiting rest states are typically not spatially homogeneous.

5 Discrete FitzHugh–Nagumo Equations

In this section we focus on several versions of the discrete FitzHugh–Nagumo equation, which can be used to describe the propagation of signal through myelinated nerve fibres [113]. In its simplest form, this system can be written as

$$\begin{aligned}\dot{v}_j(t) &= v_{j-1}(t)+v_{j+1}(t)-2v_j(t)+g\big(v_j(t)\,;\,a\big)-w_j(t),\\\dot{w}_j(t) &= \epsilon\big(v_j(t)-w_j(t)\big).\end{aligned} \tag{184}$$

For each lattice point $j \in \mathbb{Z}$, the variable v_j encodes the on-site action potential while w_j acts as a generic recovery component. The parameter $\epsilon > 0$ is related to the speed of this recovery, which is relatively slow compared to the firing process. The cubic g models the local ionic interactions. Notice that (184) reduces to the Nagumo LDE for $\epsilon = 0$ and $w = 0$.

The discrete nature of the problem is a direct consequence of the properties of the myelin coating. This coating is essential to ensure that electrical pulses travel at adequate speed, but leads to rapid signal degradation. In order to repair this, the coating admits regularly spaced gaps at the so-called nodes of Ranvier [135]. Through a process called saltatory conduction, the electrical spikes effectively jump from one node to the next [122].

By restricting the size of $\epsilon > 0$ in (184) one effectively creates a two-timescale problem that fits into the general framework of singular perturbation theory. In Sect. 5.1 we discuss how several important techniques from this area can be generalized to the discrete setting. This allows the existence and stability of travelling pulse solutions for (184) to be established.

In Sect. 5.2 we setup a bifurcation argument that allows these pulses to be extended to versions of (184) where the diffusion coefficient alternates between small and large values. Finally, in Sect. 5.3 we replace the finite-range discrete Laplacian in (184) by a class of counterparts that feature infinite-range interactions.

5.1 Singular Perturbation Theory

The LDE (184) can be seen as the nearest-neighbour discretization of the FitzHugh–Nagumo PDE

$$\begin{aligned} v_t &= v_{xx} + g(v; a) - w \\ w_t &= \epsilon(v - w), \end{aligned} \tag{185}$$

which was originally proposed [73, 74] as a simplification of the Hodgkin-Huxley equations [85] to describe the propagation of signals through the nerve fibres of giant squids. Early numerical experiments by FitzHugh [75] provided clear evidence that (185) supports travelling pulse solutions, but a rigorous analysis of these solutions turned out to be quite delicate. By now, existence and nonlinear stability results for these pulses have been obtained in various settings using a wide range of techniques such as geometric singular perturbation theory [29, 83, 109, 110], Lin's method [30, 31, 118], the variational principle [33] and the Maslov index [44, 45].

Returning to the LDE (184), substitution of the travelling wave ansatz

$$(v_j, w_j)(t) = (\phi, \psi)(j + ct) \tag{186}$$

leads naturally to the two-component MFDE

$$\begin{aligned} c\phi'(\xi) &= \phi(\xi - 1) + \phi(\xi + 1) - 2\phi(\xi) + g\big(\phi(\xi); a\big) - \psi, \\ c\psi'(\xi) &= \epsilon\big[\phi(\xi) - \psi(\xi)\big]. \end{aligned} \tag{187}$$

For certain specially tailored nonlinearities g, several ad-hoc techniques involving asymptotic expansions and Fourier transforms have been used to construct pulse solutions [37, 62, 151]. However, in order to handle the cubic and other general bistable nonlinearities, it turned out to be worthwhile to adapt several key constructions from the field of geometric singular perturbation to the infinite-dimensional setting considered here [100, 101].

To set the stage, we note that in the limiting case $\epsilon = 0$ all equilibria of (193) lie on the graph $\mathcal{M} = \big(\phi, g(\phi; a)\big)$. Choosing manifolds $\mathcal{M}_L$ and $\mathcal{M}_R$ as in Fig. 9, one can use the travelling wave solutions of the Nagumo LDE discussed in Sect. 2.2 to construct front and back connections $(\phi_f, 0)$ and (ϕ_b, w_*) between $\mathcal{M}_L$ and $\mathcal{M}_R$ that both travel at the same wave speed c_* that we assume to be positive. These

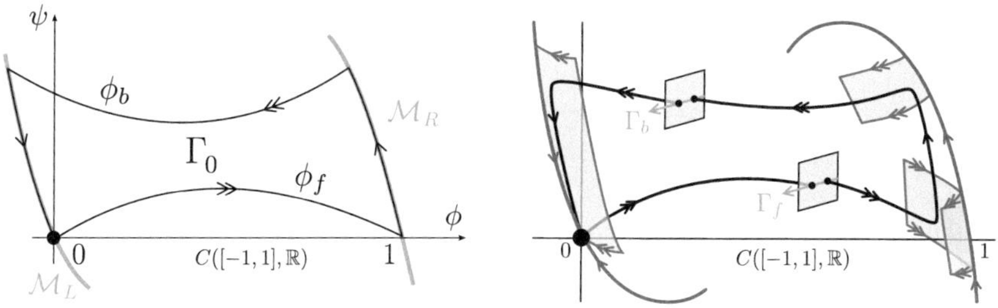

Fig. 9 The left panel describes the manifolds $\mathcal{M}_L$ and $\mathcal{M}_R$ and contains the singular orbit Γ_0. The right panel despicts a quasi-solution to (187), which has two jumps that can be captured in the one-dimensional spaces Γ_f and Γ_b

connections can be combined with segments of $\mathcal{M}_L$ and $\mathcal{M}_R$ to form the (fast) singular orbit Γ_0 depicted on the left in Fig. 9.

The main goal is to show that that this singular orbit can be continued into a branch of regular orbits $\Gamma(\epsilon)$ for small $\epsilon > 0$. In the PDE case, a typical proof proceeds by showing that the unstable manifold of $(0, 0)$ intersects the stable manifold of $\mathcal{M}_L$ in a transverse fashion, using the Exchange Lemma [110] to study the passage of these manifolds near $\mathcal{M}_R$. Such an analysis relies on the Hadamard graph transform and the Fenichel normal form to describe the dynamics near $\mathcal{M}_L$ and $\mathcal{M}_R$. Unfortunately, these tools are not readily available in infinite dimensional settings.

The main contribution in [100] is that analytical constructions are used to underpin the geometric intuition behind the ideas described above. Building on earlier work by Sakamoto [139], we used the exponential dichotomies discussed in Sect. 2.1 to construct a branch of global invariant manifolds for $\mathcal{M}_L$ and $\mathcal{M}_R$ that persist for small $\epsilon > 0$. The analytic proof of the exchange lemma developed by Krupa, Sandstede and Szmolyan in [118] could subsequently be combined with the Lin's method machinery discussed in Sect. 2.1 to construct quasi-solutions to (187). These solutions have two discontinuities that can both be captured in one-dimensional spaces and quantified by means of the Hale inner product (44); see the right panel in Fig. 9.

All that remains to construct the pulse solutions for (187) is to close these two gaps. This can be done by analyzing the associated two-component nonlinear bifurcation problem. Interestingly enough, the same setup can be used to reduce the spectral characterization of these pulses to a related two-dimensional problem involving the spectral parameter λ; see [101].

The structure of these bifurcation equations is almost identical to their counterparts for the PDE (185). In particular, this allows the existence and nonlinear stability of travelling pulse solutions for the LDE (184) to be obtained.

5.2 Periodic Diffusion

Recent developments in optical nanoscopy clearly show [49, 50, 159] that certain proteins in the cytoskeleton of nerve fibres are organized periodically. This periodicity turns out to be a universal feature of all nerve systems, not just those which are insulated with a myelin coating. Since it also manifests itself at the nodes of Ranvier, it is natural to allow the parameters in (184) to vary in a periodic fashion.

In particular, let us consider the LDE

$$\begin{aligned} \dot{v}_j &= d_j\big[v_{j-1} + v_{j+1} - 2v_j\big] + g(v_j; a) - w_j, \\ \dot{w}_j &= \rho[v_j - w_j] \end{aligned} \tag{188}$$

with 2-periodic diffusion coefficients

$$d_j = \begin{cases} \epsilon^{-2} & \text{for odd } j, \\ 1 & \text{for even } j. \end{cases} \tag{189}$$

Note that we are now using $\rho > 0$ for the recovery speed variable in order to emphasize that the relevant scale separation is now contained in the diffusion coefficients.

Searching for solutions of the form

$$(v, w)_j(t) = \begin{cases} (\phi_o, \psi_o)(j + ct), & \text{for odd } j, \\ (\phi_e, \psi_e)(j + ct), & \text{for even } j, \end{cases} \tag{190}$$

we obtain the travelling wave system

$$\begin{aligned} c\phi_o'(\xi) &= \tfrac{1}{\epsilon^2}\big[\phi_e(\xi - 1) + \phi_e(\xi + 1) - 2\phi_o(\xi)\big] + g\big(\phi_o(\xi); a\big) - \psi_o, \\ c\psi_o'(\xi) &= \rho\big[\phi_o(\xi) - \psi_o(\xi)\big], \\ c\phi_e'(\xi) &= \phi_o(\xi - 1) + \phi_o(\xi + 1) - 2\phi_e(\xi) + g\big(\phi_e(\xi); a\big) - \psi_e, \\ c\psi_e'(\xi) &= \rho\big[\phi_e(\xi) - \psi_e(\xi)\big]. \end{aligned} \tag{191}$$

Multiplying the first line of (191) by ϵ^2 and taking the limit $\epsilon \downarrow 0$, we obtain the formal relation

$$\overline{\phi}_o(\xi) = \tfrac{1}{2}\big(\overline{\phi}_e(\xi - 1) + \overline{\phi}_e(\xi + 1)\big) \tag{192}$$

between the singular waveprofiles $\overline{\phi}_o$ and $\overline{\phi}_e$. The limiting even subsystem hence decouples from the odd system and formally satisfies

$$\begin{aligned} c\overline{\phi}_e'(\xi) &= \overline{\phi}_e(\xi - 2) + \overline{\phi}_e(\xi + 2) - 2\overline{\phi}_e(\xi) + g\big(\overline{\phi}_e(\xi); a\big) - \overline{\psi}_e, \\ c\overline{\psi}_e'(\xi) &= \rho\big[\overline{\phi}_e(\xi) - \overline{\psi}_e(\xi)\big]. \end{aligned} \tag{193}$$

This is simply a spatially rescaled version of the travelling wave equation for the FitzHugh–Nagumo LDE (184) with constant diffusion. In particular, the results in Sect. 5.1 guarantee that this even limiting system has pulse solutions that are spectrally stable for small $\rho > 0$.

The main results in [142] state that *any* spectrally stable solution to the limiting system (193) can be extended to a branch of nonlinearly stable travelling wave solutions (190) to the full 2-periodic system (188) for small values of $\epsilon > 0$. This is hence another singular perturbation result, but the underlying bifurcation at $\epsilon = 0$ is of a completely different nature than in Sect. 5.1.

The key ingredient behind the proof of this result is the realization that the spirit of the spectral convergence technique discussed in Sect. 2.2 can be used to study a wide range of singular limits for families of differential operators. The main technical complication that needs to be overcome in the current context is that the four components of (191) and their derivatives need to be rescaled with various powers of ϵ before taking the appropriate limits.

5.3 *Infinite-Range Interactions*

Recently, an active interest has arisen in non-local equations that feature infinite-range interactions. Neural field models for example aim to describe the dynamics of large networks of neurons, which interact with each other by exchanging signals across long distances through their interconnecting nerve axons [22, 23, 133, 147]. One model that has been proposed [22, Eq. (3.31)] to describe these complex interactions features a FitzHugh–Nagumo type system with infinite-range interactions.

Motivated by the above, the results in [69, 143] consider a class of infinite-range FitzHugh–Nagumo LDEs that includes the prototype

$$\begin{aligned} \dot{v}_j &= \frac{1}{h^2} \sum_{k>0} e^{-k^2} [v_{j-k} + v_{j+k} - 2v_j] + g(v_j; a) - w_j, \\ \dot{w}_j &= \rho [v_j - w_j], \end{aligned} \tag{194}$$

in which $h > 0$ represents the grid-spacing. By replacing the phase-space approach in Sect. 5.1 with Fredholm-based functional-analytic considerations, Faye and Scheel were able [69] to obtain the existence of travelling pulses to (194) for arbitrary $h > 0$, provided that $\rho > 0$ is sufficiently small.

In [143] the nonlinear stability of these pulses was established in the continuum regime where $h > 0$ is sufficiently small. On the other hand, the only requirement on $\rho > 0$ is that a spectrally stable travelling pulse exists for the FitzHugh–Nagumo PDE that arises as the formal $h \downarrow 0$ limit of (194). In particular, the two main insights obtained in [143] are that the spectral convergence method discussed in Sect. 2.2 can be generalized to multi-component systems and that it can also be used to study the spectral properties of the constructed solutions.

The latter point turns out to be rather subtle. To see this, let us write $\mathcal{L}_{\text{tw}}^{(h)}$ for the linearization of (194) around the pulse solution corresponding to the grid size $h > 0$. For a fixed $\lambda \neq 0$ with $\Re\lambda \geq 0$, the spectral convergence method can be used to show that the operator $\mathcal{L}_{\text{tw}}^{(h)} - \lambda$ is invertible for small $h > 0$. However, for a spectral stability argument one must consider all such λ simultaneously for a fixed small $h > 0$. This requires careful control on all the parameter dependencies appearing in the problem.

6 Applications to Numerical Analysis

The primary motivation behind the papers discussed in this section is to contribute to the on-going systematic approach to understand the impact of discretization schemes on the dynamics that they are designed to capture. Of course, there is a tremendous amount of literature concerning the accuracy of numerical schemes, but these studies typically focus on *finite time* error bounds [117, 120]. Our concern here is more related to the persistence of structures that exist for all time. Interesting pioneering work on this topic can be found in [19, 77], where the authors discuss the impact of discretization on attractors for ODEs and PDEs.

We take one step further here and study in what sense individual stable travelling structures such as waves survive common discretization techniques. Building on previous work in [59–61], we study full spatial-temporal discretizations of the Nagumo PDE in Sect. 6.1, restricting our attention to grids that are fixed in space and time. We loosen this restriction in Sect. 6.2, where we allow the spatial grid to adjust itself to the shape of the solution.

6.1 Spatial-Temporal Discretizations

In [104] we revisited the spatially-discrete Nagumo LDE (21) and set out to replace the remaining temporal derivative by an appropriate discretized operator. We considered the family of six Backward Differentiation Formula (BDF) methods, which are multistep methods that admit good numerical stability properties for parabolic equations. They are referred to as regular methods, which means that they don't create or destroy equilibrium solutions. For our purposes here they are relatively easy to analyze, since they do not involve any of the intermediate stage values that are required by many popular one-step methods. Related results for the standard forward Euler scheme can be found in [39], but the smaller stability region severely limits the time-steps that can be used.

The first order BDF method coincides with the backward Euler scheme. Applying this to the LDE (21) with $d = 1$ and time-step Δt, we arrive at the fully discretized system

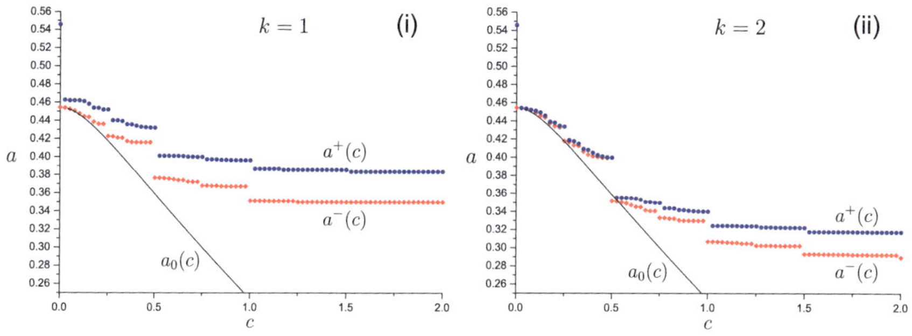

Fig. 10 Range of a for which fully discrete traveling wave solutions exist for BDF methods of first and second order, compared to the corresponding curve $a_0(c)$ for the spatially discrete problem. For these computations we used $\Delta t = 2$ and we replaced g by the nonlinearity $\frac{121}{12}u(1-u)(u-a)$; see [104]

$$\frac{1}{\Delta t}\big[U_j(n\Delta t) - U_j\big((n-1)\Delta t\big)\big] = U_{j-1}(n\Delta t) + U_{j+1}(n\Delta t) - 2U_j(n\Delta t) + g\big(U_j(n\Delta t); a\big). \tag{195}$$

Systems of this type are often referred to as coupled map lattices (CMLs). They are of independent interest and have been used for a wide range of purposes, including the construction of hash functions [157] and the study of population dynamics [48].

The main contribution in [104] is the construction of fully discretized travelling wave solutions

$$U_j(n\Delta t) = \Phi(j + nc\Delta t), \qquad \Phi(-\infty) = 0, \qquad \Phi(+\infty) = 1 \tag{196}$$

to (195). Such fronts must satisfy the difference equation

$$\frac{1}{\Delta t}\big[\Phi(\xi) - \Phi(\xi - c\Delta t)\big] = \Phi(\xi - 1) + \Phi(\xi + 1) - 2\Phi(\xi) + g\big(\Phi(\xi); a\big). \tag{197}$$

By a peculiar coincidence, this equation fits into the global homotopy framework of [126] that we discussed in Sect. 2.2, after adding an artificial term $\tilde{c}\Phi'(\xi)$ to the left hand side. Indeed, solutions to (197) live inside the parameter region of this artificially enhanced system where $\tilde{c} = 0$. As we have seen, such pinning regions generically encompass an open set of detuning parameters a.

As a consequence, it is natural to conjecture that the $a(c)$ relation for (197) can be multi-valued even for $c \neq 0$. This is confirmed by the numerical results in Fig. 10, which shows the upper and lower bounds $a^{\pm}(c)$ of the detuning parameter a for which (196)–(197) admits a solution Φ. For comparison, the curve $a_0(c)$ gives the corresponding $a(c)$ relation for the semi-discrete problem (22) with $d = 1$. We note that monostable KPP systems [132] in the presence of inhomogeneities have been found to exhibit similar behaviour.

Let us write $(\overline{c}, \overline{\Phi})$ for the solution to the semi-discrete travelling wave MFDE (22) with $d = 1$. The key technical ingredient in our construction of solutions to (196)–(197) is the understanding of the fully discrete operator

$$[\mathcal{L}_{\mathrm{fd}}v](\xi) = -\frac{1}{\Delta t}[v(\xi) - v(\xi - \overline{c}\Delta t)] + v(\xi - 1) + v(\xi + 1) - 2v(\xi) + g'(\overline{\Phi}(\xi); a)v(\xi), \tag{198}$$

for ξ in an appropriate subset of $\mathbb{R}$. This operator is associated to the linearization of (195) around the pair $(\overline{c}, \overline{\Phi})$. The main question is whether this operator inherits properties from its semi-discrete counterpart $\mathcal{L}_{\mathrm{tw}}$ defined in (24). This highly singular transition can be analyzed by a variant of the spectral convergence technique described in Sect. 2.2. The main complication in this setting is that the natural domain for ξ varies from the whole line to a subset of the line, which required the use of a delicate interpolation procedure.

6.2 Adaptive Grids

Most efficient modern solvers concentrate their meshpoints in areas where the solution under consideration fluctuates the most. This often leads to a significant increase in performance compared to fixed-grid algorithms. In order to explore this, let us write $\{x_j(t)\}$ for the positions of the grid points and introduce the approximants

$$U_j(t) \approx u(x_j(t), t). \tag{199}$$

Instead of approximating the second derivative in (3) by the standard second difference

$$u_{xx}(x) \approx \frac{1}{h^2}\big(u(x - h) + u(x - h) - 2u(x)\big), \tag{200}$$

we can now apply the non-uniform discretization

$$u_{xx}(x_j) \approx \frac{2}{x_{j+1} - x_{j-1}}\left(\frac{U_{j-1} - U_j}{x_j - x_{j-1}} + \frac{U_{j+1} - U_j}{x_{j+1} - x_j}\right). \tag{201}$$

In particular, by differentiating (199) and inspecting (3), we arrive at the LDE

$$\dot{U}_j = \left[\frac{U_{j+1} - U_{j-1}}{x_{j+1} - x_{j-1}}\right]\dot{x}_j + \frac{2}{x_{j+1} - x_{j-1}}\left(\frac{U_{j-1} - U_j}{x_j - x_{j-1}} + \frac{U_{j+1} - U_j}{x_{j+1} - x_j}\right) + g(U_j; a). \tag{202}$$

In order to close this system, we need to specify a rule for computing the grid-point velocity $\dot{x}_j(t)$. This is usually achieved by formulating an elliptic or parabolic equation that incorporates the local behaviour of the solution U. The approach taken in many codes to solve two point boundary value problems is to distribute a suitable prediction for the numerical error equally along each mesh-interval. Indeed, for accuracy and efficiency a fine mesh is desired where the solution changes rapidly and a course mesh is sufficient where the solution is relatively constant.

In [102] we considered moving mesh equations that aim to equidistribute the arclength of the numerical solution. In particular, we use the MMPDE5 [90] grid update scheme, which can be written as

$$\tau \dot{x}_j = \sqrt{(x_{j+1} - x_j)^2 + (U_{j+1} - U_j)^2} - \sqrt{(x_{j-1} - x_j)^2 + (U_{j-1} - U_j)^2} \quad (203)$$

for some tunable coefficient $\tau \geq 0$ that determines the speed of the gridpoints. We studied the limiting case $\tau = 0$, which means that the mesh instantaneously adjusts itself to equalize the arclength of the solution U between neighbouring gridpoints. This allows us to pick a constant $h > 0$ for which

$$h^2 = (x_{j+1} - x_j)^2 + (U_{j+1} - U_j)^2 \quad (204)$$

holds for all $j \in \mathbb{Z}$ and $t \geq 0$. The boundary condition $x_j(t) \to jh$ as $j \to -\infty$ subsequently fixes the mesh in a unique fashion. The quantity $h > 0$ can hence be seen as the 'background' grid size that is observed in regions where U is flat.

The main results in [102] state that for each sufficiently small $h > 0$, the resulting coupled system admits a travelling wave solution

$$U_j(t) = \phi(x_j(t) + ct), \qquad \phi(-\infty) = 0, \qquad \phi(\infty) = 1 \quad (205)$$

that is unique up to translation. Upon introducing the wave variable $\xi = x_j(t) + ct$, we note that the grid spacing can be recovered from such wave solutions by writing

$$x_{j+1}(t) - x_j(t) = h_+[\xi; \phi], \;\; x_j(t) - x_{j-1}(t) = h_-[\xi; \phi] \quad (206)$$

for a pair of suitably chosen functions $h_\pm$. The second difference (201) for this wave can now be written as

$$\frac{2}{h_+[\xi; \varphi] + h_-[\xi; \varphi]} \left(\frac{\phi(\xi - h_-[\xi; \phi]) - \phi(\xi)}{h_-[\xi; \phi]} + \frac{\phi(\xi - h_+[\xi; \phi]) - \phi(\xi)}{h_+[\xi; \phi]} \right). \quad (207)$$

The full wave equation for the pair (c, ϕ) will hence become an MFDE with state-dependent advances and shifts, for which hardly any theory is available.

Fortunately, this issue can be avoided by using the computational coordinates $\tau = jh + ct$ instead of the physical coordinates ξ. However, it is a delicate and cumbersome task to switch between the two points of view. An interesting consequence is that the correct linear operator that needs to be understood for $h = 0$ is not the standard linearization (61), but rather a spatially stretched version that takes into account the arclength equidistribution of the PDE waveprofile.

We are specially interested to uncover if and why the size of the pinning interval is smaller here when compared to stationary grids. In particular, this will give additional insight into the *theoretical* benefits of adaptive grids compared to the *practical* benefits of increased performance. Preliminary results pertaining to this issue can be found in [89].

Acknowledgements HJH, LM and WSS acknowledge support from the Netherlands Organization for Scientific Research (NWO) (grants 639.032.612 and 613.001.304). EVV was supported in part by NSF grant # DMS-1714195.

References

1. Abell, K.A., Elmer, C.E., Humphries, A.R., Van Vleck, E.S.: Computation of mixed type functional differential boundary value problems. SIAM J. Appl. Dyn. Syst. **4**, 755–781 (2005)
2. Alfaro, M., Droniou, J., Matano, H.: Convergence rate of the Allen-Cahn equation to generalized motion by mean curvature. J. Evol. Equ. **12**(2), 267–294 (2012)
3. Allen, S.M., Cahn, J.W.: A microscopic theory for antiphase boundary motion and its application to antiphase domain coarsening. Acta Metall. **27**(6), 1085–1095 (1979)
4. Anderson, T., Faye, G., Scheel, A., Stauffer, D.: Pinning and unpinning in nonlocal systems. J. Dyn. Differ. Equ. **28**(3–4), 897–923 (2016)
5. Aronson, D.G., Weinberger, H.F.: Nonlinear diffusion in population genetics, combustion, and nerve pulse propagation. Partial differential equations and related topics (Program, Tulane Univ., New Orleans, La., 1974). Lecture Notes in Mathematics, vol. 446, pp. 5–49. Springer, Berlin (1975)
6. Aronson, D.G., Weinberger, H.F.: Multidimensional nonlinear diffusion arising in population genetics. Adv. Math. **30**(1), 33–76 (1978)
7. Bakker, B.: Nonlinear waves in local and nonlocal media: a topological approach. Ph.D. Thesis (2019)
8. Bakker, B., Scheel, A.: Spatial Hamiltonian identities for nonlocally coupled systems. Forum of Mathematics, Sigma, vol. 6. Cambridge University Press, Cambridge (2018)
9. Barashenkov, I., Oxtoby, O., Pelinovsky, D.: Translationally invariant discrete kinks from one-dimensional maps. Phys. Rev. E **72**, 035602 (2005)
10. Bates, P.W., Chen, F.: Spectral analysis and multidimensional stability of traveling waves for nonlocal Allen-Cahn equation. J. Math. Anal. Appl. **273**(1), 45–57 (2002)
11. Bates, P.W., Chen, X., Chmaj, A.J.J.: Traveling waves of bistable dynamics on a lattice. SIAM J. Math. Anal. **35**(2), 520–546 (2003)
12. Bates, P.W., Chmaj, A.: A discrete convolution model for phase transitions. Arch. Ration. Mech. Anal. **150**, 281–305 (1999)
13. Beck, M., Hupkes, H.J., Sandstede, B., Zumbrun, K.: Nonlinear stability of semidiscrete shocks for two-sided schemes. SIAM J. Math. Anal. **42**, 857–903 (2010)
14. Benzoni-Gavage, S., Huot, P.: Existence of semi-discrete shocks. Discret. Contin. Dyn. Syst. **8**, 163–190 (2002)
15. Benzoni-Gavage, S., Huot, P., Rousset, F.: Nonlinear stability of semidiscrete shock waves. SIAM J. Math. Anal. **35**, 639–707 (2003)
16. Berendsen, J.: Horizontal travelling waves on the lattice. Ph.D. thesis, Masters thesis, Leiden University (2015)
17. Berestycki, H., Hamel, F., Matano, H.: Bistable traveling waves around an obstacle. Commun. Pure Appl. Math. **62**(6), 729–788 (2009)
18. Beyn, W.J.: The numerical computation of connecting orbits in dynamical systems. IMA J. Numer. Anal. **9**, 379–405 (1990)

19. Beyn, W.J., Pilyugin, S.Y.: Attractors of reaction diffusion systems on infinite lattices. J. Dyn. Diff. Equ. **15**, 485–515 (2003)
20. Beyn, W.-J., Thümmler, V.: Freezing solutions of equivariant evolution equations. SIAM J. Appl. Dyn. Syst. **3**(2), 85–116 (2004)
21. Bhattacharya, K.: Microstructure of Martensite: Why it Forms and How it Gives Rise to the Shape-Memory Effect, Vol. 2. Oxford University Press, Oxford (2003)
22. Bressloff, P.C.: Spatiotemporal dynamics of continuum neural fields. J. Phys. A: Math. Theor. **45**, 3 (2011)
23. Bressloff, P.C.: Waves in Neural Media: From Single Neurons to Neural Fields. Lecture Notes on Mathematical Modeling in the Life Sciences. Springer, Berlin (2014)
24. Brucal-Hallare, M., Van Vleck, E.: Traveling wavefronts in an antidiffusion lattice Nagumo model. SIAM J. Appl. Dyn. Syst. **10**(3), 921–959 (2011)
25. Cahn, J.W.: Theory of crystal growth and interface motion in crystalline materials. Acta Metall. **8**, 554–562 (1960)
26. Cahn, J.W., Chow, S.-N., Van Vleck, E.S.: Spatially discrete nonlinear diffusion equations. Rocky Mountain J. Math. **25**(1), 87–118 (1995). Second Geoffrey J. Butler Memorial Conference in Differential Equations and Mathematical Biology (Edmonton, AB, 1992)
27. Cahn, J.W., Mallet-Paret, J., Van Vleck, E.S.: Traveling wave solutions for systems of ODE's on a two-dimensional spatial lattice. SIAM J. Appl. Math. **59**, 455–493 (1999)
28. Cahn, J.W., Novick-Cohen, A.: Evolution equations for phase separation and ordering in binary alloys. J. Stat. Phys. **76**, 877–909 (1994)
29. Carpenter, G.: A geometric approach to singular perturbation problems with applications to nerve impulse equations. J. Differ. Equ. **23**, 335–367 (1977)
30. Carter, P., de Rijk, B., Sandstede, B.: Stability of traveling pulses with oscillatory tails in the FitzHugh-Nagumo system. J. Nonlinear Sci. **26**(5), 1369–1444 (2016)
31. Carter, P., Sandstede, B.: Fast pulses with oscillatory tails in the FitzHugh-Nagumo system. SIAM J. Math. Anal. **47**(5), 3393–3441 (2015)
32. Celli, V., Flytzanis, N.: Motion of a screw dislocation in a crystal. J. Appl. Phys. **41**(11), 4443–4447 (1970)
33. Chen, C., Choi, Y.: Traveling pulse solutions to FitzHugh-Nagumo equations. Calc. Var. Partial Differ. Equ. **54**(1), 1–45 (2015)
34. Chen, X.: Existence, uniqueness and asymptotic stability of traveling waves in nonlocal evolution equations. Adv. Differ. Equ. **2**, 125–160 (1997)
35. Chen, X., Guo, J.-S., Wu, C.-C.: Traveling waves in discrete periodic media for bistable dynamics. Arch. Ration. Mech. Anal. **189**(2), 189–236 (2008)
36. Chen, X., Guo, J.S., Wu, C.C.: Traveling waves in discrete periodic media for bistable dynamics. Arch. Ration. Mech. Anal. **189**, 189–236 (2008)
37. Chen, X., Hastings, S.P.: Pulse waves for a semi-discrete Morris-Lecar type model. J. Math. Bio. **38**, 1–20 (1999)
38. Chi, H., Bell, J., Hassard, B.: Numerical solution of a nonlinear advance-delay-differential equation from nerve conduction theory. J. Math. Bio. **24**, 583–601 (1986)
39. Chow, S.-N., Mallet-Paret, J., Shen, W.: Traveling waves in lattice dynamical systems. J. Differ. Equ. **149**(2), 248–291 (1998)
40. Chow, S.-N., Mallet-Paret, J., Van Vleck, E.S.: Pattern formation and spatial chaos in spatially discrete evolution equations. Random Comput. Dyn. **4**(2–3), 109–178 (1996)
41. Chua, L.O., Yang, L.: Cellular neural networks: applications. IEEE Trans. Circuits Syst. **35**, 1273–1290 (1988)
42. Chua, L.O., Yang, L.: Cellular neural networks: theory. IEEE Trans. Circuits Syst. **35**, 1257–1272 (1988)
43. Cook, H., de Fontaine, D., Hilliard, J.: A model for diffusion on cubic lattices and its application to the early stages of ordering. Acta Metall. **17**, 765–773 (1969)
44. Cornwell, P.: Opening the Maslov Box for Traveling Waves in Skew-Gradient Systems (2017). arXiv:1709.01908

45. Cornwell, P., Jones, C.K.R.T.: On the existence and stability of fast traveling waves in a doubly-diffusive FitzHugh-Nagumo system. SIAM J. Appl. Dyn. Syst. **17**(1), 754–787 (2018)
46. Cuevas, J., English, L.Q., Kevrekidis, P., Anderson, M.: Discrete breathers in a forced-damped array of coupled pendula: modeling, computation, and experiment. Phys. Rev. Lett. **102**(22), 224101 (2009)
47. Dauxois, T., Peyrard, M., Bishop, A.R.: Dynamics and thermodynamics of a nonlinear model for DNA denaturation. Phys. Rev. E **47**, 684–695 (1993)
48. de Camino-Beck, T., Lewis, M.: Invasion with stage-structured coupled map lattices: application to the spread of scentless chamomile. Ecol. Modell. **220**(23), 3394–3403 (2009)
49. D'Este, E., Kamin, D., Göttfert, F., El-Hady, A., Hell, S.E.: STED nanoscopy reveals the ubiquity of subcortical cytoskeleton periodicity in living neurons. Cell Rep. **10**(8), 1246–1251 (2015)
50. D'Este, E., Kamin, D., Velte, C., Göttfert, F., Simons, M., Hell, S.E.: Subcortical cytoskeleton periodicity throughout the nervous system. Sci. Rep. **6**(6), 22741 (2016)
51. Diekmann, O., van Gils, S.A., Verduyn-Lunel, S.M., Walther, H.O.: Delay Equations. Springer, New York (1995)
52. Dmitriev, S.V., Abe, K., Shigenari, T.: Domain wall solutions for EHM model of crystal: structures with period multiple of four. Phys. D: Nonlinear Phenom. **147**(1–2), 122–134 (2000)
53. Dmitriev, S.V., Kevrekidis, P.G., Yoshikawa, N.: Discrete Klein-Gordon models with static kinks free of the Peierls-Nabarro potential. J. Phys. A. **38**, 7617–7627 (2005)
54. Elmer, C.E.: Finding stationary fronts for a discrete Nagumo and wave equation; construction. Phys. D **218**, 11–23 (2006)
55. Elmer, C.E., Van Vleck, E.S.: Computation of traveling waves for spatially discrete bistable reaction-diffusion equations. Appl. Numer. Math. **20**, 157–169 (1996)
56. Elmer, C.E., Van Vleck, E.S.: Analysis and computation of traveling wave solutions of bistable differential-difference equations. Nonlinearity **12**, 771–798 (1999)
57. Elmer, C.E., Van Vleck, E.S.: Traveling wave solutions for bistable differential difference equations with periodic diffusion. SIAM J. Appl. Math. **61**, 1648–1679 (2001)
58. Elmer, C.E., Van Vleck, E.S.: A variant of Newton's method for the computation of traveling waves of bistable differential-difference equations. J. Dyn. Differ. Equ. **14**, 493–517 (2002)
59. Elmer, C.E., Van Vleck, E.S.: Anisotropy, propagation failure, and wave speedup in traveling waves of discretizations of a Nagumo PDE. J. Comput. Phys. **185**(2), 562–582 (2003)
60. Elmer, C.E., Van Vleck, E.S.: Existence of monotone traveling fronts for BDF discretizations of bistable reaction-diffusion equations. Dyn. Contin. Discret. Impuls. Syst. Ser. A Math. Anal. **10**(1–3), 389–402 (2003). Second International Conference on Dynamics of Continuous, Discrete and Impulsive Systems (London, ON, 2001)
61. Elmer, C.E., Van Vleck, E.S.: Dynamics of monotone travelling fronts for discretizations of Nagumo PDEs. Nonlinearity **18**, 1605–1628 (2005)
62. Elmer, C.E., Van Vleck, E.S.: Spatially Discrete FitzHugh-Nagumo Equations. SIAM J. Appl. Math. **65**, 1153–1174 (2005)
63. English, L.Q., Thakur, R.B., Stearrett, R.: Patterns of traveling intrinsic localized modes in a driven electrical lattice. Phys. Rev. E **77**, 066601 (2008)
64. Ermentrout, B.: Neural networks as spatio-temporal pattern-forming systems. Rep. Prog. Phys. **61**(4), 353 (1998)
65. Erneux, T., Nicolis, G.: Propagating waves in discrete bistable reaction-diffusion systems. Phys. D **67**, 237–244 (1993)
66. Evans, L.C., Soner, H.M., Souganidis, P.E.: Phase transitions and generalized motion by mean curvature. Commun. Pure Appl. Math. **45**(9), 1097–1123 (1992)
67. Fath, G.: Propagation failure of traveling waves in a discrete bistable medium. Phys. D **116**, 176–190 (1998)
68. Faye, G., Scheel, A.: Fredholm properties of nonlocal differential operators via spectral flow. Indiana Univ. Math. J. **63**, 1311–1348 (2014)

69. Faye, G., Scheel, A.: Existence of pulses in excitable media with nonlocal coupling. Adv. Math. **270**, 400–456 (2015)
70. Faye, G., Scheel, A.: Center manifolds without a phase space. Trans. Am. Math. Soc. **370**(8), 5843–5885 (2018)
71. Fiedler, B., Scheurle, J.: Discretization of homoclinic orbits, rapid forcing and "invisible" chaos. Mem. Am. Math. Soc. **119**(570) (1996)
72. Fife, P.C., McLeod, J.B.: The approach of solutions of nonlinear diffusion equations to travelling front solutions. Arch. Ration. Mech. Anal. **65**(4), 335–361 (1977)
73. FitzHugh, R.: Impulses and physiological states in theoretical models of nerve membrane. Biophys. J. **1**, 445–466 (1966)
74. FitzHugh, R.: Mathematical Models of Excitation and Propagation in Nerve. Publisher Unknown (1966)
75. Fitzhugh, R.: Motion picture of nerve impulse propagation using computer animation. J. Appl. Physiol. **25**(5), 628–630 (1968)
76. Flach, S., Zolotaryuk, Y., Kladko, K.: Moving lattice kinks and pulses: an inverse method. Phys. Rev. E **59**, 6105–6115 (1999)
77. Grüne, L.: Attraction rates, robustness, and discretization of attractors. SIAM J. Numer. Anal. **41**(6), 2096–2113 (2003)
78. Hale, J.K., Verduyn-Lunel, S.M.: Introduction to Functional Differential Equations. Springer, New York (1993)
79. Haragus, M., Scheel, A.: Almost planar waves in anisotropic media. Commun. Partial Differ. Equ. **31**(5), 791–815 (2006)
80. Haragus, M., Scheel, A.: Corner defects in almost planar interface propagation. Annales de l'Institut Henri Poincare (C) Non Linear Analysis, vol. 23, pp. 283–329. Elsevier, New York (2006)
81. Haragus, M., Scheel, A.: A bifurcation approach to non-planar traveling waves in reaction-diffusion systems. GAMM-Mitteilungen **30**(1), 75–95 (2007)
82. Härterich, J., Sandstede, B., Scheel, A.: Exponential dichotomies for linear non-autonomous functional differential equations of mixed type. Indiana Univ. Math. J. **51**(5), 1081–1109 (2002)
83. Hastings, S.: On travelling wave solutions of the Hodgkin-Huxley equations. Arch. Ration. Mech. Anal. **60**, 229–257 (1976)
84. Hillert, M.: A solid-solution model for inhomogeneous systems. Acta Metall. **9**, 525–535 (1961)
85. Hodgkin, A.L., Huxley, A.F.: A quantitative description of membrane current and its application to conduction and excitation in nerve. J. Physiol. **117** (1952)
86. Hoffman, A., Hupkes, H.J., Van Vleck, E.S.: Multi-dimensional stability of waves travelling through rectangular lattices in rational directions. Trans. Am. Math. Soc. **367**(12), 8757–8808 (2015)
87. Hoffman, A., Hupkes, H.J., Van Vleck, E.S.: Entire solutions for bistable lattice differential equations with obstacles. Mem. Am. Math. Soc. **250**(1188), 1–119 (2017)
88. Hoffman, A., Mallet-Paret, J.: Universality of crystallographic pinning. J. Dyn. Differ. Equ. **22**, 79–119 (2010)
89. Huang, W., Hupkes, H.J., Lozada-Cruz, G., Van Vleck, E.S.: Propagation failure for traveling waves of reaction-diffusion equations under moving mesh discretization. In preparation
90. Huang, W., Russell, R.D.: Adaptive mesh movement in 1D. Adaptive Moving Mesh Methods, pp. 27–135. Springer, Berlin (2011)
91. Humphries, A.R., Moore, B.E., Van Vleck, E.S.: Front solutions for bistable differential-difference equations with inhomogeneous diffusion. SIAM J. Appl. Math. **71**(4), 1374–1400 (2011)
92. Hupkes, H.J.: Invariant Manifolds and Applications for Functional Differential Equations of Mixed Type. Ph.D. Thesis(2008)
93. Hupkes, H.J., Augeraud-Véron, E.: Well-Posedness of Initial Value Problems for Vector-Valued Functional Differential Equations of Mixed Type. Preprint

94. Hupkes, H.J., Morelli, L.: Travelling Corners for Spatially Discrete Reaction-Diffusion System. Pure Appl. Anal. (2019). arXiv:1901.02319
95. Hupkes, H.J., Morelli, L., Stehlík, P.: Bichromatic travelling waves for lattice Nagumo equations. SIAM J. Appl. Dyn. Syst. **18.2**(2019), 973–1014 (2018). arXiv:1805.10977
96. Hupkes, H.J., Morelli, L., Stehlík, P., Švígler, V.: Counting and ordering periodic stationary solutions of lattice Nagumo equations. Appl. Math. Lett. (2019). arXiv:1905.06107v1
97. Hupkes, H.J., Morelli, L., Stehlík, P., Švígler, V.: Multichromatic travelling waves for lattice Nagumo equations. Appl. Math. Comput. **361**(2019), 430–452 (2019). arXiv:1901.07227
98. Hupkes, H.J., Pelinovsky, D., Sandstede, B.: Propagation failure in the discrete Nagumo equation. Proc. Am. Math. Soc. **139**(10), 3537–3551 (2011)
99. Hupkes, H.J., Sandstede, B.: Modulated Wave Trains for Lattice Differential Systems. J. Dyn. Differ. Equ. **21**, 417–485 (2009)
100. Hupkes, H.J., Sandstede, B.: Travelling pulse solutions for the discrete FitzHugh-Nagumo system. SIAM J. Appl. Dyn. Syst. **9**, 827–882 (2010)
101. Hupkes, H.J., Sandstede, B.: Stability of Pulse Solutions for the Discrete FitzHugh-Nagumo System. Trans. AMS **365**, 251–301 (2013)
102. Hupkes, H.J., Van Vleck, E.S.: Travelling Waves for Adaptive Grid Discretizations of Reaction-Diffusion Systems. Preprint
103. Hupkes, H.J., Van Vleck, E.S.: Negative diffusion and traveling waves in high dimensional lattice systems. SIAM J. Math. Anal. **45**(3), 1068–1135 (2013)
104. Hupkes, H.J., Van Vleck, E.S.: Travelling waves for complete discretizations of reaction diffusion systems. J. Dyn. Differ. Equ. **28**(3–4), 955–1006 (2016)
105. Hupkes, H.J., Verduyn-Lunel, S.M.: Analysis of Newton's method to compute travelling waves in discrete media. J. Dyn. Differ. Equ. **17**, 523–572 (2005)
106. Hupkes, H.J., Verduyn-Lunel, S.M.: Center manifold theory for functional differential equations of mixed type. J. Dyn. Differ. Equ. **19**, 497–560 (2007)
107. Hupkes, H.J., Verduyn-Lunel, S.M.: Center manifolds for periodic functional differential equations of mixed type. J. Differ. Equ. **245**, 1526–1565 (2008)
108. Hupkes, H.J., Verduyn-Lunel, S.M.: Lin's method and homoclinic bifurcations for functional differential equations of mixed type. Indiana Univ. Math. J. **58**, 2433–2487 (2009)
109. Jones, C.K.R.T.: Stability of the travelling wave solutions of the FitzHugh-Nagumo system. Trans. AMS **286**, 431–469 (1984)
110. Jones, C.K.R.T., Kopell, N., Langer, R.: Construction of the FitzHugh-Nagumo pulse using differential forms. In: Swinney, H., Aris, G., Aronson, D.G. (eds.) Patterns and Dynamics in Reactive Media. IMA Volumes in Mathematics and its Applications, vol. 37, pp. 101–116. Springer, New York (1991)
111. Kapitula, T.: Multidimensional stability of planar travelling waves. Trans. Am. Math. Soc. **349**(1), 257–269 (1997)
112. Kawasaki, K., Ohta, T.: Kinetic drumhead model of interface I. Prog. Theor. Phys. **67**(1), 147–163 (1982)
113. Keener, J., Sneed, J.: Mathematical Physiology. Springer, New York (1998)
114. Keener, J.P.: Propagation and its failure in coupled systems of discrete excitable cells. SIAM J. Appl. Math. **47**, 556–572 (1987)
115. Keener, J.P.: Propagation of waves in an excitable medium with discrete release sites. SIAM J. Appl. Math. **61**(1), 317–334 (2000)
116. Kevrekidis, P., Frantzeskakis, D., Theocharis, G., Kevrekidis, I.: Guidance of matter waves through Y-junctions. Phys. Lett. A **317**(5), 513–522 (2003)
117. Kleefeld, B., Khaliq, A., Wade, B.: An ETD Crank-Nicolson method for reaction-diffusion systems. Numer. Methods PDEs **28**(4), 1309–1335 (2012)
118. Krupa, M., Sandstede, B., Szmolyan, P.: Fast and slow waves in the FitzHugh-Nagumo equation. J. Differ. Equ. **133**, 49–97 (1997)
119. Lamb, C., Van Vleck, E.S.: Neutral mixed type functional differential equations. J. Dyn. Differ. Equ. **28**(3–4), 763–804 (2016)

120. Lax, P.D., Richtmyer, R.D.: Survey of the stability of linear finite difference equations. Commun. Pure Appl. Math. **9**(2), 267–293 (1956)
121. Lederer, F., Stegeman, G.I., Christodoulides, D.N., Assanto, G., Segev, M., Silberberg, Y.: Discrete solitons in optics. Phys. Rep. **463**(1–3), 1–126 (2008)
122. Lillie, R.S.: Factors affecting transmission and recovery in the passive iron nerve model. J. General Physiol. **7**, 473–507 (1925)
123. Lin, X.B.: Using Melnikov's method to Solve Shilnikov's problems. Proc. Roy. Soc. Edinb. **116**, 295–325 (1990)
124. Mallet-Paret, J.: Spatial patterns, spatial chaos and traveling waves in lattice differential equations. In: Stochastic and Spatial Structures of Dynamical Systems, Royal Netherlands Academy of Sciences. Proceedings, Physics Section. Series 1, Vol. 45. Amsterdam, pp. 105–129 (1996)
125. Mallet-Paret, J.: The Fredholm alternative for functional-differential equations of mixed type. J. Dyn. Differ. Equ. **11**(1), 1–47 (1999)
126. Mallet-Paret, J.: The global structure of traveling waves in spatially discrete dynamical systems. J. Dynam. Differ. Equ. **11**(1), 49–127 (1999)
127. Mallet-Paret, J.: Crystallographic Pinning: Direction Dependent Pinning in Lattice Differential Equations. Citeseer (2001)
128. Mallet-Paret, J., Verduyn-Lunel, S.M.: Exponential dichotomies and Wiener-Hopf factorizations for mixed-type functional differential equations. J. Differ. Equ. (to appear)
129. Merks, R.M., Van de Peer, Y., Inzé, D., Beemster, G.T.: Canalization without flux sensors: a traveling-wave hypothesis. Trends Plant Sci. **12**(9), 384–390 (2007)
130. Morelli, L.: Travelling Patterns on Discrete Media. Ph.D. Thesis (2019) http://pub.math.leidenuniv.nl/~morellil/Thesis.pdf
131. Mukherjee, S., Spracklen, A., Choudhury, D., Goldman, N., Öhberg, P., Andersson, E., Thomson, R.R.: Observation of a localized flat-band state in a photonic lieb lattice. Phys. Rev. Lett. **114**, 245504 (2015)
132. Nolen, J., Roquejoffre, J.-M., Ryzhik, L., Zlatoš, A.: Existence and non-existence of fisher-KPP transition fronts. Arch. Ration. Mech. Anal. **203**(1), 217–246 (2012)
133. Pinto, D.J., Ermentrout, G.B.: Spatially structured activity in synaptically coupled neuronal networks: 1. Traveling fronts and pulses. SIAM J. Appl. Math. **62**, 206–225 (2001)
134. Qin, W.-X., Xiao, X.: Homoclinic orbits and localized solutions in nonlinear Schrödinger lattices. Nonlinearity **20**, 2305–2317 (2007)
135. Ranvier, L.A.: *Lećons sur l'Histologie du Système Nerveux, par M. L. Ranvier, recueillies par M. Ed. Weber*. F. Savy, Paris (1878)
136. Roosen, A.R., McCormack, R.P., Carter, W.C.: Wulffman: a tool for the calculation and display of crystal shapes. Comput. Mater. Sci. **11**(1), 16–26 (1998)
137. Rustichini, A.: Functional-differential equations of mixed type: the linear autonomous case. J. Dyn. Differ. Equ. **1**(2), 121–143 (1989)
138. Rustichini, A.: Hopf bifurcation for functional-differential equations of mixed type. J. Dyn. Differ. Equ. **1**(2), 145–177 (1989)
139. Sakamoto, K.: Invariant manifolds in singular perturbation Problems. Proc. Roy. Soc. Edinb. A **116**, 45–78 (1990)
140. Scheel, A., Tikhomirov, S.: Depinning asymptotics in ergodic media. In: International Conference on Patterns of Dynamics, pp. 88–108. Springer (2016)
141. Scheel, A., Van Vleck, E.S.: Lattice differential equations embedded into reaction-diffusion systems. Proc. Roy. Soc. Edinb. Sect. A **139**(1), 193–207 (2009)
142. Schouten-Straatman, W.M., Hupkes, H.J.: Travelling waves for spatially discrete systems of FitzHugh-Nagumo type with periodic coefficients. SIAM J. Math. Anal. **51**(4), 3492–3532 (2018). arXiv:1808.00761
143. Schouten-Straatman, W.M., Hupkes, H.J.: Nonlinear stability of pulse solutions for the discrete FitzHugh-Nagumo equation with infinite-range interactions. Discret. Contin. Dyn. Syst. A **39**(9) (2019)

144. Sen, S., Hong, J., Bang, J., Avalos, E., Doney, R.: Solitary waves in the granular chain. Phys. Rep. **462**(2), 21–66 (2008)
145. Slavík, A.: Invariant regions for systems of lattice reaction-diffusion equations. J. Differ. Equ. **263**(11), 7601–7626 (2017)
146. Slepyan, L.I.: *Models and Phenomena in Fracture Mechanics* (2012). Springer Science & Business Media
147. Sneyd, J.: Mathematical modeling of calcium dynamics and signal transduction. Tutorials in Mathematical Biosciences II. Lecture Notes in Mathematics, vol. 187. Springer, New York (2005)
148. Speight, J.M.: Topological discrete kinks. Nonlinearity **12**, 1373–1387 (1999)
149. Stehlík, P.: Exponential number of stationary solutions for Nagumo equations on graphs. J. Math. Anal. Appl. **455**(2), 1749–1764 (2017)
150. Tagantsev, A.K., Cross, L.E., Fousek, J.: Domains in Ferroic Crystals and Thin Films. Springer, Berlin (2010)
151. Tonnelier, A.: McKean Caricature of the FitzHugh-Nagumo model: traveling pulses in a discrete diffusive medium. Phys. Rev. E **67**, 036105 (2003)
152. Vainchtein, A., Van Vleck, E.S.: Nucleation and propagation of phase mixtures in a bistable chain. Phys. Rev. B **79**, 144123 (2009)
153. Vainchtein, A., Van Vleck, E.S., Zhang, A.: Propagation of periodic patterns in a discrete system with competing interactions. SIAM J. Appl. Dyn. Syst. **14**(2), 523–555 (2015)
154. van Hal, B.: Travelling Waves in Discrete Spatial Domains. Bachelor Thesis (2017)
155. Van Vleck, E.S., Zhang, A.: Competing interactions and traveling wave solutions in lattice differential equations. Commun. Pure Appl. Anal. **15**(2), 457–475 (2016)
156. Vicencio, R.A., Cantillano, C., Morales-Inostroza, L., Real, B., Mejía-Cortés, C., Weimann, S., Szameit, A., Molina, M.I.: Observation of localized states in Lieb photonic lattices. Phys. Rev. Lett. **114**, 245503 (2015)
157. Wang, Y., Liao, X., Xiao, D., Wong, K.-W.: One-way hash function construction based on 2D coupled map lattices. Inf. Sci. **178**(5), 1391–1406 (2008)
158. Wolf, P.E., Balibar, S., Gallet, F.: Experimental observation of a third roughening transition on hcp He 4 crystals. Phys. Rev. Lett. **51**(15), 1366 (1983)
159. Xu, K., Zhong, G., Zhuang, X.: Actin, spectrin, and associated proteins form a periodic cytoskeletal structure in axons. Science **339**(6118), 452–456 (2013)
160. Zumbrun, K.: Instantaneous shock location and one-dimensional nonlinear stability of viscous shock waves. Q. Appl. Math. **69**(1), 177–202 (2011)
161. Zumbrun, K., Howard, P.: Pointwise semigroup methods and stability of viscous shock waves. Indiana Univ. Math. J. **47**(3), 741–871 (1998)

Contributed Papers

A Hilbert Space Approach to Fractional Difference Equations

Pham The Anh, Artur Babiarz, Adam Czornik, Konrad Kitzing, Michał Niezabitowski, Stefan Siegmund, Sascha Trostorff and Hoang The Tuan

Abstract We formulate fractional difference equations of Riemann–Liouville and Caputo type in a functional analytical framework. Main results are existence of solutions on Hilbert space-valued weighted sequence spaces and a condition for stability of linear fractional difference equations. Using a functional calculus, we relate the fractional sum to fractional powers of the operator $1-\tau^{-1}$ with the right shift τ^{-1} on weighted sequence spaces. Causality of the solution operator plays a crucial role for the description of initial value problems.

Keywords Computational geometry · Graph theory · Hamilton cycles

1 Introduction

1.1 Notation

We write $\mathbb{R}_{>0} := \{x \in \mathbb{R};\, x > 0\}$ and for $\mu, \varrho \in \mathbb{R}$ we define for the comprehension $\mathbb{C}_{|\cdot|<\mu} := \{z \in \mathbb{C};\, |z| < \mu\}$ and $\mathbb{C}_{|\cdot|>\mu}$, $\mathbb{C}_{|\cdot|\leq\mu}$, $\mathbb{C}_{|\cdot|\geq\mu}$ and $\mathbb{C}_{\mu\geq|\cdot|\geq\varrho}$ are defined similarly. For $\varrho > 0$ we denote the complex ball with radius ϱ centered at 0 by

P. T. Anh
Department of Mathematics, Le Quy Don Technical University, 236 Hoang Quoc Viet, Hanoi, Vietnam

A. Babiarz · A. Czornik · M. Niezabitowski
Faculty of Automatic Control, Electronics and Computer Science, Silesian University of Technology, Akademicka 16, 44-100 Gliwice, Poland

K. Kitzing · S. Siegmund (✉) · S. Trostorff
Christian-Albrechts-Universität zu Kiel, Mathematisches Seminar, Ludewig-Meyn-Straße 4, 24118 Kiel, Germany
e-mail: stefan.siegmund@tu-dresden.de

H. T. Tuan
Institute of Mathematics, Vietnam Academy of Science and Technology, 18 Hoang Quoc Viet Road, Cau Giay, Hanoi, Vietnam

M. Bohner et al. (eds.), *Difference Equations and Discrete Dynamical Systems with Applications*, Springer Proceedings in Mathematics & Statistics 312,
https://doi.org/10.1007/978-3-030-35502-9_4

$B(0, \varrho) := \{z \in \mathbb{C}; |z| < \varrho\}$ and the circle with radius ϱ centered at 0 by $S_\varrho := \partial B(0, \varrho)$. We set $\mathbb{N} := \mathbb{Z}_{\geq 0} := \{0, 1, 2, \dots\}$. For sets X, Y we denote the set of functions from Y to X by $X^Y := \{f : Y \to X\}$ and for $f \in X^Y$ we write $\operatorname{ran} f := \{f(y) \in X; y \in Y\}$ for the range of f. In particular, for any $M \subseteq \mathbb{Z}$, X^M is the space of sequences in X on M and for $u \in X^M$, $n \in M$ we write $u_n := u(n)$. The identity mapping on a vector space V is denoted by 1. For a sequence $u \in V^{\mathbb{Z}}$ we denote $\operatorname{spt} u := \{n \in \mathbb{Z}; u_n \neq 0\}$. If V is a normed vector space we denote with $\|\cdot\|_V$ the norm on V.

We recall the binomial coefficient and the binomial series including some of their properties. Proofs of the following propositions can be found in [11, 14].

Proposition 1 (Binomial coefficient [11, pp. 164–165], [14, p. 34]) *For $\alpha \in \mathbb{C}$ and $n \in \mathbb{Z}_{\geq 1}$ the binomial coefficient is defined by*

$$\binom{\alpha}{0} := 1, \quad \binom{\alpha}{n} := \frac{\alpha(\alpha-1)\cdots(\alpha-n+1)}{n!}.$$

For $\alpha \in \mathbb{C}$ and $n \in \mathbb{N}$ we have

$$(-1)^n \binom{\alpha}{n} = \binom{-\alpha+n-1}{n} \quad \text{and} \quad \sum_{k=0}^{n} (-1)^k \binom{\alpha}{k} = (-1)^n \binom{\alpha-1}{n}.$$

Proposition 2 (Binomial series [14, pp. 65, 73]) *Let $\alpha \in \mathbb{C}$. The binomial power series is defined by*

$$(1+z)^\alpha := \sum_{k=0}^{\infty} \binom{\alpha}{k} z^k.$$

The series converges absolutely in $B(0, 1)$. In particular, the mapping $\mathbb{C}_{|\cdot|>1} \to \mathbb{C}$, $z \mapsto (1 - z^{-1})^\alpha$ is holomorphic. For each $\alpha, \beta \in \mathbb{C}$ we have $(1+z)^\alpha (1+z)^\beta = (1+z)^{\alpha+\beta}$.

Binomial coefficients can be expressed with the gamma function.

Lemma 3 (Falling factorial [11, p. 164]) *With the falling factorial*

$$(x)^{(n)} := \frac{\Gamma(x+1)}{\Gamma(x-n+1)}, \qquad x \in \mathbb{C} \setminus \mathbb{Z}, \qquad n \in \mathbb{N},$$

we have for each $\alpha \in \mathbb{C} \setminus \mathbb{Z}$ and $n \in \mathbb{N}$

$$(-1)^n \binom{\alpha}{n} = \binom{-\alpha+n-1}{n} = \frac{1}{\Gamma(-\alpha)} (n - (1+\alpha))^{(-(1+\alpha))}. \tag{1}$$

Lemma 4 *Let $\alpha \in (0, 1)$ and $\varrho > 1$. Then we have for each $z \in S_\varrho$*

$$(1 - \varrho^{-1})^\alpha \leq \left|(1 - z^{-1})^\alpha\right|.$$

Proof Let $z \in S_\varrho$. For every $n \in \mathbb{Z}_{\geq 1}$ we observe that $(-1)^n\binom{\alpha}{n} < 0$ and therefore $(-1)^n\binom{\alpha}{n}z^{-n} = -\left|\binom{\alpha}{n}\right| z^{-n}$. We show by induction that for every $n \in \mathbb{N}$

$$\left|\sum_{k=0}^{n}(-1)^k\binom{\alpha}{k}z^{-k}\right| \geq \left|\sum_{k=0}^{n}(-1)^k\binom{\alpha}{k}\varrho^{-k}\right|$$

and when letting n tend to infinity the inequality follows. The induction basis is trivial. For the induction step for $n \in \mathbb{N}$ we use the lower triangle inequality to obtain

$$\begin{aligned}\left|\sum_{k=0}^{n+1}(-1)^k\binom{\alpha}{k}z^{-k}\right| &= \left|\sum_{k=0}^{n}(-1)^k\binom{\alpha}{k}z^{-k} + (-1)^{n+1}\binom{\alpha}{n+1}z^{-(n+1)}\right| \\ &\geq \left|\left|\sum_{k=0}^{n}(-1)^k\binom{\alpha}{k}z^{-k}\right| - \left|(-1)^{n+1}\binom{\alpha}{n+1}z^{-(n+1)}\right|\right| \\ &= \left|\left|\sum_{k=0}^{n}(-1)^k\binom{\alpha}{k}z^{-k}\right| + (-1)^{n+1}\binom{\alpha}{n+1}\varrho^{-(n+1)}\right| \\ &\geq \left|\sum_{k=0}^{n+1}(-1)^k\binom{\alpha}{k}\varrho^{-k}\right|.\end{aligned}$$

1.2 Fractional Difference Operators

Let V be a real or complex vector space.

The fractional sum can be motivated by the iterated sum formula and is also related to iterating the backward difference operator (see e.g. [15]). For $\alpha \in \mathbb{R}_{>0}$ the fractional sum $\nabla^{-\alpha}\colon V^{\mathbb{N}} \to V^{\mathbb{N}}$ is defined by (cf. [3, p. 3])

$$(\nabla^{-\alpha}u)_n = \sum_{k=0}^{n}\binom{n-k+\alpha-1}{n-k}u_k = \sum_{k=0}^{n}(-1)^k\binom{-\alpha}{k}u_{n-k}. \tag{2}$$

There is also a definition motivated by iterating the forward difference operator which is studied at least since [15] and can be found in [3, p. 3] as well. Note that $(\nabla^{-\alpha}u)_n$ in general depends on $u_0, \ldots, u_n$.

The approach to defining the fractional differential operators in the Riemann–Liouville and Caputo sense (cf. [8]) was applied mutatis mutandis to difference operators (see e.g. [1] and the references therein). Recall that for $\Delta : V^{\mathbb{N}} \to V^{\mathbb{N}}$, $u \mapsto (u_{n+1} - u_n)_{\mathbb{N}}$ we have $(\Delta u)_n = (\nabla u)_{n+1}$ for $n \in \mathbb{N}$. For $\alpha \in (0, 1)$ the Riemann–Liouville forward fractional difference operator is defined by (cf. [16, p. 3813])

$$\Delta^\alpha : V^{\mathbb{N}} \to V^{\mathbb{N}}, \qquad u \mapsto \Delta\nabla^{-(1-\alpha)}u. \tag{3}$$

The Caputo forward fractional difference operator is defined by (cf. [16, p. 3813])

$$\Delta_C^\alpha : V^{\mathbb{N}} \to V^{\mathbb{N}}, \qquad u \mapsto \nabla^{-(1-\alpha)} \Delta u. \tag{4}$$

In this paper we study sequences in a Hilbert space $V = H$ on $\mathbb{Z}$ and define a fractional difference sum operator using the binomial series and a functional calculus which is not purely algebraic as in the case of $\nabla^{-\alpha}$. The connection between operators defined on $H^{\mathbb{Z}}$ with those defined on $H^{\mathbb{N}}$ will be causality and we analyze how the Riemann–Liouville and the Caputo operator fit into the calculus developed for sequences in $H^{\mathbb{Z}}$. An important step for the development of the discrete, functional analytic framework which is introduced in this paper has been done in the continuous case for fractional derivatives in [19]. Lastly we study the asymptotic stability of the zero solution of a linear fractional difference equation with the Riemann–Liouville and the Caputo forward difference operator. The interest in the study of linear problems in the context of stability analysis stems from Lyapunov's first method, which has been analyzed in [6] for fractional differential equations. The results regarding asymptotic stability will be in terms of the Matignon criterion (cf. [18]), however, for bounded operators on a Hilbert space H and will be compared to those in [1, 5]. A useful tool when analyzing the asymptotic stability of linear problems is the $\mathscr{Z}$ transform which is also used in [1, 5] but which is studied here for sequences in $H^{\mathbb{Z}}$. Asymptotic stability has also been studied using the Riemann–Liouville and the Caputo backward difference operators in [4, 16].

2 Exponentially Weighted ℓ_p Spaces

We denote by $(H, \|\cdot\|_H)$ a complex and separable Hilbert space. The scalar product $\langle \cdot, \cdot \rangle_H$ on H shall be conjugate linear in the first argument and linear in the second argument. We recall several of the concepts of weighted $\ell_{p,\varrho}(\mathbb{Z}; H)$ spaces and the $\mathscr{Z}$ transform (see also [13]).

Lemma 5 (Exponentially weighted ℓ_p spaces [13]) *Let $1 \leq p < \infty$, $\varrho > 0$. Define*

$$\ell_{p,\varrho}(\mathbb{Z}; H) := \left\{ x \in H^{\mathbb{Z}};\ \sum_{k\in\mathbb{Z}} \|x_k\|_H^p\, \varrho^{-pk} < \infty \right\},$$

$$\ell_{\infty,\varrho}(\mathbb{Z}; H) := \left\{ x \in H^{\mathbb{Z}};\ \sup_{k\in\mathbb{Z}} \|x_k\|_H\, \varrho^{-k} < \infty \right\}.$$

Then $\ell_{p,\varrho}(\mathbb{Z}; H)$ and $\ell_{\infty,\varrho}(\mathbb{Z}; H)$ are Banach spaces with norms

$$\|x\|_{\ell_{p,\varrho}(\mathbb{Z};H)} := \Big(\sum_{k\in\mathbb{Z}} \|x_k\|_H^p\, \varrho^{-pk} \Big)^{\frac{1}{p}} \qquad (x \in \ell_{p,\varrho}(\mathbb{Z}; H))$$

and

$$\|x\|_{\ell_{\infty,\varrho}(\mathbb{Z};H)} := \sup_{k\in\mathbb{Z}} \|x_k\|_H\, \varrho^{-k} \qquad (x \in \ell_{\infty,\varrho}(\mathbb{Z};H)),$$

respectively. Moreover, $\ell_{2,\varrho}(\mathbb{Z};H)$ is a Hilbert space with the inner product

$$\langle x, y\rangle_{\ell_{2,\varrho}(\mathbb{Z};H)} := \sum_{k\in\mathbb{Z}} \langle x_k, y_k\rangle_H\, \varrho^{-2k} \qquad (x, y \in \ell_{2,\varrho}(\mathbb{Z};H)).$$

We write $\ell_p(\mathbb{Z};H) := \ell_{p,1}(\mathbb{Z};H)$ for $1 \le p \le \infty$.

Proposition 6 (One sided weighted sequence spaces [13]) *For $1 \le p \le \infty$, $a \in \mathbb{Z}$ and $\varrho > 0$ we define*

$$\ell_{p,\varrho}(\mathbb{Z}_{\ge a};H) := \left\{x|_{\mathbb{Z}_{\ge a}};\, x \in \ell_{p,\varrho}(\mathbb{Z};H)\right\}.$$

And for $1 \le p \le \infty$, $\varrho > 0$, $a \in \mathbb{Z}$ and for $x \in H^{\mathbb{Z}_{\ge a}}$, we define $\iota x \in H^{\mathbb{Z}}$ by

$$(\iota x)_k := \begin{cases} 0 & \text{if } k < a, \\ x_k & \text{if } k \ge a. \end{cases}$$

Then $\ell_{p,\varrho}(\mathbb{Z}_{\ge a};H)$ is a Banach space with norm $\|\cdot\|_{\ell_{p,\varrho}(\mathbb{Z}_{\ge a};H)} := \|\iota\cdot\|_{\ell_{p,\varrho}(\mathbb{Z};H)}$, and

$$\iota\colon \ell_{p,\varrho}(\mathbb{Z}_{\ge a};H) \hookrightarrow \ell_{p,\varrho}(\mathbb{Z};H)$$

is an isometric embedding. Write $\ell_{p,\varrho}(\mathbb{Z}_{\ge a};H) \subseteq \ell_{p,\varrho}(\mathbb{Z};H)$.
For $1 \le p < q \le \infty$, $\varrho, \varepsilon > 0$, $a \in \mathbb{Z}$ we have

$$\begin{aligned} &(a) \quad \ell_{p,\varrho}(\mathbb{Z}_{\ge a};H) \subsetneq \ell_{q,\varrho}(\mathbb{Z}_{\ge a};H), \\ &(b) \quad \ell_{q,\varrho}(\mathbb{Z}_{\ge a};H) \subsetneq \ell_{p,\varrho+\varepsilon}(\mathbb{Z}_{\ge a};H). \end{aligned}$$

Definition 7 For $x \in H$ and $n \in \mathbb{Z}$ we define $\delta_n x \in H^{\mathbb{Z}}$ by

$$(\delta_n x)_m := \begin{cases} x, & \text{if } m = n, \\ 0, & \text{if } m \ne n, \end{cases}$$

and $\chi_{\mathbb{Z}_{\ge n}} x \in H^{\mathbb{Z}}$ by

$$(\chi_{\mathbb{Z}_{\ge n}} x)_m := \begin{cases} x, & \text{if } m \ge n, \\ 0, & \text{if } m < n. \end{cases}$$

Note that for $\varrho > 0$, $\delta_n x \in \ell_{p,\varrho}(\mathbb{Z};H)$ and for $\varrho > 1$, $\chi_{\mathbb{Z}_{\ge n}} x \in \ell_{p,\varrho}(\mathbb{Z};H)$.

Lemma 8 (Shift operator [13]) *Let $1 \le p \le \infty$, $\varrho > 0$. Then*

$$\tau\colon \ell_{p,\varrho}(\mathbb{Z}; H) \to \ell_{p,\varrho}(\mathbb{Z}; H),$$
$$(x_k)_{k\in\mathbb{Z}} \mapsto (x_{k+1})_{k\in\mathbb{Z}},$$

is linear, bounded, invertible, and

$$\left\|\tau^n\right\|_{L(\ell_{p,\varrho}(\mathbb{Z}; H))} = \varrho^n \qquad (n \in \mathbb{Z}).$$

3 $\mathscr{Z}$ Transform

Lemma 9 (L_2 space on circle and orthonormal basis [13]) *Let* $\varrho > 0$. *Define*

$$L_2(S_\varrho; H) := \left\{ f\colon S_\varrho \to H;\ \int_{S_\varrho} \|f(z)\|_H^2 \frac{\mathrm{d}z}{|z|} < \infty \right\}.$$

Then $L_2(S_\varrho; H)$ *is a Hilbert space with the inner product*

$$\langle f, g\rangle_{L_2(S_\varrho; H)} := \frac{1}{2\pi} \int_{S_\varrho} \big\langle f(z), g(z)\big\rangle_H \frac{\mathrm{d}z}{|z|} \qquad (f, g \in L_2(S_\varrho; H)).$$

Moreover, let $(\psi_n)_{n\in\mathbb{Z}}$ *be an orthonormal basis in* H. *Then* $(p_{k,n})_{k,n\in\mathbb{Z}}$ *with*

$$p_{k,n}(z) := \varrho^k z^{-k} \psi_n \qquad (z \in S_\varrho)$$

is an orthonormal basis in $L_2(S_\varrho; H)$.

Theorem 10 ($\mathscr{Z}$ transform [13]) *Let* $\varrho > 0$. *The operator*

$$\mathscr{Z}_\varrho\colon \ell_{2,\varrho}(\mathbb{Z}; H) \to L_2(S_\varrho; H),$$
$$x \mapsto \Big(z \mapsto \sum_{k\in\mathbb{Z}} \langle \psi_n, \varrho^{-k} x_k\rangle_H\, p_{k,n}(z)\Big)$$

is well-defined and unitary. For $x \in \ell_{1,\varrho}(\mathbb{Z}; H) \subseteq \ell_{2,\varrho}(\mathbb{Z}; H)$ *we have*

$$\mathscr{Z}_\varrho(x) = \left(z \mapsto \sum_{k\in\mathbb{Z}} x_k z^{-k}\right).$$

Remark 11 ($\mathscr{Z}$ transform of $x\in\ell_{2,\varrho}(\mathbb{Z}; H) \setminus \ell_{1,\varrho}(\mathbb{Z}; H)$) Let $\varrho > 0$, $x\in\ell_{2,\varrho}(\mathbb{Z}; H) \setminus \ell_{1,\varrho}(\mathbb{Z}; H)$. Then

$$\sum_{k\in\mathbb{Z}} x_k z^{-k}$$

does not necessarily converge for all $z \in S_\varrho$. For example if $H = \mathbb{C}$, $x \in \ell_{2,\varrho}(\mathbb{Z}; H) \setminus \ell_{1,\varrho}(\mathbb{Z}; H)$ with $x_k := \frac{\varrho^k}{k}$ and $z = \varrho$.

Lemma 12 (Shift is unitarily equivalent to multiplication [13]) *Let $\varrho > 0$. Then*

$$\mathscr{Z}_\varrho \tau \mathscr{Z}_\varrho^* = \mathrm{m},$$

where m *is the multiplication-by-the-argument operator acting in $L_2(S_\varrho; H)$, i.e.,*

$$\begin{aligned} \mathrm{m}\colon L_2(S_\varrho; H) &\to L_2(S_\varrho; H), \\ f &\mapsto (z \mapsto zf(z)). \end{aligned}$$

Next, we present a Paley–Wiener type result for the $\mathscr{Z}$ transform.

Lemma 13 (Characterization of positive support [13]) *Let $\varrho > 0$, $x \in \ell_{2,\varrho}(\mathbb{Z}; H)$. Then the following statements are equivalent:*
(i) $\operatorname{spt} x \subseteq \mathbb{N}$,
(ii) $z \mapsto \sum_{k\in\mathbb{Z}} x_k z^{-k}$ *is analytic on* $\mathbb{C}_{|\cdot|>\varrho}$ *and*

$$\sup_{\mu>\varrho} \int_{S_\mu} \left\| \sum_{k\in\mathbb{Z}} x_k z^{-k} \right\|_H^2 \frac{\mathrm{d}z}{|z|} < \infty. \tag{5}$$

Definition 14 (*Causal linear operator*) We call a linear operator $B\colon \ell_{2,\varrho}(\mathbb{Z}; H) \to \ell_{2,\varrho}(\mathbb{Z}; H)$ *causal*, if for all $a \in \mathbb{Z}$, $f \in \ell_{2,\varrho}(\mathbb{Z}; H)$, we have

$$\operatorname{spt} f \subseteq \mathbb{Z}_{\geq a} \quad \Rightarrow \quad \operatorname{spt} Bf \subseteq \mathbb{Z}_{\geq a}.$$

Recall [12, VIII.3.6, p. 222] that for $A \in L(H)$ with spectrum $\sigma(A)$, the spectral radius

$$r(A) := \sup\{|z|;\, z \in \sigma(A)\}$$

of A satisfies

$$r(A) = \lim_{n\to\infty} \|A^n\|_{L(H)}^{1/n}.$$

Let $A \in L(H)$ and $\varrho > 0$. We denote the operators $\ell_{2,\varrho}(\mathbb{Z}, H) \to \ell_{2,\varrho}(\mathbb{Z}, H)$, $x \mapsto (Ax_k)$, and $L_2(S_\varrho, H) \to L_2(S_\varrho, H)$, $f \mapsto (z \mapsto Af(z))$, which have the same operator norm as A, again by A.

Proposition 15 (Convolution) *Let $c \in \ell_{1,\varrho}(\mathbb{Z}; \mathbb{C})$ and $u \in \ell_{2,\varrho}(\mathbb{Z}; H)$. Then*

$$c * u := \left(\sum_{k=-\infty}^{\infty} c_k u_{n-k} \right)_{n\in\mathbb{Z}} \in \ell_{2,\varrho}(\mathbb{Z}; H).$$

We have Young's inequality

$$\|c * u\|_{\ell_{2,\varrho}(\mathbb{Z};H)} \le \|c\|_{\ell_{1,\varrho}(\mathbb{Z};\mathbb{C})} \|u\|_{\ell_{2,\varrho}(\mathbb{Z};H)} .$$

Moreover,

$$\mathscr{Z}_\varrho(c * u) = \mathscr{Z}_\varrho c \mathscr{Z}_\varrho u.$$

Proof Let $n \in \mathbb{Z}$. With the Cauchy–Schwarz inequality we compute

$$\begin{aligned}\left(\sum_{k=-\infty}^{\infty} \|c_k u_{n-k}\|_H\right)^2 \varrho^{-2n} &= \left(\sum_{k=-\infty}^{\infty} |c_k|^{1/2}\varrho^{-k/2}|c_k|^{1/2}\varrho^{-k/2} \|u_{n-k}\|_H \varrho^{-(n-k)}\right)^2 \\ &\le \|c\|_{\ell_{1,\varrho}(\mathbb{Z};\mathbb{C})} \left(\sum_{k=-\infty}^{\infty} |c_k|\varrho^{-k} \|u_{n-k}\|_H^2 \varrho^{-2(n-k)}\right).\end{aligned}$$

Therefore using Fubini's theorem

$$\begin{aligned}\sum_{n=-\infty}^{\infty} \|(c*u)_n\|_H^2 \varrho^{-2n} &\le \sum_{n=-\infty}^{\infty} \left(\sum_{k=-\infty}^{\infty} \|c_k u_{n-k}\|_H\right)^2 \varrho^{-2n} \\ &\le \|c\|_{\ell_{1,\varrho}(\mathbb{Z};\mathbb{C})} \sum_{n=-\infty}^{\infty} \left(\sum_{k=-\infty}^{\infty} |c_k|\varrho^{-k} \|u_{n-k}\|_H^2 \varrho^{-2(n-k)}\right) \\ &= \|c\|_{\ell_{1,\varrho}(\mathbb{Z};\mathbb{C})}^2 \|u\|_{\ell_{2,\varrho}(\mathbb{Z};H)}^2 .\end{aligned}$$

This shows Young's inequality. If additionally $u \in \ell_{1,\varrho}(\mathbb{Z}; H)$ then

$$\begin{aligned}\sum_{n=-\infty}^{\infty} \|(c*u)_n\|_H \varrho^{-n} &\le \sum_{n=-\infty}^{\infty}\sum_{k=-\infty}^{\infty} \|c_k u_{n-k}\|_H \varrho^{-n} \\ &= \sum_{k=-\infty}^{\infty}\sum_{n=-\infty}^{\infty} |c_k|\varrho^{-k} \|u_{n-k}\|_H \varrho^{-(n-k)} \\ &= \|c\|_{\ell_{1,\varrho}(\mathbb{Z};\mathbb{C})} \|u\|_{\ell_{1,\varrho}(\mathbb{Z};H)} ,\end{aligned}$$

i.e., $c * u \in \ell_{1,\varrho}(\mathbb{Z}; H) \cap \ell_{2,\varrho}(\mathbb{Z}; H)$ which simplifies the $\mathscr{Z}$ transform of $c * u$. Using Fubini's theorem, we compute for $u \in \ell_{1,\varrho}(\mathbb{Z}; H) \cap \ell_{2,\varrho}(\mathbb{Z}; H)$ and $z \in S_\varrho$

$$\begin{aligned}\mathscr{Z}_\varrho(c*u)(z) &= \sum_{n=-\infty}^{\infty}\left(\sum_{k=-\infty}^{\infty} c_k u_{n-k}\right) z^{-n} = \sum_{n=-\infty}^{\infty}\left(\sum_{k=-\infty}^{\infty} c_k z^{-k} u_{n-k} z^{-(n-k)}\right) \\ &= \sum_{k=-\infty}^{\infty} c_k z^{-k}\left(\sum_{n=-\infty}^{\infty} u_{n-k} z^{-(n-k)}\right) = \mathscr{Z}_\varrho(c)\mathscr{Z}_\varrho(u).\end{aligned}$$

For $u \in \ell_{2,\varrho}(\mathbb{Z}; H)$ the formula follows by density of $\ell_{1,\varrho}(\mathbb{Z}; H) \cap \ell_{2,\varrho}(\mathbb{Z}; H) \subseteq \ell_{2,\varrho}(\mathbb{Z}; H)$.

Example 16 (The operator $(1-\tau^{-1})^\alpha$) Let $\varrho > 1$ and $\alpha \in \mathbb{C}$. For the operator $1 - \tau^{-1} : \ell_{2,\varrho}(\mathbb{Z}; H) \to \ell_{2,\varrho}(\mathbb{Z}; H)$, we compute

$$(1-\tau^{-1}) = \mathscr{Z}_\varrho^*(1-z^{-1})\mathscr{Z}_\varrho.$$

We have $|z^{-1}| < 1$ for all $z \in S_\varrho$ and therefore

$$(1-\tau^{-1})^\alpha := \mathscr{Z}_\varrho^*(1-z^{-1})^\alpha \mathscr{Z}_\varrho : \ell_{2,\varrho}(\mathbb{Z}; H) \to \ell_{2,\varrho}(\mathbb{Z}; H)$$

is well-defined. This is an application of the holomorphic functional calculus (cf. [10, pp. 13–18], [9, p. 601]).

We define $c \in \ell_{1,\varrho}(\mathbb{Z}; \mathbb{C})$ by

$$c_k := \begin{cases} (-1)^k \binom{\alpha}{k} & \text{if } k \geq 0, \\ 0 & \text{if } k < 0. \end{cases}$$

Then

$$\mathscr{Z}_\varrho c = \sum_{k=0}^{\infty} (-1)^k \binom{\alpha}{k} z^{-k} = (1-z^{-1})^\alpha.$$

Thus we compute for $u \in \ell_{2,\varrho}(\mathbb{Z}; H)$

$$\mathscr{Z}_\varrho(c * u) = \mathscr{Z}_\varrho c \mathscr{Z}_\varrho u = (1-z^{-1})^\alpha \mathscr{Z}_\varrho u.$$

Thus for $\alpha \in \mathbb{C}$ and $u \in \ell_{2,\varrho}(\mathbb{Z}; H)$ we obtain

$$(1-\tau^{-1})^\alpha u = c * u = \left(\sum_{k=0}^{\infty} (-1)^k \binom{\alpha}{k} u_{n-k}\right)_{n\in\mathbb{Z}} = \left(\sum_{k=-\infty}^{n} (-1)^{n-k} \binom{\alpha}{n-k} u_k\right)_{n\in\mathbb{Z}},$$

i.e., $(1-\tau^{-1})^\alpha$ is a convolution operator and by Young's Theorem $(1-\tau^{-1})^\alpha$ is bounded and $\left\|(1-\tau^{-1})^\alpha\right\|_{L(\ell_{2,\varrho}(\mathbb{Z};H))} = \|c\|_{\ell_{1,\varrho}(\mathbb{Z};H)}$.

If $u \in \ell_{2,\varrho}(\mathbb{Z}; H)$ with $\operatorname{spt} u \subseteq \mathbb{N}$, we have

$$(1-\tau^{-1})^\alpha u = \left(\sum_{k=0}^{n} (-1)^k \binom{\alpha}{k} u_{n-k}\right)_{n\in\mathbb{Z}}.$$

Since τ commutes with $(1-\tau^{-1})^\alpha$, we deduce that $(1-\tau^{-1})^\alpha$ is causal.

On $\ell_{2,\varrho}(\mathbb{Z}; H)$ we compute for $\alpha, \beta \in \mathbb{C}$

$$\begin{aligned}(1-\tau^{-1})^\alpha(1-\tau^{-1})^\beta &= \mathscr{Z}_\varrho^*(1-z^{-1})^\alpha \mathscr{Z}_\varrho \mathscr{Z}_\varrho^*(1-z^{-1})^\beta \mathscr{Z}_\varrho \\ &= \mathscr{Z}_\varrho^*(1-z^{-1})^{\alpha+\beta} \mathscr{Z}_\varrho = (1-\tau^{-1})^{\alpha+\beta}.\end{aligned}$$

In particular, for $\alpha \in \mathbb{C}$, $(1-\tau^{-1})^{\alpha}$ is invertible with inverse $(1-\tau^{-1})^{-\alpha}$.

4 Fractional Difference Equations on $\ell_{2,\varrho}(\mathbb{Z}; H)$

Fractional Operators

Let $\varrho > 1$ and $\alpha \in (0,1)$. We consider the operators (2), (3) and (4) defined on $V = H$. For comparing operators defined on spaces of sequences on $\mathbb{Z}$ with those defined for sequences on $\mathbb{N}$, we recall the embedding of $\ell_{2,\varrho}(\mathbb{N}; H)$ into $\ell_{2,\varrho}(\mathbb{Z}; H)$ by ι in Proposition 6. Moreover, we extend the operator Δ on $\mathbb{N}$ to $\mathbb{Z}$ by

$$\Delta : \ell_{2,\varrho}(\mathbb{Z}; H) \to \ell_{2,\varrho}(\mathbb{Z}; H), \qquad u \mapsto \chi_{\mathbb{N}}(\tau - 1)u = \chi_{\mathbb{N}}\tau(1-\tau^{-1})u.$$

Note that the left shift on $\mathbb{N}$ cuts of the first value of a sequence and embedded sequences have positive support. This is the reason for multiplying with $\chi_{\mathbb{N}}$ in the definition of Δ on $\ell_{2,\varrho}(\mathbb{Z}; H)$.

Let $v \in \ell_{2,\varrho}(\mathbb{N}; H)$ and set $u := \iota v \in \ell_{2,\varrho}(\mathbb{Z}; H)$. We compare the operator $(1-\tau^{-1})^{-\alpha}$ defined on $\ell_{2,\varrho}(\mathbb{Z}; H)$ and the fractional sum (2). We have $\operatorname{spt}\big((1-\tau^{-1})^{-\alpha}u\big) \subseteq \mathbb{N}$ and obtain

$$\iota\nabla^{-\alpha}v = (1-\tau^{-1})^{-\alpha}u.$$

Using definitions (3) and (4) of the Riemann–Liouville and Caputo difference operators, and the fact that $\Delta u = (\tau - 1)(u - \chi_{\mathbb{N}}u_0) = \tau(1-\tau^{-1})(u - \chi_{\mathbb{N}}u_0)$, we compute

$$\Delta(1-\tau^{-1})^{-(1-\alpha)}u = \chi_{\mathbb{N}}\tau(1-\tau^{-1})^{\alpha}u = \tau(1-\tau^{-1})^{\alpha}u - \delta_{-1}u_0,$$
$$(1-\tau^{-1})^{\alpha-1}\Delta u = (1-\tau^{-1})^{\alpha-1}\chi_{\mathbb{N}}\tau(1-\tau^{-1})u = \tau(1-\tau^{-1})^{\alpha}(u - \chi_{\mathbb{N}}u_0).$$

Moreover, we have

$$\iota\Delta^{\alpha}v = \chi_{\mathbb{N}}\tau(1-\tau^{-1})^{\alpha}u,$$
$$\iota\Delta_C^{\alpha}v = \tau(1-\tau^{-1})^{\alpha}(u - \chi_{\mathbb{N}}u_0).$$

In view of $\tau(1-\tau^{-1})^{\alpha}$, the Caputo and the Riemann–Liouville operators are equal whereby the Caputo operator regularizes u first. In particular, for $n \in \mathbb{N}$ by Proposition 1, we have $((1-\tau^{-1})^{\alpha}\chi_{\mathbb{N}}u_0)_n = \sum_{k=0}^{n}(-1)^k\binom{\alpha}{k}u_0 = \binom{-\alpha+n}{n}u_0$ and so

$$(\Delta^{\alpha}v)_n = (\Delta_C^{\alpha}v)_n + \binom{-\alpha+n+1}{n+1}u_0.$$

It is notable that the operator $(1-\tau^{-1})^{\alpha}$ when $H=\mathbb{C}$ maps real valued sequences to real valued sequences. We could have started with a real Hilbert space H and analyze $(1-\tau^{-1})^{\alpha}$ spectral-wise by the complexification $H\oplus H$.

Proposition 17 (Equivalence of difference equation and sequence equation) *Let* $\varrho>1$ *and* $\alpha\in(0,1)$. *Let* $x\in H$, $F:\ell_{2,\varrho}(\mathbb{Z};H)\to\ell_{2,\varrho}(\mathbb{Z};H)$ *and* $u\in\ell_{2,\varrho}(\mathbb{Z};H)$. *Let* $\operatorname{spt} u\subseteq\mathbb{N}$ *and* $\operatorname{spt} F(u)\subseteq\mathbb{N}$. *In view of the Riemann–Liouville operator, the following are equivalent:*

(i) $\tau(1-\tau^{-1})^{\alpha}u=F(u)+\delta_{-1}x$,

(ii) $u_0=x,\ ((1-\tau^{-1})^{\alpha}u)_{n+1}=F(u)_n$ *for* $n\in\mathbb{N}$,

(iii) $u_0=x,\ u_{n+1}=(-1)^{n+1}\binom{-\alpha}{n+1}u_0+\sum_{k=0}^{n}(-1)^{n-k}\binom{-\alpha}{n-k}F(u)_k$ *for* $n\in\mathbb{N}$.

In view of the Caputo operator, the following are equivalent:

(iv) $\tau(1-\tau^{-1})^{\alpha}u=F(u)+(1-\tau^{-1})^{\alpha}\chi_{\mathbb{Z}_{\geq -1}}x$,

(v) $u_0=x,\ ((1-\tau^{-1})^{\alpha}u)_{n+1}=F(u)_n+(-1)^{n+1}\binom{\alpha-1}{n+1}u_0$ *for* $n\in\mathbb{N}$,

(vi) $u_0=x,\ u_{n+1}=u_0+\sum_{k=0}^{n}(-1)^{n-k}\binom{-\alpha}{n-k}F(u)_k$ *for* $n\in\mathbb{N}$.

Proof We only proof the equivalence of (i), (ii) and (iii).
$(i)\Leftrightarrow(ii)$: If we evaluate (i) at $n\in\mathbb{Z}$ we obtain

$$(\tau(1-\tau^{-1})^{\alpha}u)_n=((1-\tau^{-1})^{\alpha}u)_{n+1}=F(u)_n+(\delta_{-1}x)_n.$$

Since $((1-\tau^{-1})^{\alpha}u)_n$ and $F(u)_n=0$ for $n\in\mathbb{Z}_{<0}$, and since $(\delta_{-1}x)_n=x$ if and only if $n=-1$ and $((1-\tau^{-1})^{\alpha}u)_0=u_0$, it follows that (i) and (ii) are equivalent.
$(i)\Leftrightarrow(iii)$: If we apply $(1-\tau^{-1})^{-\alpha}$ to (i) we see that (i) is equivalent to

$$\tau u=(1-\tau^{-1})^{-\alpha}F(u)+(1-\tau^{-1})^{-\alpha}\delta_{-1}u.$$

This equation is equivalent to (iii), since

$$(1-\tau^{-1})^{-\alpha}\delta_{-1}x=\begin{cases}0, & \text{if } n<-1,\\ (-1)^{n+1}\binom{-\alpha}{n+1}x, & \text{if } n\geq -1,\end{cases}$$

and since $\operatorname{spt} F(u)\subseteq\mathbb{N}$,

$$(1-\tau^{-1})^{-\alpha}F(u)=\sum_{k=0}^{n}(-1)^{n-k}\binom{-\alpha}{n-k}F(u)_k.$$

Remark 18 Note that the right hand side F in Proposition 17(i), (iv) maps sequences instead of values of H. If we have a function $f : H \to H$ such that for $u \in \ell_{2,\varrho}(\mathbb{Z}; H)$ we have $(f(u_n))_{n\in\mathbb{Z}} \in \ell_{2,\varrho}(\mathbb{Z}; H)$, we may set $F(u) := (f(u_n))_{n\in\mathbb{Z}}$ in Proposition 17.

Remark 19 (Grünwald–Letnikov difference operator) The Grünwald–Letnikov difference operator is defined for $h > 0$ and $\alpha \in (0, 1)$ by (c.f. [17, p. 708]):

$$\tilde{\Delta}_h^\alpha : V^{h\mathbb{N}} \to V^{h\mathbb{N}}, \qquad u \mapsto \left(t \mapsto \frac{1}{h^\alpha} \sum_{k=0}^{t/h} (-1)^k \binom{\alpha}{k} u_{t-kh} \right), \tag{6}$$

where $h\mathbb{N} = \{hn; n \in \mathbb{N}\}$. It can be shown (cf. [17, p. 708], [20, p. 43]) that for $V = \mathbb{R}$ the Grünwald–Letnikov operator can be used to approximate the Riemann–Liouville integral of sufficiently smooth functions.

Let $\alpha \in (0, 1)$. For $v \in \ell_{2,\varrho}(\mathbb{N}; H)$ and $u := \iota v$ we calculate for the Grünwald–Letnikov operator (6), $(1 - \tau^{-1})^\alpha u = \tilde{\Delta}_1^\alpha v$. Let $h > 0$, $x \in H$ and $F : H \to H$. A Grünwald–Letnikov difference equation has the form

$$(\tilde{\Delta}_h^\alpha v)(t + h) = F(v(t)), \quad v(0) = x \qquad (t \in h\mathbb{N}).$$

For $h = 1$ the Grünwald–Letnikov equation resembles the Riemann–Liouville equation of Proposition 17 and for $h \in \mathbb{R}_{>0}$, we may treat a Grünwald–Letnikov problem by considering the problem

$$\tau(1 - \tau^{-1})^\alpha u = h^\alpha F(u) + \delta_{-1}x.$$

Linear Equations on Sequence Spaces

Remark 20 Let $A \in L(H)$ and $x \in H$. In view of the Riemann–Liouville difference operator we ask whether the linear equation

$$\tau(1 - \tau^{-1})^\alpha u = Au + \delta_{-1}x \tag{7}$$

of Proposition 17 has a unique so-called causal solution that is supported in $\mathbb{N}$. In the spaces $\ell_{2,\varrho}(\mathbb{Z}; H)$, we have a unique solution of (7) for every initial value if $\tau(1 - \tau^{-1})^\alpha - A$ is invertible in $\ell_{2,\varrho}(\mathbb{Z}; H)$. In view of Proposition 17, the solution $(\tau(1 - \tau^{-1})^\alpha - A)^{-1}\delta_{-1}x$ should be causal. For the corresponding Caputo equation

$$\tau(1 - \tau^{-1})^\alpha u = Au + (1 - \tau^{-1})^\alpha \chi_{\mathbb{Z}_{\geq -1}} x, \tag{8}$$

the treatment is similar since $\chi_{\mathbb{Z}_{\geq -1}} x = \chi_{\mathbb{N}} x + \delta_{-1} x$.

Lemma 21 *Let $\alpha \in (0, 1)$ and $A \in L(H)$. We define $f : \mathbb{C}_{|\cdot|>1} \to \mathbb{C}, z \mapsto z(1 - z^{-1})^\alpha$ and set $f_\varrho := f|_{S_\varrho}$ for $\varrho > 1$. For $\varrho > 1$, the operator $\tau(1 - \tau^{-1})^\alpha - A$ is invertible in $\ell_{2,\varrho}(\mathbb{Z}; H)$ if and only if* ran $f_\varrho \cap \sigma(A) = \emptyset$. *Moreover, there is $\varrho > 1$*

such that for all $\mu > \varrho$, $\operatorname{ran} f_\mu \cap \sigma(A) = \emptyset$, *that is* $\{z(1-z^{-1})^\alpha;\ |z| > \varrho\}$ *is in the resolvent set of* A.

Proof Recall the multiplication operator m of Lemma 12. Using the $\mathcal{Z}$ transform, the operator $\tau(1-\tau^{-1})^\alpha - A$ is invertible in $\ell_{2,\varrho}(\mathbb{Z}; H)$ if and only if $\mathrm{m}(1-\mathrm{m}^{-1})^\alpha - A$ is invertible in $L_2(S_\varrho, H)$, since $\mathcal{Z}_\varrho$ is unitary. This is the case, however, if and only if $\operatorname{ran} f_\varrho \cap \sigma(A) = \emptyset$. Using Lemma 4, there is $\varrho > 1$ such that for all $\mu > \varrho$ and $z \in S_\mu$, $r(A) < \mu(1-\mu^{-1})^\alpha \leq |z(1-z^{-1})^\alpha|$. That is, for all $\mu > \varrho$, $\operatorname{ran} f_\mu \cap \sigma(A) = \emptyset$.

Proposition 22 (Causality of $(\tau(1-\tau^{-1})^\alpha - A)^{-1}$) *Let* $\varrho > 1$, $\alpha \in (0,1)$ *and* $A \in L(H)$. *Let* f_ϱ *be defined as in Lemma 21. The following are equivalent:*

$$(i)\quad (\tau(1-\tau^{-1})^\alpha - A)^{-1} \in L(\ell_{2,\varrho}(\mathbb{Z}; H)) \text{ is causal,}$$
$$(ii)\quad (\tau(1-\tau^{-1})^\alpha - A)^{-1} \in L(\ell_{2,\varrho}(\mathbb{Z}; H))$$
$$\text{and}\quad \forall x \in H\colon \operatorname{spt}(\tau(1-\tau^{-1})^\alpha - A)^{-1}\delta_{-1}x \subseteq \mathbb{N},$$
$$(iii)\quad \forall \mu \geq \varrho\colon \operatorname{ran} f_\mu \cap \sigma(A) = \emptyset.$$

Proof $(i) \Rightarrow (ii)$: Let $x \in H$ and $u := (\tau(1-\tau^{-1})^\alpha - A)^{-1}\delta_{-1}x$. Using causality assumed in (i), we obtain $\operatorname{spt} u \subseteq \mathbb{Z}_{\geq -1}$. Moreover, $u_{-1} = ((1-\tau^{-1})^{-\alpha}Au)_{-2} + ((1-\tau^{-1})^{-\alpha}\delta_{-1}x)_{-2} = 0$ so that $\operatorname{spt} u \subseteq \mathbb{N}$.
$(ii) \Rightarrow (iii)$: Suppose by contradiction that there is $\varrho' > \varrho$ with $\operatorname{ran} f_{\varrho'} \cap \sigma(A) \neq \emptyset$. The set $\{z \in \mathbb{C}_{|\cdot|\geq\varrho'};\ z(1-z^{-1})^\alpha \in \sigma(A)\}$ is closed, since $\sigma(A)$ is closed and since f is continuous, and the set is bounded, since by Lemma 21 there is a $\tilde{\varrho} > \varrho'$ such that $f(\mathbb{C}_{|\cdot|\geq\tilde{\varrho}})$ is in the resolvent set. Thus, there is $z' \in \{z \in \mathbb{C}_{|\cdot|\geq\varrho'};\ z(1-z^{-1})^\alpha \in \sigma(A)\}$ with maximal absolute value. Therefore there is a sequence $(z_n)_{n\in\mathbb{N}}$ in $\mathbb{C}$ with $|z_n| > |z'|$, that is, $z_n(1-z_n^{-1})^\alpha$ is in the resolvent set of A ($n \in \mathbb{N}$) and $\lim_{n\to\infty} z_n = z'$. Using the resolvent estimate (cf. [21, p. 378]), we have $\lim_{n\to\infty} \|(z_n(1-z_n^{-1})^\alpha - A)^{-1}\|_{L(H)} = \infty$. By applying the Banach–Steinhaus theorem (cf. [21, p. 141]), there is $x \in H$ with $\lim_{n\to\infty} \|(z_n(1-z_n^{-1})^\alpha - A)^{-1}x\|_H = \infty$. By assumption, $(\tau(1-\tau^{-1})^\alpha - A)^{-1}\delta_{-1}x \in \ell_{2,\varrho}(\mathbb{Z}; H)$ and $\operatorname{spt}(\tau(1-\tau^{-1})^\alpha - A)^{-1}\delta_{-1}x \subseteq \mathbb{N}$. Hence for $v := (\tau(1-\tau^{-1})^\alpha - A)^{-1}\delta_0 x \in \ell_{2,\varrho}(\mathbb{Z}; H)$, we have $v \in \ell_{2,\varrho}(\mathbb{Z}; H)$ and $\operatorname{spt} v \subseteq \mathbb{N}$. Applying Lemma 13, it follows that $F : \mathbb{C}_{|\cdot|>\varrho} \to H,\ z \mapsto \sum_{k=-\infty}^{\infty} v_k z^{-k}$ is analytic. Since $v \in \ell_{2,\mu}(\mathbb{Z}; H)$ for $\mu > |z'|$, it follows that for $G : \mathbb{C}_{|\cdot|>|z'|} \to H,\ z \mapsto (z(1-z^{-1})^\alpha - A)^{-1}x$, we have $G = F|_{\mathbb{C}_{|\cdot|>|z'|}}$. This means that $\lim_{n\to\infty} \|F(z_n)\|_H = \lim_{n\to\infty} \|G(z_n)\|_H = \infty$. Since F is continuous, this is a contradiction in that $\lim_{n\to\infty} \|F(z_n)\|_H \neq \infty$.
$(iii) \Rightarrow (i)$: We have $(\tau(1-\tau^{-1})^\alpha - A)^{-1} \in L(\ell_{2,\mu}(\mathbb{Z}; H))$ for $\mu > \varrho$ by Lemma 21. Since the resolvent of A is analytic, the mapping $z \mapsto (z(1-z^{-1})^\alpha - A)^{-1}$ is analytic on $\mathbb{C}_{|\cdot|>\varrho}$. Moreover the mapping $z \mapsto \|(z(1-z^{-1})^\alpha - A)^{-1}\|_{L(H)}$ is continuous and hence bounded on compact sets $\mathbb{C}_{\mu\geq|\cdot|\geq\varrho}$ where $\mu \geq \varrho$, i.e. the mapping attains its maximum on $\mathbb{C}_{\mu\geq|\cdot|\geq\varrho}$. By Lemma 4 and since A is bounded, $\sup_{z\in S_\mu} \|(z(1-z^{-1})^\alpha - A)^{-1}\|_{L(H)}$ decays to zero when μ tends to infinity. It follows that $\mu \mapsto \sup_{z\in S_\mu} \|(z(1-z^{-1})^\alpha - A)^{-1}\|_{L(H)}$ is bounded on $[\varrho, \infty)$ and there-

fore the conditions of Lemma 13(ii) are satisfied for $(\tau(1-\tau^{-1})^{\alpha}-A)^{-1}u$, where $u \in \ell_{2,\varrho}(\mathbb{Z}; H)$, $\operatorname{spt} u \subseteq \mathbb{N}$. It follows that $(\tau(1-\tau^{-1})^{\alpha}-A)^{-1}$ is causal.

Remark 23 Let $A \in L(H)$, $\varrho > 1$ and $\alpha \in (0,1)$. By Lemma 21 and Proposition 22, we can always choose ϱ large enough such that $\tau(1-\tau^{-1})^{\alpha}-A$ is invertible with causal inverse. As a consequence the linear fractional difference Eq. (7) or (8) has a unique solution $u \in \ell_{2,\varrho}(\mathbb{Z}; H)$. Moreover, from the previous Theorem it follows that (7) or (8) has a unique solution in $\ell_{2,\mu}(\mathbb{Z}; H)$ for $\mu \geq \varrho$ which coincides with the solution u, since $\ell_{2,\varrho}(\mathbb{N}; H) \subseteq \ell_{2,\mu}(\mathbb{N}; H)$. Therefore we can speak of the solution operator $(\tau(1-\tau^{-1})^{\alpha}-A)^{-1}$.

The difference equation for an initial value $x \in H$ and $A \in L(H)$

$$(\Delta^{\alpha} u)_n = A u_n, \qquad u_0 = x,$$

or

$$(\Delta_C^{\alpha} u)_n = A u_n, \qquad u_0 = x,$$

can be solved algebraically with a unique solution $u \in H^{\mathbb{N}}$ (cf. Proposition 17(iii), (vi)). Recall the embedding ι of Proposition 6. Since A has bounded spectrum, when by the previous theorem, there is $\varrho > 1$ such that $\iota u \in \ell_{2,\varrho}(\mathbb{Z}; H)$ is the unique solution of (7) or (8).

Asymptotic Stability

We discuss asymptotic stability of linear fractional difference equations. For an analysis of rates of convergence, see also [5, 7].

Definition 24 (*Asymptotic stability*) Let $A \in L(H)$. The zero equilibrium of Eq. (7) or (8), i.e., the solution $u = 0$ for the inital value 0, is said to be asymptotically stable if for every $\varrho > 1$, every solution $u \in \ell_{2,\varrho}(\mathbb{Z}; H)$ of (7) or (8) with $\operatorname{spt} u \subseteq \mathbb{N}$ satisfies $\lim_{n\to\infty} u_n = 0$ in H.

Remark 25 If a sequence $u \in H^{\mathbb{Z}}$ satisfies $\operatorname{spt} u \subseteq \mathbb{N}$ and $\lim_{n\to\infty} u_n = 0$, then necessarily for all $\varrho > 1$ we have $u \in \ell_{2,\varrho}(\mathbb{Z}; H)$. One could say that the spaces $\ell_{2,\varrho}(\mathbb{Z}; H)$, $\varrho > 1$, are large enough to look for asymptotically stable solutions of a linear sequence equation.

Proposition 26 (Necessary condition for asymptotic stability) *Let $A \in L(H)$ such that the zero equilibrium of Eq.* (7) *or* (8) *is asymptotically stable and let f_{μ} $(\mu > 1)$ be as in Lemma 21. Then for all $\mu > 1$, $\tau(1-\tau^{-1})^{\alpha}-A$ is invertible in $\ell_{2,\mu}(\mathbb{Z}; H)$ with causal inverse, i.e., for each $\mu > 1$, $\sigma(A) \cap \operatorname{ran} f_{\mu} = \emptyset$.*

Proof Assume by contradiction there is $z' \in \operatorname{ran} f_{\varrho} \cap \sigma(A) \neq \emptyset$ where $\varrho > 1$. We may assume that $\operatorname{ran} f_{\mu} \cap \sigma(A) = \emptyset$ for $\mu > |z'|$. Then there is a sequence $(z_n)_{n\in\mathbb{N}}$ with $|z_n| > |z'|$ such that $z_n(1-z_n^{-1})^{\alpha}$ is in the resolvent set of A $(n \in \mathbb{N})$ and such that $z_n \to z'$ $(n \to \infty)$. Using the resolvent estimate, we have $\lim_{n\to\infty} \left\|(z_n(1-z_n^{-1})^\alpha - A)^{-1}\right\|_{L(H)} = \infty$. Using the Banach–Steinhaus theorem, there is $x \in H$ with $\lim_{n\to\infty} \left\|(z_n(1-z_n^{-1})^\alpha - A)^{-1}x\right\|_H = \infty$. By Lemma 21 and Proposition 22, for $\mu > |z'|$, we know that $\tau(1-\tau^{-1})^\alpha - A$ is invertible in $\ell_{2,\mu}(\mathbb{Z}; H)$ and $v := (\tau(1-\tau^{-1})^\alpha - A)^{-1}\delta_0 x$ satisfies $\operatorname{spt} v \subseteq \mathbb{N}$. Since the zero equilibrium is asymptotically stable, we have $v \in \ell_{2,\varrho'}(\mathbb{Z}; H)$ for some $\varrho' \in (1, |z'|)$ by Remark 25. Then the mapping $F : \mathbb{C}_{|\cdot|>\varrho'} \to H, z \mapsto \sum_{k=-\infty}^{\infty} v_k z^{-k}$ is analytic and equals $G : \mathbb{C}_{|\cdot|>|z'|} \to H, z \mapsto (z(1-z^{-1})^\alpha - A)^{-1}\delta_0 x$ on $\mathbb{C}_{|\cdot|>|z'|}$ by Lemma 13. Therefore we have $\lim_{n\to\infty} F(z_n) < \infty$, since F is analytic which contradicts $\lim_{n\to\infty} F(z_n) = \lim_{n\to\infty} G(z_n) = \infty$.

For a sufficient condition of asymptotic stability we observe that if $u \in \ell_{2,1}(\mathbb{Z}; H)$ with $\operatorname{spt} u \subseteq \mathbb{N}$ then $\lim_{n\to\infty} u_n = 0$.

Proposition 27 (Sufficient condition for asymptotic stability) *Let $A \in L(H)$. For all $\varrho > 1$ let $\tau(1-\tau^{-1})^\alpha - A$ be invertible in $\ell_{2,\varrho}(\mathbb{Z}; H)$ with causal inverse. If for all $x \in H$ the mapping $\mathbb{C}_{|\cdot|>1} \to H, z \mapsto \sum_{k=-\infty}^{\infty}[(\tau(1-\tau^{-1})^\alpha - A)^{-1}\delta_{-1}x]_k z^{-k}$ has a continuous continuation to the unit circle S_1, then the zero equilibrium of Eq.* (7) *or* (8) *is asymptotically stable.*

Proof Let g be the continuous continuation. Then $g|_{S_1} \in L_2(S_1, H)$ and $v := \mathscr{Z}_1^{-1} g|_{S_1} \in \ell_{2,1}(\mathbb{Z}; H)$. Moreover, $u = v$ that is $u \in \ell_{2,1}(\mathbb{Z}; H)$.

Remark 28 We believe that the necessary conditions for stability in Proposition 26 are not sufficient, neither are the sufficient conditions for stability in Proposition 27 necessary. Already for semigroups the asymptotic stability can in general not be characterized by spectral conditions solely. The shift operator on continuous functions from $\mathbb{R}^+$ to $\mathbb{R}$ which decay at infinity, for example, is asymptotically stable although its spectrum consists of all complex numbers with non-positive real part [2, Example 2.5(c)]. The characterization of asymptotic stability for linear fractional difference equations is an intricate problem which still needs to be addressed.

Example 29 Let $H = \mathbb{C}$, $A : \mathbb{C} \to \mathbb{C}, z \mapsto \lambda z$ where $\lambda \in \mathbb{R}$ and $\alpha \in (0, 1)$. We study the asymptotic behavior of the linear fractional equations (7) and (8) on $\ell_{2,\varrho}(\mathbb{Z}; H)$ $(\varrho > 1)$ in view of Proposition 26 and Proposition 27 and therefore want to apply the $\mathscr{Z}$ transform to Eq. (7) and (8). In order to obtain an asymptotically stable zero equilibrium by Proposition 26, we must have $\sigma(A) \cap \operatorname{ran} f = \emptyset$ where $f : \mathbb{C}_{|\cdot|>1} \to \mathbb{C}, z \mapsto z(1-z^{-1})^\alpha$ is defined as in Lemma 21 and $\sigma(A) = \{\lambda\}$. We remark that for $z \in \mathbb{C}_{|\cdot|>1}$, $f(z) \in \mathbb{R}$ if and only if $z \in \mathbb{R}$ since f is injective and since $f(\overline{z}) = \overline{f(z)}$. Moreover $f(\mathbb{C}_{|\cdot|>1} \cap \mathbb{R}) = (-\infty, -2^\alpha) \cup (0, \infty)$ and so $\lambda \notin \operatorname{ran} f$ if and only if $\lambda \in [-2^\alpha, 0]$. By Proposition 26, we necessarily have $\lambda \in [-2^\alpha, 0]$ if the zero equilibrium of (7) or (8) is asymptotically stable. Let $\lambda \in [-2^\alpha, 0]$, and for $u \in \ell_{2,\varrho}(\mathbb{Z}; H)$ we denote $\hat{u} := \mathscr{Z}u$.

We consider (7) with $x \in \mathbb{C}$ first. Also for $z \in S_\varrho$ we have $(\mathscr{Z}\delta_{-1}x)(z) = zx$. Applying the $\mathscr{Z}$ transform to Eq. (7), we obtain for $z \in S_\varrho$

$$z(1-z^{-1})^{\alpha}\hat{u}(z) = A\hat{u}(z) + zx.$$

If $\lambda \in (-2^{\alpha}, 0)$, the mapping $\mathbb{C}_{|\cdot|>1} \to H, z \mapsto \frac{zx}{z(1-z^{-1})^{\alpha}-\lambda}$ has a continuous continuation to S_1 and by Proposition 27, we obtain that the zero equilibrium of (7) is asymptotically stable.

We now consider Eq. (8) where $x \in \mathbb{C}$. For $z \in S_{\varrho}$, we have $(\mathscr{Z}\chi_{\mathbb{Z}_{\geq 1}}x)(z) = \frac{zx}{1-z^{-1}}$. Applying the $\mathscr{Z}$ transform to Eq. (8), we obtain for $z \in S_{\varrho}$

$$z(1-z^{-1})^{\alpha}\hat{u}(z) = A\hat{u}(z) + z(1-z^{-1})^{\alpha-1}x.$$

If $\lambda \in (-2^{\alpha}, 0)$, the mapping $\mathbb{C}_{|\cdot|>1} \to H, z \mapsto \frac{z(1-z^{-1})^{\alpha-1}x}{z(1-z^{-1})^{\alpha}-\lambda}$ has a continuous continuation to S_1 and using Proposition 27, we obtain that the zero equilibrium of (8) is asymptotically stable.

The cases $\lambda = 0$ and $\lambda = -2^{\alpha}$ are discussed in [5].

Acknowledgements The research of A.B. was funded by the National Science Centre in Poland granted according to decision DEC-2015/19/D/ST7/03679. The research of A.C. was supported by the Statutory Funding of the Faculty of Automatic Control, Electronics and Computer Science, Silesian University of Technology, Gliwice, Poland, 02/990/BK_19/0121. The research of M.N. was supported by the Polish National Agency for Academic Exchange according to the decision PPN/BEK/2018/1/00312/DEC/1. The research of S.S. was partially supported by an Alexander von Humboldt Polish Honorary Research Fellowship. The work of H.T. Tuan was supported by the joint research project from RAS and VAST QTRU03.02/437 18-19.

References

1. Abu-Saris, R., Al-Mdallal, Q.: On the asymptotic stability of linear system of fractional-order difference equations. Fract. Calc. Appl. Anal. **16**(3), 613–629 (2013)
2. Arendt, W., Batty, C.J.K.: Tauberian theorems and stability of one-parameter semigroups. Trans. Amer. Math. Soc. **306**(2), 837–852 (1988)
3. Atici, F.M., Eloe, P.W.: Discrete fractional calculus with the nabla operator. Electron. J. Qual. Theory Differ. Equ. (Special Edition I) (3), 12 (2009)
4. Čermák, J., Kisela, T., Nechvátal, L.: Stability and asymptotic properties of a linear fractional difference equation. Adv. Differ. Equ. **2012**(14),122 (2012)
5. Čermák, J., Győri, I., Nechvátal, L.: On explicit stability conditions for a linear fractional difference system. Fract. Calc. Appl. Anal. **18**(3), 651–672 (2015)
6. Cong, N.D., Doan, T.S., Siegmund, S., Tuan, H.T.: Linearized asymptotic stability for fractional differential equations. Electron. J. Qual. Theory Differ. Equ., paper no. 39, 13 (2016)
7. Cong, N.D., Tuan, H.T., Trinh, H.: On asymptotic properties of solutions to fractional differential equations. Submitted (2018)
8. Diethelm, K.: The Analysis of Fractional Differential Equations: An Application-Oriented Exposition Using Differential Operators of Caputo Type. Lecture Notes in Mathematics, vol. 2004. Springer, Berlin (2010)
9. Dunford, N., Schwartz, J.T.: Linear operators. Part I. Wiley Classics Library. Wiley, New York (1988). General theory, With the assistance of William G. Bade and Robert G. Bartle, Reprint of the 1958 original, A Wiley-Interscience Publication
10. Gohberg, I., Goldberg, S., Kaashoek, M.A.: Classes of Linear Operators. Vol. I, vol. 49. Birkhäuser Verlag, Basel (1990)

11. Graham, R.L., Knuth, D.E., Patashnik, O.: Concrete Mathematics: A Foundation for Computer Science, 2nd edn. Addison-Wesley Publishing Company, Reading (1994)
12. Katznelson, Y.: An Introduction to Harmonic Analysis, 3rd edn. Cambridge University Press, Cambridge (2004)
13. Kitzing, K., Picard, R., Siegmund, S., Trostorff, S., Waurick, M.: A Hilbert space approach to difference equations. Submitted (2018)
14. Königsberger, K.: Analysis. 1. Springer-Lehrbuch, 6th edn. Springer, Berlin (2004)
15. Kuttner, B.: On differences of fractional order. Proc. London Math. Soc. **3**(7), 453–466 (1957)
16. Lizama, C.: The Poisson distribution, abstract fractional difference equations, and stability. Proc. Amer. Math. Soc. **145**(9), 3809–3827 (2017)
17. Lubich, C.: Discretized fractional calculus. SIAM J. Math. Anal. **17**(3), 704–719 (1986)
18. Matignon, D.: Stability properties for generalized fractional differential systems. In: Systèmes différentiels fractionnaires (Paris, 1998), volume 5 of ESAIM Proc., pages 145–158. Soc. Math. Appl. Indust., Paris (1998)
19. Picard, R., Trostorff, S., Waurick, M.: On evolutionary equations with material laws containing fractional integrals. Math. Methods Appl. Sci. **38**(15), 3141–3154 (2015)
20. Podlubny, I.: Fractional Differential Equations, volume 198 of Mathematics in Science and Engineering. Academic Press, Inc., San Diego, CA, 1999. An introduction to fractional derivatives, fractional differential equations, to methods of their solution and some of their applications
21. Werner, D.: Funktionalanalysis, extended edn. Springer, Berlin (2000)

Assignability of Lyapunov Spectrum for Discrete Linear Time-Varying Systems

Artur Babiarz, Irina Banshchikova, Adam Czornik, Evgeniĭ Makarov, Michał Niezabitowski and Svetlana Popova

Abstract We discuss relations between the four formulations of the problem of assignability of the Lyapunov spectrum for discrete linear time-varying systems by a time-varying feedback. For two of them: global assignability and proportional local assignability, we have already [2–4] obtained sufficient conditions in terms of uniform complete controllability and certain asymptotic properties of the free system. In the present paper we discuss the assumptions of our papers and demonstrate the use of the obtained conditions by numerical examples. We also compare our results with the classical pole placement problem. Finally, we formulate a couple of directions for further research in this area.

Keywords Discrete linear time-varying system · Lyapunov spectrum · Pole placement problem · Controllability

---

A. Babiarz · A. Czornik · M. Niezabitowski
Faculty of Automatic Control, Electronics and Computer Science, Silesian University of Technology, Akademicka 16, 44-100 Gliwice, Poland
e-mail: artur.babiarz@polsl.pl

A. Czornik
e-mail: adam.czornik@polsl.pl

I. Banshchikova · S. Popova
Department of Differential Equations, Udmurt State University, Universitetskaya 1, Izhevsk 426034, Russia
e-mail: banshhikova.irina@mail.ru

S. Popova
e-mail: udsu.popova.sn@gmail.com

E. Makarov
National Academy of Sciences of Belarus, Institute of Mathematics, Surganova 11, Minsk 220072, Belarus
e-mail: jcm@im.bas-net.by

M. Niezabitowski (✉)
Faculty of Mathematics, Physics and Chemistry, University of Silesia, Institute of Mathematics, Bankowa 14, 40-007 Katowice, Poland
e-mail: michal.niezabitowski@polsl.pl; mniezabitowski@us.edu.pl

M. Bohner et al. (eds.), *Difference Equations and Discrete Dynamical Systems with Applications*, Springer Proceedings in Mathematics & Statistics 312,
https://doi.org/10.1007/978-3-030-35502-9_5

1 Introduction

One of the main methods of designing the control strategy for linear systems with time-invariant coefficients is pole placement method [17]. This method is based on the selection of feedback in such a way that the poles of the closed-loop system are in advance given points on the complex plane. The theoretical basis of this method is the following classical theorem [8, p. 458] (see also [7]): the pair $(A, B) \in \mathbb{R}^{s\times s} \times \mathbb{R}^{s\times t}$ is completely controllable if and only if for any set $\Lambda = \{\mu_1, \mu_2, \ldots, \mu_s\}$ of arbitrary s complex numbers such that $\overline{\Lambda} = \{\overline{\mu}_1, \overline{\mu}_2, \ldots, \overline{\mu}_s\} = \Lambda$, there exists a constant matrix $U \in \mathbb{R}^{t\times s}$ such that the eigenvalues of $A + BU$ form the set Λ. At the same time, for such systems, relationships between the location of the poles and dynamic properties such as stability, stability and oscillation degrees, and the size of overshoot are well known and described in the literature [1].

This problem for time-varying systems is much more complex and less studied. In the literature most of the results are for continuous-time systems and they are summarized in [14]. For time-varying systems, there are various concepts of stability (asymptotic, uniform, exponential, etc. [13]). Similarly, the concept of controllability of such systems can be understood in several different ways (uniform, complete, output, etc. [12]). Moreover, for time-varying systems, we do not have an obvious notion which fully matches the concept of the poles of time-invariant systems. The Lyapunov exponents play, to a certain extent, the same role as the logarithms of absolute values of poles of discrete-time systems and real parts of poles of continuous-time systems.

The problems of assignability of the Lyapunov spectrum for discrete-time systems were considered by us in [2–4]. In this article, we summarize these results and compare time-invariant versions of them with pole placement theorem, give some examples and formulate directions of further research.

The paper is organized as follows. In Sect. 2, we introduce the basic notation and definitions. We present four different formulations of the assignability of the Lyapunov spectrum: global, proportional global, local and proportional local assignability. In Sect. 3 we recall our previous results from [2–4], discuss the assumptions and relations between them and present two examples to depict the relation between global and proportional local assignability, and uniform complete controllability. The work is ended with conclusions and formulations of some open questions.

2 Problem Statement

Let $\mathbb{R}^s$ be the s-dimensional Euclidean space with a fixed orthonormal basis and the Euclidean norm $\|\cdot\|$. By $\mathbb{R}^{s\times t}$ we will denote the space of all real matrices of the size $s \times t$ with the spectral norm, i.e. with the operator norm generated in $\mathbb{R}^{s\times t}$ by Euclidean norms in $\mathbb{R}^s$ and $\mathbb{R}^t$, respectively; $I \in \mathbb{R}^{s\times s}$ is the identity matrix. For any sequence $F = \big(F(n)\big)_{n\in\mathbb{N}} \subset \mathbb{R}^{s\times t}$ we define

$$\|F\|_\infty = \sup_{n\in\mathbb{N}} \|F(n)\|.$$

A bounded sequence $\left(L(n)\right)_{n\in\mathbb{N}} \subset \mathbb{R}^{s\times s}$ of invertible matrices such that $\left(L^{-1}(n)\right)_{n\in\mathbb{N}}$ is bounded, will be called the Lyapunov sequence. By $\mathbb{R}^s_\leq$ we denote the set of all nondecreasing sequences of s real numbers. For a fixed sequence $\mu = (\mu_1, \ldots, \mu_s) \in \mathbb{R}^s_\leq$ and any $\delta > 0$ let us denote by $O_\delta(\mu)$ the set of all sequences $\nu = (\nu_1, \ldots, \nu_s) \in \mathbb{R}^s_\leq$ such that $\max_{j=1,\ldots,s} |\nu_j - \mu_j| < \delta$.

We consider a discrete linear time-varying system

$$x(n+1) = A(n)x(n) + B(n)u(n), \quad n \in \mathbb{N}, \tag{1}$$

where $A = (A(n))_{n\in\mathbb{N}} \subset \mathbb{R}^{s\times s}$ is a Lyapunov sequence, $B = (B(n))_{n\in\mathbb{N}} \subset \mathbb{R}^{s\times t}$ is a bounded sequence, and $u = (u(n))_{n\in\mathbb{N}} \subset \mathbb{R}^t$ is a control sequence.

We denote the transition matrix of the free system

$$x(n+1) = A(n)x(n) \tag{2}$$

by $\Phi_A(n, m)$, $n, m \in \mathbb{N}$, and by $(x(n, x_0))_{n\in\mathbb{N}}$ its solution with the initial condition $x(1, x_0) = x_0$.

For $x_0 \in \mathbb{R}^s$, $x_0 \neq 0$, the Lyapunov exponent $\lambda(x_0)$ of the solution $x = (x(n, x_0))_{n\in\mathbb{N}}$ is defined as

$$\lambda(x_0) = \lambda[x] \doteq \limsup_{n\to\infty} \frac{1}{n} \ln \|x(n, x_0)\|$$

and $\lambda(0) \doteq -\infty$. It is well known (see [6], [9, pp. 51–52]) that if $A = (A(n))_{n\in\mathbb{N}}$ is a Lyapunov sequence, then the set of the Lyapunov exponents of all nontrivial solutions of system (2) contains at most s elements, say

$$-\infty < \Lambda_1(A) < \Lambda_2(A) < \cdots < \Lambda_q(A) < \infty,$$

where $q \leq s$. For each $i \in \{1, \ldots, q\}$, we consider the linear subspace

$$E_i = \left\{x_0 \in \mathbb{R}^s : \lambda(x_0) \leq \Lambda_i(A)\right\} \subset \mathbb{R}^s.$$

We also set $E_0 = \{0\}$. The multiplicity s_i of the Lyapunov exponent $\Lambda_i(A)$ is defined as

$$\dim E_i - \dim E_{i-1}, \quad i = 1, \ldots, q.$$

Note that $s_1 + \cdots + s_q = s$. The sequence of s numbers

$$\left(\Lambda_1(A), \ldots, \Lambda_1(A), \ldots, \Lambda_q(A), \ldots, \Lambda_q(A)\right),$$

where each Lyapunov exponent $\Lambda_i(A)$ appears s_i times, is called the Lyapunov spectrum of (2) (see [6], [9, p. 57]) and is denoted by

$$\lambda(A) = (\lambda_1(A), \lambda_2(A), \ldots, \lambda_s(A)).$$

We assume that the Lyapunov spectrum is numbered in nondecreasing order, i.e. $\lambda(A) \in \mathbb{R}^s_{\leq}$.

For any bounded sequence $U = (U(n))_{n\in\mathbb{N}} \subset \mathbb{R}^{t\times s}$, we consider a linear feedback control

$$u(n) = U(n)x(n), \quad n \in \mathbb{N}$$

for system (1). We identify this control u with the sequence U and call this sequence U a feedback control for system (1).

Definition 1 A bounded sequence

$$U = (U(n))_{n\in\mathbb{N}} \subset \mathbb{R}^{t\times s}$$

is said to be an admissible feedback control for system (1) if the sequence

$$(A(n) + B(n)U(n))_{n\in\mathbb{N}}$$

is a Lyapunov sequence.

Let $U = (U(n))_{n\in\mathbb{N}}$ be any admissible feedback control for system (1). Then, for the closed-loop system

$$x(n+1) = (A(n) + B(n)U(n))\,x(n) \tag{3}$$

we can define the Lyapunov spectrum

$$\lambda(A + BU) = \big(\lambda_1(A + BU), \ldots, \lambda_s(A + BU)\big) \in \mathbb{R}^s_{\leq}.$$

The next definition expresses one of the possible way of formulation of the Lyapunov spectrum assignability problem.

Definition 2 The Lyapunov spectrum of system (3) is called globally assignable if for each $\mu \in \mathbb{R}^s_{\leq}$ there exists an admissible feedback control U such that

$$\lambda(A + BU) = \mu. \tag{4}$$

In this definition there is in general no bound on the norm of the feedback control. In some practical applications it is desirable to have a bound on the control which tends to zero in case the placed Lyapunov spectrum tends to the Lyapunov spectrum of the free system. This requirement is the base for the following definition.

Definition 3 The Lyapunov spectrum of system (3) is called proportionally globally assignable if for all $\Delta > 0$ there exists $\ell = \ell(\Delta) > 0$ such that for any sequence $\mu = (\mu_1, \dots, \mu_s) \in O_\Delta(\lambda(A))$ there exists an admissible feedback control U, satisfying the estimate

$$\|U\|_\infty \le \ell \max_{j=1,\dots,s} |\lambda_j(A) - \mu_j| \tag{5}$$

and such that equality (4) is satisfied.

One may also consider the local version of assignability of the Lyapunov spectrum.

Definition 4 The Lyapunov spectrum of system (3) is called locally assignable if for each $\varepsilon > 0$ there exists $\delta > 0$ such that for all $\mu \in O_\delta(\lambda(A))$ there exists an admissible feedback control U such that

$$\lambda(A + BU) = \mu \text{ and } \|U\|_\infty < \varepsilon.$$

Definition 5 The Lyapunov spectrum of system (3) is called proportionally locally assignable if there exist $\ell > 0$ and $\delta > 0$ such that for all $\mu \in O_\delta(\lambda(A))$ there exists an admissible feedback control U, such that estimate (5) and equality (4) are satisfied.

All the proposed definitions of the assignability problem were formulated for continous-time systems in [14] and our Definitions 2, 4 and 5 are direct translations of their continuous counterparts. However, the direct translation of definition of proportional global assignability is as follows: the Lyapunov spectrum of system (3) is called proportionally globally assignable if there exists $\ell > 0$ such that for all $\mu = (\mu_1, \dots, \mu_s) \in \mathbb{R}^s_\le$ there exists a feedback control U, satisfying (4) and (5).

The next example justifies our modification.

Example 6 Let us consider a linear discrete-time control system

$$x(n+1) = x(n) + u(n). \tag{6}$$

Here the matrices $A(n)$, $B(n)$ have the sizes 1×1 and $A(n) = B(n) = 1$ for all n. Therefore, for the transition matrix of the free system

$$x(n+1) = x(n)$$

we have $\Phi_A(n, m) = 1$ for all n, m. Thus, system (6) is uniformly completely controllable with $K = 1$ (see Definition 7 below). Since every solution $x(n, x_0)$ of the free system is constant, it follows that the Lyapunov spectrum coincides with 0. Let us close system (6) by a feedback $u(n) = U(n)x(n)$. Then we get a system

$$x(n+1) = (1 + U(n))\,x(n). \tag{7}$$

By the Theorem 4.7 from [2] the Lyapunov spectrum of system (7) is globally assignable, so for every $\alpha \in \mathbb{R}$ we can construct a control U, such that the Lyapunov spectrum of system (7) coincides with the number α. Let us find out whether it is possible to find a number $\ell > 0$, such that for all $\alpha > 0$ there exists a control U for which we have

$$\lambda (A + BU) = \alpha, \quad \|U\|_\infty \leq \ell\alpha. \tag{8}$$

Here we restrict ourselves to the consideration of positive numbers α, since below we will prove that even for this case it is impossible. Suppose that for each $\alpha > 0$ there exists a control U for which both conditions (8) are satisfied. Then for an arbitrary nontrivial solution $x_U(n, x_0)$ of system (7) we have estimates

$$\alpha = \lambda(A + BU) = \limsup_{n\to\infty} \frac{1}{n} \ln |x_U(n, x_0)|$$
$$= \limsup_{n\to\infty} \frac{1}{n} \ln \left| \prod_{j=1}^{n-1} \big(1 + U(j)\big) x_0 \right| \leq \limsup_{n\to\infty} \frac{1}{n} \ln \prod_{j=1}^{n-1} \big(1 + |U(j)|\big)|x_0|$$
$$\leq \limsup_{n\to\infty} \frac{1}{n} \ln \prod_{j=1}^{n-1} \big(1 + \ell\alpha\big)|x_0| = \limsup_{n\to\infty} \frac{1}{n} \ln \big(1 + \ell\alpha\big)^{n-1} |x_0| = \ln(1 + \ell\alpha).$$

Thus, there exists $\ell > 0$ such that for each $\alpha > 0$ the inequality $\alpha \leq \ln(1 + \ell\alpha)$ holds, that is, $e^\alpha \leq 1 + \ell\alpha$. But this is impossible, since the exponential function grows faster than any linear function. But if we choose an arbitrary $\Delta > 0$, then there exists an $\ell = \ell(\Delta) > 0$ such that for each $\alpha \in \mathbb{R}$, $|\alpha| < \Delta$ there exists a control U for which the conditions (8) are satisfied. Here we can take $U(n) = e^\alpha - 1$. Then

$$\|U\|_\infty = |e^\alpha - 1| \leq e^{|\alpha|} - 1 \leq \ell|\alpha|,$$

where $\ell = \frac{e^\Delta - 1}{\Delta}$.

In our further consideration we will present some conditions for solvability of assignability problems of the Lyapunov spectrum for discrete-time systems. Uniform complete controllability is the first of these conditions.

Definition 7 ([10]) System (1) is called uniformly completely controllable if there exist $K \in \mathbb{N}$ and $\gamma > 0$ such that

$$W(k_0, k_0 + K) \geq \gamma I,$$

for all $k_0 \in \mathbb{N}$, where

$$W(k, n) \doteq \sum_{j=k}^{n-1} \Phi_A(k, j + 1) B(j) B^T(j) \Phi_A^T(k, j + 1)$$

is the Kalman controllability matrix.

3 Comparisons and Discussions of Assignability Problems

The next theorem presents a sufficient condition for global assignability of the Lyapunov spectrum.

Theorem 8 ([2]) *If system (1) is uniformly completely controllable, then the Lyapunov spectrum of system (3) is globally assignable.*

The next example taken from [3] shows that the global assignability of the Lyapunov spectrum does not imply in general the uniform complete controllability. Therefore, uniform complete controllability is only a sufficient, but not a necessary, condition for the global assignability.

Example 9 ([3]) Let us define a sequence $(n_k)_{k\in\mathbb{N}}$ by the recurrent formulae

$$n_1 = 1, \quad n_{2m} = mn_{2m-1}, \quad n_{2m+1} = m + n_{2m}$$

for all $m \in \mathbb{N}$, define

$$b(n) = \begin{cases} 1 \text{ for } & n = 1, \\ 1 \text{ for } n \in [n_{2m-1}, n_{2m} & -1], \\ 0 \text{ for } n \in [n_{2m}, n_{2m+1} & -1], \end{cases}$$

for $m = 2, 3, \ldots$, and consider the scalar linear control equation

$$x(n+1) = x(n) + b(n)u(n). \tag{9}$$

Equation (9) is not uniformly completely controllable. Indeed, for each $K \in \mathbb{N}$, there exists a number $m \doteq K$ such that the Kalman controllability matrix of Eq. (9) is equal to zero on the interval $[n_{2m}, n_{2m} + K]$:

$$W(n_{2m}, n_{2m} + K) = W(n_{2m}, n_{2m} + m) = W(n_{2m}, n_{2m+1}) = \sum_{j=n_{2m}}^{n_{2m+1}-1} b^2(j) = 0.$$

The closed-loop equation corresponding to Eq. (9) has the form

$$x(n+1) = \big(1 + b(n)U(n)\big)x(n), x \in \mathbb{R}, \ U \in \mathbb{R}, \ n \in \mathbb{N}. \tag{10}$$

Now, let us show that the above equation has the global assignability property of the Lyapunov spectrum. Fix any $\alpha \in \mathbb{R}$, denote $\beta = e^{\alpha} - 1$ and define $U(n) \equiv \beta$, $n \in \mathbb{N}$. The Lyapunov exponent of each nontrivial solution of Eq. (10) with the defined admissible U coincides with the upper mean value of the function $1 + \beta b(\cdot)$, that is, with the value

$$\mu \doteq \limsup_{n\to\infty} n^{-1} \sum_{j=1}^{n-1} \ln\big(1 + \beta b(j)\big).$$

Our aim is to prove that $\mu = \alpha$.

Put $\varphi(1) = 0$ and

$$\varphi(n) = \frac{1}{n} \sum_{j=1}^{n-1} \ln\bigl(1 + \beta b(j)\bigr)$$

for natural number $n > 1$.

It is clear that

$$\ln\bigl(1 + \beta b(n)\bigr) = \begin{cases} \alpha \text{ for } & n = 1, \\ \alpha \text{ for } n \in [n_{2m-1}, n_{2m} - 1], \\ 0 \text{ for } n \in [n_{2m}, n_{2m+1} - 1], \end{cases}$$

for $m = 2, 3, \ldots$.

Let $\alpha \geq 0$. Then,

$$0 \leq \ln(1 + \beta b(n)) \leq \alpha$$

and therefore, $\varphi(n) \leq \alpha$ for all $n \in \mathbb{N}$. Hence, $\mu \leq \alpha$. By the definition of the sequence $(n_k)_{k\in\mathbb{N}}$ we know that the sequence $(n_k)_{k\in\mathbb{N}}$ is strictly increasing for $k \geq 2$, tends to $+\infty$ and satisfies the relations

$$\lim_{m\to\infty} \frac{n_{2m-1}}{n_{2m}} = \lim_{m\to\infty} \frac{1}{m} = 0,$$

$$\lim_{m\to\infty} \frac{m}{n_{2m}} = \lim_{m\to\infty} \frac{1}{n_{2m-1}} = 0$$

and

$$\lim_{m\to\infty} \frac{n_{2m}}{n_{2m+1}} = \lim_{m\to\infty} \frac{1}{1 + m/n_{2m}} = 1.$$

Therefore,

$$\mu \geq \limsup_{m\to\infty} \varphi(n_{2m}) = \limsup_{m\to\infty} \frac{1}{n_{2m}} \sum_{j=1}^{n_{2m}-1} \ln\bigl(1 + \beta b(j)\bigr)$$

$$\geq \limsup_{m\to\infty} \frac{1}{n_{2m}} \sum_{j=n_{2m-1}}^{n_{2m}-1} \alpha = \alpha \lim_{m\to\infty} \frac{n_{2m} - n_{2m-1}}{n_{2m}} = \alpha.$$

Thus, $\mu = \alpha$.

Now let $\alpha \leq 0$. Then,

$$0 \geq \ln(1 + \beta b(n)) \geq \alpha$$

and therefore $0 \geq \varphi(n) \geq \alpha$ for all $n \in \mathbb{N}$. Hence, $\mu \geq \alpha$.

On the other hand, for each $k \in [n_{2m-1}, n_{2m}]$ with any natural number $m > 1$, we have

$$\begin{aligned}\varphi(k) &= \frac{1}{k}\Big(\sum_{j=1}^{n_{2m-1}-1} \ln\big(1+\beta b(j)\big) + (k-n_{2m-1})\alpha\Big)\\ &= k^{-1}\big(\varphi(n_{2m-1})n_{2m-1} + (k-n_{2m-1})\alpha\big)\\ &= \frac{n_{2m-1}}{k}\varphi(n_{2m-1}) + \alpha\frac{k-n_{2m-1}}{k} \le \varphi(n_{2m-1}).\end{aligned}$$

In addition, for $k = n_{2m}$, we obtain

$$\varphi(n_{2m}) \le \alpha\frac{n_{2m}-n_{2m-1}}{n_{2m}} = \alpha\left(1-\frac{1}{m}\right). \tag{11}$$

For each $k \in [n_{2m}, n_{2m+1}]$ with any $m \in \mathbb{N}$, we also have

$$\varphi(k) = k^{-1}\varphi(n_{2m})n_{2m} \le n_{2m+1}^{-1}\varphi(n_{2m})n_{2m} = \varphi(n_{2m+1}). \tag{12}$$

Thus, $\varphi(k) \le \varphi(n_{2m+1})$ for all $k \in [n_{2m}, n_{2m+2}]$. Moreover, from (11) and (12), we get

$$\varphi(n_{2m+1}) = \frac{n_{2m}}{n_{2m+1}}\varphi(n_{2m}) \le \alpha\frac{n_{2m}}{n_{2m+1}}\left(1-\frac{1}{m}\right),$$

so

$$\varphi(k) \le \alpha\frac{n_{2m}}{n_{2m+1}}\left(1-\frac{1}{m}\right)$$

for all $k \in [n_{2m}, n_{2m+2}]$. Note that

$$\lim_{m\to\infty} \alpha\frac{n_{2m}}{n_{2m+1}}\left(1-\frac{1}{m}\right) = \alpha.$$

Put

$$r(k) = \begin{cases} 0, & k = 1,\\ \dfrac{\alpha n_{2m}}{n_{2m+1}}\left(1-\dfrac{1}{m}\right), & k \in [n_{2m}, n_{2m+2}-1].\end{cases}$$

It is clear that $r(k) \to \alpha$ as $k \to \infty$. Since $\varphi(k) \le r(k)$ for all $k \in \mathbb{N}$, we have

$$\mu = \limsup_{k\to\infty} \varphi(k) \le \limsup_{k\to\infty} r(k) = \alpha.$$

Therefore, $\mu = \alpha$.

Thus, Eq. (10) with the defined control U has the Lyapunov spectrum consisting of α, and the Lyapunov spectrum of the Eq. (10) is globally assignable.

To present a deeper relation between global assignability and uniform complete controllability let us introduce the concept of Bebutov hull of a sequence. For any bounded sequence $F_0 = (F_0(n))_{n\in\mathbb{N}} \subset \mathbb{R}^{q\times r}$ and any $m \in \mathbb{N}$, let us consider a sequence $F_m = (F_m(n))_{n\in\mathbb{N}}$, where $F_m(n) = F_0(n+m)$ is a shift of $F_0(n)$ by m. Let us denote by $\mathfrak{R}(F_0)$ the closure in the topology of pointwise convergence on $\mathbb{N}$ of the set $\{F_m(\cdot) : m \in \mathbb{N}\}$. It is well known that $\mathfrak{R}(F_0)$ is metrizable by means of the metric

$$\varrho(F, \widehat{F}) = \sup_{n\in\mathbb{N}} \min\{\|F(n) - \widehat{F}(n)\|, n^{-1}\}.$$

The space $(\mathfrak{R}(F_0), \varrho)$ is compact [15, p. 34] and it is called the Bebutov hull of the sequence F_0 (see [11, p. 32], [16]).

Let us identify system (1) with the sequence $(A, B) = \big(A(n), B(n)\big)_{n\in\mathbb{N}} \subset \mathbb{R}^{s\times(s+t)}$. The space $\mathfrak{R}(A, B)$ will be called the Bebutov hull of system (1).

Theorem 10 ([3]) *System (1) is uniformly completely controllable if and only if for each system from $\mathfrak{R}(A, B)$ the corresponding closed-loop system has globally assignable Lyapunov spectrum.*

For a given system (1), which is not uniformly completely controllable, the problem of finding a system from $\mathfrak{R}(A, B)$ such that corresponding closed-loop system does not have assignable Lyapunov spectrum is in general a difficult task. The proof of Theorem 10 does not give a recipe to find a "bad" system from the hull, but only establishes the fact of its existence. The example below presents this "bad" system explicitly.

Example 11 ([3]) Let us consider a linear control system

$$x(n+1) = A_0(n)x(n) + B_0(n)u(n), \ x \in \mathbb{R}^2, \ u \in \mathbb{R}^2, \ n \in \mathbb{N}, \tag{13}$$

where

$$A_0(n) = I \in \mathbb{R}^{2\times 2}, \quad B_0(n) = \begin{pmatrix} 1 & 0 \\ 0 & b(n) \end{pmatrix},$$

and the sequence $b(n)$ is defined in Example 9. Since the Kalman controllability matrix of system (13) has the form

$$W_0(k, n) = \sum_{j=k}^{n-1} B_0(j)B_0^T(j) = \begin{pmatrix} 1 & 0 \\ 0 & \sum_{j=k}^{n-1} b^2(j) \end{pmatrix},$$

it follows that for each $K \in \mathbb{N}$ there exists a number $m \doteq K$ such that

$$W_0(n_{2m}, n_{2m} + K) = \begin{pmatrix} 1 & 0 \\ 0 & 0 \end{pmatrix}.$$

It means that system (13) is not uniformly completely controllable.

We will show that the hull of this system contains the system

$$x(n+1) = A_0(n)x(n) + B(n)u(n), \tag{14}$$

where

$$B(n) = \begin{pmatrix} 1 & 0 \\ 0 & 0 \end{pmatrix}.$$

In fact, let us consider the sequence $\left(n_{2m}\right)_{m\in\mathbb{N}}$ and any $n \in \mathbb{N}$. Then we have

$$\|B_{n_{2m}}(n) - B(n)\| = |b(n_{2m} + n)|.$$

For any $m > n$ the following inequalities

$$n_{2m} < n_{2m} + n < n_{2m} + m = n_{2m+1}$$

hold, therefore $b(n_{2m} + n) = 0$. It means, that

$$\lim_{m\to\infty} \|B_{n_{2m}}(n) - B(n)\| = 0$$

for any $n \in \mathbb{N}$, what implies that $\left(A_0, B\right) \in \mathfrak{R}\left(A_0, B_0\right)$.

Now, we will show that the Lyapunov spectrum of the closed-loop system

$$x(n+1) = \left(A_0 + BU(n)\right)x(n) \tag{15}$$

is not assignable. In fact, for any feedback control $U(n) = \{u_{ij}(n)\}_{i,j=1,2}$ the coefficient matrix of the closed-loop system (15) has the following form

$$A_0 + BU(n) = \begin{pmatrix} 1 + u_{11}(n) & u_{12}(n) \\ 0 & 1 \end{pmatrix}.$$

For the second coordinate $x_2(n)$ of any solution $x(n)$ of system (15) we have the equality

$$x_2(n+1) = x_2(n), \quad n \in \mathbb{N},$$

which means that the second coordinate is constant. It is clear, that every fundamental system of solutions of system (15) contains a solution with the nonzero second coordinate and for this solution we have $\lambda[x] \geq \lambda[x_2] = 0$. It means that for any admissible feedback control U the Lyapunov spectrum of system (15) contains a nonnegative number and therefore the Lyapunov spectrum of system (15) is not assignable. Moreover, the stationarity of the second coordinate of any solution of this system, when choosing the arbitrary matrix control U, implies that system (15) is not stabilizable. Thus, system (14) is the “bad” system from the hull of system (13).

To conclude the example, we show that the Lyapunov spectrum of the original system

$$x(n+1) = \big(A_0(n) + B_0(n)U(n)\big)x(n) \tag{16}$$

is assignable.

The Lyapunov spectrum of the free system

$$x(n+1) = A_0(n)x(n)$$

coincides with the sequence $(0, 0)$. Let us fix any numbers $\alpha_1 \le \alpha_2$, denote $\beta_i = e^{\alpha_i} - 1$, $i = 1, 2$, and apply to system (16) the feedback control

$$U(n) = \operatorname{diag}(\beta_1, \beta_2).$$

Then the closed-loop system (16) has the diagonal form

$$x(n+1) = \operatorname{diag}\big(1 + \beta_1, 1 + b(n)\beta_2\big)x(n), \tag{17}$$

and therefore its Lyapunov spectrum consists of upper mean values of the diagonal elements [9, p. 55], i.e. of the numbers

$$\mu_1 = \limsup_{n\to\infty} n^{-1} \sum_{j=1}^{n-1} \ln(1 + \beta_1) = \alpha_1,$$

$$\mu_2 = \limsup_{n\to\infty} n^{-1} \sum_{j=1}^{n-1} \ln\big(1 + \beta_2 b(j)\big) = \alpha_2.$$

Here the second equality follows from Example 9. It means that the spectrum of system (16) is globally assignable.

Now we will present a result about local proportional assignability of the spectrum of system (3). It will be expressed in terms of certain concepts from the asymptotic theory of linear systems, which are defined below.

Definition 12 ([9, p. 63]) System (2) is called regular (in the Lyapunov sense) if the following equality

$$\sum_{i=1}^{s} \lambda_i(A) = \liminf_{n\to\infty} \frac{1}{n} \sum_{j=1}^{n-1} \ln|\det A(j)|$$

holds.

Definition 13 (see [9, p. 100], [10, p. 15]) Let $\big(L(n)\big)_{n\in\mathbb{N}} \subset \mathbb{R}^{s\times s}$ be a Lyapunov sequence. A linear transformation

$$y = L(n)x, \quad n \in \mathbb{N}, \tag{18}$$

of the space $\mathbb{R}^s$ is called a Lyapunov transformation.

Definition 14 (*see* [10, p. 15]) We say that system (2) is dynamically equivalent to the system

$$y(n+1) = C(n)y(n), \quad n \in \mathbb{N}, \ y \in \mathbb{R}^s, \tag{19}$$

if there exists a Lyapunov transformation (18) which connects these systems, i.e. for every solution $x(n)$ of system (2) the function $y(n) = L(n)x(n)$ is a solution of system (19) and for every solution $y(n)$ of system (19) the function $x(n) = L^{-1}(n)y(n)$ is a solution of system (2).

Definition 15 System (2) is called diagonalizable if it is dynamically equivalent to a system (19) with a diagonal matrix $C = \big(C(n)\big)_{n\in\mathbb{N}}$.

Definition 16 ([5]) The Lyapunov spectrum of system (2) is called stable if for any $\varepsilon > 0$ there exists $\delta > 0$ such that for any Lyapunov sequence $R = (R(n))_{n\in\mathbb{N}} \subset \mathbb{R}^{s\times s}$ the inequality $\|R - I\|_\infty < \delta$ implies that $\lambda(AR) \in O_\varepsilon\big(\lambda(A)\big)$, where $\lambda(AR)$ is the Lyapunov spectrum of the multiplicatively perturbed system

$$z(n+1) = A(n)R(n)z(n), \quad n \in \mathbb{N}, \ z \in \mathbb{R}^s.$$

Theorem 17 ([4]) *Let system (1) be uniformly completely controllable and assume that at least one of the following conditions holds:*

1. *system (2) is regular;*
2. *system (2) is diagonalizable;*
3. *the Lyapunov spectrum of system (2) is stable.*

Then the Lyapunov spectrum of system (3) is proportionally locally assignable.

Let us compare the form of Theorems 8, 10 and 17 for time-invariant systems to the classical pole placement theorem cited in the introduction section.

The main differences are as follows:

(i) the problems are posed and solved for systems, not for a pair of matrices;
(ii) in our problem the Lyapunov spectrum is assigned, which coincides with the logarithms of the absolute values of eigenvalues of the coefficient matrix of the system, and not with the usual spectrum of this matrix as in the classical statement of the problem;
(iii) in case of Theorem 17 the assigned values of the spectrum lie in some neighborhood of the spectrum of the original system and not in the whole set of possible values of the spectrum;

(iv) in case of Theorem 17 there is a Lipschitz-type estimate of the norm of the feedback control needed to shift the spectrum by a given value from the original one;
(v) the feedback U constructed by us to assign the spectrum may depend on time;
(vi) in contrast to the classical theorem, our result provides only sufficient conditions for the assignability of the spectrum of system (3).

4 Conclusions and Open Problems

In this paper we formulated four statements of the problem of assignability of the Lyapunov spectrum for discrete linear time-varying systems: global, proportional global, local and proportional local assignability. We showed in [2] and [3] that uniform complete controllability is a sufficient but not a necessary condition for global assignability but it is a necessary and sufficient condition for global assignability of the spectrum of any system from the Bebutov hull of the original system. In [4] we also showed that diagonalizability, as well as Lyapunov regularity or stability of Lyapunov spectrum ensures the solvability of the problem of proportional local assignability for any uniformly completely controllable system. The proof of this result does not give reasons to suppose that the property of diagonalizability, regularity or stability of the Lyapunov exponents are necessary for the proportional local assignability or even close to those. The instability of the Lyapunov exponents of the original system means that the Lyapunov spectrum, considered as a function defined on the space of systems with the topology of the uniform convergence on $\mathbb{N}$, has a discontinuity at the point corresponding to the system under consideration, i.e. for arbitrarily small perturbations some of exponents may vary considerably having the so-called jumps. In this case, if the free system is neither diagonalizable nor regular, some of the corresponding control systems may not have the property of local proportional assignability of the exponents. However, even the construction of examples of such systems, not mentioning the study of the assignability of their exponents, is a difficult task that must be further investigated. The problem of the necessity of the condition of uniform complete controllability for local proportional assignability of the Lyapunov spectrum is also unsolved in general case. The problems of finding conditions for global proportional and local assignability remain open.

It is clear from the definitions that:

(1) global proportional assignability implies global assignability;
(2) local proportional assignability implies local assignability;
(3) global proportional assignability implies local proportional assignability.

The other relations between the proposed definitions of assignability are unknown.

Acknowledgements The research presented here was done by the first and third author as part of the projects funded by the National Science Centre in Poland granted according to decisions DEC-2015/19/D/ST7/03679 and DEC-2017/25/B/ST7/02888, respectively. The research of M.N.

was supported by the Polish National Agency for Academic Exchange according to the decision PPN/BEK/2018/1/00312/DEC/1. The research presented here by Banshchikova and Popova were supported by Russian Foundation for Basic Research (project no. 18–51–41005–Uzb).

References

1. Antsaklis, P., Michel, A.: Linear Systems. Birkhauser, Boston (2005)
2. Babiarz, A., Czornik, A., Makarov, E., Niezabitowski, M., Popova, S.: Pole placement theorem of discrete time-varying linear systems. SIAM J. Control Optim. **55**(2), 671–692 (2017)
3. Babiarz, A., Banshchikova, I., Czornik, A., Makarov, E., Niezabitowski, M., Popova, S.: Necessary and sufficient conditions for assignability of the Lyapunov spectrum of discrete linear time-varying systems. IEEE Trans. Autom. Control **63**(11), 3825–3837 (2018)
4. Babiarz, A., Banshchikova, I., Czornik, A., Makarov, E., Niezabitowski, M., Popova, S.: Proportional local assignability of Lyapunov spectrum of linear discrete time-varying systems. SIAM J. Control Optim., accepted for publication
5. Banshchikova, I.N., Popova, S.N.: On the spectral set of a linear discrete system with stable Lyapunov exponents. Vestnik Udmurtskogo Universiteta. Matematika. Mekhanika. Komp'yuternye Nauki **26**(1), 15–26 (2016)
6. Barreira, L., Pesin, Y.B.: Lyapunov Exponents and Smooth Ergodic Theory. American Mathematical Society, vol. 23 (2002)
7. Dickinson, B.: On the fundamental theorem of linear state variable feedback. IEEE Trans. Autom. Control **19**(5), 577–579 (1974)
8. Elaydi, S.N.: An Introduction to Difference Equations. Springer, New York (2005)
9. Gaishun, I.: Discrete-Time Systems. Natsionalnaya Akademiya Nauk Belarusi, Institut Matematiki, Minsk (2001)
10. Halanay, A., Ionescu, V.: Time-Varying Discrete Linear Systems: Input-Output Operators, Riccati Equations, Disturbance Attenuation. Springer, Berlin (1994)
11. Johnson, R., Obaya, R., Novo, S., Núñez, G., Fabbri, R.: Nonautonomous Linear Hamiltonian Systems: Oscillation, Spectral Theory and Control. Springer, Berlin (2016)
12. Klamka, J.: Controllability of Dynamical Systems. Kluwer Academic Publishers, Dordrecht (1991)
13. Ludyk, G.: Stability of Time-variant Discrete-Time Systems. Springer Fachmedien Wiesbaden GmbH, Bremen (1985)
14. Makarov, E.K., Popova, S.N.: Controllability of Asymptotic Invariants of Time-Dependent Linear Systems. Belorusskaya nauka, Minsk (2012)
15. Sell, G.R.: Topological Dynamics and Ordinary Differential Equations. Van Nostrand Reinhold Mathematical Studies (1971)
16. Sell, G.R.: The Floquet problem for almost periodic linear differential equations. Ordinary and Partial Differential Equations, vol. 239–251. Springer, Berlin (1974)
17. Sontag, E.D.: Mathematical Control Theory: Deterministic Finite Dimensional Systems, vol. 6. Springer Science & Business Media (2013)

Strongly Exponentially Separated Linear Difference Equations

Flaviano Battelli and Kenneth J. Palmer

Abstract In the study of linear differential systems, an important concept is that of exponential separation. In a previous paper, we have studied this concept for differential equations. Here we develop the theory for difference equations. Our first aim is to develop a theory which applies to unbounded systems. It turns that in order to have a reasonable theory it is necessary to add the assumption that the angle between the two separated subspaces is bounded below (note this follows automatically for bounded systems). Our second aim is to show that if a bounded linear symplectic system is exponentially separated into two subspaces of the same dimension, then it must have an exponential dichotomy. The theory follows the same lines as the differential equation case with one important difference: for the roughness theorem a different kind of perturbation is needed.

Keywords Linear difference equations · Exponential separation · Exponential dichotomy · Symplectic

1 Introduction

In this paper we study linear difference systems

$$Nowweconsidex(k+1) = A(k)x(k), \quad x \in \mathbb{R}^n \tag{1}$$

where $A(k)$ is an invertible matrix for each k. In the study of such systems, an important concept is that of exponential separation. It is closely related to the concept

F. Battelli
Department of Industrial Engineering and Mathematical Sciences, Marche Polytechnic University, Ancona, Italy
e-mail: f.battelli@univpm.it

K. J. Palmer (✉)
Department of Mathematics, National Taiwan University, Taipei, Taiwan
e-mail: palmer@math.ntu.edu.tw

M. Bohner et al. (eds.), *Difference Equations and Discrete Dynamical Systems with Applications*, Springer Proceedings in Mathematics & Statistics 312,
https://doi.org/10.1007/978-3-030-35502-9_6

of exponential dichotomy. In this paper we restrict attention to the case where the solution space is split into two exponentially separated subspaces.

Exponential separation for difference equations was first studied in [9, 10]. There it is assumed that the coefficient matrixNow we conside $A(k)$ is bounded in norm together with its inverse. Our first aim here is to develop a theory of exponential separation which applies to unbounded systems. It turns out that in order to have a reasonable theory it is necessary to add the assumption that the angle between the two separated subspaces is bounded below (note this follows automatically for bounded systems). Our second aim is to show that if a bounded linear system (1), where the $A(k)$ is symplectic, is exponentially separated into two subspaces of the same dimension, then it must have an exponential dichotomy. Note that we developed a similar theory for linear differential equations in [2]. Most of the results for invertible difference equations are analogous but there is one important difference which we mention below. Let us remark here that it was already observed in [4] that in the case of exponential separation for differential equations, an additional angle condition is needed for unbounded systems.

Now we summarize the contents of the paper. In Sect. 2, we introduce the basic definitions and examine to what extent the separated subspaces are unique. Then if a system is separated on both half-axes, we describe what must be added to ensure separation on the whole axis. In this section we do not need the additional condition on the angle. In Sect. 3 we introduce the concept of strong exponential separation, which is exponential separation plus the condition that the angle between the separated subspaces be bounded below. Then in Proposition 3, we derive a convenient necessary and sufficient condition for strong exponential separation and use it to show that the condition of strong exponential separation is preserved by the operation of taking the inverse adjoint of a system and that it is implied by exponential dichotomy. In Sect. 4, we mention the result in [9] which shows that when $A(k)$ and its inverse are bounded then exponential separation is equivalent to exponential dichotomy of a shifted equation. In Sect. 5 we show strong exponential separation is robust under small perturbation of the coefficient matrix. In fact, here we see a major difference from the theorem for differential equations and also a difference from the roughness theorem for exponential dichotomy in difference equations. Next in Sect. 6 we study block upper triangular systems. First, we show that if the block upper triangular system is strongly exponentially separated, then the corresponding block diagonal system is strongly exponentially separated. Here no boundedness assumptions are needed. Then, using the perturbation theorem in Sect. 5, we show that if the corresponding block diagonal system is strongly exponentially separated and the off-diagonal blocks are bounded in a certain sense, then the block upper triangular system is strongly exponentially separated. Here again there is a difference from the result for differential equations. Finally in Sect. 7, we show that if a bounded linear symplectic system is exponentially separated into two subspaces of the same dimension, then it must have an exponential dichotomy. An important tool here are the results in Sect. 6 about block upper triangular systems.

2 Exponential Separation in Bounded and Unbounded Systems: General Properties

In this section, we define exponential separation and examine to what extent the separated subspaces are unique. Then if a system is separated on both half-axes (that is, $\mathbb{Z}_+$ or $\mathbb{Z}_-$), we describe what must be added to ensure separation on the whole axis (that is, $\mathbb{Z}$). Note in the case of $\mathbb{Z}_-$, (1) holds for $k \in (-\infty, -1]$ but the solutions are defined for $k \in \mathbb{Z}_-$. However in the sequel we use $\mathbb{Z}_-$ in both cases where, for (1), it is to be understood that $\mathbb{Z}_-$ means $(-\infty, -1]$.

Definition 1 We say system (1) is *exponentially separated* on an infinite interval J of integers if there are nonzero invariant subspaces $V_1(k) \oplus V_2(k) = \mathbb{R}^n$ and positive constants $K \geq 1$ and α such that if $x(k)$ is a nonzero solution in $V_1(k)$ and $y(k)$ a nonzero solution in $V_2(k)$, then

$$\frac{|x(k)|\,|y(m)|}{|x(m)|\,|y(k)|} \leq Ke^{-\alpha(k-m)}, \quad k \geq m \text{ in } J.$$

$V_1(k)$ is called the *stable subspace* and $V_2(k)$ the *unstable subspace*. If we denote by $P(k)$ (note that $P(k) \neq 0, \mathbf{I}$) the projection with range $V_1(k)$ and nullspace $V_2(k)$, then this can be written as

$$|\Phi(k,m)\xi|\,|\Phi(m,k)\eta| \leq Ke^{-\alpha(k-m)}|\xi|\,|\eta|, \quad k \geq m \text{ in } J \tag{2}$$

for all $\xi \in \mathcal{R}P(m)$ and all $\eta \in \mathcal{N}P(k)$, where $\Phi(k,m)$ is the transition matrix. If $P(k)$ has rank r $(1 \leq r \leq n-1)$, or equivalently $\dim V_1(k) = r$, we say that (1) is exponentially separated with rank r. Note that $P(k)$ has the invariance property $A(k)P(k) = P(k+1)A(k)$ for all k.

Remark 1 It is easy to see that (1) is exponentially separated on $\mathbb{Z}_+$ with subspaces $V_1(k)$, $V_2(k)$ if and only if $x(k+1) = A^{-1}(-k-1)x(k)$ is exponentially separated on $\mathbb{Z}_-$ with subspaces $V_2(-k)$, $V_1(-k)$.

First we make a simple but useful observation.

Proposition 1 *If* (1) *is exponentially separated on* $[T, \infty)$ *for some* $T > 0$*, it is exponentially separated on* $\mathbb{Z}_+$.

Proof Let (1) be exponentially separated on $[T, \infty)$ with projection $P(k)$ and constants K, α. Then

$$|\Phi(k,m)\xi|\,|\Phi(m,k)\eta| \leq Ke^{-\alpha(k-m)}|\xi|\,|\eta|, \quad k \geq m \geq T \tag{3}$$

for all $\xi \in \mathcal{R}P(m)$ and all $\eta \in \mathcal{N}P(k)$.

There exists M such that $|A(k)|, |A^{-1}(k)| \le M$ for $0 \le k \le T-1$. Then

$$|\Phi(k,m)| \le M^{|k-m|} \quad \text{for} \quad 0 \le k, m \le T$$

so that if $T \ge k \ge m \ge 0$ and $\xi \in \mathcal{R}P(m)$, $\eta \in \mathcal{N}P(k)$,

$$|\Phi(k,m)\xi|\,|\Phi(m,k)\eta| \le M^{2T} e^{\alpha T} e^{-\alpha(k-m)} |\xi|\,|\eta|. \tag{4}$$

Next if $k \ge T \ge m \ge 0$ and $\xi \ne 0$ in $\mathcal{R}P(m)$, $\eta \ne 0$ in $\mathcal{N}P(k)$,

$$\begin{aligned}
&\frac{|\Phi(k,m)\xi|\,|\Phi(m,k)\eta|}{|\xi|\,|\eta|} \\
&= \frac{|\Phi(k,T)\xi_1|\,|\Phi(m,T)\eta_1|}{|\xi|\,|\eta|}, \quad \xi_1 = \Phi(T,m)\xi \in \mathcal{R}P(T),\ \eta_1 = \Phi(T,k)\eta \in \mathcal{N}P(T) \\
&= \frac{|\Phi(k,T)\xi_1|}{|\xi_1|}\frac{|\Phi(T,k)\eta|}{|\eta|} \times \frac{|\Phi(T,m)\xi|}{|\xi|}\frac{|\Phi(m,T)\eta_1|}{|\eta_1|} \\
&\le Ke^{-\alpha(k-T)} \times M^{2T} e^{\alpha T} e^{-\alpha(T-m)}, \quad \text{using} \quad (3), (4) \\
&= KM^{2T} e^{\alpha T} e^{-\alpha(k-m)}.
\end{aligned}$$

Together with (3) and (4), this proves the proposition.

Now we examine the extent to which exponentially separated subspaces are unique.

Proposition 2 *For exponentially separated systems on $\mathbb{Z}_+$ the stable subspace is uniquely defined (for a given dimension) and for exponentially separated systems on $\mathbb{Z}_-$ the unstable subspace is uniquely defined. The other subspace can be any complement.*

Proof Consider first $\mathbb{Z}_+$. So we are assuming there is an invariant projection $P(k) \ne 0, \mathbf{I}$ and positive constants K and α such that (2) holds. Let $Q(k)$ be another invariant projection with the same range as $P(k)$.

Suppose $\xi \in \mathcal{R}(Q(m)) = \mathcal{R}(P(m))$ and $\eta \ne 0 \in \mathcal{N}(Q(k))$. Note that $(\mathbf{I} - P(k))\eta \ne 0$, since otherwise $\eta \in \mathcal{R}(P(k)) = \mathcal{R}(Q(k))$ and therefore would be 0 since it is in $\mathcal{N}(Q(k))$ also. Then if $0 \le m \le k$,

$$\begin{aligned}
&|\Phi(k,m)\xi|\,|\Phi(m,k)\eta| \\
&= |\Phi(k,m)\xi|\,|\Phi(m,k)(\mathbf{I} - P(k))\eta| \frac{|\Phi(m,k)(\mathbf{I} - Q(k))\eta|}{|\Phi(m,k)(\mathbf{I} - P(k))\eta|} \\
&\le Ke^{-\alpha(k-m)} |\xi|\,|(\mathbf{I} - P(k))\eta| \frac{|\Phi(m,k)(\mathbf{I} - Q(k))\eta|}{|\Phi(m,k)(\mathbf{I} - P(k))\eta|} \\
&= Ke^{-\alpha(k-m)} |\xi|\,|\eta| \frac{|\Phi(m,k)(\mathbf{I} - Q(k))\eta|\,|(\mathbf{I} - P(k))\eta|}{|\Phi(m,k)(\mathbf{I} - P(k))\eta|\,|(\mathbf{I} - Q(k))\eta|}.
\end{aligned} \tag{5}$$

Now

$$\begin{aligned}\frac{|\Phi(m,k)(\mathbf{I}-Q(k))\eta|}{|\Phi(m,k)(\mathbf{I}-P(k))\eta|} &= \frac{|\Phi(m,0)(\mathbf{I}-Q(0))\Phi(0,k)\eta|}{|\Phi(m,0)(\mathbf{I}-P(0))\Phi(0,k)\eta|}\\ &\leq 1+\frac{|\Phi(m,0)(P(0)-Q(0))\Phi(0,k)\eta|}{|\Phi(m,0)(\mathbf{I}-P(0))\Phi(0,k)\eta|}\\ &\leq 1+Ke^{-\alpha m}\frac{|(P(0)-Q(0))\Phi(0,k)\eta|}{|(\mathbf{I}-P(0))\Phi(0,k)\eta|}\\ &\leq 1+Ke^{-\alpha m}N,\end{aligned}$$

since $(P(0)-Q(0))\Phi(0,k)\eta \in \mathcal{R}(P(0))$, $(\mathbf{I}-P(0))\Phi(0,k)\eta \in \mathcal{N}(P(0))$ and where

$$N = \sup_{\eta\in\mathcal{N}Q(0),|\eta|=1} \frac{|(P(0)-Q(0))\eta|}{|(\mathbf{I}-P(0))\eta|}.$$

Similarly,

$$\begin{aligned}\frac{|(\mathbf{I}-Q(k))\eta|}{|(\mathbf{I}-P(k))\eta|} &= \frac{|\Phi(k,0)(\mathbf{I}-Q(0))\Phi(0,k)\eta|}{|\Phi(k,0)(\mathbf{I}-P(0))\Phi(0,k)\eta|}\\ &\geq 1-\frac{|\Phi(k,0)(P(0)-Q(0))\Phi(0,k)\eta|}{|\Phi(k,0)(\mathbf{I}-P(0))\Phi(0,k)\eta|}\\ &\geq 1-Ke^{-\alpha k}\frac{|(P(0)-Q(0))\Phi(0,k)\eta|}{|(\mathbf{I}-P(0))\Phi(0,k)\eta|}\\ &\geq 1-Ke^{-\alpha k}N.\end{aligned}$$

So if $k \geq m \geq T = \alpha^{-1}\log(3KN)$,

$$|\Phi(k,m)\xi|\,|\Phi(m,k)\eta| \leq Ke^{-\alpha(k-m)}|\xi|\,|\eta|\frac{1+1/3}{1-1/3} = 2Ke^{-\alpha(k-m)}|\xi|\,|\eta|.$$

Then we get the exponential separation on $\mathbb{Z}_+$ using Proposition 1. Thus $Q(k)$ can also be used as a projection and hence any invariant complement of $\mathcal{R}(P(k))$ can be taken as the unstable subspace.

Now we prove the uniqueness of the stable subspace. Suppose $R(k)$ is an invariant projection with different range from $P(k)$ but with the same rank, with respect to which the system is exponentially separated. Let p be a vector which is in the range of $R(0)$ but not in the range of $P(0)$ and q a vector which is in the range of $P(0)$ but not in the range of $R(0)$. We can take $Q_1(k)$ as an invariant projection with the same range as $P(k)$ with p in the nullspace of $Q_1(0)$ and $Q_2(k)$ as an invariant projection with the same range as $R(k)$ with q in the nullspace of $Q_2(0)$. Now there exist positive constants K_i and α_i such that for $k \geq m \geq 0$,

$$|\Phi(k,m)\xi|\,|\Phi(m,k)\eta| \leq K_1e^{-\alpha_1(k-m)}|\xi|\,|\eta|$$

for all $\xi \neq 0$ in $\mathcal{R}Q_1(m)$ and all $\eta \neq 0$ in $\mathcal{N}Q_1(k)$ and

$$|\Phi(k,m)\xi|\,|\Phi(m,k)\eta| \leq K_2 e^{-\alpha_2(k-m)}|\xi|\,|\eta|$$

for all $\xi \neq 0$ in $\mathcal{R}Q_2(m)$ and all $\eta \neq 0$ in $\mathcal{N}Q_2(k)$. Since q is in the range of $Q_1(0)$ and $\Phi(k,0)p$ is in the nullspace of $Q_1(k)$, it follows that for $k \geq 0$,

$$|\Phi(k,0)q|\,|\Phi(0,k)\Phi(k,0)p| \leq K_1 e^{-\alpha_1 k}|q|\,|\Phi(k,0)p|,$$

and since p is in the range of $Q_2(0)$ and $\Phi(k,0)q$ in the nullspace of $Q_2(k)$, it follows that for $k \geq 0$,

$$|\Phi(k,0)p|\,|\Phi(0,k)\Phi(k,0)q| \leq K_2 e^{-\alpha_2 k}|p|\,|\Phi(k,0)q|.$$

Then, combining these inequalities,

$$|p|\,|\Phi(k,0)q| \leq K_1 e^{-\alpha_1 k} K_2 e^{-\alpha_2 k}|p|\,|\Phi(k,0)q|$$

so that for $k \geq 0$

$$1 \leq K_1 e^{-\alpha_1 k} K_2 e^{-\alpha_2 k},$$

clearly impossible. Thus the stable subspace is unique.

The proof of the Proposition for $\mathbb{Z}_-$ follows using Remark 1.

In the following corollary, we show what additional conditions are needed to ensure that a system which is exponentially separated on both half-axes is also exponentially separated on the whole axis.

Corollary 1 *System (1) is exponentially separated on $\mathbb{Z}$ if and only if it is exponentially separated on $\mathbb{Z}_+$ and $\mathbb{Z}_-$, the respective ranks are the same and the stable subspace on $\mathbb{Z}_+$ and the unstable subspace on $\mathbb{Z}_-$ intersect in $\{0\}$ at $k = 0$.*

Proof Clearly the conditions are necessary.

For the sufficiency, suppose (1) is exponentially separated on $\mathbb{Z}_+$ and $\mathbb{Z}_-$, the respective ranks are the same and the stable subspace on $\mathbb{Z}_+$ and the unstable subspace on $\mathbb{Z}_-$ intersect in $\{0\}$ at $k = 0$. According to Proposition 2, at $k = 0$, we can take the unstable subspace on $\mathbb{Z}_+$ to be the unstable subspace on $\mathbb{Z}_-$ and the stable subspace on $\mathbb{Z}_-$ to be the stable subspace on $\mathbb{Z}_+$. This means we have the same invariant projection $P(k)$ on both $\mathbb{Z}_+$ and $\mathbb{Z}_-$. Then there exist positive constants K and α such that if $\xi \in \mathcal{R}(P(m))$ and $\eta \in \mathcal{N}(P(k))$, we have

$$|\Phi(k,m)\xi|\,|\Phi(m,k)\eta| \leq K e^{-\alpha(k-m)}|\xi|\,|\eta|$$

for $k \geq m \geq 0$ and $0 \geq k \geq m$. Next if $k \geq 0 \geq m$,

$$\begin{aligned}
|\Phi(k,m)\xi|\,|\Phi(m,k)\eta| &= \frac{|\Phi(k,m)\xi|\,|\Phi(0,k)\eta|}{|\Phi(0,m)\xi|}\,\frac{|\Phi(0,m)\xi|\,|\Phi(m,k)\eta|}{|\Phi(0,k)\eta|} \\
&= \frac{|\Phi(k,0)\Phi(0,m)\xi|\,|\Phi(0,k)\eta|}{|\Phi(0,m)\xi|}\,\frac{|\Phi(0,m)\xi|\,|\Phi(m,0)\Phi(0,k)\eta|}{|\Phi(0,k)\eta|} \\
&\leq \frac{Ke^{-\alpha k}|\Phi(0,m)\xi|\,|\eta|}{|\Phi(0,m)\xi|}\,\frac{Ke^{\alpha m}|\xi|\,|\Phi(0,k)\eta|}{|\Phi(0,k)\eta|} \\
&= K^2 e^{-\alpha(k-m)}|\xi|\,|\eta|.
\end{aligned}$$

It follows that (1) is exponentially separated on $\mathbb{Z}$.

3 Strongly Exponentially Separated Systems

In this section we introduce the definition of strong exponential separation. Then we derive a simple necessary and sufficient condition for strong exponential separation and we use it to show that strong exponential separation is preserved by the operation of taking adjoints and also that exponential dichotomy implies strong exponential separation.

First we recall the definitions of kinematic similarity and reducibility.

Definition 2 Systems (1) and $y(k+1) = B(k)y(k)$ are *kinematically similar* if there exists a bounded, invertible matrix function $S(k)$ with bounded inverse such that the transformation $x = S(k)y$ takes (1) into $y(k+1) = B(k)y(k)$, where $B(k) = S^{-1}(k+1)A(k)S(k)$. We refer to the transformation $x = S(k)y$ as a *kinematic similarity*.

Definition 3 System (1) is *reducible* if it is kinematically similar to a block diagonal system

$$y(k+1) = \begin{pmatrix} A_1(k) & 0 \\ 0 & A_2(k) \end{pmatrix} y(k).$$

The following proposition follows from [3] (see also Lemma 1.5.4 in [11]). Note that a projection $P(k)$ is *invariant* for (1) if $A(k)P(k) = P(k+1)A(k)$ for all k.

Proposition 3 *System (1) is reducible if and only if* (1) *has a bounded invariant projection* $P(k) \neq 0, \mathbf{I}$.

Remark 2 Note that the boundedness of $P(k)$ is equivalent to the angle between the range and nullspace of $P(k)$ being bounded below by a positive number.

Remark 3 Here we prove the sufficiency in Proposition 3. If $P(k)$ is invariant and bounded, it follows from [3] that there is a kinematic similarity $S(k)$ such that $P(k) = S(k)PS^{-1}(k)$ with $P = \begin{pmatrix} \mathbf{I} & 0 \\ 0 & 0 \end{pmatrix}$. Then the transformation $x = S(k)y$ takes (1) into a system $y(k+1) = B(k)y(k)$, where

$$\begin{aligned}
B(k)P &= B(k)S^{-1}(k)P(k)S(k) \\
&= S^{-1}(k+1)A(k)P(k)S(k) \\
&= S^{-1}(k+1)P(k+1)A(k)S(k) \quad \text{by invariance of } P(k) \\
&= S^{-1}(k+1)P(k+1)S(k+1)B(k) \\
&= PB(k).
\end{aligned}$$

So the transformed system has the form

$$y(k+1) = \begin{pmatrix} A_1(k) & 0 \\ 0 & A_2(k) \end{pmatrix} y,$$

and the projection corresponding to $P(k)$ is $S^{-1}(k)P(k)S(k) = P$.

Remark 4 Analogously to Lemma 1 in [7], it can be proved that if $A(k)$ and $A^{-1}(k)$ are bounded and (1) is exponentially separated with corresponding projection $P(k)$, then $P(k)$ is bounded so that (1) is reducible. The example below shows that an unbounded exponentially separated system need not be reducible, in contrast to the case of bounded systems.

Example 1

$$x(k+1) = \begin{pmatrix} e & (e^2-1)e^{k+1} \\ 0 & e^2 \end{pmatrix} x(k). \tag{6}$$

This has the two solutions

$$x(k) = (e^k, 0), \quad y(k) = (e^{3k}, e^{2k}).$$

Using the maximum norm in $\mathbb{R}^2$, we see that $|x(k)| = e^k$ and $|y(k)| = e^{3k}$ for all $k \geq 0$. Then if $k \geq m \geq 0$

$$\frac{|x(k)|\,|y(m)|}{|x(m)|\,|y(k)|} = \frac{e^k\, e^{3m}}{e^m\, e^{3k}} = e^{-2(k-m)}$$

so that the system is exponentially separated on $\mathbb{Z}_+$. Now suppose (6) is reducible. Then it follows that there exists a bounded invariant projection $P(k)$ of rank 1. By direct calculation, it can be shown that there is no such $P(k)$. Indeed if

$$P(0) = \begin{pmatrix} a & b \\ c & d \end{pmatrix}, \quad P^2(0) = P(0),$$

then with

$$\Phi(k,0) = \begin{pmatrix} e^k & e^{3k} - e^k \\ 0 & e^{2k} \end{pmatrix},$$

we find that the (2, 1) entry in $P(k)$, where by invariance $P(k) = \Phi(k,0)P(0)\Phi(0,k)$, is ce^k. So, if $P(k)$ is bounded, then $c = 0$. Next we find that the (1, 2) entry in $P(k)$ is

$(d - a)e^k + (a + b - d)e^{-k}$. So $a = d$ and it follows that $P(0) = 0$ or the identity and hence cannot have rank 1.

Now we give the definition of strong exponential separation.

Definition 4 System (1) is said to be *strongly exponentially separated* if it is exponentially separated with corresponding subspaces $V_1(k) \oplus V_2(k) = \mathbb{R}^n$ where the angle between $V_1(k)$ and $V_2(k)$ is bounded below by a positive number, or equivalently the projection $P(k)$ on to $V_1(k)$ along $V_2(k)$ is bounded.

Remark 5 Kinematic similarity preserves both exponential separation and strong exponential separation. Indeed, suppose the kinematic similarity $x = S(k)y$ takes (1) into the system $y(k+1) = B(k)y(k)$. Moreover suppose (1) is exponentially separated on an interval J with projection $P(k)$ so that

$$|\Phi(k, m)\xi|\,|\Phi(m, k)\eta| \le Ke^{-\alpha(k-m)}|\xi|\,|\eta|, \quad k \ge m \text{ in } J$$

for all $\xi \in \mathcal{R}P(m)$ and all $\eta \in \mathcal{N}P(k)$. The transition matrix for $y(k+1) = B(k)y(k)$ is $\Psi(k, m) = S^{-1}(k)\Phi(k, m)S(m)$. So if we define the projection $Q(k) = S^{-1}(k)P(k)S(k)$, then $Q(k)$ is invariant with respect to $y(k+1) = B(k)y(k)$ and if $\xi \in \mathcal{R}Q(m)$ and $\eta \in \mathcal{N}Q(k)$, then $S(m)\xi \in \mathcal{R}P(m)$ and $S(k)\eta \in \mathcal{N}P(k)$ so that if $k \ge m$ in J,

$$\begin{aligned}
|\Psi(k, m)\xi|\,|\Psi(m, k)\eta| &= |S^{-1}(k)\Phi(k, m)S(m)\xi|\,|S^{-1}(m)\Phi(m, k)S(k)\eta| \\
&\le N^2|\Phi(k, m)S(m)\xi|\,|\Phi(m, k)S(k)\eta|, \quad \text{with} \quad N = \sup|S^{-1}(k)| \\
&\le N^2Ke^{-\alpha(k-m)}|S(m)\xi|\,|S(k)\eta| \\
&\le M^2N^2Ke^{-\alpha(k-m)}|\xi|\,|\eta|, \quad \text{where} \quad M = \sup|S(k)|.
\end{aligned}$$

Moreover, if $P(k)$ is bounded then $Q(k)$ is bounded also.

Remark 6 In view of the remark before the example, if $A(k)$ and $A^{-1}(k)$ are bounded, exponential separation implies strong exponential separation.

Now we derive a simple criterion for strong exponential separation which is similar to one given in [5] and was proved for the bounded case in [9].

Proposition 4 *A system $x(k+1) = A(k)x(k)$ is strongly exponentially separated on an infinite interval J with projection $P(k)$ if and only if there exist positive constants K and α such that*

$$|\Phi(k, m)P(m)|\,|\Phi(m, k)(\mathbf{I} - P(k))| \le Ke^{-\alpha(k-m)}, \quad m \le k \in J. \tag{7}$$

Proof To prove the sufficiency, note that (7) with $k = m$ implies that

$$|P(k)|\,|\mathbf{I} - P(k))| \le K, \quad k \in J.$$

Since $\mathbf{I} - P(k)$ is a nonzero projection, $|\mathbf{I} - P(k)| \geq 1$ and so

$$|P(k)| \leq K, \quad k \in J.$$

Next observe that if $\xi \in \mathcal{R}P(m)$ and $\eta \in \mathcal{N}P(k)$

$$\begin{aligned}|\Phi(k,m)\xi|\,|\Phi(m,k)\eta| &= |\Phi(k,m)P(m)\xi|\,|\Phi(m,k)(\mathbf{I} - P(k))\eta| \\ &\leq |\Phi(k,m)P(m)|\,|\xi|\,|\Phi(m,k)(\mathbf{I} - P(k))|\,|\eta| \\ &\leq Ke^{-\alpha(k-m)}|\xi|\,|\eta|.\end{aligned}$$

for $k \geq m$. So the sufficiency is proved.

Now we prove the necessity. We are supposing that there are positive constants K and α such that if $\xi \in \mathcal{R}P(m)$ and $\eta \in \mathcal{N}P(k)$, then for $k \geq m$

$$|\Phi(k,m)\xi|\,|\Phi(m,k)\eta| \leq Ke^{-\alpha(k-m)}|\xi|\,|\eta|, \quad |P(k)| \leq K.$$

Then for all ξ and η,

$$\begin{aligned}|\Phi(k,m)P(m)\xi|\,|\Phi(m,k)(\mathbf{I} - P(k))\eta| &\leq Ke^{-\alpha(k-m)}|P(m)\xi|\,|(\mathbf{I} - P(k))\eta| \\ &\leq Ke^{-\alpha(k-m)}|P(m)|\,|\xi|\,|(\mathbf{I} - P(k))|\,|\eta| \\ &\leq K^2(1+K)e^{-\alpha(k-m)}|\xi|\,|\eta|.\end{aligned}$$

Hence

$$|\Phi(k,m)P(m)|\,|\Phi(m,k)(\mathbf{I} - P(k))| \leq K^2(1+K)e^{-\alpha(k-m)}, \quad k \geq m.$$

We use the criterion just derived to show that strong exponential separation is preserved by the operation of taking adjoints.

Corollary 2 *If a system $x(k+1) = A(k)x(k)$ is strongly exponentially separated on an interval J with projection $P(k)$, then so also is its adjoint $x(k+1) = [A^*(k)]^{-1}x(k)$ with projection $\mathbf{I} - P^*(k)$.*

Proof By Proposition 4, there exist an invariant projection $P(k)$ and positive constants K and α such that

$$|\Phi(k,m)P(m)|\,|\Phi(m,k)(\mathbf{I} - P(k))| \leq Ke^{-\alpha(k-m)}, \quad k \geq m,$$

which, using invariance $P(k)\Phi(k,m) = \Phi(k,m)P(m)$, can be rewritten as

$$|P(k)\Phi(k,m)|\,|(\mathbf{I} - P(m))\Phi(m,k)| \leq Ke^{-\alpha(k-m)}, \quad k \geq m.$$

Taking adjoints and using the Euclidean norm, we get

$$|\Phi^*(k,m)P^*(k)|\,|\Phi^*(m,k)(\mathbf{I} - P^*(m))| \leq Ke^{-\alpha(k-m)}, \quad k \geq m.$$

Now the transition matrix for the adjoint system is $\Psi(k, m) = \Phi^*(m, k)$ and so we have

$$|\Psi(k, m)(\mathbf{I} - P^*(m))|\,|\Psi(m, k)P^*(k)| \leq Ke^{-\alpha(k-m)}, \quad k \geq m.$$

Thus the adjoint system is strongly exponentially separated with projection $\mathbf{I} - P^*(k)$.

Remark 7 In 2 dimensions this also holds without "strong". For let $x(k) = (x_1(k), x_2(k))^T$ and $y(k) = (y_1(k), y_2(k))^T$ be exponentially separated solutions. Set $X(k) = (x(k), y(k))$. Then $X^*(k)^{-1}$ is a matrix solution of the adjoint system. But

$$X^*(k)^{-1} = (x_1(k)y_2(k) - x_2(k)y_1(k))^{-1}\begin{pmatrix} y_2(k) & -x_2(k) \\ -y_1(k) & x_1(k) \end{pmatrix}.$$

We see that the columns of this are exponentially separated solutions for the adjoint system. However this does not extend to higher dimensions. We give an example of an exponentially separated discrete equation in three dimensions for which the corollary about the adjoint system does not hold.

The system is

$$x(k+1) = \begin{pmatrix} e^3 & 0 & (1-e)e^{k+2} \\ 0 & e^2 & e(1-e) \\ 0 & 0 & e \end{pmatrix} x(k), \quad k \geq 0, \tag{8}$$

for which the transition matrix is

$$\Phi(k, 0) = \begin{pmatrix} e^{3k} & 0 & e^{2k}(1-e^k) \\ 0 & e^{2k} & e^k(1-e^k) \\ 0 & 0 & e^k \end{pmatrix}.$$

If P is the projection

$$P := \begin{pmatrix} 0 & 0 & 1 \\ 0 & 1 & 0 \\ 0 & 0 & 1 \end{pmatrix},$$

then

$$\Phi(k, 0)P = \begin{pmatrix} 0 & 0 & e^{2k} \\ 0 & e^{2k} & e^k(1-e^k) \\ 0 & 0 & e^k \end{pmatrix}, \quad \Phi(k, 0)(\mathbf{I} - P) = \begin{pmatrix} e^{3k} & 0 & -e^{3k} \\ 0 & 0 & 0 \\ 0 & 0 & 0 \end{pmatrix}.$$

The space spanned by the columns of $\Phi(k, 0)P$ is the

$$\text{span}\{x_1(k), x_2(k)\} \text{ with } x_1(k) = e^{2k}\begin{pmatrix} 0 \\ 1 \\ 0 \end{pmatrix} \text{ and } x_2(k) = e^k\begin{pmatrix} e^k \\ 1 \\ 1 \end{pmatrix}.$$

With the ℓ^1-norm we have

$$|x_1(k)| = e^{2k} \qquad e^{2k} \le |x_2(k)| \le 3e^{2k}.$$

Next the space spanned by the columns of $\Phi(k,0)(\mathbf{I}-P)$ is

$$\text{span}\{y(k)\} \text{ with } y(k) = e^{3k}\begin{pmatrix}1\\0\\0\end{pmatrix}.$$

Let $x(k) = ax_1(k) + bx_2(k)$ with $a^2+b^2 \neq 0$. Then $(|a|+|b|)e^{2k} \le |x(k)| \le (|a| + 3|b|)e^{2k}$. Then for $k \ge 0$, $m \ge 0$,

$$\frac{|x(k)|}{|x(m)|} \le \frac{|a|+3|b|}{|a|+|b|}e^{2(k-m)}.$$

Next

$$\frac{|y(m)|}{|y(k)|} = e^{3(m-k)}.$$

Hence for $k \ge 0$, $m \ge 0$,

$$\frac{|x(k)|}{|x(m)|}\frac{|y(m)|}{|y(k)|} \le \frac{|a|+3|b|}{|a|+|b|}e^{-(k-m)} \le 3e^{-(k-m)}.$$

As a consequence the discrete system (8) is exponentially separated on $\mathbb{Z}_+$ with projection

$$P(k) = \Phi(k,0)P\Phi(0,k) = \begin{pmatrix}0 & 0 & e^k(e^k+1)\\ 0 & 1 & 0\\ 0 & 0 & 1\end{pmatrix},$$

which is not bounded so that (8) is not strongly exponentially separated.

We prove that the adjoint system is not exponentially separated with projection (at $k=0$)

$$\mathbf{I} - P^*(0) = \mathbf{I} - P^* = \begin{pmatrix}1 & 0 & 0\\ 0 & 0 & 0\\ -1 & 0 & 0\end{pmatrix}.$$

If we denote by $\Psi(k,m)$ the transition matrix for the adjoint system, then

$$\Psi(k,0) = \Phi^*(0,k) = \begin{pmatrix}e^{-3k} & 0 & 0\\ 0 & e^{-2k} & 0\\ e^{-k}-e^{-2k} & e^{-k}-e^{-2k} & e^{-k}\end{pmatrix}$$

so that

$$\Psi(k,0)P^* = \begin{pmatrix} 0 & 0 & 0 \\ 0 & e^{-2k} & 0 \\ e^{-k} & e^{-k} - e^{-2k} & e^{-k} \end{pmatrix}, \quad \Psi(k,0)(\mathbf{I} - P^*) = \begin{pmatrix} e^{-3k} & 0 & 0 \\ 0 & 0 & 0 \\ -e^{-2k} & 0 & 0 \end{pmatrix}.$$

Thus the stable subspace for the adjoint system should be spanned by the solution

$$u(k) = \begin{pmatrix} e^{-3k} \\ 0 \\ -e^{-2k} \end{pmatrix}$$

and we can take the unstable space as the one spanned by the two solutions:

$$v_1(k) = \begin{pmatrix} 0 \\ 0 \\ e^{-k} \end{pmatrix}, \quad v_2(k) = \begin{pmatrix} 0 \\ e^{-2k} \\ -e^{-2k} \end{pmatrix}.$$

Then

$$e^{-2k} \le |u(k)| = e^{-2k}(e^{-k} + 1) \le 2e^{-2k} \quad \text{and} \quad |v_2(k)| = 2e^{-2k}.$$

Now if there was an exponential separation, there would exist positive constants K and α such that for $k \ge m \ge 0$,

$$\frac{|u(k)|}{|u(m)|}\frac{|v_2(m)|}{|v_2(k)|} \le Ke^{-\alpha(k-m)}.$$

However then for $k \ge m \ge 0$,

$$\frac{1}{2} = \frac{e^{-2k}}{2e^{-2m}}\frac{2e^{-2m}}{2e^{-2k}} \le \frac{|u(k)|}{|u(m)|}\frac{|v_2(m)|}{|v_2(k)|} \le Ke^{-\alpha(k-m)}.$$

This is impossible. We conclude that the adjoint system cannot be exponentially separated with projection $\mathbf{I} - P^*$ at $k = 0$.

Next we show that exponential dichotomy implies strong exponential separation. First we give the defintion of exponential dichotomy.

Definition 5 We say system (1) has an *exponential dichotomy* on an infinite interval J of integers if there is an invariant projection $P(k)$ and positive constants $K \ge 1$ and α such that

$$|\Phi(k,m)P(m)| \le Ke^{-\alpha(k-m)}, \quad |\Phi(m,k)(\mathbf{I} - P(k))| \le Ke^{-\alpha(k-m)} \quad k \ge m \text{ in } J,$$

where $\Phi(k,m)$ is the transition matrix. If $P(k)$ has rank r $(0 \le r \le n)$, we say that (1) has an exponential dichotomy with rank r.

Corollary 3 *If a system* $x(k+1) = A(k)x(k)$ *has an exponential dichotomy on an interval* J *with projection not equal to* 0 *or* $\mathbf{I}$*, then it is strongly exponentially separated on* J *with the same projection.*

Proof The proof follows at once from Proposition 4.

Finally we show that if a system is strongly exponentially separated on $\mathbb{Z}_+$ (resp. $\mathbb{Z}_-$) with a certain invariant projection, then it is strongly exponentially separated with respect to an invariant projection with the same range (resp. nullspace).

Proposition 5 *If system* (1) *is strongly exponentially separated on* $\mathbb{Z}_+$ *(resp.* $\mathbb{Z}_-$*) with projection* $P(k)$*, then it is strongly exponentially separated with respect to an invariant projection* $Q(k)$ *with the same range (resp. nullspace) as* $P(k)$*.*

Proof We just prove it for $\mathbb{Z}_+$. By Proposition 2, we know that (1) is exponentially separated with respect to $Q(k)$. All we need to show is that $Q(k)$ is bounded. As in Remark 3, there is a kinematic similarity $x = S(k)y$ taking (1) into a system of the form

$$y(k+1) = \begin{pmatrix} A_1(k) & 0 \\ 0 & A_2(k) \end{pmatrix} y(k),$$

where the projection for the exponential separation is the constant $P = \begin{pmatrix} \mathbf{I} & 0 \\ 0 & 0 \end{pmatrix} = S^{-1}(k)P(k)S(k)$ and so by Proposition 4, there exist positive constants K and α such that

$$|\Phi_1(k,m)|\,|\Phi_2(m,k)| \le Ke^{-\alpha(k-m)}, \quad k \ge m,$$

$\Phi_1(k,m)$ and $\Phi_2(k,m)$ being the respective transition matrices for $y_1(k+1) = A_1(k)y_1(k)$ and $y_2(k+1) = A_2(k)y_2(k)$. If $Q(k)$ is a projection with the same range as $P(k)$, $\tilde{Q}(k) = S^{-1}(k)Q(k)S(k)$ has the same range as $S^{-1}(k)P(k)S(k) = P$ so that

$$\tilde{Q}(k) = \begin{pmatrix} \mathbf{I} & C(k) \\ 0 & 0 \end{pmatrix}$$

for some matrix function $C(k)$, where by the invariance of $\tilde{Q}(k)$,

$$C(k) = \Phi_1(k,0)C(0)\Phi_2(0,k)$$

so that

$$|C(k)| \le |\Phi_1(k,0)|\,|C(0)|\,|\Phi_2(0,k)| \le Ke^{-\alpha k}|C(0)|.$$

So $\tilde{Q}(k)$ is bounded and hence also $Q(k) = S(k)\tilde{Q}(k)S^{-1}(k)$. Note also that we have shown that $|Q(k) - P(k)| \to 0$ as $k \to \infty$ exponentially fast.

4 Exponential Separation and Dichotomy

In Proposition 3 in [9], Papaschinopoulos showed that for a system (1) where $A(k)$ and its inverse are bounded, strong exponential separation is equivalent to the existence of a bounded sequence $p(k) > 0$, where $1/p(k)$ is also bounded, such that the shifted equation

$$x(k+1) = \frac{1}{p(k)} A(k)x(k)$$

has an exponential dichotomy with the same projection. A similar result, in the context of diffeomorphisms on a compact manifold, has been given in [5]. There exponential separation is referred to as "dominated splitting" and exponential dichotomy as "hyperbolicity". It is not known whether or not this result can be extended to the case where $A(k)$ and its inverse are not bounded.

5 Roughness of Strong Exponential Separation

We would like to show, as for differential equations (see [2]), that strong exponential separation is preserved under small perturbations of the coefficient matrix. Now Kalkbrenner [6] has shown that if (1) has an exponential dichotomy and $|B(k)|$ is uniformly small, then $x(k+1) = [A(k) + B(k)]x(k)$ has an exponential dichotomy with projection near that for the unperturbed system. (Such a result was also proved in [8], but under the additional assumptions that $A(k)$ be invertible and $A(k)$ and its inverse be bounded.) However the following example shows that the analogous result is not true, in general, for strong exponential separation.

Example 2 Consider the equation

$$x(k+1) = \begin{pmatrix} e^{-\alpha}a(k) + \delta & 0 \\ 0 & a(k) \end{pmatrix} x(k) \tag{9}$$

with $k \geq 0$, $\delta > 0$ and $a(k) > 0$. Two independent solutions are:

$$u(k) = \left(e^{-\alpha k}\sigma(k) \prod_{j=0}^{k-1} \left(1 + \frac{\delta e^{\alpha}}{a(j)}\right), 0 \right), \quad v(k) = (0, \sigma(k)),$$

where $\sigma(0) = 1$ and

$$\sigma(k) = \prod_{j=0}^{k-1} a(j) \quad \text{for } k \geq 1. \tag{10}$$

Then

$$\frac{|u(k)|}{|u(m)|}\frac{|v(m)|}{|v(k)|} = e^{-\alpha(k-m)}\prod_{j=m}^{k-1}\left(1+\frac{\delta e^{\alpha}}{a(j)}\right).$$

We define $a(k)$ as follows. Let T_i, $i \geq 0$, be a sequence of positive integers such that $T_{i+1} \geq T_i + 2$ (e.g. $T_i = 2i$). Then we define

$$a(k) = \begin{cases} A_i & (i \geq 1,\ T_{2i-1} \leq k < T_{2i}), \\ B_i & (i \geq 0,\ T_{2i} \leq k < T_{2i+1}), \end{cases}$$

where $A_i \to 0$ and $B_i \to \infty$ (e.g. $A_i = 1/i$, $B_i = i$). Then

$$\frac{|u(k)|}{|u(m)|}\frac{|v(m)|}{|v(k)|} = e^{-\alpha(k-m)}\left(1+\frac{\delta e^{\alpha}}{A_i}\right)^{k-m} = e^{[\log(1+\delta e^{\alpha}/A_i)-\alpha](k-m)}. \tag{11}$$

when $T_{2i-1} \leq m \leq k < T_{2i}$. This implies that it is not possible that there exist positive constants K and β such that

$$\frac{|u(k)|}{|u(m)|}\frac{|v(m)|}{|v(k)|} \leq Ke^{-\beta(k-m)}, \quad k \geq m$$

for otherwise, setting $m = T_{2i-1}$ and $k = T_{2i} - 1$ and taking logs,

$$[\log(1+\delta e^{\alpha}/A_i) - \alpha](T_{2i} - 1 - T_{2i-1}) \leq \log(K) - \beta(T_{2i} - 1 - T_{2i-1}) \leq \log(K) - \beta,$$

which is impossible since the left side $\to \infty$ as $i \to \infty$. On the other hand, if $T_{2i} \leq m \leq k < T_{2i+1}$ we have

$$\frac{|u(k)|}{|u(m)|}\frac{|v(m)|}{|v(k)|} = e^{-\alpha(k-m)}(1+\delta e^{\alpha}/B_i)^{k-m} = e^{[\log(1+\delta e^{\alpha}/B_i)-\alpha](k-m)}.$$

This implies that it is not possible that there exist positive constants K and β such that

$$\frac{|v(k)|}{|v(m)|}\frac{|u(m)|}{|u(k)|} \leq Ke^{-\beta(k-m)}, \quad k \geq m$$

for otherwise, setting $m = T_{2i}$ and $k = T_{2i+1} - 1$ and taking logs,

$$-(\log(1+\delta e^{\alpha}/B_i) - \alpha)(T_{2i+1} - 1 - T_{2i}) \leq \log(K) - \beta(T_{2i+1} - 1 - T_{2i}) \leq \log(K) - \beta,$$

which is impossible since the left side $\to \infty$ as $i \to \infty$.

As a consequence system (9) is not exponentially separated with, at $k = 0$, projection $\begin{pmatrix} 1 & 0 \\ 0 & 0 \end{pmatrix}$ or with projection $\begin{pmatrix} 0 & 0 \\ 0 & 1 \end{pmatrix}$. This also covers all cases where the stable subspace at $k = 0$ is the x-axis, because we may take the unstable subspace at $k = 0$ as the y-axis so that at $k = 0$ the projection is $\begin{pmatrix} 1 & 0 \\ 0 & 0 \end{pmatrix}$.

This leaves the case that system (9) is exponentially separated with stable subspace at $k = 0$ different from the x-axis. Then we can take the x-axis as the unstable subspace at $k = 0$, that is, we can assume that (9) is exponentially separated with projection $\begin{pmatrix} 0 & \gamma \\ 0 & 1 \end{pmatrix}$ at $k = 0$. Set

$$u(k) = e^{-\alpha k}\sigma(k)\prod_{j=0}^{k-1}\left(1 + \frac{\delta e^{\alpha}}{a(j)}\right), \quad v(k) = \sigma(k),$$

with $\sigma(k)$ as in (10). Then the stable and unstable solutions are, respectively

$$x_s(k) = (\gamma u(k), v(k)), \quad x_u(k) = (u(k), 0)$$

and there should exist positive constants K and β such that

$$\max\left\{|\gamma|, \frac{|v(k)|}{|u(k)|}\right\} = \frac{|x_s(k)|}{|x_u(k)|} \leq Ke^{-\beta k}, \quad k \geq 0,$$

where we are using the maximum norm in $\mathbb{R}^2$. This implies that $\gamma = 0$. However we have shown the system is not exponentially separated with projection $\begin{pmatrix} 0 & 0 \\ 0 & 1 \end{pmatrix}$ at $k = 0$. This completes the proof that system (9) is not exponentially separated.

In view of this example, we must choose a different kind of perturbation.

Theorem 1 *Suppose* (1) *is strongly exponentially separated on an interval J with projection $P(k)$. Then if $|B(k)| \leq \delta$, where δ is sufficiently small, the perturbed system*

$$x(k+1) = A(k)[\mathbf{I} + B(k)]x(k) \tag{12}$$

is also strongly exponentially separated with projection $Q(k)$ of the same rank. Also there exists a constant N such that

$$|Q(k) - P(k)| \leq N\delta.$$

First we prove the perturbation theorem for the special case where the system has been split into its stable and unstable parts and the perturbation is uncoupled. Note that this proof is rather more complicated than the proof of a corresponding lemma for dichotomy would be.

Lemma 1 *Consider the system*

$$x_1(k+1) = A_1(k)[\mathbf{I} + C_1(k)]x_1(k), \quad x_2(k+1) = A_2(k)[\mathbf{I} + C_2(k)]x_2(k).$$

Suppose there exist positive constants K and α *such that*

$$|\Phi_1(k,m)|\,|\Phi_2(m,k)| \le Ke^{-\alpha(k-m)}, \quad k \ge m \in J,$$

$\Phi_i(k,m)$ *being the respective transition matrices for* $x_i(k+1) = A_i(k)x_i(k)$. *Then if* $|C_i(k)| \le \delta$, *where* $K\delta < 1$,

$$|\tilde{\Phi}_1(k,m)|\,|\tilde{\Phi}_2(m,k)| \le Ke^{-(\alpha-\log[(1+K\delta)/(1-K\delta)])(k-m)}, \quad k \ge m \in J,$$

where $\tilde{\Phi}_i(k,m)$ *are the respective transition matrices for* $x_i(k+1) = A_i(k)[\mathbf{I} + C_i(k)]x_i(k)$.

Proof Let $x_2(k)$ be a nonzero solution of $x_2(k+1) = A_2(k)x_2(k)$. Then

$$|\Phi_1(k,m)|\,|x_2(m)| = |\Phi_1(k,m)|\,|\Phi_2(m,k)x_2(k)| \le Ke^{-\alpha(k-m)}|x_2(k)|, \quad k \ge m.$$

Next if $x_1(k)$ is a solution of $x_1(k+1) = A_1(k)[\mathbf{I} + C_1(k)]x_1(k)$, then

$$x_1(k) = \Phi_1(k,m)x_1(m) + \sum_{p=m}^{k-1} \Phi_1(k,p)C_1(p)x_1(p)$$

so that

$$|x_1(k)| \le Ke^{-\alpha(k-m)}\frac{|x_2(k)|}{|x_2(m)|}|x_1(m)| + \sum_{p=m}^{k-1} Ke^{-\alpha(k-p)}\frac{|x_2(k)|}{|x_2(p)|}\delta|x_1(p)|.$$

Writing $z(k) = e^{\alpha k}|x_1(k)|/|x_2(k)|$, we get

$$z(k) \le Kz(m) + K\delta\sum_{p=m}^{k-1} z(p)$$

so that for $k \ge m$,

$$z(k) \le K(1+K\delta)^{k-m}z(m).$$

Therefore

$$|x_1(k)| \le Ke^{-(\alpha-\log(1+K\delta))(k-m)}\frac{|x_2(k)|}{|x_2(m)|}|x_1(m)|$$

and hence

$$|\tilde{\Phi}_1(k,m)| \le Ke^{-(\alpha-\log(1+K\delta))(k-m)}\frac{|x_2(k)|}{|x_2(m)|}, \quad k \ge m.$$

That is,

$$|\tilde{\Phi}_1(k,m)|\,|\Phi_2(m,k)x_2(k)| \le Ke^{-(\alpha-\log(1+K\delta))(k-m)}|x_2(k)|, \quad k \ge m.$$

Since $x_2(k)$ is arbitrary, it follows that

$$|\tilde{\Phi}_1(k,m)|\,|\Phi_2(m,k)| \le Ke^{-(\alpha-\log(1+K\delta))(k-m)}, \quad k \ge m.$$

Next let $x_1(k)$ be a solution of $x_1(k+1) = A_1(k)[\mathbf{I} + C_1(k)]x_1(k)$. Then

$$|x_1(k)|\,|\Phi_2(m,k)| = |\tilde{\Phi}_1(k,m)x_1(m)|\,|\Phi_2(m,k)| \le Ke^{-(\alpha-\log(1+K\delta))(k-m)}|x_1(m)|, \quad k \ge m.$$

Now if $x_2(k)$ is a solution of $x_2(k+1) = A_2(k)[\mathbf{I} + C_2(k)]x_2(k)$, then for $k \le m$

$$x_2(k) = \Phi_2(k,m)x_2(m) - \sum_{p=k}^{m-1} \Phi_2(k,p)C_2(p)x_2(p)$$

so that

$$|x_2(k)| \le Ke^{-(\alpha-\log(1+K\delta))(m-k)}\frac{|x_1(k)|}{|x_1(m)|}|x_2(m)| + \sum_{p=k}^{m-1} Ke^{-(\alpha-\log(1+K\delta))(p-k)}\frac{|x_1(k)|}{|x_1(p)|}\delta|x_2(p)|.$$

Writing $z(k) = e^{-(\alpha-\log(1+K\delta))k}|x_2(k)|/|x_1(k)|$, we get

$$z(k) \le Kz(m) + K\delta\sum_{p=k}^{m-1} z(p), \quad k \le m, \quad \text{so that} \quad z(k) \le Kz(m)(1-K\delta)^{k-m}, \quad k \le m$$

and so

$$|x_2(k)| \le Ke^{-(\alpha-\log[(1+K\delta)/(1-K\delta)])(m-k)}\frac{|x_1(k)|}{|x_1(m)|}|x_2(m)|, \quad k \le m.$$

Thus

$$|\tilde{\Phi}_2(k,m)| \le Ke^{-(\alpha-\log[(1+K\delta)/(1-K\delta)])(m-k)}\frac{|x_1(k)|}{|x_1(m)|}, \quad k \le m.$$

That is,

$$|\tilde{\Phi}_1(m,k)x_1(k)|\,|\tilde{\Phi}_2(k,m)| \le Ke^{-(\alpha-\log[(1+K\delta)/(1-K\delta)])(m-k)}|x_1(k)|, \quad k \le m.$$

Since $x_1(k)$ is arbitrary, it follows that

$$|\tilde{\Phi}_1(k,m)|\,|\tilde{\Phi}_2(m,k)| \le Ke^{-(\alpha-\log[(1+K\delta)/(1-K\delta)])(k-m)}, \quad m \le k.$$

Now we prove the general perturbation theorem, using the special case just proved and also Lemma 2. Here we follow the method in the proof on p. 42 of [3]. First we prove a lemma which is essentially well-known (see, for example, Proposition 2.8 in [8]) but we give the proof for the sake of completeness.

Lemma 2 *Let $A(k)$ be an invertible matrix function on $J = \mathbb{Z}$, $\mathbb{Z}_+$ or $\mathbb{Z}_-$. Suppose there exist positive constants K and α such that the transition matrix $\Phi(k,m)$ for (1) satisfies*

$$|\Phi(k,m)| \le Ke^{-\alpha(k-m)}$$

for $k \ge m$ in J. Next let $f(k,x)$ be a function satisfying

$$|f(k,0)| \le \mu, \quad |f(k,x_1) - f(k,x_2)| \le \theta|x_1 - x_2|$$

for $k \in J$ and $|x_1|$, $|x_2| \le \Delta$. Then if

$$2K\mu \le (1-e^{-\alpha})\Delta, \quad 2K\theta \le 1 - e^{-\alpha},$$

the equation

$$x(k+1) = A(k)x(k) + f(k,x(k)), \quad k \in J$$

has a solution $x(k)$ such that $|x(k)| \le 2K\mu/(1-e^{-\alpha})$.

Proof Let E be the Banach space of bounded sequences $x(k)$ defined for $k \in J$ with the supremum norm $\|\cdot\|_\infty$ and let S be the ball of radius $2K\mu/(1-e^{-\alpha})$ in E. Then we define a mapping T on S by

$$(Tx)(k) = \sum_{m=b}^{k-1} \Phi(k,m+1)f(m,x(m)), \quad k \in J,$$

where $b = -\infty$ when $J = \mathbb{Z}, \mathbb{Z}_-$ and $b = 0$ when $J = \mathbb{Z}_+$. Note that if $k \in J$,

$$\begin{aligned} |(Tx)(k)| &\le \sum_{m=b}^{k-1} Ke^{-\alpha(k-m-1)}[\mu + \theta\, 2K\mu/(1-e^{-\alpha})] \\ &\le K(1-e^{-\alpha})^{-1}[\mu + \theta\, 2K\mu/(1-e^{-\alpha})] \\ &\le 2K\mu(1-e^{-\alpha})^{-1}. \end{aligned}$$

Hence T maps S into itself. Next if $x_1(k)$ and $x_2(k)$ are two sequences in S, then for $k \in J_1$,

$$\begin{aligned} |(Tx_1)(k) - (Tx_2)(k)| &\le \sum_{m=b}^{k-1} Ke^{-\alpha(k-m-1)}\theta\|x_1 - x_2\|_\infty \\ &\le K(1-e^{-\alpha})^{-1}\theta\|x_1-x_2\|_\infty \le \frac{1}{2}\|x_1 - x_2\|_\infty. \end{aligned}$$

Thus T is a contraction on S and has a unique fixed point $x(k)$ which satisfies $|x(k)| \leq 2K\mu/(1-e^{-\alpha})$ for all $k \in J$ and also

$$x(k) = \sum_{m=b}^{k-1} \Phi(k, m+1) f(m, x(m)), \quad k \in J,$$

from which it follows that for $k \in J$,

$$\begin{aligned} x(k+1) &= \sum_{m=b}^{k} \Phi(k+1, m+1) f(m, x(m)) \\ &= A(k) \sum_{m=b}^{k-1} \Phi(k, m+1) f(m, x(m)) + f(k, x(k)) = A(k)x(k) + f(k, x(k)). \end{aligned}$$

Now we continue with the proof of the theorem.

Proof As in Remark 3, there is a kinematic similarity $x = T(k)y$, $k \in J$, taking (1) into a block diagonal system

$$x(k+1) = \begin{pmatrix} A_1(k) & 0 \\ 0 & A_2(k) \end{pmatrix} x(k).$$

This coefficient matrix commutes with $P = \begin{pmatrix} \mathbf{I} & 0 \\ 0 & 0 \end{pmatrix}$. Also the block diagonal system is strongly exponentially separated with constant projection P so that by Proposition 4 there exist positive constants K and α such that

$$|\Phi_1(k, m)|\, |\Phi_2(m, k)| \leq K e^{-\alpha(k-m)}, \quad m \leq k, \ m, k \in J$$

$\Phi_i(k, m)$ being the respective transition matrices. If we apply the transformation $x = T(k)y$ to the perturbed system (12), we obtain a system with coefficient matrix

$$\begin{aligned} T^{-1}(k+1)A(k)[\mathbf{I} + B(k)]T(k) &= T^{-1}(k+1)A(k)T(k) \\ &\quad + T^{-1}(k+1)A(k)T(k)T^{-1}(k)B(k)T(k) \\ &= \begin{pmatrix} A_1(k) & 0 \\ 0 & A_2(k) \end{pmatrix} + \begin{pmatrix} A_1(k)C_{11}(k) & A_1(k)C_{12}(k) \\ A_2(k)C_{21}(k) & A_2(k)C_{22}(k) \end{pmatrix}, \end{aligned}$$

where

$$T^{-1}(k)B(k)T(k) = \begin{pmatrix} C_{11}(k) & C_{12}(k) \\ C_{21}(k) & C_{22}(k) \end{pmatrix} \quad \text{so that} \quad |C_{ij}(k)| \leq M\delta,$$

M being an upper bound on $|T^{-1}(k)|\,|T(k)|$. We assume

$$2KM\delta < 1.$$

The transformed equation is

$$\begin{aligned} y_1(k+1) &= A_1(k)[y_1(k) + C_{11}(k)y_1(k) + C_{12}(k)y_2(k)], \\ y_2(k+1) &= A_2(k)[y_2(k) + C_{21}(k)y_1(k) + C_{22}(k)y_2(k)]. \end{aligned} \tag{13}$$

Then we use the transformation

$$\begin{pmatrix} y_1 \\ y_2 \end{pmatrix} = S(k)\begin{pmatrix} w_1 \\ w_2 \end{pmatrix} = \begin{pmatrix} \mathbf{I} & H_{12}(k) \\ H_{21}(k) & \mathbf{I} \end{pmatrix}\begin{pmatrix} w_1 \\ w_2 \end{pmatrix} \tag{14}$$

where, provided $H_{12}(k)$, $H_{21}(k)$ are bounded and $|H_{12}(k)H_{21}(k)| \le \gamma < 1$, $S(k)$ is invertible with bounded inverse given by

$$\begin{pmatrix} [\mathbf{I} - H_{12}(k)H_{21}(k)]^{-1} & -[\mathbf{I} - H_{12}(k)H_{21}(k)]^{-1}H_{12}(k) \\ -[\mathbf{I} - H_{21}(k)H_{12}(k)]^{-1}H_{21}(k) & [\mathbf{I} - H_{21}(k)H_{12}(k)]^{-1} \end{pmatrix}.$$

The transformation (14) takes (13) into a block diagonal system

$$\begin{aligned} w_1(k+1) &= A_1(k)[\mathbf{I} + C_{11}(k) + C_{12}(k)H_{21}(k)]w_1(k), \\ w_2(k+1) &= A_2(k)[\mathbf{I} + C_{22}(k) + C_{21}(k)H_{12}(k)]w_2(k) \end{aligned} \tag{15}$$

provided for $k \in J$,

$$H_{12}(k+1)A_2(k)[\mathbf{I} + C_{22}(k) + C_{21}(k)H_{12}(k)] = A_1(k)[(\mathbf{I} + C_{11}(k))H_{12}(k) + C_{12}(k)], \tag{16}$$

$$H_{21}(k+1)A_1(k)[\mathbf{I} + C_{11}(k) + C_{12}(k)H_{21}(k)] = A_2(k)[C_{21}(k) + (\mathbf{I} + C_{22}(k))H_{21}(k)]. \tag{17}$$

First we solve (16). With H_{12} replaced by H, (16) can be rewritten as

$$\begin{aligned} H(k+1) = A_1(k)[\mathbf{I} + C_{11}(k)]H(k)[\mathbf{I} + C_{22}(k) + C_{21}(k)H(k)]^{-1}A_2^{-1}(k) \\ + A_1(k)C_{12}(k)[\mathbf{I} + C_{22}(k) + C_{21}(k)H(k)]^{-1}A_2^{-1}(k), \end{aligned} \tag{18}$$

assuming $\mathbf{I} + C_{22}(k) + C_{21}(k)H(k)$ is invertible, which it will be as long as δ is sufficiently small. Now we know the equation

$$H(k+1) = \mathcal{A}(k)H(k) = A_1(k)H(k)A_2^{-1}(k)$$

is uniformly asymptotically stable because it has transition operator

$$H \to \Psi(k,m)H = \Phi_1(k,m)H\Phi_2(m,k),$$

where

$$|\Psi(k,m)| \le Ke^{-\alpha(k-m)}, \quad k \ge m.$$

We write (18) as

$$H(k+1) = A_1(k)H(k)A_2^{-1}(k) + f(k, H(k)),$$

where

$$f(k,H) = A_1(k)g(k,H)A_2^{-1}(k) \quad \text{with} \quad g(k,H) = g_1(k,H)g_2^{-1}(k,H),$$

and

$$\begin{aligned} g_1(k,H) &= C_{11}(k)H - HC_{22}(k) - HC_{21}(k)H + C_{12}(k), \\ g_2(k,H) &= \mathbf{I} + C_{22}(k) + C_{21}(k)H. \end{aligned}$$

We see that $|g(k,0)| \le (1-M\delta)^{-1}M\delta$ so that

$$|f(k,0)| \le Ke^{-\alpha}(1-M\delta)^{-1}M\delta$$

in view of the fact that

$$|A_1(k)||A_2^{-1}(k)| = |\Phi_1(k,k-1)||\Phi_2(k-1,k)| \le Ke^{-\alpha}.$$

Next note that when $|H| \le 1$,

$$|g_1(k,H)| \le 4M\delta, \quad |g_2^{-1}(k,H)| \le (1-2M\delta)^{-1}$$

and if $|H_1|, |H_2| \le 1$,

$$\begin{aligned} |g_1(k,H_1) - g_1(k,H_2)| &\le 4M\delta|H_1 - H_2|, \\ |g_2^{-1}(k,H_1) - g_2^{-1}(k,H_2)| &= |g_2^{-1}(k,H_1)[g_2(k,H_1) - g_2(k,H_2)]g_2^{-1}(k,H_2)| \\ &\le (1-2M\delta)^{-2}M\delta|H_1 - H_2| \end{aligned}$$

so that

$$\begin{aligned} |g(k,H_1) - g(k,H_2)| &\le |g_1(k,H_1)|\,|g_2^{-1}(k,H_1) - g_2^{-1}(k,H_2)| \\ &\quad + |g_2^{-1}(k,H_2)|\,|g_1(k,H_1) - g_1(k,H_2)| \\ &\le [4M\delta(1-2M\delta)^{-2}M\delta + (1-2M\delta)^{-1}4M\delta]|H_1 - H_2| \\ &= 4M\delta(1-M\delta)(1-2M\delta)^{-2}|H_1 - H_2|. \end{aligned}$$

Then if $|H_1|, |H_2| \le 1$,

$$|f(k,H_1) - f(k,H_2)| \le 4MKe^{-\alpha}(1-M\delta)(1-2M\delta)^{-2}\delta|H_1 - H_2|.$$

Now we apply Lemma 2 with $\Delta = 1, \mu = Ke^{-\alpha}(1 - M\delta)^{-1}M\delta, \theta = 4MKe^{-\alpha}(1 - M\delta)(1 - 2M\delta)^{-2}\delta$. Then if

$$2K^2M(1 - M\delta)^{-1}\delta \le e^\alpha - 1, \quad 8K^2M(1 - M\delta)(1 - 2M\delta)^{-2}\delta \le e^\alpha - 1,$$

equation (16) has a solution $H_{12}(k)$ such that $|H_{12}(k)| \le 2K^2M(e^\alpha - 1)^{-1}(1 - M\delta)^{-1}\delta$ for all t.

Now we consider Eq. (17). If we define $\tilde{H}(k) = H_{21}(-k+1)$, with $k \in 1 - J = \{1 - \ell : \ell \in J\}$, (17) reads:

$$\begin{aligned}&\tilde{H}(k)A_1(-k)[\mathbf{I} + C_{11}(-k) + C_{12}(-k)\tilde{H}(k+1)]\\&\quad = A_2(-k)[C_{21}(-k) + (\mathbf{I} + C_{22}(-k))\tilde{H}(k+1)]\end{aligned}$$

for $k \in 1 - J$, that is,

$$\begin{aligned}&\tilde{H}(k+1)\\&= [A_2(-k)(\mathbf{I} + C_{22}(-k)) - \tilde{H}(k)A_1(-k)C_{12}(-k)]^{-1}\\&\quad\{\tilde{H}(k)A_1(-k)[\mathbf{I} + C_{11}(-k)] - A_2(-k)C_{21}(-k)\}\\&= A_2(-k)^{-1}\tilde{H}(k)A_1(-k) + f(k, \tilde{H}(k)), \quad k \in 1 - J,\end{aligned} \tag{19}$$

where

$$f(k, H) = g_2^{-1}(k, H)g_1(k, H),$$

with

$$\begin{aligned}&g_1(k, H) = p(k, H)C_{11}(-k) - C_{21}(-k) - C_{22}(-k)p(k, H) + p(k, H)C_{12}(k)p(k, H),\\&g_2(k, H) = \mathbf{I} + C_{22}(-k) - p(k, H)C_{12}(-k), \quad p(k, H) = A_2^{-1}(-k)HA_1(-k).\end{aligned}$$

Note that if $|H|, |H_1|, |H_2| \le 1$,

$$\begin{aligned}&p(k, 0) = 0, \quad |p(k, H_1) - p(k, H_2)| \le Ke^{-\alpha}|H_1 - H_2|\\&|g_1(k, H)| \le M\delta(1 + Ke^{-\alpha})^2, \quad |g_2^{-1}(k, H)| \le [1 - M\delta(1 + Ke^{-\alpha})]^{-1}\\&|g_1(k, H_1) - g_1(k, H_2)| \le 2MKe^{-\alpha}\delta[1 + Ke^{-\alpha}]|H_1 - H_2|,\\&|g_2^{-1}(k, H_1) - g_2^{-1}(k, H_2)| \le [1 - M\delta(1 + Ke^{-\alpha})]^{-2}M\delta Ke^{-\alpha}|H_1 - H_2|\end{aligned}$$

so that if $|H_1|, |H_2| \le 1$,

$$|f(k, H_1) - f(k, H_2)| \le \theta|H_1 - H_2|,$$

where

$$\theta = [1 - M\delta(1 + Ke^{-\alpha})]^{-1} 2MKe^{-\alpha}\delta[1 + Ke^{-\alpha}] \\ + M\delta(1 + Ke^{-\alpha})^2[1 - M\delta(1 + Ke^{-\alpha})]^{-2} M\delta K e^{-\alpha} \\ \leq (1 - 2KM\delta)^{-2} 4K^2 e^{-\alpha} M\delta.$$

Also we have

$$|f(k, 0)| = |[\mathbf{I} + C_{22}(-k)]^{-1}|\,|C_{21}(-k)| \leq \frac{M\delta}{1 - M\delta}.$$

The transition operator of equation

$$H(k + 1) = A_2(-k)^{-1} H(k) A_1(-k), \quad k \in 1 - J,$$

is $\Psi(k, m)H = \Phi_2(1 - k, 1 - m)H\Phi_1(1 - m, 1 - k)$ and satisfies

$$|\Psi(k, m)| \leq Ke^{-\alpha(k-m)}$$

for $m \leq k$. Then, applying Lemma 2 with $\mu = \frac{M\delta}{1-M\delta}$, $\theta = \frac{4K^2e^{-\alpha}M\delta}{(1-2KM\delta)^2}$ and $\Delta = 1$ we see that if

$$M\delta \leq \frac{e^\alpha - 1}{2Ke^\alpha + e^\alpha - 1} \quad \text{and} \quad \frac{8K^3M\delta}{(1 - 2KM\delta)^2} \leq e^\alpha - 1,$$

equation (19) has a solution bounded on $1 - J$ and hence (17) has a solution bounded on J.

In fact, $|H_{21}(k)| \leq 2KM(1 - e^{-\alpha})^{-1}(1 - M\delta)^{-1}\delta$. Then provided

$$|H_{12}(k)H_{21}(k)| \leq 4K^3e^{-\alpha}M^2(1 - e^{-\alpha})^{-2}(1 - M\delta)^{-2}\delta^2 < 1,$$

$x = S(k)w$ is a kinematic similarity taking (13) into (15).

Next we show that the transformed Eq. (15) is strongly exponentially separated. Note that

$$|C_{11}(k) + C_{12}(k)H_{21}(k)| \leq \delta_1, \quad |C_{22}(k) + C_{21}(k)H_{12}(k)| \leq \delta_1,$$

where

$$\delta_1 = M\delta(1 + 2KM \max\{1, Ke^{-\alpha}\}(1 - e^{-\alpha})^{-1}(1 - M\delta)^{-1}\delta).$$

Then it follows from Lemma 1 that

$$|\tilde{\Phi}_1(k, m)|\,|\tilde{\Phi}_2(m, k)| \leq Ke^{-(\alpha - \log[(1+K\delta_1)/(1-K\delta_1)])(k-m)}, \quad m \leq k,$$

$\tilde{\Phi}_i(k, m)$ being the respective transition matrices. Provided that

$$\log[(1 + K\delta_1)/(1 - K\delta_1)] < \alpha,$$

it follows that (15) is strongly exponentially separated with constant projection $P = \begin{pmatrix} \mathbf{I} & 0 \\ 0 & 0 \end{pmatrix}$. Hence (12) is strongly exponentially separated with projection $Q(k) = T(k)S(k)PS^{-1}(k)T^{-1}(k)$. Now the projection for the unperturbed system is $P(k) = T(k)PT^{-1}(k)$. It follows that

$$\begin{aligned} |Q(k) - P(k)| &\leq |T(k)|\,|S(k)PS^{-1}(k) - P|\,|T^{-1}(k)| \\ &\leq |T(k)|\left[|S(k) - \mathbf{I}|\,|P|\,|S^{-1}(k)| + |P|\,|S^{-1}(k) - \mathbf{I}|\right]|T^{-1}(k)| \\ &\leq M\left[|S(k) - \mathbf{I}|\,|P|\,|S^{-1}(k)| + |P|\,|S^{-1}(k) - \mathbf{I}|\right] \\ &\leq N\delta \end{aligned}$$

for some constant N, since $|S(k) - \mathbf{I}| = O(\delta)$, $|S^{-1}(k) - \mathbf{I}| = O(\delta)$.

6 Exponential Separation in Block Upper Triangular Systems

In this section we consider a block upper triangular system

$$u(k+1) = A(k)u(k) = \begin{pmatrix} A_{11}(k) & A_{12}(k) & \cdots & A_{1p}(k) \\ 0 & A_{22}(k) & \cdots & A_{2p}(t) \\ \cdot & \cdot & \cdots & \cdot \\ 0 & 0 & \cdots & A_{pp}(k) \end{pmatrix} u(k), \tag{20}$$

where $A_{ii}(t)$ is $n_i \times n_i$, and its associated block diagonal system

$$u(k+1) = \begin{pmatrix} A_{11}(k) & 0 & \cdots & 0 \\ 0 & A_{22}(k) & \cdots & 0 \\ \cdot & \cdot & \cdots & \cdot \\ 0 & 0 & \cdots & A_{pp}(k) \end{pmatrix} u(k). \tag{21}$$

In Proposition 6, we show if (20) is strongly exponentially separated on a half-axis, then (21) is strongly exponentially separated also. Then in Proposition 7, we show the converse of this statement on a half-axis or whole axis but with a boundedness condition. At the end we give some results for diagonal systems and upper triangular systems, where the blocks are scalars.

Proposition 6 *If the block upper triangular system (20) is strongly exponentially separated on a half-axis J, then the projection for the exponential separation of (20) at $k = 0$ can be taken as*

$$P = \begin{pmatrix} P_1 & P_{12} & P_{13} & \cdots & \cdot & P_{1p} \\ 0 & P_2 & P_{23} & \cdots & \cdot & P_{2p} \\ 0 & 0 & P_3 & \cdots & \cdot & \cdot \\ 0 & 0 & 0 & \cdots & \cdot & \cdot \\ 0 & 0 & 0 & \cdots & P_{p-1} & P_{p-1,p} \\ 0 & 0 & 0 & \cdots & \cdot & P_p \end{pmatrix}, \tag{22}$$

where the P_i are projections. Moreover, the block diagonal system (21) is strongly exponentially separated with projection at $k = 0$ given by

$$P = \begin{pmatrix} P_1 & 0 & 0 & \cdots & \cdot & 0 \\ 0 & P_2 & 0 & \cdots & \cdot & 0 \\ 0 & 0 & P_3 & \cdots & \cdot & \cdot \\ 0 & 0 & 0 & \cdots & \cdot & \cdot \\ 0 & 0 & 0 & \cdots & P_{p-1} & 0 \\ 0 & 0 & 0 & \cdots & \cdot & P_p \end{pmatrix}.$$

In the unbounded case, "strongness" is necessary here to get exponential separation. Consider the following example.

Example 3

$$u(k+1) = \begin{pmatrix} 1 & e^k(e-1) \\ 0 & 1 \end{pmatrix} u.$$

This upper triangular system is exponentially separated on $\mathbb{Z}_+$, since we have the two solutions $(1, 0)$ and $(e^k - e, 1)$. However the diagonal system is clearly not exponentially separated. The reason for this is that the exponential separation is not strong.

In order to prove the proposition, we use the following lemma from [2].

Lemma 3 *Consider the Cartesian product of vector spaces*

$$U = U_1 \times U_2 \times \cdots \times U_k,$$

where $k \geq 2$. If V is a subspace of U, there is a projection P on U with range V, which has the form

$$P = \begin{pmatrix} P_1 & P_{12} & P_{13} & \cdots & \cdot & P_{1k} \\ 0 & P_2 & P_{23} & \cdots & \cdot & P_{2k} \\ 0 & 0 & P_3 & \cdots & \cdot & \cdot \\ 0 & 0 & 0 & \cdots & \cdot & \cdot \\ 0 & 0 & 0 & \cdots & P_{k-1} & P_{k-1,k} \\ 0 & 0 & 0 & \cdots & 0 & P_k \end{pmatrix},$$

where P_i is a projection on U_i for $i = 1, \ldots, k$.

Now we prove Proposition 6.

Proof Denote by V the stable (resp. unstable) subspace for (20) at $k = 0$. Then it follows from Lemma 3 that there is a projection P of the form given in the statement of the Proposition which has range V; in the case $\mathbb{Z}_-$, we replace P by $\mathbf{I} - P$ so that the new P has nullspace V. It follows from Proposition 2 that we may take P as the projection for the exponential separation at $k = 0$. The transition matrix for (21) is

$$\tilde{U}(k,m) = \begin{pmatrix} \Phi_1(k,m) & 0 & \cdots & 0 \\ 0 & \Phi_2(k,m) & \cdots & 0 \\ \cdot & \cdot & \cdots & \cdot \\ 0 & 0 & \cdots & \Phi_p(k,m) \end{pmatrix},$$

whereas that for (20) is

$$U(k,m) = \begin{pmatrix} \Phi_1(k,m) & W_{12}(k,m) & \cdots & W_{1p}(k,m) \\ 0 & \Phi_2(k,m) & \cdots & W_{2m}(k,m) \\ \cdot & \cdot & \cdots & \cdot \\ 0 & 0 & \cdots & \Phi_p(k,m) \end{pmatrix},$$

where $\Phi_i(k,m)$ is the transition matrix for $x_i(k+1) = A_{ii}(k)x_i(k)$. We take

$$\tilde{P} = \begin{pmatrix} P_1 & 0 & 0 & \cdots & \cdot & 0 \\ 0 & P_2 & 0 & \cdots & \cdot & 0 \\ 0 & 0 & P_3 & \cdots & \cdot & \cdot \\ 0 & 0 & 0 & \cdots & \cdot & \cdot \\ 0 & 0 & 0 & \cdots & P_{p-1} & 0 \\ 0 & 0 & 0 & \cdots & \cdot & P_p \end{pmatrix}.$$

Then $\tilde{P}(k) = \tilde{U}(k,0)\tilde{P}\tilde{U}(0,k)$ is an invariant projection for (21) and $P(k) = U(k,0)PU(0,k)$ is a projection for the exponential separation of (20). We see that $U(k,0)PU(0,m) = U(k,m)P(m)$ and $\tilde{U}(k)\tilde{P}\tilde{U}^{-1}(m) = \tilde{U}(k,m)\tilde{P}(m)$ only differ in the (i,j)-entries, $i < j$, with those in $\tilde{U}(k,m)\tilde{P}(m)$ being zero. It follows that

$$|\tilde{U}(k,m)\tilde{P}(m)| \le |U(k,m)P(m)|.$$

(Note we use a matrix norm with the property that if $A = [A_{ij}]$ and $B = [B_{ij}]$ are partitioned matrices with $|A_{ij}| \le |B_{ij}|$ for all i, j, then $|A| \le |B|$.) Similarly

$$|\tilde{U}(m,k)(\mathbf{I} - \tilde{P}(k))| \le |U(m,k)(\mathbf{I} - P(k))|.$$

Since by Proposition 4 the strong exponential separation of (20) implies the existence of positive constants K and α such that

$$|U(k,m)P(m)|\,|U(m,k)(\mathbf{I} - P(k))| \le Ke^{-\alpha(k-m)}, \quad m \le k,$$

we conclude that

$$|\tilde{U}(k,m)\tilde{P}(m)|\,|\tilde{U}(m,k)(\mathbf{I}-\tilde{P}(k))| \le Ke^{-\alpha(k-m)}, \quad m \le k.$$

Then the proposition follows from Proposition 4.

Now we prove the converse.

Proposition 7 *Consider system* (20) *on* $J = \mathbb{Z}_+$, $\mathbb{Z}_-$ *or* $\mathbb{Z}$, *where* $A_{ii}^{-1}(k)A_{ij}(k)$ *is bounded for* $i < j$. *Then if the diagonal system* (21) *is strongly exponentially separated, system* (20) *is also strongly exponentially separated.*

Proof We define the matrix

$$S = \begin{pmatrix} \mathbf{I}_{n_1} & 0 & \cdots & 0 \\ 0 & \beta\mathbf{I}_{n_2} & \cdots & 0 \\ \cdot & \cdot & \cdots & \cdot \\ 0 & 0 & \cdots & \beta^{p-1}\mathbf{I}_{n_p} \end{pmatrix},$$

where $A_{ii}(k)$ is $n_i \times n_i$ and $\beta > 0$. A simple calculation shows that with $A(k)$ as in (20),

$$S^{-1}A(k)S = A_\beta(k) = \begin{pmatrix} A_{11}(k) & \beta A_{12}(k) & \cdots & \beta^{p-1}A_{1p}(k) \\ 0 & A_{22}(k) & \cdots & \beta^{p-2}A_{2p}(k) \\ \cdot & \cdot & \cdots & \cdot \\ 0 & 0 & \cdots & A_{pp}(k) \end{pmatrix}.$$

This means the constant kinematic similarity $u = Sv$ takes (20) into the system $v(k+1) = A_\beta(k)v(k)$, where

$$A_\beta(k) = \begin{pmatrix} A_{11}(k) & 0 & \cdots & 0 \\ 0 & A_{22}(k) & \cdots & 0 \\ \cdot & \cdot & \cdots & \cdot \\ 0 & 0 & \cdots & A_{pp}(k) \end{pmatrix} \left[\mathbf{I} + \begin{pmatrix} 0 & \beta A_{11}^{-1}(k)A_{12}(k) & \cdots & \beta^{p-1}A_{11}^{-1}(k)A_{1p}(k) \\ \cdot & \cdot & \cdots & \cdot \\ \cdot & \cdot & \cdots & \beta A_{p-1,p-1}^{-1}(k)A_{p-1,p}(k) \\ 0 & 0 & \cdots & 0 \end{pmatrix} \right]$$

However, when $A_{ii}^{-1}(k)A_{ij}(k)$ is bounded for $i < j$, the last matrix is small when β is small. Then it follows from Theorem 1 that $v(k+1) = A_\beta(k)v(k)$ is strongly exponentially separated if β is sufficiently small. From this it follows that (20) is strongly exponentially separated.

Example 4 These examples show that in Proposition 7 the boundedness condition on $A_{ii}^{-1}(k)A_{ij}(k)$ for $i < j$ is necessary. In both examples the diagonal system is strongly exponentially separated. In the first example the triangular system is not exponentially separated but in the second example it is exponentially separated but not strongly so. Note we use the maximum norm in $\mathbb{R}^2$.

(i) Consider the system

$$u(k+1) = \begin{pmatrix} e & e^k \\ 0 & 1 \end{pmatrix} u(k).$$

Clearly the diagonal system is strongly exponentially separated on $\mathbb{Z}_+$ having the solutions

$$u_s(k) = \begin{pmatrix} 0 \\ 1 \end{pmatrix}, \quad u_u(k) = \begin{pmatrix} e^k \\ 0 \end{pmatrix}, \quad k \geq 0.$$

However we will show that the upper triangular system is not exponentially separated. The solution with $u(0) = (a, b)$ is $u(k) = (e^{k-1}(ea + bk), b)$. When $b \neq 0$, we see that $|u_u(k)|/|u(k)| \to 0$ as $t \to \infty$. So if the system is exponentially separated on $\mathbb{Z}_+$, $u_u(k)$ must be the unique stable solution and we may take $w(k) = (ke^{k-1}, 1)$ as the unstable solution. Then there must exist positive constants K and α such that

$$\frac{|u_u(k)|\,|w(m)|}{|u_u(m)|\,|w(k)|} \leq Ke^{-\alpha(k-m)}, \quad k \geq m.$$

That is,

$$\frac{m}{k} \leq Ke^{-\alpha(k-m)}, \quad \text{for any } k \geq m \geq 1.$$

But, with $k = 2m$, $m \geq 1$, this gives

$$\frac{1}{2} \leq Ke^{-\alpha m} \to 0 \quad \text{as } m \to \infty,$$

a contradiction. Hence the system is not exponentially separated.

(ii) Next consider the system:

$$u(k+1) = \begin{pmatrix} \frac{2}{k+1} & \delta \\ 0 & \frac{1}{k+1} \end{pmatrix} u(k).$$

When $\delta = 0$ it is strongly exponentially separated on $\mathbb{Z}_+$ with constant projection $\begin{pmatrix} 0 & 0 \\ 0 & 1 \end{pmatrix}$. The stable solution is $\left(0, \frac{1}{k!}\right)$ and the unstable is $\left(\frac{2^k}{k!}, 0\right)$. When $\delta \neq 0$, the transition matrix is

$$\Phi(k, 0) = \begin{pmatrix} \frac{2^k}{k!} & \frac{2^{k+1}\delta}{k!}\left(1 - \frac{k+2}{2^{k+1}}\right) \\ 0 & \frac{1}{k!} \end{pmatrix}.$$

Let $u(k)$ be the solution $\left(-\delta\frac{k+2}{k!}, \frac{1}{k!}\right)$ and let $v(k)$ be the solution $\left(\frac{2^k}{k!}, 0\right)$. Then, for $k \geq m \geq \delta^{-1} - 2$,

$$\frac{|u(k)|\,|v(m)|}{|u(m)|\,|v(k)|} = \frac{k+2}{m+2} 2^{-(k-m)} \leq e^{(k-m)/(m+2)} 2^{-(k-m)} \leq e^{-(\log 2 - \frac{1}{2})(k-m)}.$$

So this system is exponentially separated. However it is not strongly exponentially separated. To see this, note that taking $u(k) = \left(-\frac{k+2}{k!}, \frac{1}{\delta k!}\right)$ as the stable solution and $v(k) = \left(\frac{2^k}{k!}, 0\right)$ as the unstable corresponds to taking the projection $P(0)$ as

$$P(0) = \begin{pmatrix} 0 & -2\delta \\ 0 & 1 \end{pmatrix}.$$

Then, by invariance, a projection for the exponential separation is

$$P(k) = \Phi(k,0)P(0)\Phi(0,k) = \begin{pmatrix} 0 & -\delta(k+2) \\ 0 & 1 \end{pmatrix}.$$

Since $|P(k)| \to \infty$ as $k \to \infty$, it follows from Proposition 5 that the system is not strongly exponentially separated.

We finish the section with two propositions about upper triangular systems, that is, block upper triangular systems where the blocks are scalars. We need the following lemma, which gives information about the stable subspace of (21) when it is strongly exponentially separated. Actually we only need it for scalar diagonal systems but prove it for the block diagonal case.

Lemma 4 *Assume (21) is exponentially separated on $\mathbb{Z}+$ (resp. $\mathbb{Z}_-$). Then the stable (resp. unstable) subspace is a Cartesian product $V_1 \times V_2 \times \cdots \times V_p$, where V_i is a subspace of $\mathbb{R}^{n_i}$. Moreover, if $1 \le \dim V_i \le n_i - 1$, $u_i(k+1) = A_{ii}(k)u_i(k)$ is exponentially separated with stable (resp. unstable) subspace V_i and strongly exponentially separated if (21) is.*

Proof First we prove the case $p = 2$. Define V_1 as the subspace of $\mathbb{R}^{n_1}$ consisting of those ξ for which $(\Phi_1(k,0)\xi, 0)$ is in the stable (resp. unstable) subspace for (21) and V_2 as the subspace of $\mathbb{R}^{n_2}$ consisting of those η for which $(0, \Phi_2(k,0)\eta)$ is in the stable (resp. unstable) subspace, where $\Phi_i(k,m)$ is the transition matrix for $u_i(k+1) = A_{ii}(k)u_i(k)$. Then $V_1 \times V_2$ is a subspace of the stable (resp. unstable) subspace for (21) at $k = 0$.

Suppose $z(k) = (u_1(k), u_2(k))$ is a solution of (21) (still with $p = 2$) in the stable (resp. unstable) subspace with $u_1(0) \ne 0, u_2(0) \ne 0$ and suppose there exists α and β with $\alpha\beta \ne 0$ for which $w(k) = (\alpha u_1(k), \beta u_2(k))$ is not in the stable (resp. unstable) subspace. Then we can choose the unstable (resp. stable) space so as to include $w(k)$. Then, in the $\mathbb{Z}_+$ case, there exist positive constants K and γ such that

$$\frac{|z(k)|\,|w(m)|}{|z(m)|\,|w(k)|} \le Ke^{-\gamma(k-m)}, \quad k \ge m.$$

However note that $w(k) = Dz(k)$, where $D = \begin{pmatrix} \alpha\mathbf{I}_{n_1} & 0 \\ 0 & \beta\mathbf{I}_{n_2} \end{pmatrix}$, so that

$$\frac{|w(m)|}{|w(k)|} \ge \frac{|z(m)|}{|z(k)|}\frac{1}{|D|\,|D^{-1}|}.$$

This implies that

$$\frac{1}{|D|\,|D^{-1}|} \le Ke^{-\gamma(k-m)}, \quad k \ge m,$$

which is clearly absurd. In the $\mathbb{Z}_-$ case, there exist positive constants K and γ such that

$$\frac{|w(k)|\,|z(m)|}{|w(m)|\,|z(k)|} \le Ke^{-\gamma(k-m)}, \quad k \ge m.$$

However

$$\frac{|w(k)|}{|w(m)|} \ge \frac{|z(k)|}{|z(m)|}\frac{1}{|D|\,|D^{-1}|}$$

which, as before, leads to an absurdity. So we have proved that if $z(k) = (u_1(k), u_2(k))$ is a solution in the stable (resp. unstable) subspace, then all solutions $(\alpha u_1(k), \beta u_2(k))$ with $\alpha\beta \neq 0$ are in the stable (resp. unstable) subspace. Then we can let $(\alpha, \beta) \to (1, 0)$ and $(0, 1)$ and conclude, by continuity, that $(u_1(k), 0)$ and $(0, u_2(k))$ are in the stable (resp. unstable) subspace. It follows that $z(0) \in V_1 \times V_2$. Thus we have shown that the stable (resp. unstable) subspace at $k = 0$ is $V_1 \times V_2$.

Then we let W_i be subspaces such that $\mathbb{R}^{n_i} = V_i \oplus W_i$. Then by Proposition 2 we may take $W = W_1 \times W_2$ as the unstable (resp. stable) subspace at $k = 0$ for the whole system. Then the conclusion about exponential separation of $u_i(k+1) = A_{ii}(k)u_i(k)$ follows easily. Note also that when (21) is strongly exponentially separated, it follows from Proposition 5 that the angle between $V_1(k) \times V_2(k)$ and $W_1(k) \times W_2(k)$ (here $V_i(k) = \Phi_i(k, 0)V_i$, etc.) is bounded below by a positive number. Hence the angles between the $V_i(k)$ and $W_i(k)$ are bounded below by positive numbers also, and so the systems $u_i(k+1) = A_{ii}(k)u_i(k)$ are strongly exponentially separated also.

The proof for general $p \ge 2$ follows by induction on p using the $p = 2$ case.

Now we give a necessary and sufficient condition that a diagonal system be strongly exponentially separated. As we shall see, strong exponential separation and exponential separation are equivalent for diagonal systems.

Proposition 8 *The diagonal system*

$$u_i(k+1) = a_i(k)u_i(k), \quad i = 1, \dots, n \tag{23}$$

is exponentially separated on $J = \mathbb{Z}, \mathbb{Z}_+$ *or* $\mathbb{Z}_-$ *with rank* r *if and only if there exists* $I \subset \{1, \dots, n\}$, *where* $\#I = r$, *and there exist constants* K *and* $\alpha > 0$ *such that*

$$\sum_{p=m}^{k-1} [\log |a_i(p)| - \log |a_j(p)|] \le K - \alpha(k-m) \tag{24}$$

for $i \in I$, $j \notin I$ *and* $k \ge m$ *in* J. *Morever, if* (23) *is exponentially separated, it is also strongly exponentially separated.*

Proof Suppose (23) is exponentially separated on $\mathbb{Z}_+$ (resp. $\mathbb{Z}_-$). Then by Lemma 4, there exists I such that the stable (resp. unstable) subspace is $\text{span}\{e_i : i \in I\}$ (resp. $\text{span}\{e_i : i \notin I\}$), where the e_i form the standard basis in $\mathbb{R}^n$. Then we may take the unstable (resp. stable) subspace as span$\{e_i : i \notin I\}$ (resp. span$\{e_i : i \in I\}$). If (23) is exponentially separated on $\mathbb{Z}$, then these must of course be the stable and unstable subspaces. Then if $x_i(k) = \prod_{p=0}^{k-1} a_i(p)e_i$, there exist positive constants K and α such that

$$\frac{|x_i(k)||x_j(m)|}{|x_i(m)||x_j(k)|} \leq Ke^{-\alpha(k-m)}, \quad k \geq m \text{ in } J,$$

for $i \in I$, $j \notin I$, from which the inequalities (24) follow with $\log K$ instead of K.

Suppose conversely that inequalities (24) hold. Let P be the projection with range span$\{e_i : i \in I\}$ and nullspace span$\{e_i : i \notin I\}$. Then, for $k \geq m$, if $\Phi(k, m)$ is the transition matrix for (23), $\Phi(k, m)P$ is a diagonal matrix with $\prod_{p=m}^{k-1} a_i(p)$ in the ith position when $i \in I$ and 0 otherwise and $\Phi(m, k)(\mathbf{I} - P)$ is a diagonal matrix with $\prod_{p=m}^{k-1}(1/a_j(p))$ in the jth position when $j \notin I$ and 0 otherwise. If we use the maximum norm for matrices, then for all $k \geq m$ there exist $i \in I$ and $j \notin I$ such that

$$|\Phi(k, m)P|\,|\Phi(m, k)(\mathbf{I} - P)| = \prod_{p=m}^{k-1} |a_i(p)| \prod_{p=m}^{k-1} (1/|a_j(p)|) \leq e^K e^{-\alpha(k-m)}.$$

The strong exponential separation follows.

Finally we examine the relation between the strong exponential separation of an upper triangular system and its associated diagonal system.

Proposition 9 *If the upper triangular system*

$$u(k+1) = A(k)u(k) = \begin{pmatrix} a_{11}(k) & a_{12}(k) & \cdots & a_{1n}(k) \\ 0 & a_{22}(k) & \cdots & a_{2n}(k) \\ \cdot & \cdot & \cdots & \cdot \\ 0 & 0 & \cdots & a_{nn}(k) \end{pmatrix} u$$

is strongly exponentially separated on a half-axis with rank r, then there exists $I \subset \{1, \ldots, n\}$, where $\#I = r$, and there exist constants K and $\alpha > 0$ such that

$$\sum_{p=m}^{k-1} [\log |a_{ii}(p)| - \log |a_{jj}(p)|] \leq K - \alpha(k-m)$$

for $i \in I$, $j \notin I$ and $k \geq m$. The converse holds if $\frac{a_{ij}(k)}{a_{ii}(k)}$ is bounded for $i < j$ and on the whole axis also.

Proof Immediate from Propositions 8, 6 and 7.

7 Exponential Separation in Linear Symplectic Systems

In this section we restrict to bounded systems so that exponential separation is equivalent to strong exponential separation. We begin with the observation that, in general, exponential separation does not imply that the *dichotomy spectrum* (or Sacker–Sell spectrum) has at least two components. For example, consider the diagonal system

$$x(k+1) = \begin{pmatrix} a(k) & 0 \\ 0 & e^{\gamma}a(k) \end{pmatrix} x(k)$$

on $\mathbb{Z}_+$, with $0 < \gamma < \beta - \alpha$, where $a(k)$ is a bounded real sequence such that $1/a(k)$ is also bounded satisfying

$$\alpha = \liminf_{k-m\to\infty} \sum_{p=m}^{k-1} \log|a(p)| < \beta = \limsup_{k-m\to\infty} \sum_{p=m}^{k-1} \log|a(p)|.$$

This is clearly exponentially separated. However for any $\lambda > 0$,

$$x(k+1) = \begin{pmatrix} \lambda^{-1}a(k) & 0 \\ 0 & \lambda^{-1}e^{\gamma}a(k) \end{pmatrix} x(k)$$

has an exponential dichotomy if and only if each scalar equation has (see Theorem 3.1 in [12]) and hence if and only if (see Proposition 2.4 in [12])

$$1 \notin [e^{\alpha}/\lambda, e^{\beta}/\lambda], \text{ that is, } \lambda \notin [e^{\alpha}, e^{\beta}]$$

and

$$1 \notin [e^{\alpha+\gamma}/\lambda, e^{\beta+\gamma}/\lambda], \text{ that is, } \lambda \notin [e^{\alpha+\gamma}, e^{\beta+\gamma}].$$

So the dichotomy spectrum of the system is the single interval $[e^{\alpha}, e^{\beta}] \cup [e^{\alpha+\gamma}, e^{\beta+\gamma}] = [e^{\alpha}, e^{\beta+\gamma}]$.

However in this section, we show when the system is bounded linear symplectic and it is exponentially separated with subspaces of the same dimension, the system has an exponential dichotomy with stable and unstable subspaces having the same dimension and hence the dichotomy spectrum has at least two components.

Let

$$x(k+1) = A(k)x(k), \quad x(k) \in \mathbb{R}^{2n} \tag{25}$$

be a linear system, where $A(k)$ is bounded together with its inverse, and is symplectic, that is,

$$A^*(k)\mathcal{J}A(k) = \mathcal{J}, \quad \text{where } \mathcal{J} = \begin{pmatrix} 0 & \mathbf{I}_n \\ -\mathbf{I}_n & 0 \end{pmatrix}.$$

Theorem 2 *Let* (25) *be a linear symplectic system, where* $A(k)$ *is bounded together with its inverse. Suppose* (25) *is exponentially separated on an interval* $J = \mathbb{Z}$, $\mathbb{Z}_+$ *or* $\mathbb{Z}_-$ *such that the stable and unstable subspaces have the same dimension* n. *Then* (25) *has an exponential dichotomy on* J *with stable subspace of dimension* n.

Proof The transition matrix $\Phi(k, m)$ associated with (25) is symplectic. So, as in [2], we have the Iwasawa decomposition

$$\Phi(k, 0) = G(k)R(k),$$

where $G(k)$ is orthogonal and

$$R(k) = \begin{pmatrix} R_{11}(k) & R_{12}(k) \\ 0 & (R_{11}^*(k))^{-1} \end{pmatrix},$$

$R_{11}(k)$ being upper triangular with positive diagonal entries. Note that

$$A(k) = \Phi(k+1, k) = \Phi(k+1, 0)\Phi(0, k) = G(k+1)R(k+1)R^{-1}(k)G^{-1}(k).$$

The transformation $x(k) = G(k)y(k)$ takes system (25) to

$$y(k+1) = B(k)y(k) = G^{-1}(k+1)A(k)G(k)y(k) = R(k+1)R^{-1}(k)y(k). \tag{26}$$

It is clear that $B(k)$ and its inverse are bounded. Of course, it is trivial that $G(k)$ and its inverse are bounded. It follows that (25) and (26) share those dynamical properties preserved by kinematic similarity.

$B(k)$ is a block upper triangular matrix

$$B(k) = R(k+1)R^{-1}(k) = \begin{pmatrix} B_{11}(k) & B_{12}(k) \\ 0 & B_{22}(k) \end{pmatrix},$$

where $B_{11}(k) = R_{11}(k+1)R_{11}^{-1}(k)$ is upper triangular and

$$B_{22}(k) = (R_{11}^*(k+1))^{-1}R_{11}^*(k) = (B_{11}^*)^{-1}(k).$$

Then $B_{22}(k)$ is lower triangular with diagonal entries the reciprocal of those in $B_{11}(k)$.

Since (25), and therefore also (26), is exponentially separated with rank n on an interval $J = \mathbb{Z}$, $\mathbb{Z}_+$ or $\mathbb{Z}_-$, we conclude using Proposition 6 that the block diagonal system

$$y(k+1) = \begin{pmatrix} B_{11}(k) & 0 \\ 0 & B_{22}(k) \end{pmatrix} y(k)$$

is exponentially separated with rank n on J also. We can turn this into an upper triangular system by reversing the order of the last n variables. Then we can conclude, by Proposition 9, that on each half-axis (that is, on J when $J = \mathbb{Z}_+$ or $\mathbb{Z}_-$ and on

$\mathbb{Z}_+$ and $\mathbb{Z}_-$ when $J = \mathbb{Z}$), its diagonal part is also strongly exponentially separated with rank n. Returning to the original variables, this diagonal system has the form

$$y_i(k+1) = a_i(k)y(k), \quad i = 1, \dots, 2n,$$

where $a_{i+n}(k) = 1/a_i(k)$ for $i = 1, \dots, n$. It follows from Proposition 8 that there exist constants K and $\alpha > 0$ such that the set $\{1, 2, \dots, 2n\}$ can be divided into two disjoint subsets I_1 and I_2 of cardinality n with

$$\sum_{p=m}^{k-1} [\log |a_i(p)| - \log |a_j(p)|] \le K - \alpha(k-m), \quad k \ge m$$

when $i \in I_1$ and $j \in I_2$. Suppose we have $i \in I_1$ and $i + n \in I_1$ for some i. We subtract n from those numbers in I_1 or I_2 which exceed n. So there must be a number j, $1 \le j \le n$, which after this subtraction does not appear in I_1. This means $j \in I_2$ and $j + n \in I_2$. Then

$$\sum_{p=m}^{k-1} [\log |a_i(p)| - \log |a_j(p)|] \le K - \alpha(k-m), \quad k \ge m$$

and

$$\sum_{p=m}^{k-1} [\log(|a_{i+n}(p)| - \log |a_{j+n}(p)|] \le K - \alpha(k-m), \quad k \ge m.$$

The latter means that

$$\sum_{p=m}^{k-1} [-\log(|a_i(p)| + \log |a_j(p)|] \le K - \alpha(k-m), \quad k \ge m.$$

and we conclude that

$$-K + \alpha(k-m) \le \sum_{p=m}^{k-1} [\log |a_i(p)| - \log |a_j(p)| \le K - \alpha(k-m)$$

so that

$$\alpha(k-m) \le K, \quad k \ge m,$$

which is clearly impossible. Hence i and $i + n$ cannot both belong to I_1. Similar reasoning shows that they cannot both belong to I_2.

Suppose $i \in I_1$. Then $i + n \in I_2$, where we interpret $i + n$ as $i - n$ when $i > n$. Then

$$\sum_{p=m}^{k-1}[\log|a_i(p)| - (-\log|a_i(p)|] \le K - \alpha(k-m), \quad k \ge m$$

so that

$$\sum_{p=m}^{k-1}\log|a_i(p)| \le K/2 - (\alpha/2)(k-m), \quad k \ge m.$$

Next suppose $i \in I_2$. Then $i + n \in I_1$, where we interpret $i + n$ as $i - n$ when $i > n$. Then

$$\sum_{p=m}^{k-1}[-\log|a_i(p)| - \log|a_i(p)|]u \le K - \alpha(k-m), \quad k \ge m$$

so that

$$\sum_{p=m}^{k-1}\log|a_i(p)| \ge -K/2 + (\alpha/2)(k-m), \quad k \ge m.$$

Hence the diagonal system has an exponential dichotomy on J with projection of rank n when J is a half-axis and on each half-axis when $J = \mathbb{Z}$. Then, by Theorem 4.1 in [12], system (26) and hence (25) has an exponential dichotomy on J with projection of rank n when J is a half-axis and on both half-axes when J is the whole axis. This finishes the proof when $J = \mathbb{Z}_+$ or $\mathbb{Z}_-$.

When $J = \mathbb{Z}$, we know from Corollary 1 that on $\mathbb{Z}_+$ the stable subspace (for the exponential separation) on $\mathbb{Z}_+$ and the unstable subspace on $\mathbb{Z}_-$ intersect in 0 and each has dimension n. However, by Corollary 3, since (25) has an exponential dichotomy with projection not equal to zero or the identity, it is exponentially separated with the same stable and unstable subspaces, the stable (resp. unstable) subspace being unique when $J = \mathbb{Z}_+$ (resp. $\mathbb{Z}_-$). Hence, by the discrete analogue of Proposition 1 in [1], system (25) has an exponential dichotomy on $\mathbb{Z}$ with projection of rank n.

Acknowledgements F.B. was partially supported by GNAMPA-INdAM (Italy) and K.P. received support from Marche Polytechnic University (Ancona, Italy).

References

1. Battelli, F., Palmer, K.J.: Criteria for exponential dichotomy for triangular systems. J. Math. Anal. Appl. **428**, 525–543 (2015)
2. Battelli, F., Palmer, K.J.: Strongly exponentially separated linear systems. J. Dyn. Differ. Equ. **31**, 573–600 (2019)
3. Coppel, W.A.: Dichotomies and Stability. Lecture Notes. Springer, Berlin (1978)
4. Cong, N.D., Son, D.T.: An open set of unbounded cocycles with simple Lyapunov spectrum and no exponential separation. Stoch. Dyn. **7**, 335–355 (2007)
5. Fang, C., Gyllenberg, M., Liu, S.: Characterizations of dominated splitting and its relation to hyperbolicity. arXiv: 1209.5595 [math.DS]

6. Kalkbrenner, J.: Exponentielle Dichotomie und chaotische Dynamik nichtinvertierbarer Differenzengleichungen, Dissertation, Universität Augsburg (1994)
7. Palmer, K.J.: Exponential dichotomy, exponential separation and spectral theory for linear systems of ordinary differential equations. J. Differ. Equ. **46**, 324–345 (1982)
8. Palmer, K.J.: Exponential dichotomies, the shadowing lemma and transversal homoclinic points. In: Kirchgraber, U., Walther, H.-O. (eds.) Dynamics Reported, vol. 1, pp. 265–306. B.G. Teubner/Wiley, Stuttgart/Chichester (1988)
9. Papaschinopoulos, G.: Exponential separation, exponential dichotomy and almost periodicity of linear difference equations. J. Math. Anal. Appl. **120**, 276–287 (1986)
10. Papaschinopoulos, G.: A characterization of exponential separation in terms of reducibility for linear difference equations. J. Math. Anal. Appl. **133**, 231–248 (1988)
11. Pötzsche, C.: Nichtautonome Differenzengleichungen mit stationären und invarianten Mannigfaltigkeiten. Diplomarbeit, Universität Augsburg (1998)
12. Pötzsche, C.: Dichotomy spectra of triangular equations. Discret. Contin. Dyn. Syst. Ser. A **36**, 423–450 (2016)

An Integrable SIS Model on Time Scales

Martin Bohner and Sabrina Streipert

Abstract In this work, we generalize the dynamic model introduced in Bohner and Streipert (Pliska Stud. Math. 26:11–28, 2016, [5]) in the context of epidemiology. This model exhibits many similarities to the continuous susceptible-infected-susceptible model and is therefore of particular interest to formulate a generalization of a continuous model on time scales. In this work, we extend the results in Bohner and Streipert (Pliska Stud. Math. 26:11–28, 2016, [5]) for time-dependent coefficients rather than constant parameters and derive an explicit solution. We further discuss the stability of periodic solutions for the corresponding discrete model with periodic coefficients. We conclude the analysis of the SIS model by considering time-dependent vital dynamics and derive its explicit solution on a general time scale.

Keywords Time scales · Dynamic equations · Difference equations · Epidemiology · Periodic solution · Stability

1 Introduction

The susceptible-infected-removed model introduced by Kermack and McKendrick remains an important mathematical tool to investigate diseases, see for example [8, 9]. Not only diseases can be modelled with these epidemic models but also the spreading of ideas. For example in [12], the authors apply such a classic disease model to investigate the development of ideas on social media. A similar idea was applied in [7] and in [11] to investigate the success of viral marketing.

M. Bohner (✉)
Missouri University of Science and Technology, Rolla, MO 65409, USA
e-mail: bohner@mst.edu

S. Streipert
Centre for Applications in Natural Resource Mathematics, University of Queensland, St Lucia, QLD 4072, Australia
e-mail: s.streipert@uq.edu.au

M. Bohner et al. (eds.), *Difference Equations and Discrete Dynamical Systems with Applications*, Springer Proceedings in Mathematics & Statistics 312,
https://doi.org/10.1007/978-3-030-35502-9_7

The susceptible-infected-susceptible SIS model describes diseases, where, after recovery, no immunity is manifested, so that individuals can become infected right after recovery. The model then reads as

$$\begin{aligned} S' &= -\beta S I + \delta I, \\ I' &= \beta S I - \delta I, \end{aligned} \tag{1}$$

where S represents the group of susceptible individuals that can get infected, I is the group of infected individuals that transmit the disease with transmission rate $\beta > 0$ and recover with recovery rate $\delta > 0$. A critical property is that in this model, the total population remains constant, as $(S + I)' = 0$.

The interest in discrete and time scales analogues of this model mounts on several reasons, besides the mathematical interest of generalizing mathematical models to any time domain. A practical reason is that computational methods analysing the spread of diseases rely upon a discrete analogue behaving similar to the continuous model. From a modelling perspective, the interest is caused by describing diseases that spread at specific time ranges, for example, within a certain temperature range.

Some authors did formulate discrete versions of these traditional epidemic models, but there has been no attempt, besides [5, 6], of generalizing it to any time scales. Further, the discrete formulations often exhibit undesirable behaviour, see [1]. In [5], the authors introduced a discrete epidemic model of the SIS-form with constant transmission rate β and constant recovery rate γ that displays similar characteristics as the continuous model (1). Further, the authors also generalized the model in [5] to any time domain that is a nonempty subset of the real numbers.

One of many modifications of the classical model (1) is the implementation of a birth and death rate that allows the total population to change over time. To consider vital dynamics, a constant birth rate aK and a proportionate death rate a was added to the model, which is then

$$\begin{aligned} S' &= -\beta S I + \delta I + aK - aS, \\ I' &= \beta S I - \delta I - aI. \end{aligned} \tag{2}$$

The explicit solution of (2) is given in [10] as

$$\begin{aligned} S(t) = {}& \frac{a+\delta}{\beta} + \frac{1}{\beta}\frac{A\exp\{-(a+\delta-\beta K)t + Be^{-at}\}}{Aa\int \exp\{-(a+\delta-\beta K)s + Be^{-as}\}\mathrm{d}s + C} \\ & \times\Bigg\{a+\delta-\beta K - aBe^{-at} \\ & + \frac{1}{\beta}\frac{A\exp\{-(a+\delta-\beta K)t + Be^{-at}\}}{Aa\int \exp\{-(a+\delta-\beta K)s + Be^{-as}\}\mathrm{d}s + C}\Bigg\}, \end{aligned} \tag{3}$$

$$I(t) = \frac{1}{\beta}\frac{A\exp\{-(a+\delta-\beta K)t + Be^{-at}\}}{Aa\int \exp\{-(a+\delta-\beta K)s + Be^{-as}\}\mathrm{d}s + C}. \tag{4}$$

In [5], the SIS model on time scales was extended using vital dynamics and an analogue of the results in [10] was obtained.

In this work, we generalize the SIS-model with and without vital dynamics, considering a time-dependent transmission and time-dependent recovery rate. This assists in describing diseases, where the probability of infection during an interaction between a susceptible and infected individual is time-dependent. This allows to model for example an increasing awareness of the disease, rising cautious behaviour, resulting in a reduction of the transmission rate, as well as medical advances that lead to an increase in the recovery rate.

Before we introduce the formulation of the SIS model on time scales, we provide a brief introduction of time scales, giving a short summary of time scales definitions and results used in this paper.

2 Time Scales

A time scale $\mathbb{T}$ is an arbitrary nonempty closed subset of the real numbers $\mathbb{R}$, see [3, p. 1].

Definition 1 (See [3, Definition 1.1]) For $t \in \mathbb{T}$, the forward jump operator $\sigma : \mathbb{T} \to \mathbb{T}$ is

$$\sigma(t) := \inf\{s \in \mathbb{T} : s > t\}.$$

For any function f, we put $f^\sigma = f \circ \sigma$.

For the definition of an *rd-continuous* function $f : \mathbb{T} \to \mathbb{R}$, denoted by $f \in \mathrm{C}_{\mathrm{rd}}$, see [3, Definition 1.24].

Definition 2 (See [3, Definition 2.25]) A function $p : \mathbb{T} \to \mathbb{R}$ is called *regressive* provided

$$1 + \mu(t)p(t) \neq 0 \quad \text{for all} \quad t \in \mathbb{T}, \quad \text{where} \quad \mu(t) = \sigma(t) - t.$$

The set of regressive and rd-continuous functions is denoted by $\mathcal{R}$. Moreover, $p \in \mathcal{R}$ is called *positively regressive*, denoted by $\mathcal{R}^+$, if

$$1 + \mu(t)p(t) > 0 \quad \text{for all} \quad t \in \mathbb{T}.$$

For the definition of the *delta derivative* f^Δ of a function $f : \mathbb{T} \to \mathbb{R}$, we refer to [3, Definition 1.10]. For the reader unfamiliar with time scales, it helps to recall that $f^\Delta = f'$ if $\mathbb{T} = \mathbb{R}$ and $f^\Delta = \Delta f$ if $\mathbb{T} = \mathbb{Z}$, where $\Delta f(t) = f(t+1) - f(t)$ for $t \in \mathbb{Z}$ is the usual forward difference operator. We will use the product rule for $f, g : \mathbb{T} \to \mathbb{R}$ that reads as

$$(fg)^\Delta = f^\Delta g + f^\sigma g^\Delta = f^\Delta g^\sigma + f g^\Delta. \tag{5}$$

Theorem 1 (See [3, Theorem 2.33]) *Let $p \in \mathcal{R}$ and $t_0 \in \mathbb{T}$. Then*

$$y^{\Delta} = p(t)y, \quad y(t_0) = 1$$

possesses a unique solution, denoted by $e_p(\cdot, t_0)$, called the time scales exponential function.

Useful properties of the time scales exponential function are the following.

Theorem 2 (See [3, Theorem 2.36]) *If $p \in \mathcal{R}$, then*

1. $e_0(t, s) = 1$, *and* $e_p(t, t) = 1$,
2. $e_p(t, s) = \frac{1}{e_p(s,t)}$,
3. *the semigroup property holds:* $e_p(t, r)e_p(r, s) = e_p(t, s)$.

Theorem 3 (See [3, Theorem 2.44]) *If $p \in \mathcal{R}^+$ and $t_0 \in \mathbb{T}$, then $e_p(t, t_0) > 0$ for all $t \in \mathbb{T}$.*

Definition 3 (See [3, p. 58]) Define the "circle plus" addition on $\mathcal{R}$ as

$$(p \oplus q)(t) = p(t) + q(t) + \mu(t)p(t)q(t)$$

and the "circle minus" subtraction as

$$(p \ominus q)(t) = \frac{p(t) - q(t)}{1 + \mu(t)q(t)}.$$

The following identities hold.

Corollary 1 (See [3, Corollary 2.36]) *If $p, q \in \mathcal{R}$, then*

(a) $e_{p\oplus q}(t, s) = e_p(t, s)e_q(t, s)$,
(b) $e_{\ominus p}(t, s) = e_p(s, t) = \frac{1}{e_p(t,s)}$,
(c) $e_{p\ominus q}(t, s) = \frac{e_p(t,s)}{e_q(t,s)}$.

Theorem 4 (Variation of Constants, see [4, Theorem 2.1]) *Let $p \in \mathcal{R}$, $f \in C_{rd}$, $t_0 \in \mathbb{T}$, and $y_0 \in \mathbb{R}$. The unique solution of the IVP*

$$y^{\Delta} = p(t)y + f(t), \quad y(t_0) = y_0$$

is given by

$$y(t) = e_p(t, t_0)y_0 + \int_{t_0}^{t} e_p(t, \sigma(s))f(s)\Delta s.$$

The unique solution of the IVP

$$y^{\Delta} = -p(t)y^{\sigma} + f(t), \quad y(t_0) = y_0$$

is given by

$$y(t) = e_{\ominus p}(t, t_0)y_0 + \int_{t_0}^{t} e_{\ominus p}(t, s)f(s)\Delta s.$$

Lemma 1 (See [3, Theorem 2.39]) *If* $p \in \mathcal{R}$ *and* $a, b, c \in \mathbb{T}$, *then*

$$\int_{a}^{b} p(t)e_p(t, c)\Delta t = e_p(b, c) - e_p(a, c)$$

and

$$\int_{a}^{b} p(t)e_p(c, \sigma(t))\Delta t = e_p(c, a) - e_p(c, b).$$

3 Dynamic SIS Model

In this section, we introduce a generalization of the classical susceptible-infected-susceptible model on time scales, as presented in [5], by considering a time-dependent transmission and time-dependent recovery rate.

Let us consider the susceptible-infected-susceptible model of the form

$$S^{\Delta} = -\beta(t)S^{\sigma}I + \gamma(t)I, \tag{SISa}$$
$$I^{\Delta} = \beta(t)S^{\sigma}I - \gamma(t)I, \tag{SISb}$$

where $S, I : \mathbb{T} \to \mathbb{R}^{+}$ are the susceptible and infected population, respectively. The functions $\beta, \gamma : \mathbb{T} \to \mathbb{R}^{+}$ are the transmission and recovery rate, respectively.

Remark 1 For the reader concerned about the presence of the factor $S(\sigma(t))I(t)$ on the right-hand sides of (SISa) and (SISb), we remark that it is easy to see that (SISa) and (SISb) may be, using the "simple useful formula" $S^{\sigma} = S + \mu S^{\Delta}$, rewritten as

$$S^{\Delta} = -\frac{\beta(t)SI}{1 + \mu(t)\beta(t)I} + \frac{\gamma(t)I}{1 + \mu(t)\beta(t)I},$$
$$I^{\Delta} = \frac{\beta(t)SI}{1 + \mu(t)\beta(t)I} - \frac{\gamma(t)I}{1 + \mu(t)\beta(t)I}.$$

Theorem 5 *If* $\beta \in \mathrm{C_{rd}}$ *and* $p := \beta N - \gamma \in \mathcal{R}^{+}$ *for* $N = I_0 + S_0$, *then the unique solution to* (SISa)–(SISb) *with initial conditions* $S_0 = S(t_0) > 0$ *and* $I_0 = I(t_0) > 0$ *is given by*

$$\begin{aligned} S(t) &= N - \frac{(N - S_0)e_p(t, t_0)}{1 + (N - S_0)\int_{t_0}^{t} e_p(s, t_0)\beta(s)\Delta s}, \\ I(t) &= \frac{I_0 e_p(t, t_0)}{1 + I_0 \int_{t_0}^{t} e_p(s, t_0)\beta(s)\Delta s}. \end{aligned} \tag{6}$$

Proof First, if S and I solve (SISa)–(SISb), then we realize that $(S + I)^\Delta = 0$ and therefore $S + I = N = S_0 + I_0$. Replacing $S = N - I$ in (SISb), we have

$$I^\Delta = \beta(t)(N - I)^\sigma I - \gamma(t)I = \beta(t)(N - I^\sigma)I - \gamma(t)I.$$

For $I \neq 0$, we apply the transformation $v = 1/I$. We then have

$$v^\Delta = -\beta(t)N\frac{1}{I^\sigma} + \beta(t) + \gamma(t)\frac{1}{I^\sigma} = v^\sigma(-\beta(t)N + \gamma(t)) + \beta(t) = -p(t)v^\sigma + \beta(t),$$

where $p(t) = \beta(t)N - \gamma(t)$. By assumption, $p \in \mathcal{R}$, and therefore, the solution of this first-order dynamic equation in v is given by Theorem 4 as

$$v(t) = e_{\ominus p}(t, t_0)v_0 + \int_{t_0}^{t} e_{\ominus p}(t, s)\beta(s)\Delta s, \tag{7}$$

and therefore

$$I(t) = \frac{I_0}{e_{\ominus p}(t, t_0) + I_0 \int_{t_0}^{t} e_{\ominus p}(t, s)\beta(s)\Delta s}.$$

Using Corollary 1 and Theorem 2, we can express this in the form of (6).

Conversely, assume that I is defined by (6) and let $S = N - I$. Hence

$$\beta(t)S^\sigma - \gamma(t) = \beta(t)N - \gamma(t) - \beta(t)I^\sigma = p(t) - \beta(t)I^\sigma. \tag{8}$$

Using (6), we have

$$I\left(1 + I_0\int_{t_0}^{t} e_p(s, t_0)\beta(s)\Delta s\right) = I_0 e_p(t, t_0).$$

By (5), we get

$$I^\Delta\left(1 + I_0\int_{t_0}^{t} e_p(s, t_0)\beta(s)\Delta s\right) + I^\sigma I_0 e_p(t, t_0)\beta(t) = I_0 p(t)e_p(t, t_0),$$

and thus, again by (6),

$$I^\Delta = I\left[p(t) - \beta(t)I^\sigma\right] \overset{(9)}{=} \beta(t)S^\sigma I - \gamma(t)I,$$

completing the proof.

We begin the discussion about the stability of the disease-free equilibrium by recalling the following lemma.

Lemma 2 (See [2, Lemma 2]) *If f is nonnegative and $-f \in \mathcal{R}^+$, then*

$$1 - \int_s^t f(\tau)\Delta\tau \leq e_{-f}(t, s) \leq e^{-\int_s^t f(\tau)\Delta\tau}.$$

Theorem 6 *Consider* (SISa)–(SISb) *with* $\beta, \gamma : \mathbb{T} \to \mathbb{R}^+$, $\beta \in \mathrm{C}_{\mathrm{rd}}$, *for a time scale* $\mathbb{T}$ *that is unbounded above and initial conditions* $I_0, S_0 > 0$ *such that* $I_0 + S_0 = N$. *If* $p := \beta N - \gamma \in \mathcal{R}^+$ *is eventually negative and* $\int_{t_0}^{\infty} |p(\tau)|\Delta\tau = \infty$, *then the disease-free equilibrium is globally asymptotically stable.*

Proof Note that $p = \beta N - \gamma \in \mathcal{R}^+$ implies by Theorem 3 that $e_p(t, s) > 0$ for all $t, s \in \mathbb{T}$. By Theorem 5, the solution to (SISb) is given by

$$I(t) = \frac{I_0 e_p(t, t_0)}{1 + I_0 \int_{t_0}^t e_p(s, t_0)\beta(s)\Delta s} \leq I_0 e_p(t, t_0) = I_0 e_p(t^*, t_0) e_p(t, t^*).$$

Applying Lemma 2 to $f = -p$ and using $p \in \mathcal{R}^+$ and $p(t) < 0$ for $t \geq t^*$, we have

$$0 \leq I(t) \leq I_0 e_p(t^*, t_0) e_p(t, t^*) \leq I_0 e_p(t^*, t_0) e^{\int_{t^*}^t (\beta(s)N - \gamma(s))\Delta s} \overset{t\to\infty}{\longrightarrow} 0,$$

which completes the proof.

Remark 2 Let us now analyse in more detail the special case of the discrete space, i.e., $\mathbb{T} = \mathbb{Z}$. Theorem 5 provides the solution to the discrete SIS model with time-dependent parameters

$$\begin{aligned} \Delta S_t &= -\beta_t S_{t+1} I_t + \gamma_t I_t, \\ \Delta I_t &= \beta_t S_{t+1} I_t - \gamma_t I_t. \end{aligned} \tag{9}$$

The derivation of this nonclassical model formulation follows the idea that the group of susceptible at time $t + 1$ (the end of period $(t, t + 1]$/start of $[t + 1, t + 2)$) is given by a proportion φ_t of the susceptible at the beginning of time-period $(t, t + 1]$ and the proportion of recovered individuals that were infected at the start of the time period $(t, t + 1]$. That is,

$$S_{t+1} = \varphi_t (S_t + \gamma_t I_t),$$

where φ is a multiplicative factor determining how many of the group of susceptible remain susceptible. This percentage φ_t should not only be within $[0, 1]$ but also decrease as I_t increases. Hence, one immediate choice is $\varphi_t = \frac{1}{1+\beta_t I_t}$, which results in the model stated above.

Considering a periodic transmission and periodic recovery rate, we analyse the existence and stability of a periodic solution to the discrete SIS model. The results are summarized in the following theorem.

Theorem 7 *Let* $p_t := \beta_t N - \gamma_t > -1$ *for all* $t \in \mathbb{Z}$, *where* $N = I_0 + S_0$ *and* I_0, $S_0 > 0$. *Further, let* $\beta, \gamma : \mathbb{Z} \to \mathbb{R}^+$ *be* ω*-periodic, that is,* $\beta_t = \beta_{t+\omega}$ *and* $\gamma_t = \gamma_{t+\omega}$ *for all* $t \in \mathbb{Z}$. *Then the unique* ω*-periodic solution to* (9), *i.e.,*

$$S_{t+1} = \frac{S_t + \gamma_t I_t}{1 + \beta_t I_t},$$
$$I_{t+1} = (1 - \gamma_t) I_t + \beta_t \frac{S_t + \gamma_t I_t}{1 + \beta_t I_t} I_t,$$

is given by

$$\bar{I}_t = (e_p(\omega, 0) - 1) \left[\sum_{i=t}^{t+\omega-1} e_p(i, t) \beta_i \right]^{-1},$$
$$\bar{S}_t = N - (e_p(\omega, 0) - 1) \left[\sum_{i=t}^{t+\omega-1} e_p(i, t) \beta_i \right]^{-1}.$$

If additionally $e_p(\omega, 0) > 1$, *then the* ω*-periodic solution is globally asymptotically stable.*

Proof By Theorem 5, the solution to (9) is given by

$$S_t = N - \frac{(N - S_0) e_p(t, t_0)}{1 + (N - S_0) \sum_{i=t_0}^{t-1} e_p(i, t_0) \beta_i},$$
$$I_t = \frac{I_0 e_p(t, t_0)}{1 + I_0 \sum_{i=t_0}^{t-1} e_p(i, t_0) \beta_i},$$

where

$$p_t = \beta_t N - \gamma_t \quad \text{and} \quad e_p(t, s) = \prod_{i=s}^{t-1} (1 + p_i), \quad s < t.$$

We first note that since β, γ are ω-periodic, so is p and $\ominus p$. For $I \neq 0$, we let $v = 1/I$. Then an ω-periodic solution $\bar{v}$ satisfies

$$\begin{aligned}
\bar{v}_t = \bar{v}_{t+\omega} &\overset{(8)}{=} e_{\ominus p}(t + \omega, t_0) \bar{v}_0 + \sum_{i=t_0}^{t+\omega-1} e_{\ominus p}(t + \omega, i) \beta_i \\
&= e_{\ominus p}(t + \omega, t) e_{\ominus p}(t, t_0) \bar{v}_0 + e_{\ominus p}(t + \omega, t) \sum_{i=t_0}^{t-1} e_{\ominus p}(t, i) \beta_i \\
&\quad + e_{\ominus p}(t + \omega, t) \sum_{i=t}^{t+\omega-1} e_{\ominus p}(t, i) \beta_i \\
&= e_{\ominus p}(\omega, 0) \bar{v}_t + e_{\ominus p}(\omega, 0) \sum_{i=t}^{t+\omega-1} e_{\ominus p}(t, i) \beta_i,
\end{aligned}$$

where we used that if f is ω-periodic, then $e_f(t+\omega,t) = e_f(\omega,0)$. This can be verified by

$$e_f(t+\omega,t) = \prod_{i=t}^{t+\omega-1}(1+f_i) = \prod_{i=0}^{\omega-1}(1+f_i) = e_f(\omega,0).$$

Therefore, a solution that is ω-periodic, i.e., $\bar{v}_t = \bar{v}_{t+\omega}$ for all $t \in \mathbb{T}$, is given by

$$\bar{v}_t = \frac{1}{e_p(\omega,0)-1} \sum_{i=t}^{t+\omega-1} e_p(i,t)\beta_i,$$

which yields the ω-periodic solution $\bar{I}$ as

$$\bar{I}_t = (e_p(\omega,0)-1)\left[\sum_{i=t}^{t+\omega-1} e_p(i,t)\beta_i\right]^{-1}.$$

The corresponding ω-periodic solution $\bar{S}$ is given by

$$\bar{S}_t = N - (e_p(\omega,0)-1)\left[\sum_{i=t}^{t+\omega-1} e_p(i,t)\beta_i\right]^{-1}.$$

Conversely, it is not difficult to show that $\bar{I}$ and $\bar{S}$ solve (9).

The uniqueness implies that there exist unique initial conditions $\bar{I}_0$, $\bar{S}_0$ such that the solutions are ω-periodic. Let, w.l.o.g. $t_0 = 0$. Then

$$\begin{aligned}|I_t - \bar{I}_t| &= \left|\frac{I_0 e_p(t,0)}{1+I_0\sum_{s=0}^{t-1} e_p(s,0)\beta(s)} - \frac{\bar{I}_0 e_p(t,0)}{1+\bar{I}_0\sum_{s=0}^{t-1} e_p(s,0)\beta(s)}\right| \\ &= |I_0 - \bar{I}_0|\left|\frac{1}{1+I_0\sum_{s=0}^{t-1} e_p(s,0)\beta(s)}\right|\left|\frac{e_p(t,0)}{1+\bar{I}_0\sum_{s=0}^{t-1} e_p(s,0)\beta(s)}\right|.\end{aligned}$$

The first and the last term, which is simply $\bar{I}_t/\bar{I}_0$, are bounded. We note further that, for all $t \geq 1$, it exists $k \in \{0,1,2,\ldots\}$ and $r \in \{0,1,2,\ldots,\omega-1\}$ such that $t-1 = k\omega + r$ and hence

$$\begin{aligned}1+I_0\sum_{s=0}^{t-1} e_p(s,0)\beta(s) &\geq 1+I_0\beta_{\min}\sum_{s=0}^{k\omega+r} e_p(s,0) \geq 1+I_0\beta_{\min}\sum_{j=0}^{k} e_p(j\omega,0) \\ &= 1+I_0\beta_{\min}\sum_{j=0}^{k} e_p^j(\omega,0) \overset{k\to\infty}{\longrightarrow} \infty,\end{aligned}$$

where we have used that $\beta, I_0 \in \mathbb{R}^+$, and $e_p(t,s) > 0$ since $p \in \mathcal{R}^+$. This completes the claim.

It should be pointed out that the following theorem can be proven in a similar fashion as Theorem 5.

Theorem 8 *Let* $\beta \in \mathrm{C}_{\mathrm{rd}}$ *and* $p := \gamma - \beta N \in \mathcal{R}^+$. *Then the unique solution to*

$$S^\Delta = -\beta(t)SI^\sigma + \gamma(t)I^\sigma, \tag{SIS2a}$$

$$I^\Delta = \beta(t)SI^\sigma - \gamma(t)I^\sigma, \tag{SIS2b}$$

with initial conditions $S_0 = S(t_0) > 0$ *and* $I_0 = I(t_0) > 0$, *satisfying* $N = S_0 + I_0$, *is given by*

$$S(t) = N - \frac{N - S_0}{e_p(t,t_0) + (N - S_0)\int_{t_0}^t e_p(t,\sigma(s))\beta(s)\Delta s},$$

$$I(t) = \frac{I_0}{e_p(t,t_0) + I_0\int_{t_0}^t e_p(t,\sigma(s))\beta(s)\Delta s}.$$

4 Dynamic SIS Model with Vital Dynamics

In this section, we generalize (2) by considering the dynamic time scale system

$$S^\Delta = -\beta(t)S^\sigma I - \nu(t)S + \gamma(t)I + \nu(t)\kappa, \tag{SISVDa}$$

$$I^\Delta = \beta(t)S^\sigma I - \gamma(t)I - \nu(t)I, \tag{SISVDb}$$

where $S, I, \beta, \gamma, \nu : \mathbb{T} \to \mathbb{R}^+$ represent the group of susceptible, group of infected, transmission rate, recovery rate, birth rate/death rate, respectively. $\kappa \in \mathbb{R}^+$ is a constant representing a constant reproduction density. It is assumed that newborns join the group of susceptible.

This model is a generalization of the SIS model with vital dynamics discussed in [5], where (SISVDa)–(SISVDb) was considered with constant parameters. This model not only generalizes the results in [10] to any time scale but also extends it by considering time-dependent parameters.

Theorem 9 *Let* $\beta \in \mathrm{C}_{\mathrm{rd}}$, $-\nu \in \mathcal{R}$, *and* $p := \beta A^\sigma - (\gamma + \nu) \in \mathcal{R}^+$. *Then the unique solution to* (SISVDa)–(SISVDb) *with initial conditions* $S(t_0) = S_0 > 0$ *and* $I(t_0) = I_0 > 0$ *is given by*

$$\begin{aligned} S(t) &= A(t) - \frac{I_0 e_p(t,t_0)}{1 + I_0\int_{t_0}^t e_p(s,t_0)\beta(s)\Delta s}, \\ I(t) &= \frac{I_0 e_p(t,t_0)}{1 + I_0\int_{t_0}^t e_p(s,t_0)\beta(s)\Delta s}, \end{aligned} \tag{10}$$

where

$$A(t) = e_{-\nu}(t, t_0)(I_0 + S_0 - \kappa) + \kappa.$$

Proof First, if S and I solve (SISVDa)–(SISVDb), then we realize that for $v = S + I$, we have

$$v^{\Delta} = (S + I)^{\Delta} = -\nu(t)(S + I) + \nu(t)\kappa = -\nu(t)v + \nu(t)\kappa.$$

This first-order dynamic equation has a solution given by Theorem 4 as

$$\begin{aligned} v(t) &= e_{-\nu}(t, t_0)v_0 + \int_{t_0}^{t} e_{-\nu}(t, \sigma(s))\nu(s)\kappa\Delta s \\ &= e_{-\nu}(t, t_0)v_0 + \kappa(e_{-\nu}(t, t) - e_{-\nu}(t, t_0)) = e_{-\nu}(t, t_0)(v_0 - \kappa) + \kappa \\ &= e_{-\nu}(t, t_0)(I_0 + S_0 - \kappa) + \kappa =: A(t). \end{aligned}$$

Plugging $S = A - I$ into (SISVDb) and defining $w := 1/I$ for $I \neq 0$, we have

$$\begin{aligned} w^{\Delta} &= \frac{-I^{\Delta}}{II^{\sigma}} = -\beta(t)A(\sigma(t))w^{\sigma} + \beta(t) + \gamma(t)w^{\sigma} + \nu(t)w^{\sigma} \\ &= -p(t)w^{\sigma} + \beta(t), \end{aligned}$$

where

$$p(t) = \beta(t)A(\sigma(t)) - (\gamma(t) + \nu(t)).$$

This first-order dynamic equation has the solution given by Theorem 4 as

$$w(t) = e_{\ominus p}(t, t_0)w_0 + \int_{t_0}^{t} e_{\ominus p}(t, s)\beta(s)\Delta s,$$

which yields (10).

Conversely, assume that I is defined by (10) and let $S = A - I$. Then, as in the proof of Theorem 5, we get

$$I^{\Delta} = I\left[p(t) - \beta(t)I^{\sigma}\right].$$

However, this time, we have

$$\beta(t)S^{\sigma} - \gamma(t) - \nu(t) = \beta(t)A^{\sigma} - \beta(t)I^{\sigma} - \gamma(t) - \nu(t) = p(t) - \beta(t)I^{\sigma},$$

so that

$$I^{\Delta} = I\left[\beta(t)S^{\sigma} - \gamma(t) - \nu(t)\right],$$

i.e., (SISVDb) follows. Recalling $S = A - I$ completes the proof.

Corollary 2 *If* $\beta \in \mathrm{C_{rd}}$, $-\nu \in \mathcal{R}^+$, $\beta A^\sigma - (\gamma+\nu) \in \mathcal{R}^+$, *and* $S_0 + I_0 < \kappa$, *then the solutions of* (SISVDa)–(SISVDb) *are nonnegative for positive initial conditions.*

Proof First, we note that $I(t) = \frac{I_0 e_p(t,t_0)}{1+I_0\int_{t_0}^t e_p(s,t_0)\beta(s)\Delta s} \geq 0$ for $p \in \mathcal{R}^+$. It is left to show that $S(t) \geq 0$, i.e., $I(t) \leq A(t)$. We note that $I_0 < I_0 + S_0 = A(t_0)$ and that A is in fact increasing, since

$$A^\Delta(t) = -\nu(t)e_{-\nu}(t,t_0)(S_0 + I_0 - \kappa) > 0.$$

To show that

$$I(t) = \frac{I_0 e_p(t,t_0)}{1+I_0\int_{t_0}^t e_p(s,t_0)\beta(s)\Delta s} \leq A(t) \quad \text{for all} \quad t \in \mathbb{T},$$

we further note that since $p \in \mathcal{R}^+$, implying $e_p(t,s) > 0$, it suffices to show that

$$I_0 g(t) \leq A(t) \quad \text{for all} \quad t \in \mathbb{T}, \tag{11}$$

where

$$g(t) = e_p(t,t_0) - A(t)\int_{t_0}^t e_p(s,t_0)\beta(s)\Delta s.$$

At $t = t_0$, the inequality (11) is satisfied since $g(t_0) = 1$ and $A(t_0) = S_0 + I_0 > I_0$. We further realize that

$$\begin{aligned}
g^\Delta(t) &= p(t)e_p(t,t_0) - A^\Delta(t)\int_{t_0}^t e_p(s,t_0)\beta(s)\Delta s - A^\sigma(t)e_p(t,t_0)\beta(t)\\
&= (\beta(t)A^\sigma(t) - \gamma(t) - \nu(t))e_p(t,t_0) - A^\Delta(t)\int_{t_0}^t e_p(s,t_0)\beta(s)\Delta s\\
&\quad - A^\sigma(t)e_p(t,t_0)\beta(t)\\
&= -(\gamma(t)+\nu(t))e_p(t,t_0) - A^\Delta(t)\int_{t_0}^t e_p(s,t_0)\beta(s)\Delta s
\end{aligned}$$

and since A is increasing, $A^\Delta > 0$, $g^\Delta < 0$ and therefore, g is decreasing. Thus, since the right-hand-side of (11) is increasing, the inequality remains true for all $t \in \mathbb{T}$.

Remark 3 For $\mathbb{T} = \mathbb{R}$ and $t_0 = 0$, Theorem 9 reads as follows:
The unique solution to

$$\begin{aligned}
S' &= -\beta(t)SI - \nu(t)S + \gamma(t)I + \nu(t)\kappa,\\
I' &= \beta(t)SI - \gamma(t)I - \nu(t)I,
\end{aligned}$$

with initial conditions $S(0) = S_0 > 0$ and $I(0) = I_0 > 0$ is given by

$$S(t) = A(t) - \frac{I_0 e^{\int_0^t p(\tau)\mathrm{d}\tau}}{1 + I_0 \int_0^t e^{\int_0^s p(\tau)\mathrm{d}\tau} \beta(s)\mathrm{d}s},$$
$$I(t) = \frac{I_0 e^{\int_0^t p(s)\mathrm{d}s}}{1 + I_0 \int_0^t e^{\int_0^s p(\tau)\mathrm{d}\tau} \beta(s)\mathrm{d}s},$$

where

$$A(t) = e^{-\int_0^t \nu(\tau)\mathrm{d}\tau}(I_0 + S_0 - \kappa) + \kappa$$

and

$$p(t) = \beta(t)\left(e^{-\int_0^t \nu(\tau)\mathrm{d}\tau}(I_0 + S_0 - \kappa) + \kappa\right) - (\gamma(t) + \nu(t)).$$

If β, γ, ν are constant, this solution coincides with (3) and (4) in [10].

Example 1 If $\mathbb{T} = h\mathbb{Z}$, then (SISVDa)–(SISVDb) read as

$$\Delta S_n = -\beta_n S_{n+h} I_n + \gamma_n I_n - \nu_n S_n + \nu_n \kappa,$$
$$\Delta I_n = \beta_n S_{n+h} I_n - \gamma_n I_n - \nu_n I_n.$$

By Theorem 9, the solution for $t_0 = 0$ is given by

$$S_n = A_n - \frac{I_0 \prod_{i=0}^{n-h}(1 + hp_i)}{1 + I_0 \sum_{i=0}^{n-h}\left(\prod_{j=0}^{i-h}(1 + hp_j)\right)\beta_i},$$
$$I_n = \frac{I_0 \prod_{i=0}^{n-h}(1 + hp_i)}{1 + I_0 \sum_{i=0}^{n-h}\left(\prod_{j=0}^{i-h}(1 + hp_j)\right)\beta_i},$$

where

$$A_n = \left(\prod_{i=0}^{n-h}(1 - h\nu_i)\right)(I_0 + S_0 - \kappa) + \kappa$$

and

$$p_n = \beta_n\left(\prod_{i=0}^{n}(1 - h\nu_i)\right)(I_0 + S_0 - \kappa) + (\beta_n\kappa - \gamma_n - \nu_n).$$

References

1. Allen, L.J.S.: Some discrete-time SI, SIR, and SIS epidemic models. Math. Biosci. **124**, 83–105 (1994)
2. Bohner, M.: Some oscillation criteria for first order delay dynamic equations. Far East J. Appl. Math. **18**, 289–304 (2005)

3. Bohner, M., Peterson, A.: Dynamic Equations on Time Scales. Birkhäuser, Boston (2001)
4. Bohner, M., Peterson, A.: Advances in Dynamic Equations on Time Scales. Birkhäuser, Boston (2003)
5. Bohner, M., Streipert, S.: The SIS-model on time scales. Pliska Stud. Math. **26**, 11–28 (2016)
6. Bohner, M., Streipert, S., Torres, D.F.M.: Exact solution to a dynamic SIR model. Nonlinear Anal. Hybrid Syst. **32**, 228–238 (2019)
7. Kandhway, K., Kuri, J.: How to run a campaign: optimal control of SIS and SIR information epidemics. Appl. Math. Comput. **231**, 79–92 (2014)
8. Keeling, M.J., Rohani, P.: Modeling Infectious Diseases in Humans and Animals, p. xiii. Princeton University Press, Princeton (2008)
9. Mkhatshwa, T., Mummert, A.: Modeling super-spreading events for infectious diseases: case study SARS. IAENG Int. J. Appl. Math. **41**, (2011)
10. Nucci, M.C., Leach, P.G.L.: An integrable SIS model. J. Math. Anal. Appl. **290**, 506–518 (2004)
11. Rodrigues, H.S., Fonseca, M.J.: Viral marketing as epidemiological model. In: Proceedings of the 15th International Conference on Computational and Mathematical Methods in Science and Engineering, pp. 946–955 (2015)
12. Wang, J., Wang, Y.Q.: SIR rumor spreading model with network medium in complex social networks. Chinese J. Phys. **53**, Art. ID 020702, 21 (2015)

Equilibrium Stability and the Geometry of Bifurcation Graphs for a Class of Nonlinear Leslie Models

J. M. Cushing

Abstract For nonlinear scalar difference equations that arise in population dynamics the geometry of the graph obtained by plotting the population growth rate as a function of inherent fertility leads to information about the number of positive equilibria and about the local stability of positive equilibria. Specifically, equilibria on decreasing segments of this graph are always unstable. Equilibria on increasing segments are stable in two circumstances: when the equilibrium is sufficiently close either to 0 or to a critical point on the graph. These geometric criteria are shown to hold for a class of nonlinear Leslie models in which (age-specific) survival rates are population density independent and fertilities are dependent on a weighted total population size. Examples are given to show how this geometric method can be used to identity strong Allee and hysteresis effects in these models.

Keywords Leslie matrix models · Bifurcation · Stability · Allee effects · Hysteresis

2010 Mathematics Subject Classification 92D25 · 92D15 · 37G35 · 39A30

1 Nonlinear Leslie Matrix Equations

Difference equations arise as models of population dynamics by means of a straightforward accounting of the individuals present at time $t+1$, the total of which consists of new individuals who were not present at time t plus those who were present at time t and survived to time $t+1$. If no immigration or emigration occurs, the former are newborns, so that the population at time $t+1$ consists simply of newborns plus survivors. If $\hat{x}(t)$ is demographic m-dimensional column vector of population densities as categorized by a classification scheme such as age, size, life history stage,

---

J. M. Cushing (✉)
Department of Mathematics and Interdisciplinary Program in Applied Mathematics, University of Arizona, Tucson, AZ 85721, USA
e-mail: cushing@math.arizona.edu

M. Bohner et al. (eds.), *Difference Equations and Discrete Dynamical Systems with Applications*, Springer Proceedings in Mathematics & Statistics 312,
https://doi.org/10.1007/978-3-030-35502-9_8

disease classifications, etc., then

$$\hat{x}(t+1) = F\hat{x}(t) + T\hat{x}(t)$$

where $F\hat{x}(t)$ is the vector of newborns and $T\hat{x}(t)$ the vector of survivors. Here $F = [f_{ij}]$ is an $m \times m$ non-negative matrix of class specific fertilities f_{ij} and $T = [s_{ij}]$ is an $m \times m$ non-negative matrix of survival and between class transition probabilities s_{ij}, which if dependent on population density become functions of $\hat{x}(t)$ and create a nonlinear difference equation

$$\hat{x}(t+1) = F(\hat{x}(t))\,\hat{x}(t) + T(\hat{x}(t))\,\hat{x}(t). \tag{1}$$

In the non-structured population case $m = 1$, we have a scalar difference equation

$$x(t+1) = f(x(t))\,x(t) + s(x(t))\,x(t).$$

Denoting the inherent (or intrinsic) fertility by b (i.e. fertility in the absence of density effects), we re-write the scalar equation as

$$x(t+1) = b\beta(x(t))\,x(t) + s(x(t))\,x(t)$$

where $\beta(0) = 1$. Besides the extinction equilibrium $x = 0$, we are interested in positive equilibria, which are roots of the algebraic equation

$$1 = b\beta(x) + s(x)$$

or

$$b = \frac{1 - s(x)}{\beta(x)}. \tag{2}$$

To have a feasible population model, we require $\beta(x) > 0$ and $0 \leq s(x) < 1$ on some interval of positive x values. We assume these inequalities hold for $x \in I$ where I is an open interval containing $x = 0$. In applications I is often a half line, as is the case when β and/or s have one of the commonly used forms

$$\frac{1}{1+cx} \text{ or } \exp(-cx), \quad c > 0.$$

The graph, G, of $b = b(x)$ as a function of x defined by (2), together with the horizontal line $x = 0$, plotted in the positive quadrant geometrically displays the existence of relevant equilibria for the population model. From this graph the number of equilibria for any value of b can be determined as well as any bifurcation points where that number changes. The following facts are straightforwardly derived from the linearization principle, provided β and s are twice continuously differentiable on I. See [9].

1. The extinction equilibrium loses stability as b increases through the critical value $b_0 := 1 - s(0) > 0$.
2. A positive equilibrium $x = x_e$ corresponding to $b = b_e$ is unstable if the point (b_e, x_e) lies on a decreasing segment of the graph G.
3. A positive equilibrium $x = x_e$ corresponding to $b = b_e$ is stable if the point (b_e, x_e) lies on a increasing segment of the graph G provided $x_e \approx 0$ or x_e is near a critical point of $b(x)$, i.e. a point for which $b'(x_e) = 0$.

Thus, by simply graphing $b(x)$ given by (2) one can not only determine the existence and number of positive equilibria, but also learn a considerable amount about their local stability properties.

Here we will extend these conclusions to a class of matrix models (1) when $m > 1$. Specifically, we consider Leslie age-structured models under two assumptions: that only fertility is density dependent and that the density dependence is by means of a weighted total population size

$$w(t) = \hat{w}^{\tau}\hat{x}(t)$$

where $\hat{w} \in R_{+}^{m}\backslash\{\hat{0}\}$. Here R_{+}^{m} denotes the non-negative cone in Euclidean space R^{m} and the superscript τ denotes the vector transpose. Thus, in (1) we have

$$F(w) = \begin{pmatrix} 0 \cdots 0 & f_n(w) & \cdots & f_m(w) \\ 0 \cdots 0 & 0 & \cdots & 0 \\ 0 \cdots 0 & 0 & \cdots & 0 \\ \vdots \quad \vdots & \vdots & & \vdots \\ 0 \cdots 0 & 0 & \cdots & 0 \end{pmatrix}, \quad T = \begin{pmatrix} 0 & 0 & \cdots & 0 & 0 \\ s_1 & 0 & \cdots & 0 & 0 \\ 0 & s_2 & \cdots & 0 & 0 \\ \vdots & \vdots & & \vdots & \vdots \\ 0 & 0 & \cdots & s_{m-1} & s_m \end{pmatrix} \tag{3}$$

where n is the first adult (reproducing) age class and $0 < s_i \leq 1$ for $i = 1, ..., m-1$ and $0 \leq s_m < 1$. If $s_m = 0$ then $F(w) + T$ is a standard Leslie age-structured projection matrix, designed so that no individual lives past age class m. If $s_m > 0$ then the matrix is an extended Leslie matrix in which no maximal age is specified and the m-age class consists of all individuals older than age m.

In summary, we consider the nonlinear Leslie matrix equation

$$\hat{x}(t+1) = F(w(t))\hat{x}(t) + T\hat{x}(t) \tag{4}$$

with $F(w)$ and T given by (3). We assume

A1: $f_i(w) \geq 0$ are twice continuously differentiable functions of $w \in I$.

In order that n is the youngest and m is the oldest fertile age class we make the following assumption:

A2: $f_n(w) > 0$ and $f_m(w) > 0$ on I.

Under this assumption $F(w)+T$ is non-negative and irreducible for $w \in I$. Finally, we will use the inherent fertility of the youngest fertility class as a bifurcation parameter.

A3: Let λ denote the inherent fertility of the youngest adult class n and write

$$f_n(w) = \lambda\beta(w) \quad \text{where} \quad \beta(0) = 1.$$

In the $m = 1$ dimensional case discussed above, λ is the inherent fertility rate b.

2 Equilibria

By Perron–Frobenius theory $F(w)+T$ has a simple, dominant eigenvalue $r(w) > 0$ which possesses a positive eigenvector (i.e. an eigenvector in the interior $int\left(R_+^m\right)$ of the positive cone R_+^m in R^m) and no other eigenvalue has a nonnegative eigenvector (i.e. no other eigenvector in $R_+^m \backslash \{\hat{0}\}$). The equilibrium equation

$$\hat{x} = (F(w)+T)\hat{x} \tag{5}$$

has, of course, the solution $\hat{x} = \hat{0}$ (the extinction equilibrium). If $\hat{x} \in R_+^m \backslash \{\hat{0}\}$ is a solution of (5), then by Perron–Frobenius theory $r(w) = 1$ and $\hat{x}$ is in fact a positive vector, i.e. $\hat{x} \in int\left(R_+^m\right)$. Thus, *the only non-negative equilibria of the nonlinear Leslie model* (4) *are positive equilibria.*

Suppose $\hat{x} \in int\left(R_+^m\right)$ solves the equilibrium equation (5). Since $r(w) = 1$ it follows from well-known theorems [4, 12] that the net reproduction number $R_0(w)$ of $F(w)+T$ also equals 1. While there is no analytic formula for $r(w)$, there is a formula for $R_0(w)$ for Leslie matrices [5, 6], namely

$$R_0(w) = p_n\lambda\beta(w) + \sum_{i=n+1}^{m-1} p_i f_i(w) + p_m \frac{1}{1-s_m} f_m(w) \tag{6}$$

where

$$p_1 = 1, \quad p_i = \prod\nolimits_{j=1}^{i-1} s_j$$

is the probability a newborn lives to reach age class i. Thus, we conclude that to each positive equilibrium $\hat{x}_e \in int\left(R_+^m\right)$ of the nonlinear Leslie equation (4) there corresponds a positive solution $w_e = \hat{w}^{\top}\hat{x}_e > 0$ of the equation

$$p_n\lambda\beta(w) + \sum_{i=n+1}^{m-1} p_i f_i(w) + p_m \frac{1}{1-s_m} f_m(w) = 1 \tag{7}$$

which we can re-write as

$$\lambda = \frac{1}{p_n \beta(w)} \left(1 - \sum_{i=n+1}^{m-1} p_i f_i(w) - p_m \frac{1}{1 - s_m} f_m(w) \right). \tag{8}$$

Conversely, suppose $w_e > 0$ solves this equation. Then $R_0(w_e) = 1$ which in turn implies $r(w_e) = 1$ [4, 12]. By Perron–Frobenius theory, the eigen-space of $F(w_e) + T$ associated with 1 is spanned by a positive eigenvector $\hat{v} \in int(R_+^m)$. There is, then, a unique positive eigenvector $\hat{x}_e$ of $F(w_e) + T$ such that $\hat{w}^{\tau}\hat{x}_e = w_e$, namely, $\hat{x}_e = (w_e/\hat{w}^{\tau}\hat{v})\,\hat{v}$, which is therefore a unique positive equilibrium of (5).

In conclusion, we have shown that *there is a one-one correspondence between the positive equilibria of the nonlinear Leslie matrix equation* (3)–(4) *and positive solutions* w *of* (7), *or equivalently of* (8). This allows us to study the existence of positive equilibria by investigating the graph G of $\lambda = \lambda(w)$, as defined by (8), in the (λ, w)-plane, as we did in the $m = 1$ case above. To be biologically meaningful, λ must be positive and so we are interested in the graph of G only in the positive quadrant of the (λ, w)-plane.

3 Equilibrium Stability

The Jacobian of (4) evaluated at the extinction equilibrium $\hat{x} = \hat{0}$ is (the non-negative and irreducible matrix) $F(\hat{0}) + T$. The extinction equilibrium is locally asymptotically stable (LAS) if the dominant eigenvalue $r(\hat{0}) < 1$ and unstable if $r(\hat{0}) > 1$. It follows, then, from well-known theorems [4, 12] that the extinction equilibrium is stable if $R_0(\hat{0}) < 1$ and unstable if $R_0(\hat{0}) > 1$. Using formula (6) we conclude by the linearization principle [11] that the extinction equilibrium is LAS if $\lambda < \lambda_0$ and unstable if $\lambda > \lambda_0$ where

$$\lambda_0 := \frac{1}{p_n} \left(1 - \sum_{i=n+1}^{m-1} p_i f_i(0) + p_m \frac{1}{1 - s_m} f_m(0) \right). \tag{9}$$

If $\lambda_0 \leq 0$ then the extinction equilibrium is unstable for all biologically meaningful values of $\lambda > 0$ and the population is not threatened by extinction. Therefore, in so far as a study of extinction versus survival is concerned, we are interested in the case when $\lambda_0 > 0$. We have proved part (a) of the following Theorem.

Theorem 1 *Assume A1, A2, and A3 and that* $\lambda_0 > 0$.
(a) The extinction equilibrium of the nonlinear Leslie matrix equation (4) is LAS if $\lambda < \lambda_0$ *and unstable if* $\lambda > \lambda_0$.
(b) A positive equilibrium $\hat{x}_e \in int(R_+^m)$ *of (4) is unstable if the point* $(\lambda, w_e) = (\lambda, \hat{w}^{\tau}\hat{x}_e)$ *lies on a strictly decreasing segment of the graph G, i.e. if* $\lambda'(w_e) < 0$.

(c) A positive equilibrium $\hat{x}_e \in int\left(R_+^m\right)$ of (4) is LAS if the point $(\lambda, w_e) = \left(\lambda, \hat{w}^\tau \hat{x}_e\right)$ lies on a strictly increasing segment of the graph G, i.e. $\lambda'(w_e) > 0$, and is either near the bifurcation point $(\lambda_0, 0)$ or near a critical point (λ_c, w_c) on G, i.e. a point where $\lambda'(w_c) = 0$.

Proof We have only to prove (b) and (c) From (6) we calculate

$$R_0'(w) = p_n \lambda \beta'(w) + \sum_{i=n+1}^{m-1} p_i f_i'(w) + p_m \frac{1}{1-s_m} f_m'(w). \tag{10}$$

At a positive equilibrium point (λ, w_e), with λ given by (8), we have

$$R_0'(w_e) = \frac{1}{\beta(w_e)}\left(1 - \sum_{i=n+1}^{m-1} p_i f_i(w_e) - p_m \frac{1}{1-s_m} f_m(w_e)\right)\beta'(w_e) + \sum_{i=n+1}^{m-1} p_i f_i'(w_e) + p_m \frac{1}{1-s_m} f_m'(w_e).$$

From the definition (8) of $\lambda = \lambda(w)$ we calculate

$$\lambda'(w) = -\frac{1}{p_n \beta(w)}\left[\frac{1}{\beta(w)}\left(1 - \sum_{i=n+1}^{m-1} p_i f_i(w) - p_m \frac{1}{1-s_m} f_m(w)\right)\beta'(w) + \sum_{i=n+1}^{m-1} p_i f_i'(w) + p_m \frac{1}{1-s_m} f_m'(w)\right].$$

Thus, we find that

$$\lambda'(w_e) = -\frac{1}{p_n \beta(w_e)} R_0'(w_e). \tag{11}$$

and that, when nonzero, $\lambda'(w_e)$ and $R_0'(w_e)$ have opposite signs. (This is related to general results concerning $r(w)$ and $R_0(w)$ in [7].)

(b) If $\lambda'(w_e) < 0$ then $R_0'(w_e) > 0$ and hence by Theorem 1 in [13] the positive equilibrium $\hat{x}_e$ is unstable.

(c) If $\lambda'(w_e) > 0$ then $R_0'(w_e) < 0$ then, by Theorem 3 in [13] the positive equilibrium $\hat{x}_e = \text{col}(x_i)$ is LAS provided $\left|F'(w_e)\hat{x}_e\right|$ is sufficiently small. From the equilibrium equation (5) it is easily seen that any positive equilibrium has the form

$$\hat{x}_e = \begin{pmatrix} 1 \\ p_1 \\ \vdots \\ p_{m-1} \\ p_m \frac{1}{1-s_m} \end{pmatrix} x_1$$

and hence (using the vector norm $\left|\hat{x}\right| = \Sigma_{i=1}^{m} |x_i|$) we find that $\left|F'(w_e)\hat{x}_e\right| = \left|R_0'(w_e)\right| x_1$. Thus, $\left|F'(w_e)\hat{x}_e\right|$ is small if $x_1 \approx 0$, which is true if the equilibrium is near the bifurcation point. However, $\left|R_0'(w_e)\right| x_1$ is also small if the equilibrium point (λ, w_e) lies near a point (λ_c, u_c) where $R_0'(w_c) = 0$, i.e. by (11) where $\lambda'(w_c) = 0$. ■

The graph G connects to the point $(\lambda, w) = (\lambda_0, 0)$ since this point satisfies equation (8), i.e. this point is a transcritical bifurcation point where the branches of extinction and positive equilibria intersect. By Theorem 1 the bifurcating positive equilibria near this bifurcation point are LAS (respectively, unstable) if the bifurcation is forward (respectively, backward), i.e. if the graph of G is increasing (respectively, decreasing) at $(\lambda_0, 0)$. That is to say, near the bifurcation point, the bifurcating positive equilibria are stable (respectively, unstable) if $\lambda'(0) > 0$ (respectively, $\lambda'(0) < 0$) or equivalently $R_0'(0) < 0$ (respectively $R_0'(0) > 0$). From

$$R_0'(0) = p_n\lambda\beta'(0) + \sum_{i=n+1}^{m-1} p_i f_i'(0) + p_m\frac{1}{1-s_m}f_m'(0)$$

we see that a forward bifurcation of stable positive equilibria occurs if all density effects are negative (at low population levels $w \approx 0$), i.e. if $\beta'(0) \leq 0$, $f_i'(0) \leq 0$ and not all equal 0. The density effects used in most population models are in fact negative effects, i.e. $\beta'(w) \leq 0$, $f_i'(w) \leq 0$ and not all equal to 0 for all values of $w > 0$. By (10) this implies $R_0'(w) < 0$ and, by (11), $\lambda'(w) > 0$ for all population levels $w \geq 0$. Thus, there are no critical points nor decreasing segments on the graph G. As is well known, the stability of the bifurcating positive equilibria is not necessarily maintained along the entire graph G. Positive equilibrium destabilization can occur and result in periodic orbits, invariant loops, and cascades to chaos.

Positive density effects at low population levels are called *component Allee effects* [3]. If present and if they are of sufficient magnitude so as to give $R_0'(0) > 0$, then component Allee effects result in the backward bifurcation of positive equilibria from the extinction equilibrium $(\lambda_0, 0)$. Backward bifurcations, when coupled with negative density effects at high population levels, generally lead to a multiple attractor scenario called a *strong Allee effect* [8]. The importance of strong Allee effects in population dynamics has been emphasized since the early work of W. C. Allee [1]; see for example [3, 10]. For the nonlinear Leslie models (4) considered here, the geometry of the graph G, as a straightforward plotting exercise, can easily detect strong Allee effects (and other tipping points and hysteresis effects). We give an example in the next section.

4 An Application

Consider (4) with $m = 4$ and $n = 2$, i.e. a Leslie matrix model with four age classes, the first of which consists of juveniles:

$$F(w) = \begin{pmatrix} 0 & f_2(w) & f_3(w) & f_4(w) \\ 0 & 0 & 0 & 0 \\ 0 & 0 & 0 & 0 \\ 0 & 0 & 0 & 0 \end{pmatrix}, \quad T = \begin{pmatrix} 0 & 0 & 0 & 0 \\ s_1 & 0 & 0 & 0 \\ 0 & s_2 & 0 & 0 \\ 0 & 0 & s_3 & 0 \end{pmatrix}. \tag{12}$$

We consider density terms $f_i(w)$ in which two factors $\exp(-cw)$ and $\exp\left(1 - e^{-dw}\right)$ are present. The first is of the famous Ricker-type used to model negative density effects and the second, which is increasing in w, is one often used to model component Allee effects [3, 10]. We take

$$f_i(w) = b_i \exp(-c_i w) \exp\left(1 - e^{-d_i w}\right) \tag{13}$$

for $c_i \geq 0$ and $d_i \geq 0$. A rationale for this model is that fertility $f_i(w)$ is the product of a per capita birth rate times the probability that a newborn survives to the next census. In this model, the birth rates are negatively affected by increased population density w while newborn survival is positively affected by increased population density (for example, because of herding protection of newborns from predators [3, 10]).

The bifurcation parameter is $\lambda = b_2$ and

$$\beta(w) = \exp(-c_2 w) \exp\left(1 - e^{-d_2 w}\right).$$

The graph G in the (λ, w)-plane, is that of $\lambda = \lambda(w)$ given by (8), i.e.

$$\lambda(w) = \frac{1}{s_1 \beta(w)} \left(1 - s_1 s_2 f_3(w) - s_1 s_2 s_3 f_4(w)\right). \tag{14}$$

The bifurcation point $\lambda_0 = \lambda(0)$, which we assume is positive, is given by (9)

$$\lambda_0 = \frac{1 - s_1 s_2 b_3 - s_1 s_2 s_3 b_4}{s_1} > 0.$$

The direction of bifurcation is determined by the sign of

$$\lambda'(0) = c_2 \lambda_0 + b_3 c_3 s_2 + b_4 c_4 s_2 s_3 - \lambda_0 d_2 - b_3 s_2 d_3 - b_4 s_2 s_3 d_4.$$

Thus, we see that in the absence of component Allee effects, i.e. when all $d_i = 0$, the bifurcation is forward since $\lambda'(0) > 0$ and the graph of G is increasing at the bifurcation point $(\lambda, w) = (\lambda_0, 0)$. If, on the other hand, component Allee effects d_i are sufficiently large so that

$$\lambda_0 d_2 + b_3 s_2 d_3 + b_4 s_2 s_3 d_4 > c_2 \lambda_0 + b_3 c_3 s_2 + b_4 c_4 s_2 s_3$$

i.e., so that $\lambda'(0) < 0$, then the bifurcation is backward.

Figure 1a shows an example of the graph G when component Allee effects are absent. By Theorem 1 we know the positive equilibria are LAS at least near the

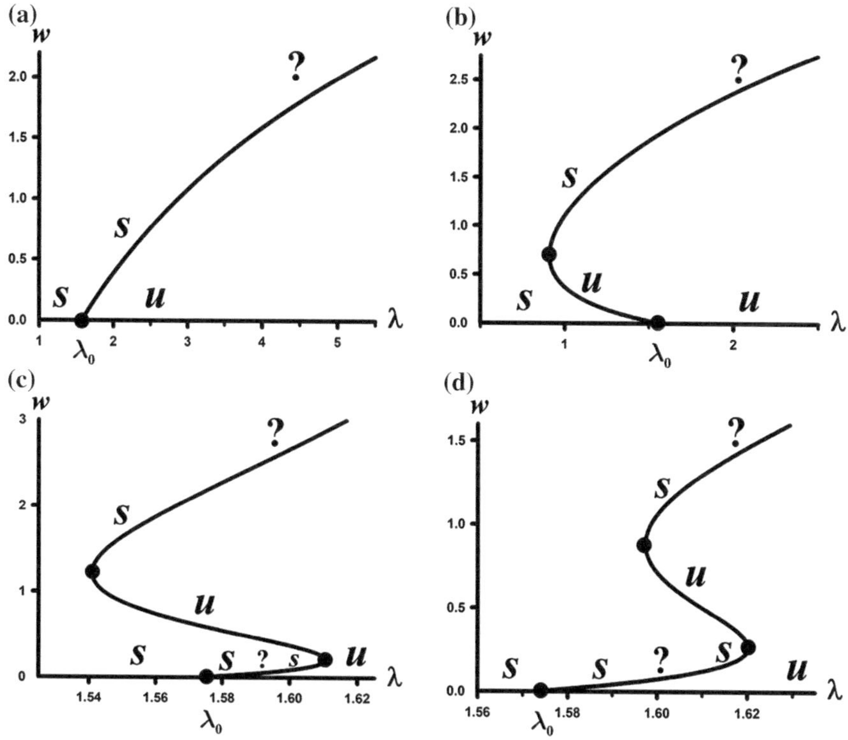

Fig. 1 In all plots of the graph G defined by (14)–(13) used parameter values $b_2 = b_3 = 1/4$, $s_1 = 1/2$, $s_2 = s_3 = 9/10$. **a Forward bifurcation:** $c_1 = 1/2$, $c_2 = 1/2$, $c_3 = 1/2$, $d_1 = 0$, $d_2 = 0$, $d_3 = 0$; **b Backward bifurcation and strong Allee effect:** $c_1 = 1/2$, $c_2 = 1/2$, $c_3 = 1/2$, $d_1 = 2$, $d_2 = 2$, $d_3 = 2$; **c Forward bifurcation and strong Allee effect:** $c_1 = 0$, $c_2 = 1/5$, $c_3 = 9/2$, $d_1 = 1/100$, $d_2 = 2$, $d_3 = 1/5$; **d Hysteresis:** $c_1 = 0$, $c_2 = 1/3$, $c_3 = 9/2$, $d_1 = 1/100$, $d_2 = 2$, $d_3 = 1/5$. The letters u stands for unstable equilibria of the nonlinear matrix model (4)–(12). The letter s stands for LAS equilibria, but only in neighborhoods of critical (bifurcation) points, indicated by solid circles, as guaranteed by Theorem 1. Outside such neighborhoods, equilibrium stability is not assured, as indicated by the question marks

bifurcation point. Figure 1b shows an example when the bifurcation is backward. In this case, Theorem 1 tells us that the positive equilibria corresponding to points (λ, w) on the increasing segment of G are LAS at least near the critical point (a saddle node bifurcation point). This is an example of a backward bifurcation producing a strong Allee effect, i.e. an interval of parameter values $\lambda < \lambda_0$ for which there exists both a stable positive equilibrium and a stable extinction equilibrium (and hence survival is initial condition dependent). Figure 1c shows an interesting case when a strong Allee effect occurs in an example when the bifurcation is forward. Finally, Fig. 1d shows a case of hysteresis and multiple stable positive equilibria, which illustrates the complicated geometry that can arise in a nonlinear Leslie model (4) when component Allee effects are present in each of the fertility terms.

5 Some Concluding Remarks

We have considered only Leslie matrix models in this paper. The geometric analysis of equilibria and their stability properties, as given in Theorem 1, are straightforwardly extendable to any model in which the transition matrix T is density independent and has spectral radius less than 1 and the fertility matrix depends on a single total weighted population size w provided there is only one newborn class (i.e. all but one row of F is a row of zeroes). This is because for such matrices there is an explicit formula for R_0 [4, 5, 12]. An example is an Usher, or standard population size structured, matrix [2, 5, 6]. It is an open question, however, whether this method can be successfully used when the transition matrix T also depends on w. Other open problems involve the construction of geometric methods for matrix models in which F and T have other types of nonlinear dependencies, for example, a dependency on more than one weighted population size. Examples in [13] suggest that this will not be, in general, straightforward.

Acknowledgements The author gratefully acknowledges the support of the National Science Foundation (Mathematical Biology Program in the Division of Mathematical Sciences and the Population and Community Ecology Program in the Division of Environmental Biology) under NSF grant DMS 1407564. The author is also grateful to two anonymous reviewers for their careful reading of the manuscript and for their suggestions.

References

1. Allee, W.C.: Animal Aggregations, a Study in General Sociology. University of Chicago Press, Chicago (1931)
2. Caswell, H.: Matrix Population Models: Construction, Analysis and Interpretation, 2nd edn. Sinauer Associates, Inc. Publishers, Sunderland (2001)
3. Courchamp, F., Berec, L., Gascoigne, J.: Allee Effects in Ecology and Conservation. Oxford University Press, Oxford (2008)
4. Cushing, J.M., Yicang, Z.: The net reproductive value and stability in matrix population models. Nat. Resour. Modell. **8**, 297–333 (1994)
5. Cushing, J.M.: An Introduction to Structured Population Dynamics, Conference Series in Applied Mathematics, vol. 71. SIAM, Philadelphia (1998). ISBN 0-89871-417-6
6. Cushing, J.M.: Matrix models and population dynamics. In: Lewis, M., Chaplain, A.J., Keener, J.P., Maini, P.K. (eds.) Mathematical Biology. IAS/Park City Mathematics Series, vol. 14, pp. 47–150. American Mathematical Society, Providence (2009)
7. Cushing, J.M.: On the relationship between r and R_0 and its role in the bifurcation of equilibria of Darwinian matrix models. J. Biol. Dyn. **5**, 277–297 (2011)
8. Cushing, J.M.: Backward bifurcations and strong Allee effects in matrix models for the dynamics of structured populations. J. Biol. Dyn. **8**, 57–73 (2014)
9. Cushing, J.M.: One dimensional maps as population and evolutionary dynamic models. In: Cushing, J.M., Saleem, M., Srivastava, H.M., Khan, M.A., Merajuddin, M. (eds.) Applied Analysis in Biological and Physical Sciences. Springer Proceedings in Mathematics & Statistics, vol. 186. Springer, India (2016)
10. Dennis, B.: Allee effects: population growth, critical density, and the chance of extinction. Nat. Resour. Model. **3**, 481–538 (1989)

11. Elaydi, S.N.: An Introduction to Difference Equations, 3rd edn. Springer, New York (2005)
12. Li, C.-K., Schneider, H.: Applications of Perron-Frobenius theory to population dynamics. J. Math. Biol. **44**, 450–462 (2002)
13. Yicang, Z., Cushing, J.M.: Stability conditions for equilibria of nonlinear matrix population models. J. Differ. Equ. Appl. **4**, 95–126 (1996)

Non-monotone Behavior of the Heavy Ball Method

Marina Danilova, Anastasiia Kulakova and Boris Polyak

Abstract We focus on the solutions of second-order stable linear difference equations and demonstrate that their behavior can be non-monotone and exhibit peak effects depending on initial conditions. The results are applied to the analysis of the accelerated unconstrained optimization method—the Heavy Ball method. We explain non-standard behavior of the method discovered in practical applications. In addition, such non-monotonicity complicates the correct choice of the parameters in optimization methods. We propose to overcome this difficulty by introducing new Lyapunov function which should decrease monotonically. By use of this function convergence of the method is established under less restrictive assumptions (for instance, with the lack of convexity). We also suggest some restart techniques to speed up the method's convergence.

Keywords Difference equations · Optimization methods · Non-monotone behavior · The Heavy Ball method · Lyapunov function · Global convergence

1 Introduction

It is well known that n-th order scalar linear difference equations

$$x_k + a_1 x_{k-1} + \cdots + a_n x_{k-n} = 0,\ k = n,\ n+1,\ \ldots;\ a_i \in \mathbb{R}$$

M. Danilova · B. Polyak
Laboratory of Adaptive and Robust Systems, Institute of Control Sciences RAS, Profsoyuznaya, 65, Moscow 117342, Russia
e-mail: Danilovamarina15@gmail.com

B. Polyak
e-mail: boris@ipu.ru

A. Kulakova (✉)
Department of Control and Applied Mathematics, Moscow Institute of Physics and Technology, Institutskiy per., 9, Dolgoprudniy 141700, Russia
e-mail: anastasiya.kulakova@phystech.edu

M. Bohner et al. (eds.), *Difference Equations and Discrete Dynamical Systems with Applications*, Springer Proceedings in Mathematics & Statistics 312,
https://doi.org/10.1007/978-3-030-35502-9_9

with initial conditions

$$x^{(0)} = (x_0, \dots, x_{n-1}) \in \mathbb{R}^n$$

are stable, i.e. $\lim_{k\to\infty} x_k = 0$, if and only if the moduli of the roots λ_i of the characteristic polynomial

$$p(\lambda) = \lambda^n + a_1\lambda^{n-1} + \cdots + a_{n-1}\lambda + a_n$$

are less than 1 [1].

However, the convergence to zero can be non-monotone. This effect can be described by the following quantity:

$$\eta(x^{(0)}) = \max_{k=n,n+1,\dots} |x_k|,$$

which will be referred to as peak of the solution (provided that $\eta(x^{(0)}) > 1$) for a given root location λ and initial condition $x^{(0)}$. Without loss of generality we assume that $\|x^{(0)}\|_\infty \le 1$. The estimates of $\eta(x^{(0)})$ are demonstrated in the recent paper [2].

The main objective of the present paper is to link the peak effects in linear difference equations with the non-monotone behavior of such unconstrained optimization methods as the Heavy Ball method [3, 4], Nesterov's accelerated gradient method [5] and, for example, the recently proposed triple momentum method [6]. When applied to a quadratic function, momentum methods are described by second-order linear matrix difference equations. In numerous simulations it was found out that the methods demonstrate a non-monotone convergence to a minimum [7].

Below we restrict ourselves with the analysis of the Heavy Ball method, but similar techniques can be extended to other accelerated optimization methods. It is worth mentioning that such methods are currently widespread and implemented to minimization problems occurring in neural networks. That is why a detailed study of these methods is of a great importance. In addition, the numerous restart techniques [8] gain their popularity. To design the restart technique it is important to know how large the deviations from the solution should be to make an assumption about incorrect parameters' choice and start the method from the beginning at the current point.

The paper is organized as follows. In Sect. 2 we provide the results on peak effects in second-order linear difference equations. Next Sect. 3 contains applications of these results to the Heavy Ball method and demonstrates the dependence of peak value on initial conditions. In Sect. 4 a new Lyapunov function is constructed; it decreases monotonically in contrast with the objective function or distance to the solution. By use of this function we are also able to prove global convergence for non-convex functions (under less restrictive Polyak-Łojasiewicz condition). Conclusions and future directions for research are summarized in Sect. 5.

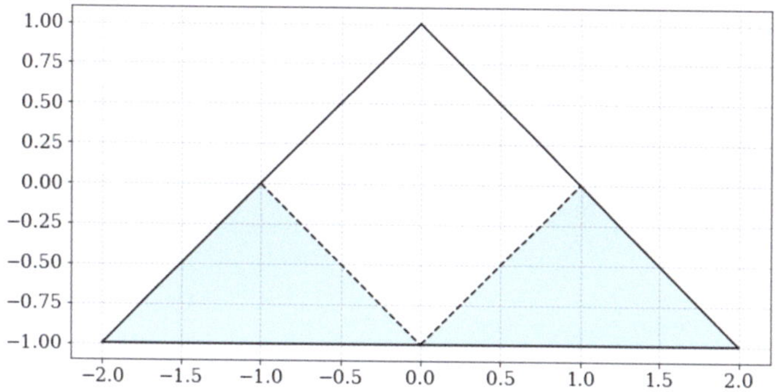

Fig. 1 The stability domain of the second-order difference equation, the shaded area corresponds to peak effect

2 Peak Effect for Second-Order Difference Equations

We consider the case of second-order difference equation ($n = 2$), which can be written down in the form

$$x_k = a_1 x_{k-1} + a_2 x_{k-2},\ k = 2,\ 3,\ \ldots;\ a_1,\ a_2 \in \mathbb{R} \tag{1}$$

with initial conditions

$$x^{(0)} = (x_0, x_1) \in \mathbb{R}^2$$

and the characteristic polynomial

$$p(\lambda) = \lambda^2 - a_1\lambda - a_2. \tag{2}$$

For the second-order difference equation the following stability domain (Fig. 1) in the space (a_1, a_2) is well known [1].

Our aim is to describe possible non-monotone behavior of stable solutions and, in particular, to measure $\eta(x^{(0)})$.

Two cases can be considered: the first corresponding to equal roots of the characteristic polynomial and the second with different roots.

In the case of equal roots, i.e. $\lambda^2 - a_1\lambda - a_2 = (\lambda - \rho)^2 = \lambda^2 - 2\lambda\rho + \rho^2$, $0 < \rho < 1$, and (1) reads:

$$x_k = 2\rho x_{k-1} - \rho^2 x_{k-2}.$$

The following expression for solution can be easily derived:

$$x_k = x_1 k\rho^{k-1} - x_0(k-1)\rho^k. \tag{3}$$

In this case the non-monotone behavior can be observed (Fig. 2).

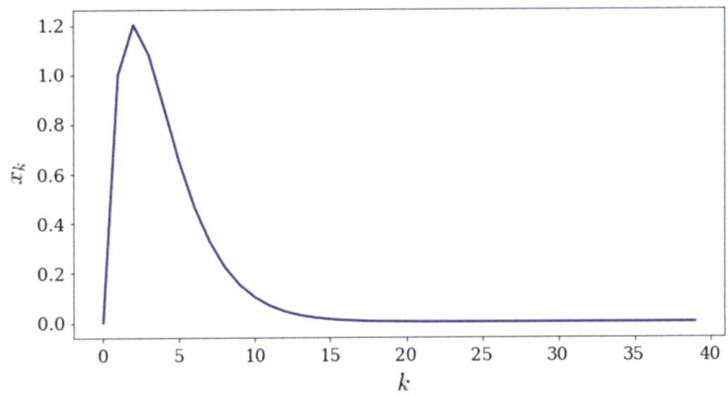

Fig. 2 The iterative process with $\rho = 0.6$ and $x^{(0)} = (0, 1)$

We conclude that for all $k \geq 2$

$$\max_{\|x^{(0)}\|_\infty \leq 1} x_k = k\rho^{k-1} + (k-1)\rho^k. \tag{4}$$

This maximum is achieved for $x^{(0)} = (-1, 1)$.

Now let's derive $k_{max} = \text{argmax}\left(k\rho^{k-1} + (k-1)\rho^k\right)$ and $\eta(x^{(0)})$. By differentiation and setting the derivative value equal to zero we obtain an expression for k_{max}:

$$k_{max} = \left\lceil \frac{\rho \ln \rho - \rho - 1}{\ln \rho(1+\rho)} \right\rceil. \tag{5}$$

The value of the peak can be obtained by a substitution (5) in the formula (4).

For $\rho \to 1$ we get

$$\eta(x^{(0)}) \approx \frac{2}{e(1-\rho)}.$$

Thus large deviations can arise for some initial conditions. However, some pairs (x_0, x_1) do not imply peak effects. For instance, for $x_0 = 1, x_1 = 1$ we obtain that $|x_k| < 1,\ k = 2, \ldots$.

Considering the case of different real roots λ_1, λ_2 of the characteristic polynomial (2), we notice the following:

• If $\lambda_2 > \lambda_1 > \rho,\ 0 < \rho < 1$, than there exist such initial conditions $x^{(0)}$, that the trajectory x_k will behave non-monotonically and its peak will be located higher than in the case of equal roots (Fig. 3).

• If $\lambda_1 < \lambda_2 < \rho,\ 0 < \rho < 1$, the peak of x_k (if any) will be located lower than in the case of equal roots ρ for all initial conditions (Fig. 4).

The proof of both statements is given in [2].

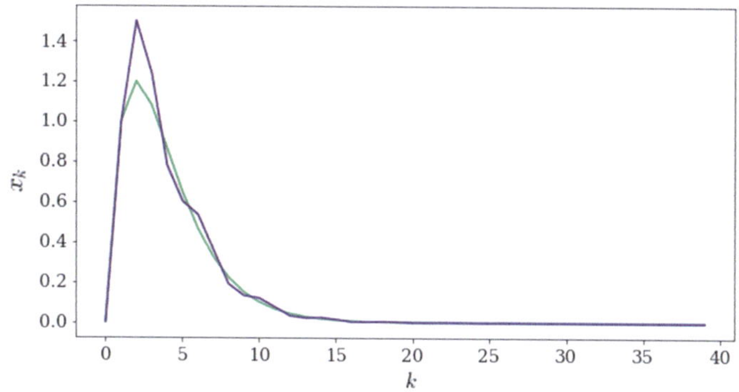

Fig. 3 The trajectories of the iterative processes with initial conditions $x^{(0)} = (0, 1)$ and $\rho = 0.6$ (green) and $\lambda_1 = 0.7,\ \lambda_2 = 0.8$ (purple)

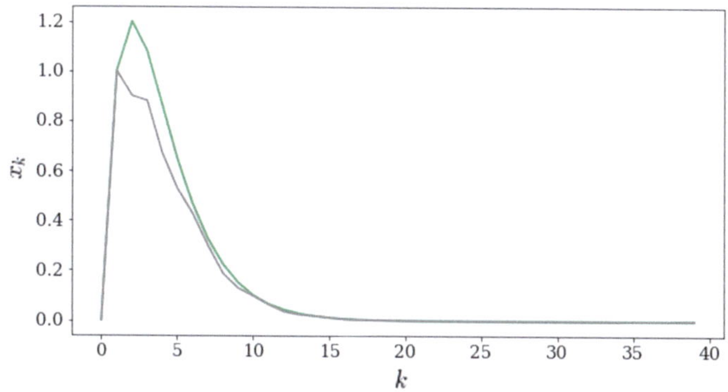

Fig. 4 The trajectories of the iterative processes with initial conditions $x^{(0)} = (0, 1)$ and $\rho = 0.6$ (green) and $\lambda_1 = 0.4,\ \lambda_2 = 0.5$ (gray)

3 Analysis of the Heavy Ball Method

3.1 *The Heavy Ball Method*

We consider the simplest unconstrained optimization problem

$$\min_{x \in \mathbb{R}^n} f(x), \tag{6}$$

where $f(x) : \mathbb{R}^n \to \mathbb{R}$ is a smooth objective function, x^*—the minimum point. We restrict our analysis with the case of quadratic objective function:

$$\min \frac{1}{2}(Ax, x) - (b, x),\ \ x,\ b \in \mathbb{R}^n$$

with $A \in \mathbb{R}^{n\times n}$ being positive-definite matrix $A \succ 0$, $\nabla f(x) = Ax - b$, $x^* = A^{-1}b$, L and $\mu > 0$—the maximum and minimum eigenvalues of A respectively. It means that we focus on strongly convex, smooth case of the objective function.

Without loss of generality we can assume that after substitution $\hat{x} = x - x^*$ objective function f has the following form:

$$f(x) = \frac{1}{2}(A\hat{x}, \hat{x})$$

So, the optimal point is $x^* = 0$, $f^* = 0$.

There are numerous iterative methods to solve this problem; gradient methods are among the most popular, see e.g. [4, 5]. The behavior of gradient methods is simple enough: they exhibit monotone convergence both for objective function and distance to the minimum point. The situation with accelerated first-order methods is much more complicated. We will focus on one of them—so called Heavy Ball method proposed in [3]:

$$x_{k+1} = x_k - \alpha\nabla f(x_k) + \beta(x_k - x_{k-1}) = x_k - \alpha Ax_k + \beta(x_k - x_{k-1}) \tag{7}$$

Here a momentum term was added to the classic gradient method, which accelerated the convergence and made the trajectory look like a smooth descent to the bottom of the ravine, rather than zigzag. The traditional choice of initial condition is

$$x_1 = x_0, \tag{8}$$

i.e. the first iteration is just a gradient step. It is known from [3] that for

$$0 \le \beta < 1, \quad 0 < \alpha < \frac{2(1+\beta)}{L}$$

there is a convergence to the solution with linear rate, $||x_k|| = O(q^k),\ q < 1$. The optimal parameters α and β providing the fastest convergence $q = \frac{\sqrt{L}-\sqrt{\mu}}{\sqrt{L}+\sqrt{\mu}}$ are also known:

$$\alpha = \frac{4}{(\sqrt{L}+\sqrt{\mu})^2}, \quad \beta = \left(\frac{\sqrt{L}-\sqrt{\mu}}{\sqrt{L}+\sqrt{\mu}}\right)^2 = q^2. \tag{9}$$

However, the non-asymptotic behavior of the method with optimal parameters for a simple example is shown on Fig. 5 and with parameters very close to optimal—on Fig. 6. In all examples below matrix A is taken diagonal.

We conclude that the method exhibits strongly non-monotone behavior.

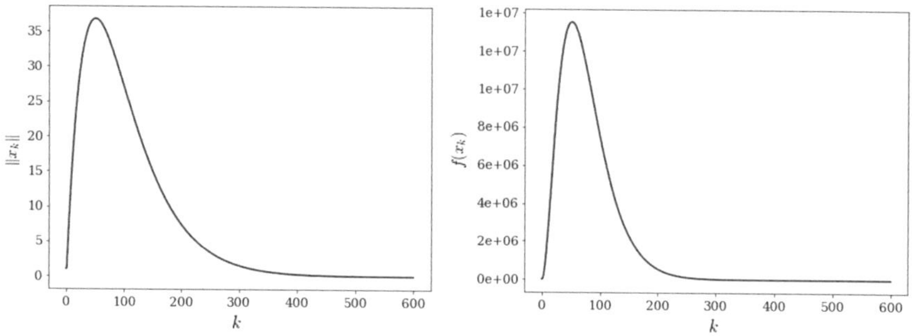

Fig. 5 Non-monotone behavior of the Heavy Ball method with $x_0 = [0, 0, 0, 1]$, $\mu = 1$, $L = 10^4$, α, β—optimal

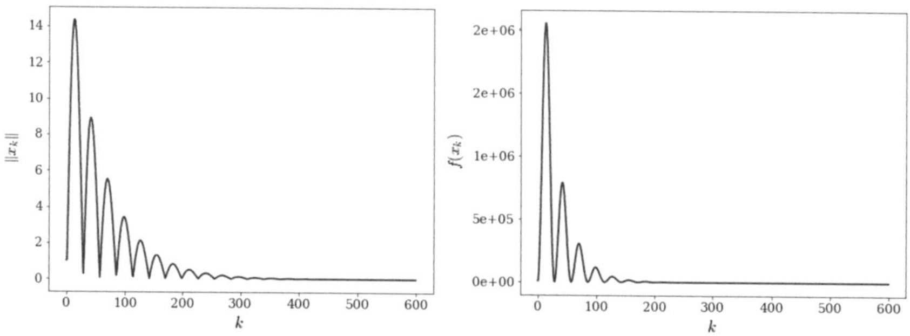

Fig. 6 Non-monotone behavior of the Heavy Ball method with $x_0 = [0, 0, 0, 1]$, $\mu = 1$, $L = 10^4$, α, β—close to optimal

3.2 Convergence Analysis

To explain the behavior of the Heavy Ball method (7) with optimal α and β (9) written in the form of second-order difference equation we consider it component-wise.

Let's start with the coordinate $x^1 = (x, e_1)$, which corresponds to the minimal eigenvalue $\lambda_{\min} = \mu$, $Ae_1 = \mu e_1$. The method for this coordinate in the form of scalar linear difference equation along with its characteristic polynomial is written down below:

$$x^1_{k+1} = (1 - \alpha\mu + \beta)x^1_k - \beta x^1_{k-1},$$

$$\rho^2 - (1 - \alpha\mu + \beta)\rho + \beta = 0.$$

It is easily determined that a characteristic polynomial has both roots equal to q, meaning that $\rho = q = \left(\frac{\sqrt{L}-\sqrt{\mu}}{\sqrt{L}+\sqrt{\mu}}\right)$. So the general solution is given by expression (3), while maximum provided by formula (4) with obvious change of notation.

Moving on to the coordinate $x^n = (x, e_n),\ Ae_n = Le_n$, which corresponds to the maximum eigenvalue $\lambda_{\max} = L$, we notice that the only difference from the previous case is in the sign of roots of the characteristic polynomial, i.e. $\rho = -q$. However this implies different behavior of solutions, even for $x_0 = x_1 = 1$ the trajectory is oscillating with possible large deviations. From formula (3) we conclude that initial conditions $x_0 = x_1 = 1$ cause the largest (in absolute value) peak effect equal to (4).

Now we consider a more general case of the coordinate $x^i = (x, e_i),\ 2 \le i \le n-1,\ Ae_i = \lambda e_i,\ \mu < \lambda < L$. The characteristic polynomial

$$\rho^2 - (1 - \alpha\lambda + \beta)\rho + \beta = 0$$

has complex roots

$$\rho_{1,2} = \frac{\left(\sqrt{L-\lambda} \pm i\sqrt{\lambda-\mu}\right)^2}{(\sqrt{L}+\sqrt{\mu})^2}, \quad |\rho| = \frac{\sqrt{L}-\sqrt{\mu}}{\sqrt{L}+\sqrt{\mu}} = q.$$

The general solution is written down below:

$$x_k^i = [C_1 \cos(\omega k) + C_2 \sin(\omega k)]\, q^k,$$

where

$$\sin(\omega) = \frac{2\sqrt{\lambda-\mu}\sqrt{L-\lambda}}{(L-\mu)}, \quad \cos(\omega) = \frac{L+\mu-2\lambda}{(L-\mu)}$$

$$C_1 = x_0^2, \quad C_2 = \frac{x_1^2(\sqrt{L}+\sqrt{\mu})^2 - x_0^2(L+\mu-2\lambda)}{2\sqrt{\lambda-\mu}\sqrt{L-\lambda}}.$$

The trajectory (for $n = 3$) demonstrates oscillations (Fig. 7).

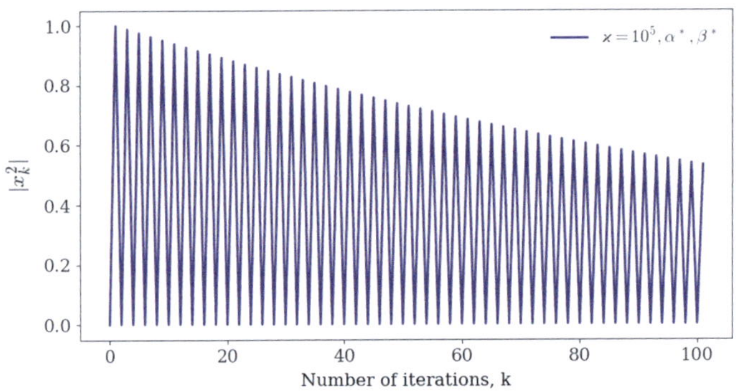

Fig. 7 Dependence of the coordinate $|x^2|$ on the number of iterations k under conditions $x_0 = 0_n,\ x_1 = 1_n$

3.3 Peak Effect

From our previous observations in Sect. 2 we note that peak effects in linear difference equations are common and depend on initial conditions. For the worst case the following proposition holds (we provide the estimates for the most important case of large condition number $\kappa = L/\mu$).

Theorem 1 *Assume that* $f(x) = \frac{1}{2}(Ax, x)$, *where* $A \in \mathbb{R}^{n\times n}$, $A = A^\top \succ 0$. *We have strong convexity parameter* $\mu = \lambda_{\min} > 0$ *and* $L = \lambda_{\max}$, *where* $\lambda_{\min}$ *and* $\lambda_{\max}$ *are the minimum and maximum eigenvalues of A, respectively. Then initial conditions* $x_0 = -e_1,\ x_1 = e_1 \in \mathbb{R}^n,\ \|x_0\| = \|x_1\| = 1$ *cause peak effect in the Heavy Ball method with optimal parameters* α^*, β^*:

$$\max_k ||x_k|| \geq \frac{\sqrt{\kappa}}{2e},$$

while standard (8) initial conditions $x_0 = x_1 = e_n \in \mathbb{R}^n,\ \|x_0\| = \|x_1\| = 1$ *cause the same peak effect combined with oscillating behavior.*

The proof follows from the estimates obtained in the previous section. Of course, other initial conditions also may lead to non-monotone behavior of iterations; we indicate the ones which provide the largest deviations from the minimum point.

The figures below show that the Heavy Ball method exhibits the non-monotone behavior on the test problem $n = 3,\ \kappa = 10^4$, namely, a sharp increase in the function under various initial conditions (Figs. 8 and 9).

To sum up, we applied the results on linear difference equations to the Heavy Ball method analysis. It was shown that even the choice of optimal parameters and standard initial conditions can not guarantee the monotone convergence.

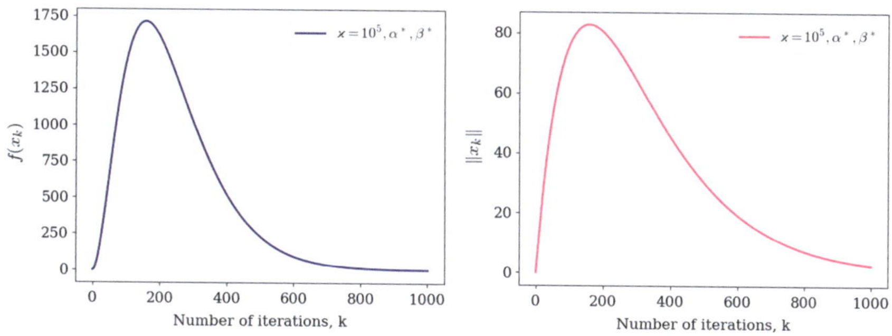

Fig. 8 Dependence of the objective function $f(x_k)$ and $\|x_k\|$ on the number of iterations k under the initial conditions $x_0 = 0_n,\ x_1 = 1_n$

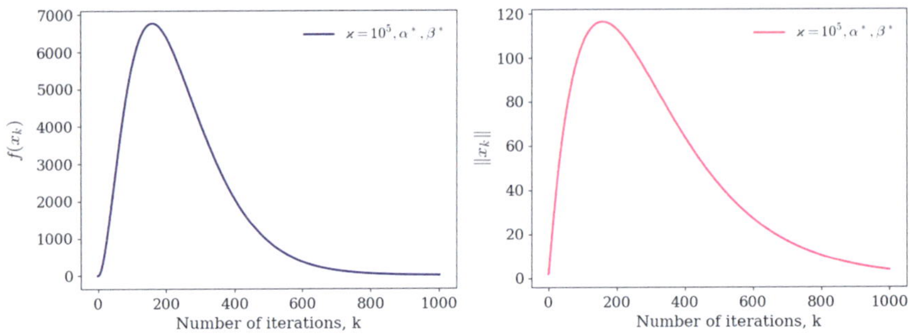

Fig. 9 Dependence of the objective function $f(x_k)$ and $\|x_k\|$ on the number of iterations k under the initial conditions $x_0 = x_1 = 1_n$

4 Lyapunov Function for the Heavy Ball Method

In this section we extend the analysis of the Heavy Ball method (both for continuous and discrete versions) to non-quadratic objective functions via Lyapunov function technique. Thus we treat the unconstrained optimization problem

$$\min_{x \in \mathbb{R}^n} f(x),$$

where $f(x)$ is differentiable function bounded from below: $f(x) \geq f^*$. Notice that here we do not assume neither convexity nor strong convexity of $f(x)$.

4.1 Construction of the Lyapunov Function

Lyapunov functions is a common tool for proving the stability of nonlinear systems described by differential or difference equations. The Lyapunov function is a scalar function that decreases monotonically in stable system.

The Heavy Ball method, as we verified above, does not exhibit monotone behavior even for the simplest case of quadratic function. Thus neither $f(x)$ nor $\|x - x^*\|$ can be used as the Lyapunov function. We suggest new Lyapunov function that can help to select the parameters of the method and overcome the difficulties related to its non-monotone transient process.

Continuous case. Before moving to the construction of the Lyapunov function for discrete case, we would like to consider continuous case from [3]:

$$\ddot{x} + a\dot{x} + b\nabla f(x) = 0. \tag{10}$$

In mechanical interpretation, $x(t) \in \mathbb{R}^n$ is the trajectory of the body ("heavy ball"), $\dot{x}, \ddot{x}$ are its velocity and acceleration, $a > 0$, $b > 0$ are scalar parameters while

$f(x) \geq f^*$ is the potential energy and $\nabla f(x)$ is its gradient. It is known from [3, 9] that $\nabla f(x(t))$ tends to zero, however the convergence can be non-monotone. Our goal is to obtain the upper bounds for the convergence.

Firstly, let's rewrite (10):

$$\dot{x} = y,$$
$$\dot{y} = -ay - bf'(x)$$

with some initial conditions $x(0),\ y(0)$. According to paper [9], $V(x, y)$ can be chosen from mechanical analogies. Consequently, it is possible to represent the function $V(x, y)$ in the form of the total energy of the system:

$$V(x, y) = f(x) + \frac{1}{2b}\|y\|^2.$$

For the time derivative we have

$$\dot{V}(x, y) = \big(f'(x), y\big) + \frac{1}{2b}2\big(y, -ay - bf'(x)\big) = -\frac{a}{b}\|y\|^2 \leq 0.$$

Thus we get an upper bound for $f(x(t))$:

$$f(x(t)) - f^* \leq f(x(0)) - f^* + \frac{1}{2b}\|y(0)\|^2.$$

In particular, for zero initial velocity $f(x(t)) - f^* \leq f(x(0)) - f^*$.

Discrete case. Now we proceed to discrete-time version of the Heavy Ball method

$$x_{k+1} = x_k - \alpha\nabla f(x_k) + \beta(x_k - x_{k-1}), \tag{11}$$

and assume additionally that f is L-smooth:

$$||\nabla f(x) - \nabla f(y)|| \leq L||x - y||.$$

Theorem 2 *Assume that*

$$0 < \alpha < \frac{1}{L},\ 0 \leq \beta \leq \sqrt{1 - \alpha L}. \tag{12}$$

Then for any initial conditions $x_0,\ x_1 \in \mathbb{R}^n$ *the function*

$$V_k = f(x_k) - f^* + \frac{1 - \alpha L}{2\alpha}\|x_k - x_{k-1}\|^2 \tag{13}$$

is the Lyapunov function for the discrete case of the Heavy Ball method, that is $V_k \le V_{k-1}$.

Proof For iterations (11), we have

$$x_k - x_{k-1} = -\alpha \nabla f(x_{k-1}) + \beta(x_{k-1} - x_{k-2}), \tag{14}$$

$$\begin{aligned} \|x_k - x_{k-1}\|^2 = \alpha^2 \|\nabla f(x_{k-1})\|^2 + \beta^2 \|x_{k-1} - x_{k-2}\|^2 \\ -2\alpha\beta \left\langle \nabla f(x_{k-1}), x_{k-1} - x_{k-2} \right\rangle. \end{aligned} \tag{15}$$

Since function $f \in \mathscr{F}_L^{1,1}$, it has the Lipschitz gradient, so following equation from [5] can be applied

$$f(x_k) \le f(x_{k-1}) + \langle \nabla f(x_{k-1}), x_k - x_{k-1} \rangle + \frac{L}{2} \|x_k - x_{k-1}\|^2. \tag{16}$$

Adding (14), (15) to (16), we get

$$\begin{aligned} & f(x_k) \le f(x_{k-1}) - \alpha \|\nabla f(x_{k-1})\|^2 + \\ & \beta \left\langle \nabla f(x_{k-1}), x_{k-1} - x_{k-2} \right\rangle + \frac{\alpha^2 L}{2} \|\nabla f(x_{k-1})\|^2 + \\ & \frac{\beta^2 L}{2} \|x_{k-1} - x_{k-2}\|^2 - L\alpha\beta \left\langle \nabla f(x_{k-1}), x_{k-1} - x_{k-2} \right\rangle. \end{aligned} \tag{17}$$

Multiplying (15) by $\frac{1-\alpha L}{2\alpha}$ and adding to (17) we obtain

$$f(x_k) + \frac{1-\alpha L}{2\alpha} \|x_k - x_{k-1}\|^2 \;\le\; f(x_{k-1}) - \alpha \|\nabla f(x_{k-1})\|^2 +$$

$$\beta \left\langle \nabla f(x_{k-1}), x_{k-1} - x_{k-2} \right\rangle + \frac{\alpha^2 L}{2} \|\nabla f(x_{k-1})\|^2 + \frac{\beta^2 L}{2} \|x_{k-1} - x_{k-2}\|^2 -$$

$$L\alpha\beta \left\langle \nabla f(x_{k-1}), x_{k-1} - x_{k-2} \right\rangle + \frac{1-\alpha L}{2\alpha} \cdot$$

$$\left(\alpha^2 \|\nabla f(x_{k-1})\|^2 + \beta^2 \|x_{k-1} - x_{k-2}\|^2 - 2\alpha\beta \left\langle \nabla f(x_{k-1}), x_{k-1} - x_{k-2} \right\rangle\right).$$

Collecting terms yields

$$f(x_k) + \frac{1-\alpha L}{2\alpha} \|x_k - x_{k-1}\|^2 \;\le\; f(x_{k-1}) + \left(\frac{\beta^2 L}{2} + \frac{(1-\alpha L)\beta^2}{2\alpha} \right) \cdot$$

$$\|x_{k-1} - x_{k-2}\|^2 + \left(\frac{\alpha^2 L}{2} - \alpha + \frac{(1-\alpha L)\alpha}{2} \right) \|\nabla f(x_{k-1})\|^2 +$$

$$(\beta - L\alpha\beta - \beta + \alpha L\beta)\,\langle\nabla f(x_{k-1}), x_{k-1} - x_{k-2}\rangle\,.$$

As a result,

$$f(x_k) + \frac{1-\alpha L}{2\alpha}\|x_k - x_{k-1}\|^2 \;\leq\; f(x_{k-1}) + \left(\frac{\beta^2 L}{2} + \frac{(1-\alpha L)\beta^2}{2\alpha}\right)\cdot$$

$$\|x_{k-1} - x_{k-2}\|^2 + \left(\frac{\alpha^2 L}{2} - \alpha + \frac{(1-\alpha L)\alpha}{2}\right)\|\nabla f(x_{k-1})\|^2$$

- $$\frac{\beta^2 L}{2} + \frac{(1-\alpha L)\beta^2}{2\alpha} \;=\; \frac{\beta^2 L\alpha + \beta^2 - \beta^2\alpha L}{2\alpha} \;=\; \frac{\beta^2}{2\alpha} \;>\; 0$$

- $$\frac{\alpha^2 L}{2} - \alpha + \frac{(1-\alpha L)\alpha}{2} = \frac{\alpha^2 L - 2\alpha + \alpha - \alpha^2 L}{2} = -\frac{\alpha}{2} \;<\; 0$$

$$f(x_k) + \frac{1-\alpha L}{2\alpha}\|x_k - x_{k-1}\|^2 \;\leq\; f(x_{k-1}) + \left(\frac{\beta^2}{2\alpha}\right)\|x_{k-1} - x_{k-2}\|^2 + \left(-\frac{\alpha}{2}\right)\|\nabla f(x_{k-1})\|^2 \qquad (18)$$

Since $\left(-\frac{\alpha}{2}\right)\|\nabla f(x_{k-1})\|^2 \;<\; 0$ and having (12) we arrive to the desired inequality $V_k \leq V_{k-1}$.

Theorem 2 provides the upper bound of $f(x_k)$ for the Heavy Ball method:

$$f(x_k) - f^* \leq f(x_0) - f^* + \frac{1-\alpha L}{2\alpha}||x_1 - x_0||^2$$

and for standard initial condition $x_1 = x_0$ we obtain

$$f(x_k) - f^* \leq f(x_0) - f^*.$$

Notice the lack of such bound for more narrow class of strongly convex quadratic functions, on the other hand this estimate holds for more restrictive conditions on parameters $\alpha,\ \beta$, see (12).

Numerical experiments. We have obtained the following Lyapunov function:

$$V_k = f(x_k) + \frac{1-\alpha L}{2\alpha}\|x_k - x_{k-1}\|^2$$

with the restriction on the parameters (12).

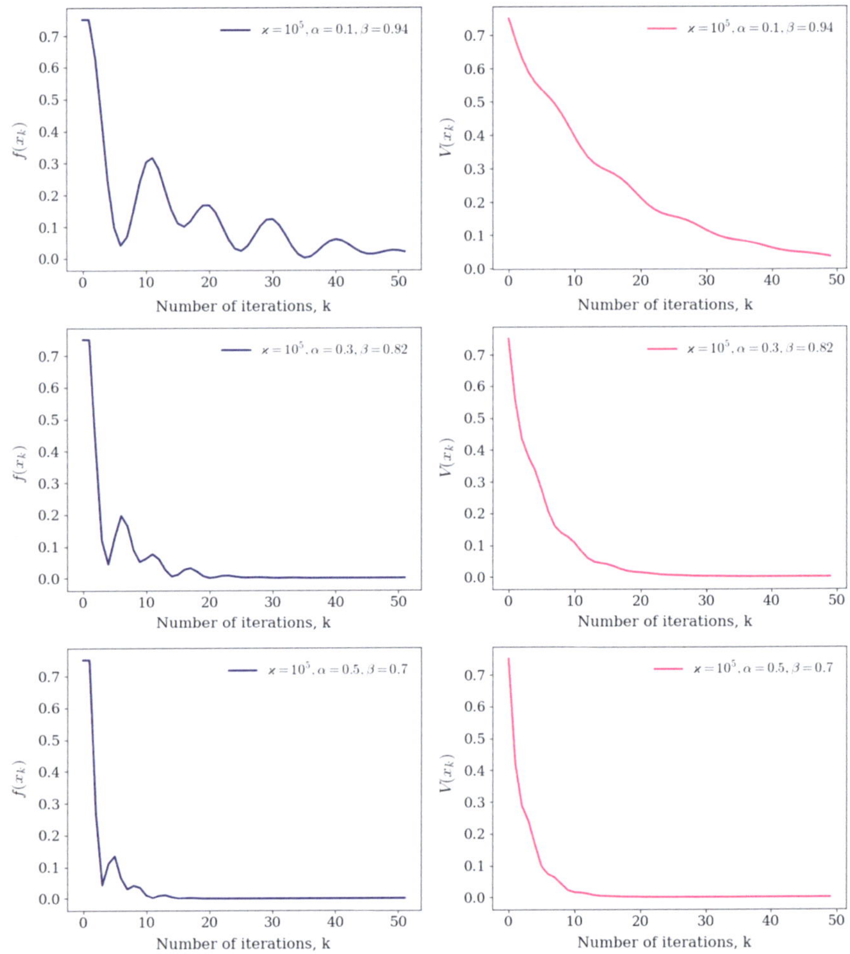

Fig. 10 The behavior of objective function $f(x_k)$ and Lyapunov function $V(x_k)$ with different parameters α, β and the condition number $\kappa = 10^5$

We would like to show numerically that the function proposed above is indeed the Lyapunov function. As a particular example a quadratic function $f(x)$, $n = 20$, $\kappa = 10^5$ is taken along with the Heavy Ball method with a set of parameters α, β. It can be observed (Fig. 10) that the Lyapunov function decreases monotonically on the trajectory of the method.

4.2 Global Convergence

In the statement above we did not prove neither convergence of $f(x_k)$ to f^* nor convergence V_k to zero. To obtain such results further assumptions on $f(x)$ are needed. The least restrictive condition is

$$||\nabla f(x)||^2 \geq 2\mu(f(x) - f^*), \mu > 0 \tag{19}$$

for all $x \in \mathbb{R}^n$. This inequality is satisfied for strongly convex functions [4, 10], but in general it does not require convexity. The condition has been proposed in [11], sometimes it is called Polyak-Łojasiewicz condition [12].

Theorem 3 *If (19) holds and*

$$\alpha \in \left(0, \frac{1}{L}\right), \ \beta \in \left[0, \sqrt{(1-\alpha L)(1-\alpha\mu)}\right].$$

then for any initial conditions $x_0, x_1 \in \mathbb{R}^n$ Lyapunov function converges linearly

$$V_k \leq V_0 q^k, \quad q = 1 - \alpha\mu < 1,$$

while for $x_0 = x_1$ objective function converges linearly

$$f(x_k) - f^* \leq (f(x_0) - f^*)q^k.$$

Proof It was shown above (18), that

$$f(x_k) - f^* + \frac{1-\alpha L}{2\alpha}\|x_k - x_{k-1}\|^2 \leq f(x_{k-1}) - f^* + \frac{\beta^2}{2\alpha}\|x_{k-1} - x_{k-2}\|^2 - \frac{\alpha}{2}\|\nabla f(x_{k-1})\|^2.$$

Due to (19) we get

$$f(x_k) - f^* + \frac{1-\alpha L}{2\alpha}\|x_k - x_{k-1}\|^2 \leq f(x_{k-1}) - f^* + \frac{\beta^2}{2\alpha}\|x_{k-1} - x_{k-2}\|^2 -$$

$$\alpha\mu(f(x_{k-1}) - f^*) = (1-\alpha\mu)\left((f(x_{k-1}) - f^*) + \frac{\beta^2}{2\alpha(1-\alpha\mu)}\|x_{k-1} - x_{k-2}\|^2\right).$$

Provided that

$$\frac{\beta^2}{2\alpha(1-\alpha\mu)} \leq \frac{1-\alpha L}{2\alpha}; \qquad \beta \leq \sqrt{(1-\alpha L)(1-\alpha\mu)},$$

we obtain

$$V(x_{k+1}) \le qV(x_k) \le q^k V(x_1), \quad q = (1 - \alpha\mu) < 1.$$

For convenience, we denote $\gamma = \frac{1-\alpha L}{2\alpha}$ and thus

$$f(x_{k+1}) - f^* \le f(x_{k+1}) - f^* + \gamma\|x_{k+1} - x_k\|^2 = V_{k+1} \le$$

$$q^k V_1 = q^k \left(f(x_1) - f^* + \gamma\|x_1 - x_0\|^2 \right) = q^k \left(f(x_0) - f^* \right).$$

Global convergence of the Heavy Ball method is established here for function under condition (19) through Lyapunov function (13). The similar results on global convergence for more narrow class of strongly convex functions were obtained in paper [13].

4.3 Adaptive Algorithm

In this section we will consider a general idea of choosing the optimal parameters for the accelerated gradient method.

We consider the smooth and strongly convex problem (6) focusing on the Heavy Ball method (7). In order to ensure the convergence of the method, it is necessary to select the parameters (9) correctly. Therefore, the values of strong convexity constant μ and the Lipschitz constant L are required. Unfortunately, these constants are difficult and time-consuming to compute for a real problem. Moreover, as already discussed earlier, accelerated first-order methods are not guaranteed to be monotonic.

To sum up, the following situations should be distinguished:

- non-monotone behavior (natural situation arising in first-order methods);
- mistake in parameter values.

Wide range of papers offer different options for adaptive restarting schemes dedicated to avoiding non-monotone behavior. By restart we denote starting the method from the very beginning with new parameters, where new initial condition is the current iteration. It is suggested to restart after a fixed number of iterations (fixed restart) or more efficiently after checking certain conditions of restart (adaptive restart). The papers [7, 8, 14] propose the different adaptive restart schemes and analysis of these techniques. For example, [8] offers the following two adaptive restart techniques. They restart whenever:

1. function scheme

$$f(x_k) > f(x_{k-1});$$

2. gradient scheme

$$\nabla f(x_{k-1})^T (x_k - x_{k-1}) > 0.$$

In this work, we offer an alternative to restart techniques in the form of the Lyapunov function $V(x)$. This function strictly decreases at each iteration. It assures that the method is stable. The algorithm converges to optimal value. We propose to monitor the values of the Lyapunov function $V(x)$ instead of the objective function $f(x)$ at each iteration. We will restart the algorithm with new parameters, only if the value of the Lyapunov function starts to increase.

Lyapunov scheme

$$V(x_k) > V(x_{k-1}).$$

We propose construction of the Lyapunov function (13) for discrete case:

$$V(x_k) = f(x_k) + \frac{1-\alpha L}{2\alpha}\|x_k - x_{k-1}\|^2.$$

Note that the correct choice of the parameters implies only the knowledge of Lipschitz constant, which can be determined iteratively according to [15]. It means, that we will check inequalities (16) with additional term $\frac{\epsilon}{2}$, where $\epsilon > 0$—the required accuracy.

$$f(x_k) \leq f(x_{k-1}) + \langle \nabla f(x_{k-1}), x_k - x_{k-1}\rangle + \frac{L}{2}\|x_k - x_{k-1}\|^2 + \frac{\epsilon}{2}$$

As a result, we obtain the following adaptive algorithm:

Algorithm 1 Adaptive Heavy ball method with Lyapunov function

Input: $f \in \mathscr{F}_L^{1,1}$, $x_0 = y_0 \in R^n$, $L_0 > 0$, $\alpha_0 \in \left(0, \frac{1}{L_0}\right)$, $\beta_0 \in \left(0, \sqrt{1-\alpha_0 L_0}\right)$.
for $k \geq 0$ **do**
    $x_{k+1} = x_k - \alpha_k \nabla f(x_k) + \beta_k (x_{k+1} - x_k)$
    if $V(x_{k+1}) > V(x_k)$
        then $L_{k+1} = 2L_k$, $\alpha_{k+1} \in \left(0, \frac{1}{L_{k+1}}\right)$, $\beta_{k+1} \in \left(0, \sqrt{1-\alpha_{k+1}L_{k+1}}\right)$
        else $L_{k+1} = L_k$, $\alpha_{k+1} = \alpha_k$, $\beta_{k+1} = \beta_k$

5 Conclusion

In this work, attention has been paid to behavior of second-order difference equations' solutions and their connection with accelerated gradient methods' convergence. Firstly, considering linear difference equations we derived the initial conditions causing peaking effects. In the next section these results were applied to the analysis of the Heavy Ball method in case of a quadratic objective function and optimal parameters α and β. Then, we moved on to discussing the non-monotonic behaviour of the method for strongly convex and smooth functions $f(x) \in \mathscr{F}_{L,\mu}^{1,1}$. Finally, the concept of the Lyapunov function for method's control was suggested.

The future work implies expanding the notion of Lyapunov function to other classes of objective functions and developing an adaptive algorithm with a better convergence rate. The obtained results are supposed to be applied to numerous unconstrained optimization problems, arising in power system engineering, deep-learning and other fields.

Funding Financial support for this work was provided by the Russian Science Foundation, project no. 16-11-10015.

References

1. Elaydi, S.: An Introduction to Difference Equations. Springer, Berlin (2005)
2. Polyak, B.T., Shcherbakov, P.S., Smirnov, G.: Peak effects in stable linear difference equations. J. Differ. Equ. Appl. **24**(9), 1488–1502 (2018). https://doi.org/10.1080/10236198.2018.1504930
3. Polyak, B.: Some methods of speeding up the convergence of iteration methods. USSR Comput. Math. Math. Phys. **4**(5), 1–17 (2004). https://doi.org/10.1016/0041-5553(64)90137-5
4. Polyak, B.: Introduction to Optimization. Optimization Software (1987)
5. Nesterov, Y.: Introductory Lectures on Convex Optimization. Kluwer, Netherlands (2004)
6. Van Scoy, B., Freeman, R., Lynch, K.: The fastest known globally convergent first-order method for the minimization of strongly convex functions. IEEE Control Syst. Lett. **2**, 49–54 (2018)
7. Giselsson, P., Boyd, S.: Monotonicity and restart in fast gradient methods. IEEE Conf. Decis. Control **53**, 5058–5063 (2014)
8. O'donoghue, B., Candes, E.: Adaptive restart for accelerated gradient schemes. Found. Comput. Math. **15**, 715–732 (2015)
9. Polyak, B., Shcherbakov, P.: Lyapunov functions: an optimization theory perspective. IFAC-PapersOnLine **50**(1), 7456–7461 (2017). https://doi.org/10.1016/j.ifacol.2017.08.1513
10. Nesterov, Y.: A method for unconstrained convex minimization problem with the rate of convergence $O(1/k^2)$. Sov. Math. Dokl. **27**, 372–376 (1983)
11. Polyak, B.: Gradient methods for the minimization of functionals. USSR Comp. Math. and Math. Phys. **3**(4), 643–653 (1963)
12. Karimi, H., Nutini, J., Schmidt, M.: Linear convergence of gradient and proximal-gradient methods under the Polyak-Łojasiewicz condition. In: Joint European Conference on Machine Learning (2016)
13. Ghadimi, E., Feyzmahdavian, H.R., Johansson, M.: Global convergence of the heavy-ball method for convex optimization. In: 2015 European Control Conference (ECC), pp. 310–315 (2015). https://doi.org/10.1109/ECC.2015.7330562
14. Donghwan, K., Fessler, J.A.: Adaptive restart of the optimized gradient method for convex optimization. J. Optim. Theory Appl. **178**(1), 240–263 (2018). https://doi.org/10.1007/s10957-018-1287-4
15. Nesterov, Yu.: Universal gradient methods for convex optimization problems. Math. Program. **152**(1–2), 381–404 (2015). https://doi.org/10.1007/s10107-014-0790-0

Global Asymptotic Stability in a Non-autonomous Difference Equation

Anatoli F. Ivanov

Abstract Non-autonomous first order difference equation of the form $x_{n+1} = x_n + a_n f(x_n)$, $n \in \mathbf{N}_0$, is considered where $f : \mathbf{R} \to \mathbf{R}$ is a continuous function satisfying the negative feedback assumption $xf(x) < 0, x \neq 0$, and $a_n \geq 0$ is a non-negative sequence. Sufficient conditions for the global asymptotic stability of the zero solution are derived in terms of the attractivity of the fixed point $x_* = 0$ under the iterations of distinct maps of the family of one-dimensional maps $F_\lambda(x) = x + \lambda f(x), \lambda \geq 0$. The principal motivation for consideration of the difference equation and the corresponding family of interval maps comes from a problem of asymptotic behavior in differential equations with piece-wise constant argument (DEPCA).

Keywords Parametric family of interval maps · Successive iterations under distinct maps · Global attractivity of fixed points · DEPCA · Reduction to difference equations

1 Introduction

This paper studies the global asymptotic stability in a class of non-autonomous first order difference equations. The problem is equivalent to the global attractivity of a fixed point under the iterations of distinct maps within a fixed family of interval maps. Both the difference equations and the family of interval maps appear as exact reductions from certain differential equations with piece-wise constant argument (DEPCA). The theoretical basics are developed here for a class of parameter dependent interval maps that are unimodal, have a single fixed point which domain of immediate attraction is the entire real axis $\mathbf{R}$. Others and more general possibilities require modifications and expansions of the approaches and techniques of the current paper; they are expected to be derived in the forthcoming paper [1].

A. F. Ivanov (✉)
Department of Mathematics, Pennsylvania State University,
Dallas, PA 18612, USA
e-mail: aivanov@psu.edu

M. Bohner et al. (eds.), *Difference Equations and Discrete Dynamical Systems with Applications*, Springer Proceedings in Mathematics & Statistics 312,
https://doi.org/10.1007/978-3-030-35502-9_10

The structure of this paper is as follows. Section 2 describes a formal statement of the principal problem under study. Section 3 is made up of two components. Section 3.1 provides a motivation and a background for us to consider difference equations of the form (1). They naturally appear as reductions of solutions of initial value problems for specific DEPCA. A brief exposition of well-known facts of such reductions is given. Section 3.2 contains some basic facts on interval maps which are necessary for related exposition in the paper. Section 4 contains several statements on global asymptotic stability of the fixed point $x = 0$ under the iterations of distinct maps of the family F_λ. They are Theorems 1–3. Those results are translated to the corresponding global asymptotical stability result for a respective DEPCA (Corollary 1). Several examples demonstrating the applicability of our results to both families of interval maps and DEPCA are given. Section 5 contains a discussion of additional aspects of this work, such as subcases not worked out in full detail in this paper, further generalizations of obtained results, and conjectures. The current work initiates a novel approach to study the global asymptotic stability in a class of DEPCA through the reduction to the problem of global attractivity of a fixed point under the action of a family of one-dimensional maps.

2 Statement of the Problem

Consider the scalar non-autonomous difference equation of the first order

$$x_{n+1} = x_n + a_n f(x_n), \quad n \in \mathbf{N}_0 = \mathbf{N} \cup \{0\}, \tag{1}$$

under the following assumptions:

($\mathbf{H}_1$) Function $f : \mathbf{R} \to \mathbf{R}$ is continuous for all $x \in \mathbf{R}$ and satisfies the negative feedback condition

$$x \cdot f(x) < 0 \quad \text{for} \quad x \neq 0; \tag{2}$$

($\mathbf{H}_2$) Sequence a_n, $n \in \mathbf{N}_0$, is non-negative, $a_n \geq 0$;
($\mathbf{H}_3$) Function $f(x)$ is continuously differentiable in a neighborhood of $x = 0$ with $f'(0) = f_0 < 0$;
($\mathbf{H}_4$) For every $\lambda > 0$ each function of the family of interval maps

$$F_\lambda(x) = x + \lambda f(x), \ \lambda \geq 0, \tag{3}$$

is unimodal for $x \in \mathbf{R}$;

Given arbitrary initial value $x_0 \in \mathbf{R}$ the corresponding solution $\{x_n, n \in \mathbf{N}_0\}$, to Eq. (1) is found by successive iterations through the composite function $\mathcal{F}_n$ as follows

$$x_n = F_{a_{n-1}} \circ F_{a_{n-2}} \circ \ldots \circ F_{a_1} \circ F_{a_0}(x_0) := \mathcal{F}_n(x_0) = \mathcal{F}_n(x_0; a_0, \ldots, a_{n-1}). \tag{4}$$

In view of ($\mathbf{H}_1$) the difference Eq. (1) has the only constant (trivial) solution $\{x_n \equiv 0, n \in \mathbf{N}_0\}$. As usual the constant solution is called (locally) *stable* if for arbitrary $\varepsilon > 0$ there exists $\delta > 0$ such that for every initial value x_0 with $|x_0| < \delta$ one has that the corresponding solution x_n satisfies $|x_n| < \varepsilon$ for all $n \in \mathbf{N}_0$. The trivial solution is called *asymptotically stable* if it is stable and $\lim_{n\to\infty} x_n = 0$ for every solution with $|x_0| < \delta$. In the case when the zero solution is not stable, it is called *unstable*. When the zero solution is asymptotically stable, and there is a set $D \subseteq \mathbf{R}$, $0 \in D$, independent of ε, δ, such that $\lim_{n\to\infty} x_n = 0$ for all $x \in D$ then we shall call the zero solution to be *globally asymptotically stable* (on set D).

We are interested in sufficient conditions when the zero solution of difference Eq. (1) is globally asymptotically stable. Such criteria are derived in Sect. 4 it terms of iterations $\mathcal{F}_n$ of distinct maps F_λ of the family (3) as $n \to \infty$.

3 Preliminaries

In this section we provide a motivation and a justification for consideration of families of difference equations of the form (1). Such difference equations appear as exact reduction for solutions of initial value problems for a particular type of differential equations with piece-wise constant argument (DEPCA). More background information on DEPCA and on the procedure of reduction are given in Sect. 3.1. A comprehensive description of fundamental properties of DEPCA can be found in e.g. [2, 3]. Section 3.2 contains necessary definitions and basic knowledge on interval maps that are used in the remainder of the paper. A complete theory of the interval maps can be found in multiple sources including the monographs [4, 5].

3.1 DEPCA: Reduction to Discrete Maps

The principal motivation for consideration of difference Eq. (1) is the study of asymptotic behavior of solutions of scalar first order differential equations with piecewise continuous argument (DEPCA) of the form

$$x'(t) = f(t, x(t), x([t - N])), \tag{5}$$

where f is a continuous function, $f(t, u, v) \in C(\mathbf{R}_+ \times \mathbf{R}^2, \mathbf{R})$, and the $[\cdot]$ stands for the greatest integer value function.

We briefly describe here some well-known basics about DEPCA as they relate to Eq. (5). Those facts as well as comprehensive additional information on the subject can be found in multiple sources, see e.g. [2, 3, 6–8] and further references therein. In order to solve Eq. (5) for $t \geq t_0 = 0$ one needs an initial function defined on the interval $[-N, 0]$. The set $\mathbf{X} := C([-N, 0], \mathbf{R})$, $N \geq 1$ can be chosen as the formal phase space for Eq. (5) (when it is viewed as a non-autonomous dynamical system).

Given arbitrary $\varphi(s) \in \mathbf{X}$ Eq. (5) is solved for $t \geq 0$ by the step method. On the interval $[0, 1)$ the solution $x(t, \varphi)$ is found by solving the initial value problem for an ordinary differential equation:

$$x'(t) = f(t, x(t), \varphi(-N)), \quad x(0) = \varphi(0), \quad t \in [0, 1)$$

By the continuity, one sets $x(1) := \lim_{t\to 1^-} x(t, \varphi) =: x_1$. It is also clear that the value $x(1) = x_1$ only depends on $x(-N) = \varphi(-N)$ and does not depend on the values $\varphi(s)$ when $s \in (-N, -N+1)$. On the interval $t \in [1, 2]$ the above integration procedure is repeated so to obtain the solution $x(t)$ by using the value $\varphi(-N+1)$ only, and so on:

$$x'(t) = f(t, x(t), x(-N+1)), \quad x(1) = x_1, \quad t \in [1, 2),$$

with $x(2) = \lim_{t\to 2^-} x(t) =: x_2$. Thus, the asymptotic behavior of solutions to Eq. (5) at $t \to \infty$ is determined by that for the successive iterations of the following family of non-autonomous maps:

$$F_n : (u_0, u_1, \ldots, u_{N-1}, u_N) \mapsto (v_0, v_1, \ldots, v_{N-1}, v_N),$$

where $u_0 = x(n)$, $u_1 = x(n-1), \ldots, u_{N-1} = x(n-N+1)$, $u_N = x(n-N)$, $v_k = u_{k-1}$, $1 \leq k \leq N$, and $v_0 := x(n+1)$ with $x(t)$ being the solution of the initial value problem

$$x'(t) = f(t, x(t), x(n-N)), \quad x(n) = u_0, \quad t \in [n, n+1), \tag{6}$$
$$x(n+1) = \lim_{t\to(n+1)^-} x(t).$$

Consider now the case of Eq. (5) when function $f(t, u, v)$ does not depend on u and has the form:

$$x'(t) = a(t)f(x([t-N])), \quad t \geq 0. \tag{7}$$

The asymptotic behavior of its solutions as $t \to \infty$ is reduced to the successive iterations of the family of maps F_n, $n \in \mathbf{N}$, on the $(N+1)$-dimensional Euclidean space $\mathbf{R}^{N+1}$:

$$F_n : (x_0, x_1, \ldots, x_{N-1}, x_N) \mapsto (x_0 + a_n f(x_N), x_0, \ldots, x_{N-2}, x_{N-1}),$$

where $a_n = \int_n^{n+1} a(t)\,dt$. Of particular interest is the case of $N = 0$ when Eq. (7) becomes

$$x'(t) = a(t)f(x([t])), \quad t \geq 0. \tag{8}$$

The family F_n then becomes a family of one dimensional maps of the form $F_n(x) = x + a_n f(x)$, $n \in \mathbf{N}$. Several specific cases of Eq. (8) were studied by others, see e.g. [8, 9] and further references therein.

Another special case of Eq. (5) is the following DEPCA

$$x'(t) = a(t)\,x(t)\,f(x([t-N])), \quad t \geq 0. \tag{9}$$

If the range of $x(t)$ is in $\mathbf{R}_+$, $x(t) > 0$, $\forall t \geq 0$, then the substitution $z(t) = \ln[x(t)]$ reduces it to the form

$$z'(t) = a(t)\, f(\exp\{z([t-N])\}), \quad t \geq 0,$$

which is Eq. (8).

As it is transparent from the reasoning above, the asymptotic properties of the solution as $t \to \infty$ of the initial value problem

$$x'(t) = a(t)\, f(x([t]), \quad x(0) = x_0, \tag{10}$$

are equivalent to the asymptotic properties of the solution of the iterative sequence

$$x_{n+1} = F_{a_n}(x_n) =: x_n + a_n\, f(x_n), \quad a_n = \int_n^{n+1} a(t)\,dt, \quad n \in \mathbf{N}_0. \tag{11}$$

Asymptotic properties of solutions of Eq. (11) as $n \to \infty$ are determined by the iterative properties of the family of one-dimensional maps

$$F_\lambda(x) : x \mapsto x + \lambda f(x), \quad \lambda \in \mathbf{R}, \tag{12}$$

when successive iterations are applied for generally different values of the parameter λ. This way iterations on $\mathbf{R}$ under the family (12) are viewed as a non-autonomous dynamical system.

In the prior studies of autonomous DEPCA of the form (8) the nonlinearity f is assumed to satisfy the negative feedback assumption with respect to the unique equilibrium [8, 9]. This is a fairly standard assumption for many systems with delay modeling biological applications among others. Therefore, the assumption (2) appears naturally in this context for both differential equation (8) and the family of maps (12). In particular, for an extension of the well-known Wright equation [10] to the case of DEPCA the nonlinearity f has the form $f(x) = 1 - e^x$ (see Example 1 in Sect. 4.4).

3.2 Related Interval Maps

In this subsection we recall some basic notions and definitions related to interval maps. Comprehensive expositions on the theory of one-dimensional maps can be found in monographs [4, 5].

Definition 1 (i) A fixed point $x = x_*$ of a continuous map F of an interval $I \subseteq \mathbf{R}$ into itself is called *attracting* if there exists an open interval $J \subseteq I$ such that $x_* \in J$, $f(J) \subseteq J$, and for every point $x \in J$ one has that $\lim_{n\to\infty} F^n(x) = x_*$ holds.
(ii) The largest connected interval $J \subseteq I$ with this property is called the domain of immediate attraction of the fixed point x_*.

As customary F^n stand for the nth iteration of F, $F^n = F \circ \cdots \circ F$.

The following statement is a well-known simple fact in the theory of interval maps. Its proof easily follows from related facts of Sect. 2.4 in [5].

Proposition 1 *For arbitrary point $x \in J$ in the domain of immediate attraction of the fixed point x_* there always exists a closed finite interval $I_0 = I_0(x) \subset J$ such that $x \in I_0$, $F(I_0) \subseteq I_0$, and $\cap_{n\geq 0} F^n(I_0) = x_*$.*

Definition 2 Let x_* be an attracting fixed point of a continuous map F. An infinite set of intervals $\{I_n, n \in \mathbf{N}_0\}$ will be called a *squeezing sequence of imbedded intervals* if the following holds:

$$I_{k+1} \subseteq I_k,\ F(I_k) \subseteq I_{k+1},\ \text{and}\ \cap_{k\geq 0} I_k = x_*.$$

It is evident that the sequence of intervals $I_k = F^k(I_0)$, $n \in \mathbf{N}_0$, in Proposition 1 is a squeezing imbedded sequence. Given an initial point x_0 in the domain of immediate attracting it is also clear that a squeezing imbedded sequence containing its iterations in not uniquely defined in general.

For C^3-smooth interval maps an important role is played by the Schwartz derivative:

$$SF(x) = \frac{F'''(x)}{F'(x)} - \frac{3}{2}\left[\frac{F''(x)}{F'(x)}\right]^2. \tag{13}$$

The Schwartz derivative exists everywhere in the phase space $I \subseteq \mathbf{R}$ except the critical points x_{cr} where the first derivative is zero, $F'(x_{cr}) = 0$. The following statement is a well-known fact for unimodal interval maps that relates local and global attractivity of a fixed point. It follows from [11].

Proposition 2 *Suppose that a unimodal map $F \in C^3(I, I)$ has a unique attracting fixed point $x_* \in I$. If the Schwartz derivative satisfies $SF(x) < 0$ for all $x \in I$, $x \neq x_{cr}$, then the fixed point x_* is also globally attracting on I: $\lim_{n\to\infty} F^n(x) = x_*$, $\forall x \in I$.*

4 Main Results

We assume throughout this section that the hypotheses $(\mathbf{H}_1)$–$(\mathbf{H}_4)$ are satisfied. Since $f'(0) = f_0 < 0$ holds the unique fixed point $x = 0$ of each of the maps $F_\lambda(x) = x + \lambda f(x)$ is locally attracting for every $0 < \lambda < -2/f_0$, and it is repelling for any

$\lambda > -2/f_0$. The value $\lambda_{cr} = -2/f_0 > 0$ is critical as the derivative is $F'_{\lambda_{cr}}(0) = -1$. The fixed point $x_* = 0$ at this parameter value can typically be either attracting or repelling, depending on further local properties of function $f(x)$ in the vicinity of $x = 0$. An important subcase for us will be when it is attracting; this happens in particular when $SF(x) < 0$ in some neighborhood of $x = 0$. Throughout this section we always assume that the parameter λ belongs to the interval $\Lambda_* = [0, -2/f_0] = [0, \lambda_{cr}]$.

We further assume that for each $\lambda \in \Lambda_* \setminus \{0\}$ the map F_λ is unimodal with the following additional specification:

($\mathbf{H}_5$) $F_\lambda(x)$ has a unique critical point $x = c_\lambda$; it is increasing on $(-\infty, c_\lambda)$ and decreasing on $(c_\lambda, +\infty)$ with

$$\lim_{x\to-\infty} F_\lambda(x) = \lim_{x\to+\infty} F_\lambda(x) = -\infty. \tag{14}$$

Note that the symmetric case of the unimodality, when the limits in (14) are $+\infty$, or when one or both of the limits are finite, is completely analogous to the case of (14) for corresponding mathematical reasoning of this section; its details are not given in this paper, due to space considerations; however, they are presented in the forthcoming paper [1]. Example 3 in Sect. 4.4 deals with such particular case.

4.1 Squeezing Sequences of Imbedded Intervals

In this subsection we shall construct and describe related properties for specific squeezing sequences of imbedded intervals for unimodal interval maps satisfying the hypothesis ($\mathbf{H}_5$). We also assume that the domain of immediate attraction of the fixed point $x_* = 0$ for each of the maps F_λ is the entire real axis $\mathbf{R}$.

Let F stand for an arbitrary representative of the family F_λ for which $x = 0$ is an attractive fixed point, and let $x = c = c_\lambda$ be its critical point. We shall distinguish between two possibilities how the critical point c is located with respect to the fixed point $x_* = 0$. If it is to the right of the fixed point, $c \geq 0$, we call the respective map F to be of type I. If it is to the left, $c < 0$, then the map F is said to be of type II. This distinction by the type is important for the structure of the squeezing sequences of imbedded intervals that we describe next.

(**A**). Suppose first that map F is of type I, i.e. the corresponding $c = c_\lambda$ satisfies $c \geq 0$. This is the case when the parameter λ satisfies $0 < \lambda \leq -1/f_0 = \lambda_{cr}/2$, since $1 > F'(0) \geq 0$ then. There is a unique value $z_0 \geq c$ such that $F(z_0) = 0$. The function $F(x)$ is monotone increasing on $(-\infty, c] \ni 0$ with $0 > F(x) > x \ \forall x < 0$ and $0 < F(x) < x \ \forall x \in (0, z_0)$. Therefore, every interval I of the form $[a, b]$, $a < 0 < b \leq z_0$, is mapped into itself, $F(I) \subset I$, with the following properties $F^{n+1}(I) \subset F^n(I)$, $n \in \mathbf{N}$, and $\cap_{n\geq 0} F^n(I) = \{0\}$ satisfied. Given an initial point x_0 a squeezing sequence $\{I_n, n \in \mathbf{N}_0\}$ of imbedded intervals now can be chosen accordingly as follows:

(i) If $x_0 \le 0$ set $I_0 = [x_0, 0]$ and $I_n = F^n(I_0)$;
(ii) If $x_0 \in (0, c]$ set $I_0 = [0, c]$ and $I_n = F^n(I_0)$;
(iii) If $x_0 \in (c, z_0]$ set $I_0 = [0, z_0]$ and $I_n = F^n(I_0)$;
(iv) If $x_0 > z_0$ set $I_0 = [F(x_0), x_0]$ and $I_n = F^n(I_0)$.

(**B**). Suppose next that F is of type II, i.e. $c = c_\lambda < 0$. This is the case when the parameter λ satisfies $-1/f_0 = \lambda_{cr}/2 < \lambda \le -2/f_0 = \lambda_{cr}$ since then $-1 \le F'(0) < 0$ holds. Then one also has that the inequalities $0 < F(c)$ and $0 > F^2(c) > c$ are satisfied. Indeed, if one assumes that the opposite $F^2(c) \ge c$ holds then the map F has a periodic point of period 2 on the interval $[c, F(c)]$. This contradicts the global attractivity of the fixed point $x = 0$. Therefore, the interval $[c, F(c)] := I_0$ is mapped into itself, and the sequence of intervals $I_n = F^n(I_0)$ forms a squeezing sequence of imbedded intervals for every initial point $x_0 \in I_0$:

$$I_0 \supset F(I_0) \supset F^2(I_0) \supset \cdots \supset F^n(I_0) \supset F^{n+1}(I_0) \supset \ldots, \quad \bigcap_{n\ge 0} F^n(I_0) = 0.$$

Since $F(x)$ is increasing and $F(x) > x$ on $(-\infty, c]$ there is a unique value $c_1 < c$ such that $F(c_1) = c$. Likewise, there is a unique value $c_2 < c_1$ such that $F(c_2) = c_1$. By induction, one builds a sequence of pre-images of the critical point c by setting $c_{n+1} < c_n$ and $F(c_{n+1}) = c_n$, $n \in \mathbf{N}$. One easily concludes that $\lim_{n\to\infty} c_n = -\infty$. Otherwise $x_* = \lim_{n\to\infty} c_n$ would be an additional fixed point of the map F. Set $J_n = [c_{n+1}, c_n]$, $n \in \mathbf{N}_0$ (with $c_0 := c$). Given an initial point x_0, a squeezing sequence I_n, $n \in \mathbf{N}$, of imbedded intervals now can be chosen as follows:

(i) For $x_0 \in I_0$ set $I_n = F^n(I_0)$;
(ii) For $x_0 \in J_m$ for some $m \ge 0$ set $I_0 = [c_{m+1}, F(c)]$ and $I_n = F^n(I_0)$;
(iii) For $x_0 > F(c)$ let $m \in \mathbf{N}$ be such that $F(x_0) \in J_m$. Set $I_0 = [c_{m+1}, x_0]$ and $I_n = F^n(I_0)$;

4.2 Global Attractivity Under Unimodal Families

The following Theorems 1–3 represent the principal results of this paper on the global attractivity of the zero fixed point under the iterations of the entire family of maps F_λ.

Given $\lambda \in \Lambda_*$ let D_λ denote the domain of immediate attraction of the fixed point $x_* = 0$ under the map F_λ. Let parameter values α, β with $0 < \alpha < \beta \le \lambda_{cr}$ be arbitrary but fixed, and let $\Lambda_{\alpha\beta}$ stands for the closed interval $[\alpha, \beta] \subset \Lambda_*$.

Theorem 1 *For arbitrary sequence* $\{\lambda_n, n \in \mathbf{N}_0\} \subseteq \Lambda_{\alpha\beta}$ *and for any initial point* $x_0 \in D := \bigcap_{\lambda\in\Lambda_{\alpha\beta}} D_\lambda$ *one has that* $\lim_{n\to\infty} \mathcal{F}_n(x_0) = 0$ *holds.*

Here we provide an outline of the proof for the case when the domain D_λ of immediate attraction of the fixed point $x_* = 0$ for each of the maps F_λ is the entire real axis $\mathbf{R}$. Therefore $D = \mathbf{R}$. The general case involving other possibilities for D_λ will be treated in paper [1].

We first make several simple general observations about the family F_λ which are due to the negative feedback assumption (2) on function $f(x)$ and the unimodality of the entire family F_λ. The following holds:

(a)
$$x < F_\alpha(x) < F_\lambda(x) < F_\beta(x) \quad \text{for all } x < 0 \text{ and every } \lambda \in (\alpha, \beta);$$

(b)
$$x > F_\alpha(x) > F_\lambda(x) > F_\beta(x) \quad \text{for all } x > 0 \text{ and every } \lambda \in (\alpha, \beta);$$

(c) For every $\lambda \in [0, \lambda_{cr}/2]$ function $F_\lambda(x)$ is increasing for $x \in (-\infty, 0]$;
(d) For every $\lambda \in [\lambda_{cr}/2, \lambda_{cr}]$ function $F_\lambda(x)$ is decreasing for $x \in [0, \infty)$;

These properties will be used multiple times in the remainder of the proof, without repeated mentioning.

In order to prove Theorem 1 we need to establish first the following result.

Lemma 1 *Assume hypotheses of Theorem 1. There exists a finite closed interval I_*, independent of the sequence $\{\lambda_n\}$, such that it is mapped into itself under each map of the family: $F_\lambda(I_*) \subseteq I_*$, $\lambda \in \Lambda_*$. Moreover, for arbitrary sequence $\{\lambda_n, n \in \mathbf{N}_0\}$ and any initial point $x_0 \in \mathbf{R}$ there exists a finite positive integer $N_0 = N_0(x_0, \{\lambda_n\})$) such that $\mathcal{F}_n(x_0) \in I_*$ for all $n \geq N_0$.*

Proof We shall construct the interval I_* explicitly. The function $u(x, \lambda) = F_\lambda(x) = x + \lambda f(x)$ is continuous for all $x \in \mathbf{R}$, $\lambda \in \Lambda_*$, and it is uniformly bounded from above. It achieves its absolute maximum value at some finite point (x_0^*, λ_0^*): $\sup\{u(x, \lambda); x \in \mathbf{R}, \lambda \in \Lambda_*\} = u_* = u(x_0^*, \lambda_0^*) = x_0^* + \lambda_0^* f(x_0^*)$. Since F_λ is a unimodal family, x_0^* is the critical point of the map $F_{\lambda_0^*}$; it is actually the point of absolute maximum for the function $y = F_\beta(x) = x + \lambda_\beta f(x)$, $x_0^* = c_\beta$. Define $b_* = u_* = F_\beta(c_\beta)$ to be the right endpoint of the interval I_*. Consider next the function $y = F_\beta(x) = x + \beta f(x)$ and define $a_* = F_\beta(b_*) = F_\beta(c_\beta)$. Since $f(x) < 0\ \forall x > 0$, this implies that $F_\lambda(x) \geq a_*$ for all $x \in [0, b_*]$ and every $\lambda \in \Lambda_*$. On the other hand, $F_\lambda(x) > x\ \forall x \in [a_*, 0]$ and every $\lambda \in \Lambda_*$ since $f(x) > 0\ \forall x < 0$. This shows that the interval $[a_*, b_*] := I_* = [c_\beta, F_\beta(c_\beta)]$ is invariant under any of the maps F_λ: $F_\lambda(I_*) \subseteq I_*$, $\lambda \in \Lambda_*$.

Suppose now that $x_0 \notin I_*$. There are two possibilities, either $x_0 < 0$ or $x_0 > 0$. Assume first that $x_0 < 0$ holds. We shall show next that the point x_0 gets mapped into the invariant interval I_* after a finite number of iterations of the map F_α, $F_\alpha^{N_0}(x_0) \in I_*$ for some positive integer N_0. Indeed, if $F_\alpha^{N_0}(x_0) \geq 0$ is satisfied for some N_0 then the claim is true. Suppose on the contrary that $F_\alpha^n(x_0) < 0\ \forall n \in \mathbf{N}_0$. Then $x_n = F_\alpha^n(x_0)$ is an increasing sequence with the finite limit $\lim_{n\to\infty} x_n = \bar{x} \leq 0$. If one assumes that $\bar{x} < 0$ then it is a fixed point of F_α different from $x = 0$, a contradiction. Therefore, $\bar{x} = 0$. This means that there exists N_0 such that $F_\alpha^{N_0}(x_0) \in I_*$. Now note that for any other map F_λ, $\lambda \in \Lambda_*$, $\lambda > \alpha$, of the family one has that the inequality $F_\lambda(x) > F_\alpha(x)$ holds for all $x < 0$. Therefore, $\mathcal{F}_n(x_0) = F_{\lambda_{n-1}} \circ F_{\lambda_{n-2}} \circ \cdots \circ F_{\lambda_1} \circ F_{\lambda_0}(x_0) \geq F_\alpha^n(x_0)$ holds as long as both $\mathcal{F}_n(x_0) < 0$ and $F_\alpha^n(x_0) < 0$ are satisfied. Thus, there exists $\hat{N}_0 \leq N_0$ such that $\mathcal{F}_{\hat{N}_0}(x_0) \in I_*$. The considerations in the case $x_0 > 0$ are analogous, and details are left out. This completes the proof of the lemma.

Outline of the proof of Theorem 1. Let the sequence $\{\lambda_n, n \in \mathbf{N}_0\} \subseteq \Lambda_{\alpha\beta}$ and the initial value $x_0 \in \mathbf{R}$ be given. Lemma 1 allows us to assume that $x_0 \in I_0^*$ is satisfied. We consider several different possibilities depending on how the composite map $\mathcal{F}_n = F_{\lambda_{n-1}} \circ F_{\lambda_{n-1}} \circ \cdots \circ F_{\lambda_1} \circ F_{\lambda_0}$ is made up. We shall show that $\lim_{n\to\infty} \mathcal{F}_n(x_0) = 0$ in all cases.

(i) Assume first that all the maps F_{λ_n}, $n \in \mathbf{N}_0$, have their critical point c_{λ_n} on the right from the fixed point $x_* = 0$, $c_{\lambda_n} \geq 0$, $\forall n \in \mathbf{N}_0$ (type I).

Proposition 3 *Suppose that* $x_0 < 0$. *Then the sequence* $x_n = \mathcal{F}_n(x_0), n \in \mathbf{N}_0$, *is increasing with* $\lim_{n\to\infty} x_n = 0$.

Indeed, this is easily seen from the inequalities $F_\alpha(x) < F_\lambda(x) < F_{\lambda_{cr}/2}(x)$ valid for all $x < 0$ and the fact that every function $F_\lambda(x), \alpha \leq \lambda \leq \lambda_{cr}/2$ is strictly increasing on $(-\infty, 0]$. It is straightforward to see that the double inequalities $F_\alpha^n(x_0) \leq \mathcal{F}_n(x_0) \leq F_{\lambda_{cr}/2}^n(x_0)$ hold for all $n \in \mathbf{N}$. Since $\lim_{n\to\infty} F_\alpha^n(x_0) = \lim_{n\to\infty} F_{\lambda_{cr}/2}^n(x_0) = 0$ the claim of the proposition follows.

Proposition 3 can also be proved by means of a squeezing sequence of imbedded intervals. Indeed, one can build the sequence following the standard steps by setting $I_0 = [x_0, 0]$ and $I_n = F_\alpha^n(I_0), n \in \mathbf{N}_0$ (see part (**A**) above). Then in view of properties (a) and (c) one has that the inclusion $\mathcal{F}_n(I_0) \subseteq F_\alpha^N(I_0)$ holds, implying $\lim_{n\to\infty} \mathcal{F}_n(x_0) = 0$.

Proposition 4 *Suppose that* $x_0 > 0$ *and the sequence* $x_n = \mathcal{F}_n(x_0) > 0$ *is positive for all* $n \in \mathbf{N}_0$. *Then* x_n *is necessarily decreasing with* $\lim x_n = 0$.

Indeed, since $f(x) < 0$, $\forall x > 0$ and $\lambda_{cr}/2 \geq \lambda_0 \geq \alpha$ then $0 < x_1 = F_{\lambda_0}(x_0) \leq F_\alpha(x_0)$ holds. By induction, one concludes $0 < x_n = \mathcal{F}_n(x_0) \leq F_\alpha^n(x_0)$. However, $F_\alpha^n(x_0) \to 0$ as $n \to \infty$ for any $x_0 > 0$ such that $F_\alpha(x_0) > 0$. Again, an associated squeezing sequence of imbedded intervals can be constructed here as $I_0 = [0, x_0]$ and $I_n = F_\alpha^n(I_0), n \in \mathbf{N}$.

The remaining possibility for the subcase (i) is that $x_0 > 0$, $x_n > 0$ for all $1 \leq n \leq N_0$, and $x_{N_0+1} < 0$ for some $N_0 \geq 1$. Then $x_n < 0$ holds for all $n \geq N_0 + 1$, and this situation is reduced to the subcase of Proposition 3. An associated squeezing sequence of imbedded intervals can be chosen here as $I_0 = [x_{N_0}, x_0]$ and $I_n = F_\alpha^n(I_0), n \in \mathbf{N}$.

(ii) Assume next that all the maps F_{λ_n} have their critical point c_{λ_n} on the left from the fixed point $x_* = 0$, $c_{\lambda_n} < 0$ (type II).

For any $\lambda \in (\lambda_{cr}/2, \lambda_{cr}]$ let c_λ be the critical point of the map F_λ. Denote the interval $[c_\lambda, F_\lambda(c_\lambda)] = I_\lambda$.

It is readily seen from the proof of Lemma 1 that for arbitrary $x_0 \in \mathbf{R}$ there exists a positive integer N_0 such that $\mathcal{F}_n(x_0) \in I_0^\beta = [c_\beta, F_\beta(c_\beta)]$ for all $n \geq N_0$. Indeed, for every $\lambda_0 \in [\lambda_{cr}/2, \lambda_{cr}]$ if $x_0 > 0$ then $x_1 = F_{\lambda_0}(x_0) < 0$. Therefore, one can assume that the initial point satisfies $x_0 < 0$. If the corresponding sequence x_n is such that $x_n < 0$ $\forall n \in \mathbf{N}$ then $\lim_{n\to\infty} x_n = 0$. Therefore, in the latter case $0 > x_{N_0} \geq c_\beta$ holds for some N_0 (and then for all $n \geq N_0$). If there exists positive integer N_0 such that $x_n < 0 \ \forall \ 1 \leq n \leq N_0$ and $x_{N_0+1} > 0$ then $x_{N_0+1} = F_{\lambda_n}(x_{N_0}) \leq F_\beta(x_{N_0}) \leq F_\beta(c_\beta)$. In either case, $c_\beta \leq x_{N_0+1} \leq F_\beta(c_\beta)$.

The interval I_β is mapped into itself under any map F_λ, $\lambda \in [\lambda_{cr}/2, \beta.]$. This is easily seen from two facts: (i) each $F_\lambda(x)$ is decreasing for $x > 0$ with the inequalities $0 > F_\lambda(x) > F_\beta(x)$ satisfied, and (ii) $F_\beta(x) > F_\lambda(x) > x$ for all $x \in (c_\beta, 0)$. This means that the inclusion $F_{\lambda_n}(x_0) \in I_\beta$ holds for all $n \in \mathbf{N}$ provided $x_0 \in I_\beta$.

In view of the above we can assume that the initial point satisfies $x_0 \in I_\beta$.

We shall construct now a squeezing sequence of imbedded intervals associated with the interval I_β and the entire family of maps F_λ for $\lambda \in [\lambda_{cr}/2, \beta]$ (type II). Note that the following inequalities hold for each such λ:

$$F_\beta(x) \geq F_\lambda(x) \geq F_{\lambda_{cr}/2}(x) \quad \text{for all} \quad x \in [c_\beta, 0] \tag{15}$$

and

$$F_\beta(x) \leq F_\lambda(x) \leq F_{\lambda_{cr}/2}(x) \quad \text{for all} \quad x \in [0, \infty). \tag{16}$$

Lemma 2 *There exists a squeezing sequence of imbedded intervals* $\{I_n, n \in \mathbf{N}_0\}$ *such that for every map* F_λ, $\lambda \in [\lambda_{cr}/2, \beta]$, *one has:*

$$F_\lambda(I_n) \subseteq I_{n+1}, \ \ I_{n+1} \subseteq I_n, \quad \text{and} \quad \bigcap_{n\geq 0} I_n = \{0\}.$$

Proof Consider the image of the interval $I_\beta := I_0 =: [a_0, b_0]$ under any representative F_λ, $\lambda \in [\lambda_{cr}/2, \beta]$, of the family. In view of the inequalities (15) and (16) it is easily seen that for any $x_0 \in (c_\beta, 0)$ one has $F_\lambda(x_0) \leq F(c_\beta)$. Also the other inequality holds for such x_0: $F_\lambda(x_0) \geq F_{\lambda_{cr}/2}(x_0) \geq F_{\lambda_{cr}/2}(c_\beta)$. Since all $F_\lambda(x)$ are decreasing for $x \geq c_\lambda$, one has that for every initial value $x_0 \in [0, F_\beta(c_\beta)]$ the inequality holds $F_\beta(c_\beta) \leq F_\lambda(x_0) \leq 0$. Therefore, $F_\lambda(I_0) \subseteq F_\beta(I_0)$ for every $\lambda \in [\lambda_{cr}/2, \beta]$. Note that more generally, in view of inequalities (15) and (16), the following holds: for every subinterval $J = [\xi, \eta] \subseteq I_\beta$ such that $J \ni 0$ the inclusion $F_\lambda(J) \subseteq F_\beta(J)$ is valid for every parameter value $\lambda \in [\lambda_{cr}/2, \beta]$.

Next we shall compare the values $F_{\lambda_{cr}/2}(c_\beta)$ and $F^2_\beta(c_\beta)$ and construct two intervals I_1 and I_2 accordingly.

(1) If $F_{\lambda_{cr}/2}(c_\beta) \geq F^2_\beta(c_\beta)$ we set $I_1 := F_\beta(I_0) = [F^2_\beta(c_\beta), F_\beta(c_\beta)] =: [a_1, b_1]$ and $I_2 := F_\beta(I_1) = [F^2_\beta(c_\beta), F^3_\beta(c_\beta)] =: [a_2, b_2]$. Due to the inequalities (15) and (16) one sees that the inclusions hold: $F_\lambda(I_k) \subseteq I_{k+1}, k = 0, 1,$ for every $\lambda \in [\lambda_{cr}/2, \beta]$.

(2) If $F_{\lambda_{cr}/2}(c_\beta) < F^2_\beta(c_\beta)$ we set $I_1 := [F_{\lambda_{cr}/2}(c_\beta), F_\beta(c_\beta)] =: [a_1, b_1]$ and $I_2 := [a_1, F_\beta(a_1)] =: [a_2, b_2] \subseteq I_1$. Then one easily derives the inclusion $F_\beta(I_1) = [F^2_\beta(c_\beta), F_\beta(a_1)] \subseteq I_2$ since $F^2_\beta(c_\beta) > a_1 = F_{\lambda_{cr}/2}(c_\beta)$. Due to the construction of intervals I_0, I_1, I_2 in this subcase one sees that the inclusions hold: $I_0 \supset I_1 \supset I_2$ and $F_\beta(I_0) \subset I_1$, $F_\beta(I_1) \subset I_2$. It is also an immediate conclusion that in this subcase the inclusions $F_\lambda(I_k) \subseteq I_{k+1}, k = 0, 1,$ hold for every $\lambda \in [\lambda_{cr}/2, \beta]$.

We proceed now iteratively by using the interval I_2 as the initial interval and repeat the above construction. In either subcase (1) and (2) above $b_2 = F_\beta(a_2)$ holds. Consider the value $F_\beta^2(a_2)$ and compare it with $F_{\lambda_{cr}/2}(a_2)$. If the inequality $F_{\lambda_{cr}/2}(a_2) \geq F_\beta^2(a_2)$ holds, then we build intervals I_3 and I_4 exactly as in part (1) above: $I_3 = F_\beta(I_2)$ and $I_4 = F_\beta(I_3)$. If the opposite inequality holds, $F_{\lambda_{cr}/2}(a_2) < F_\beta^2(a_2)$, then we proceed to build intervals I_3 and I_4 exactly as in part (2) above: $I_3 = [F_{\lambda_{cr}/2}(a_2), F_\beta(a_2)] =: [a_3, b_3]$ and $I_4 = [a_3, F_\beta(a_3)] \subseteq I_3$. In both cases the inclusions $I_2 \supseteq I_3 \supseteq I_4$ and $F_\beta(I_2) \subseteq I_3$, $F_\beta(I_3) \subseteq I_4$ hold.

By continuing this procedure by induction indefinitely we will have built a sequence of imbedded intervals $\{I_n, n \in \mathbf{N}_0\}$ such that $I_{n+1} \subseteq I_n$ and $F_\beta(I_n) \subseteq I_{n+1}$ for all $n \in \mathbf{N}_0$. This is also a squeezing sequence of intervals with $\bigcap_{k\geq 0} I_k = \{0\}$. The latter is seen from the fact that each left end point a_{2n}, $n \in \mathbf{N}$, of the interval I_{2n} is defined as a minimum of two functions on the previous values, $F_{\lambda_{cr}/2}(a_{2n-2})$ and $F_\beta^2(a_{2n-2})$. More precisely, it follows from the following proposition.

Proposition 5 *Suppose two continuous maps G_1, G_2 of the negative semi-axis $\mathbf{R}_- = (-\infty, 0]$ into itself are increasing and satisfying the following additional conditions: $G_i(0) = 0$ and $0 > G_i(x) > x$ for all $x < 0, i = 1, 2$. Then any solution of the difference equation $x_{n+1} = \min\{G_1(x_n), G_2(x_n)\}$, $x_0 \in \mathbf{R}_-$, satisfies $\lim_{n\to\infty} x_n = 0$.*

Proof Define two continuous functions H_1, H_2 by $H_1(x) = \max\{G_1(x), G_2(x)\}$, $H_2(x) = \min\{G_1(x), G_2(x)\}$. Since both are increasing with $0 > H_i(x) > x\ \forall x < 0$, then $H_i^n(x_0) \to 0$ as $n \to \infty$ for every $x_0 > 0$ $(i = 1, 2)$. On the other hand, the inequalities $H_2^n(x_0) \leq x_n \leq H_1^n(x_0)$ hold for all $n \in \mathbf{N}$, which proves the proposition.

Note that in the symmetric case when functions G_1, G_2 are defined on the positive semi-axis $\mathbf{R}_+ = [0, \infty)$, are monotone increasing and satisfying the assumptions $G_i(0) = 0$ and $0 < G_i(x) < x$, $i = 1, 2$, for all $x > 0$, the conclusion of Proposition 5 is also valid for the following difference equation: $x_{n+1} = \max\{G_1(x_n), G_2(x_n)\}$, $x_0 \in \mathbf{R}_+$,

An immediate implication of Lemma 2 is the following

Corollary 1 *For arbitrary sequence $\lambda_n \in [\lambda_{cr}, \beta]$, $n \in \mathbf{N}_0$, and for any initial value $x_0 \in I_\beta$, one has that $\mathcal{F}_n(x_0) \in I_n$ holds. Therefore, $\lim_{n\to\infty} \mathcal{F}_n(x_0) = 0$.*

The corollary follows from the fact that $x_n = \mathcal{F}_n(x_0) \subseteq \mathcal{F}_n(I_0) \subseteq F_\beta^n(I_0) \subseteq I_n$ since the inclusions $F_\lambda(I_k) \subseteq F_\beta(I_k) \subseteq I_{k+1}$ hold for every $\lambda \in [\lambda_{cr}/2, \beta]$ and all $k \in \mathbf{N}_0$. This completes the proof for subcase (ii).

(iii) In the case when the composite map $\mathcal{F}_n$ is such that all its components are either of type I or type II eventually is easily reducible to one of the subcases (i) or (ii) above. One has simply to choose $\mathcal{F}_{N_0}(x_0)$ as a new initial value for some sufficiently large N_0.

(iv) Assume finally that $\mathcal{F}_n$ has unbounded numbers of components of both types (I) and (II) as $n \to \infty$.

Proposition 6 *For arbitrary $x_0 \in I_0$ one of the following two possibilities can happen:*

(i) The iterative sequence $x_n = \mathcal{F}_n(x_0)$ *is eventually monotone with* $\lim x_n = 0$*;*
(ii) For any iterative sequence $\{x_n\}$ *that is not eventually monotone there exists a positive integer* N_0 *such that* $\mathcal{F}_n(x_0) \in [c_\beta, F_\beta(c_\beta)]$ *for all* $n \geq N_0$.

This proposition is similar to Propositions 3 and 5 with the following two differences. (1) The lower bound for the iterative sequence x_n with a negative initial value x_0 is given by the map F_α. If $x_0 < 0$ then $F_\lambda(x_0) \geq F_\alpha(x_0)$ for every $\lambda \in \Lambda_{\alpha\beta}$. This results in the estimate $x_n = \mathcal{F}_n(x_0) \geq F_\alpha^n(x_0)$ as long as $x_k < 0$, $\forall\, 0 \leq k \leq n$. (2) The sequence x_n does not have to be monotone. Therefore, either $x_n < 0\ \forall n \in \mathbf{N}_0$ with $\lim_{n\to\infty} x_n = 0$ holds, as in Proposition 3; or there exists first positive integer $N_0 \geq 1$ such that $x_{N_0} > 0$. In both cases $x_n \in I_\beta\ \forall n \geq N_0$.

Proposition 6 allows us to start with initial points x_0 from the interval I_β.

We shall build next a squeezing sequence of imbedded intervals similar to that in Lemma 2 for the subcase (ii).

Lemma 3 *There exists a squeezing sequence of imbedded intervals* $\{I_n, n \in \mathbf{N}_0\}$ *such that for the map* F_β *one has:*

$$F_\beta(I_n) \subseteq I_{n+1}, \quad I_{n+1} \subseteq I_n, \quad \text{and} \quad \bigcap_{n\geq 0} I_n = \{0\}.$$

Proof The construction of the sequence $\{I_n, n \in \mathbf{N}_0\}$ is very similar to that in the proof of Lemma 2. We start with $I_0 = I_\beta = [c_\beta, F_\beta(c_\beta)]$ and proceed with relevant step (1) or (2) to construct I_1 and I_2. The major difference now is that in step (2) we use the function F_α rather than $F_{\lambda_{cr}/2}$, since it is now the lower bound for the entire family F_λ consisting of both types (I) and (II). We repeat the procedure to get intervals I_3 and I_4, and apply it inductively to have the entire sequence of intervals $\{I_n = [a_n, b_n], n \in \mathbf{N}_0\}$. The fact that $\{I_n\}$ is a squeezing sequence follows from Proposition 5 for which F_α and F_β are now used as the lower and upper bound functions, respectively: $a_{2n} = \min\{F_\alpha(a_{2n-2}), F_\beta^2(a_{2n-2})\}, n \in \mathbf{N}$.

Finally, consider the iterative maps $\mathcal{F}_n, n \in \mathbf{N}_0$, applied to the interval I_0, $\mathcal{F}_n(I_0) = F_{\lambda_{n-1}} \circ F_{\lambda_{n-2}} \circ \cdots \circ F_{\lambda_1} \circ F_{\lambda_0}(I_0)$. For the given sequence λ_n of the parameter values let $\{\lambda_{n_1}, \lambda_{n_2}, \lambda_{n_3}, \ldots, \lambda_{n_k}, \ldots\}$ be its complete subsequence corresponding to all the maps of type II in the composition $\mathcal{F}_n$ as $n \to \infty$. In view of the inequalities (15)–(16) for any interval I_n of the squeezing sequence the inclusion $F_\lambda \circ F_{\lambda_{n_k}}(I_n) \subseteq F_{\lambda_{n_k}}(I_n)$ holds for any map F_λ of type I. This is because any such F_λ satisfies: $x < F_\alpha(x) \leq F_\lambda(x) < F_{c_{cr}/2}(x)$ for $x \in I_0 \cap \mathbf{R}_-$ and $x > F_\alpha(x) \geq F_\lambda(x) > F_{c_{cr}/2}(x)$ for $x \in I_0 \cap \mathbf{R}_+$. Therefore, $\mathcal{F}_n(I_0) \subseteq F_{\lambda_{n_1}} \circ F_{\lambda_{n_2}} \circ \cdots \circ F_{\lambda_{n_k}}(I_0) \subseteq I_k$, where n_k is the largest positive integer of the subsequence such that $n_k \leq n$. Since $k \to \infty$ as $n \to \infty$ this implies that $\lim_{n\to\infty} \mathcal{F}_n(x_0) = 0$ for every $x_0 \in I_0$.

This completes the proof of Theorem 1.

4.3 Critical Case $\lambda = 0$

We now consider the iterative maps $\mathcal{F}_n = F_{\lambda_{n-1}} \circ F_{\lambda_{n-2}} \circ \cdots \circ F_{\lambda_1} \circ F_{\lambda_0}$ for the range of the parameter values λ_k, $k \in \mathbf{N}_0$, that are close to zero.

Lemma 4 *Given an arbitrary initial point $x_0 \in \mathbf{R}$, there exists a parameter value $\lambda = \mu_0 = \mu_0(x_0)$ such that the family of maps $F_\lambda(x) = x + \lambda f(x)$ satisfies one of the following two statements:*

(1) $F_\lambda(x) > 0$ for all $x \in (0, x_0)$ and for every $\lambda \in (0, \mu_0]$ in the case when $x_0 > 0$, or

(2) $F_\lambda(x) < 0$ for all $x \in (x_0, 0)$ and for every $\lambda \in (0, \mu_0]$ in the case when $x_0 < 0$.

Proof Consider first the case when $x_0 > 0$. Since $f(x)$ is continuously differentiable in some neighborhood of $x = 0$ with $f'(0) = f_0 < 0$ there exists $\delta_0 > 0$ and $\mu_0^1 > 0$ such that $F_\lambda(x)$ satisfies $x > F_\lambda(x) > 0$ for all $x \in (0, \delta_0]$ and every $\lambda \in (0, \mu_0^1]$. Since $f(x)$ is continuous on the closed interval $[\delta_0, x_0]$ and it is negative there one has the two-sided estimate $-M \le f(x) \le -m$ for all $[\delta_0, x_0]$. If one chooses μ_0 such that $\mu_0 > \delta_0/M$ then the inequality $G_\lambda(x) > 0$ holds for all $x \in (0, x_0]$ and arbitrary $0 < \lambda \le \mu_0$. Since $G_\lambda(x) < x$ for all $x > 0$ this complete the proof when $x_0 > 0$. The symmetric case $x_0 < 0$ is proved analogously.

An easy implication of Lemma 4 is the following

Corollary 2 *Given arbitrary initial point x_0 there exists $\mu_0 > 0$ such that the iterative sequence x_n defined by (1) is monotone for arbitrary sequence of the coefficients $\{\lambda_0, \lambda_1, \lambda_2, \ldots\} \subset [0, \mu_0]$. Moreover, x_n is decreasing when $x_0 > 0$, and it is increasing when $x_0 < 0$.*

The corollary is obvious, since the inequality $0 < F_\lambda(x) < x$, $\forall x \in (0, x_0]$ is satisfied in the case $x_0 > 0$, implying $x_n > x_{n+1} > 0$, $\forall n \in \mathbf{N}_0$. The other case $x_0 < 0$ yields $x_n < x_{n+1} < 0$, $\forall n \in \mathbf{N}_0$.

Theorem 2 *Suppose the sequence $\{\lambda_n\} \subset \Lambda_*$ of parameter values satisfies $\lim_{n\to\infty} \lambda_n = 0$. Then the iterative sequence $\{x_n\}$ defined by difference Eq. (1) is monotone for large n, and the limit $\lim_{n\to\infty} x_n = x^*$ is finite. Moreover, $x^* = 0$ if and only if the following series diverges: $\sum_{n=0}^{\infty} \lambda_n = +\infty$.*

Proof Note that $D_\lambda = \mathbf{R}$ for all $\lambda \in (0, \lambda_{cr}/2)$. This follows from the fact that F_λ is unimodal, $0 < F_\lambda'(0) < 1$, and that the critical point is to the right from the unique fixed point, $c_\lambda > 0$.

Since $\lim_{n\to\infty} \lambda_n = 0$ we can assume without loosing a generality that all the values $\lambda_n \in \mathbf{N}$ are sufficiently small, so that we are under the assumptions of Lemma 1. To be definite let $x_0 > 0$. Then the sequence x_n is monotone decreasing with $\lim_{n\to\infty} x_n = x^* \ge 0$.

Assume first that $x^* > 0$. Then the following quantities are well defined on the interval $[x^*, x_0]$:

$$\max_{x\in[x^*,x_0]}\{f(x)\}=-m,\qquad \min_{x\in[x^*,x_0]}\{f(x)\}=-M,\quad \text{with}\quad 0<m\le M.$$

Therefore, one has the following estimate: $x_1=x_0+a_0f(x_0)\le x_0-ma_0$. Likewise, $x_2\le x_1-a_1m\le x_0-a_0m-a_1m$. By induction, one derives the estimate $x_n\le x_0-m\sum_{k=0}^{n}a_k$. The latter leads to a contradiction $x_n\to-\infty$ when the series $\sum_{n=0}^{\infty}a_n=\infty$ diverges.

Suppose next that $x^*=0$. One can also assume that in this case all members of the sequence x_n are sufficiently small (this is true for large n). By using the Taylor expansion $f(t)=f(t_0)+f'(\theta)(t-t_0)$, $\theta\in(t_0,t)$, in a small neighborhood of $t_0=0$, one derives the following two sided estimate: $-k_1x\le f(x)\le -k_2x$, $\forall x\in(-\delta,\delta)$ for some $0<k_2<k_1$ and $\delta>0$. This in turn implies that $x_1=x_0+a_0f(x_0)$ satisfies $x_0(1-k_1a_0)\le x_1\le x_0(1-k_2a_0)$. Likewise, the second iteration x_2 satisfies:

$$x_0(1-k_1a_0)(1-k_1a_1)\le x_1\le x_0(1-k_2a_0)(1-k_2a_1).$$

By induction, one derives the following two-sided inequalities:

$$x_0\prod_{i=0}^{n}(1-a_ik_1)\le x_n\le x_0\prod_{i=0}^{n}(1-a_ik_2),\ n\in\mathbf{N}. \tag{17}$$

By applying the natural logarithmic function to the latter one arrives at

$$\ln x_0+\sum_{i=0}^{n}\ln(1-a_ik_1)\le\ln x_n\le\ln x_0+\sum_{i=0}^{n}\ln(1-a_ik_2). \tag{18}$$

By the limit comparison test both series $\sum_{i=0}^{\infty}\ln(1-a_ik_1)$ and $\sum_{i=0}^{\infty}\ln(1-a_ik_2)$ converge if and only if the series $\sum_{i=0}^{\infty}a_i$ converges. Since $\lim_{n\to\infty}\ln(x_n)=-\infty$ holds, by taking the limit on both sides of the inequalities (18) one sees that the series $\sum_{i=0}^{\infty}a_i$ is necessarily divergent. This completes the proof when $x_0>0$. The reasoning and details of the other case $x_0<0$ are completely analogous and are left out.

Let an initial point $x_0<0$ be arbitrary but fixed, and let x^* be any other given number such that $0>x^*>x_0$. We shall show that there exists a sequence of the parameter values μ_n, $n\in\mathbf{N}_0$, such that the recursive sequence $x_{n+1}=x_n+\mu_n f(x_n)$, $n\in\mathbf{N}_0$, is monotone increasing and $\lim x_n=x^*$. Indeed, this follows from the fact that the value $x_1=x_1(\mu)=x_0+\mu f(x_0)$ is continuously dependent on μ, and satisfies $x_1(0)=x_0$ and $x_1(\mu)>x_0$ for all small $\mu>0$. Therefore, the value x_1 can be assigned arbitrarily to some small right neighborhood of $x=x_0$, by appropriately choosing the parameter value $\mu_0>0$. Likewise, x_2 can be put arbitrarily in its sufficiently small but fixed right neighborhood, by an appropriate choice of μ_2. By induction, a sequence $\{x_n\}$ can be constructed, by an appropriate choice of the corresponding sequence $\{\mu_n\}$, such that

$$x_0 < x_1 < x_2 < x_3 < \cdots < x_n < x_{n+1} < \cdots \longrightarrow x^*.$$

Theorems 1 and 2 yield the following result:

Theorem 3 *Let λ_*, $0 < \lambda_* \le -2/f_0$, be a parameter value such that $x = 0$ is an attracting fixed point for every map F_λ, $\lambda \in (0, \lambda_*]$. Set $D := \bigcap_{\lambda \in (0,\lambda_*]} D_\lambda$ where D_λ is the domain of immediate attraction of $x = 0$ under F_λ. Then for any $x \in D$ and for arbitrary sequence $\{\lambda_n\} \in (0, \lambda_*]$ such that $\sum_0^\infty \lambda_n = \infty$ one has that $\lim_{n\to\infty} \mathcal{F}_n(x) = 0$ holds.*

Proof If the sequence $\{\lambda_n, n \in \mathbf{N}_0\}$ contains an infinite subsequence λ_{n_k} such that $\lambda_{n_k} \ge \alpha$ for some $\alpha > 0$ then the reasoning of Theorem 1 applies. Otherwise $\lambda_n \to 0$ as $n \to \infty$, thus Theorem 2 invokes.

Corollary 3 *Assume the hypotheses of Theorem 3. Any solution of the corresponding initial value problem for the DEPCA:*

$$x'(t) = a(t)\, f(x([t])), \quad x(0) = x_0 \in D$$

converges to zero as $t \to \infty$, $\lim_{t\to\infty} x(t) = 0$, in the case when $0 \le a(t) \le \lambda_*$ and $\int_0^\infty a(t)\, dt = \infty$ are satisfied.

This immediately follows from the relationship between the initial value problem (10) and the family of maps (12) through the difference Eq. (11).

4.4 Examples

Example 1 The following differential equation

$$x'(t) = -a(t)x([t])[1 + x(t)], \qquad a(t) \ge 0, \tag{19}$$

can be viewed as a modification for the case of DEPCA of the well-known continuous time differential delay equation introduced by E. M. Wright. Similarly to the reasoning in [10] it can be shown that its solutions satisfy $x(t) > -1$ $\forall t \ge 0$ provided $x(0) > -1$. The equation is reducible to the form

$$x'(t) = a(t)\ (1 - \exp\{x([t]))], \quad t \ge 0, \tag{20}$$

through the substitution $1 + x = e^z$ (in the latter we revert back to the x-notation from z).

The family F_λ of the maps has the form

$$F_\lambda(x) = x + \lambda f(x) = x + \lambda\left(1 - e^x\right), \quad \lambda > 0.$$

It is easily verifiable that the following statements are true:

(1) For any $\lambda > 0$ map F_λ has unique fixed point $x = 0$;
(2) Function $f(x) = 1 - e^x$ satisfies the negative feedback condition $x \cdot f(x) < 0,\ \forall x \neq 0$;
(3) $F_\lambda(x)$, $\lambda > 0$, is a family of unimodal maps with the critical point $c_\lambda = -\ln \lambda$. F_λ is increasing for $x \in (-\infty, c_\lambda)$ and F_λ is decreasing for $x \in (c_\lambda, \infty)$ with $\lim_{x\to-\infty} F_\lambda(x) = \lim_{x\to+\infty} F_\lambda(x) = -\infty$;
(4) The Schwartzian derivative is negative outside the critical point:

$$(S F_\lambda)(x) = \frac{-\lambda e^x - \frac{1}{2}\lambda^2 e^{2x}}{(1 - \lambda e^x)^2} < 0 \quad \forall x \in \mathbf{R}, x \neq c_\lambda.$$

(5) Derivative at zero is $F'_\lambda(0) = 1 - \lambda$. Fixed point $x = 0$ is globally attracting for $0 < \lambda \leq 2$; it is repelling for $\lambda > 2$.

Theorem 3 applies to conclude the following

Proposition 7 *Suppose that the coefficient $a(t)$ of Eq. (19) satisfies $a(t) \geq 0$, $\int_n^{n+1} a(t)\,dt \leq 2, n \in \mathbf{N}$, and $\int_0^\infty a(t)\,dt = \infty$. Then its constant solution $x(t) \equiv 0$ is globally asymptotically stable: $\lim_{t\to\infty} x(t) = 0$ for any solution $x(t)$ with the initial condition satisfying $x(0) > -1$.*

Paper [9] considers DEPCA of the form $y'(t) = r(t)y(t)\{1 - y([t])/K\}$, which is an equivalent form to Eq. (19) in Example 1. It is shown that any solution $y(t)$ with the initial condition $y(0) = y_0 > 0$ satisfies $\lim_{t\to\infty} y(t) = K$ provided $0 \leq r(t) \leq 2$ and $\int_0^\infty r(t)\,dt = \infty$. This result is largely overlapping with our Proposition 7. The methods and techniques used in [9], however, are completely different from those developed in this work.

Example 2 Consider the following equation

$$x'(t) = a(t) \arctan(x([t])), \quad \text{where} \quad a(t) \leq 0. \tag{21}$$

The corresponding family of maps in this example is given by $F_\lambda(x) = x - \lambda \arctan x$, $\lambda > 0$. For $0 < \lambda \leq 1$ all functions are monotone increasing on $\mathbf{R}$. For every $\lambda > 1$ the corresponding map F_λ has two critical points, $c_1 = -\sqrt{\lambda - 1}$ and $c_2 = \sqrt{\lambda - 1}$ with $F_\lambda(x)$ increasing on $(-\infty, c_1) \cup (c_2, +\infty)$ and decreasing on (c_1, c_2). The only fixed point $x = 0$ is globally attracting on $\mathbf{R}$ for every $\lambda \in [-2, 0)$, and it is repelling for $\lambda < -2$.

Though the structure of the family of maps F_λ in this case does not formally fit the unimodality assumption $(\mathbf{H}_5)$ their dynamics are very similar to those described by Theorems 1–3. This is due to the symmetric shape of the maps with respect to the diagonal $y = x$, since $y = \arctan(x)$ is an odd function.

In the case $0 < \lambda \leq 1$ the dynamics of F_λ are very simple since it is a monotone increasing function for all $x \in \mathbf{R}$. The sequence $x_n = F^n_\lambda(x_0)$ converges to zero as $n \to \infty$ for every initial point x_0. The rate of convergence is faster when the parameter value λ is larger. This is easily seen from the fact that when $0 < \lambda_1 < \lambda_2 \leq 1$ then the inequalities hold $x < F_{\lambda_1}(x) < F_{\lambda_2}(x) < 0$ for $x < 0$, and $x > F_{\lambda_1}(x) > F_{\lambda_2}(x) > 0$ for $x > 0$. Therefore, if the parameter sequence $\{\lambda_n, n \in \mathbf{R}\}$ is such that $1 \geq \lambda_n \geq \alpha > 0$ holds for all n then the inequalities $0 > \mathcal{F}_n(x_0) \geq F_\alpha^n(x_0)$ are valid for $x_0 < 0$, and $0 < \mathcal{F}_n(x_0) \leq F_\alpha^n(x_0)$ are satisfied for $x_0 > 0$. Thus, $\lim_{n\to\infty} \mathcal{F}_n(x_0) = 0$ in either case.

In the case of the parametric values $1 < \lambda \leq 2$ the dominant map for the family is $F_2(x) = x - 2\arctan(x)$, in the sense that it provides the slowest convergence to zero for the entire family F_λ, $1 < \lambda \leq 2$. Indeed, in this case the following inequalities hold $F_2(x) \geq F_\lambda(x) \geq F_1(x)$ for $x < 0$, and $F_2(x) \leq F_\lambda(x) \leq F_1(x)$ for $x > 0$.

We first note that the fixed point $x_* = 0$ is globally attracting for $F_2(x) = x - 2\arctan(x)$. This follows from the fact that the following inequalities are satisfied $x < F_2(x) < -x$ for $x < 0$, and $x > F_2(x) > -x$ for $x > 0$. There are no cycles of period two, and the sequence $F_2^{2n}(x_0)$ is monotone and convergent to zero for any initial point $x_0 \in [-1, 1]$. Denote $I_0 = [-1, 1]$, the closed interval between two critical points of the map $F_2(x)$. Exactly as in the proof of Lemma 1 it can be shown that for arbitrary parameter sequence $\{\lambda_n, n \in \mathbf{N}_0\}$ such that $1 < \lambda_n \leq 2$ and every initial value $x_0 \in \mathbf{R}$ there exists a finite positive integer $N_0 = N_0(x_0, \{\lambda_n\})$ such that $\mathcal{F}_n(x_0) \in I_0$, $\forall n \geq N_0$. This allows us to consider initial values x_0 from the interval I_0 only. Finally, one can build the squeezing sequence of imbedded intervals exactly the same way as in the proof of Theorem 1 in Sect. 4.2, to show that $\lim_{n\to\infty} \mathcal{F}_n(x_0) = 0$ for every $x_0 \in I_0$. In essence, Theorems 1–3 hold for the family of maps $F_\lambda(x) = x - \lambda\arctan(x)$, $0 < \lambda \leq 2$ with minor modifications and without the assumption of unimodality.

The following statement holds true.

Proposition 8 *Suppose that the coefficient $a(t)$ of Eq. (21) satisfies $a(t) \leq 0$, $\int_n^{n+1} a(t)\,dt \geq -2, n \in \mathbf{N}$, and $\int_0^\infty a(t)\,dt = -\infty$. Then its zero solution is globally asymptotically stable:* $\lim_{t\to\infty} x(t) = 0$ *for any initial condition* $x(0) = x_0 \in \mathbf{R}$.

Example 3 The conclusions derived for DEPCA (20) can be easily generalized to the equation

$$x'(t) = a(t)g((x([t]))), \quad x \cdot g(x) < 0, x \neq 0, \ a(t) \geq 0,$$

where the resulting family of maps F_λ is unimodal, however, the opposite limits hold to those in (14): $\lim_{x\to\infty} F_\lambda(x) = \lim_{x\to-\infty} F_\lambda(x) = +\infty$. To achieve this type of unimodality one has to choose the nonlinearity $g(x)$ such that it grows to infinity faster than a power $|x|^p$, $p > 1$, when $x \to -\infty$, and it decays at positive infinity not faster than say a linear function $y = -mx$, $0 < m < 1$ (more precise conditions for this type of unimodality can be stated). A symmetric case to that in Example 1 is given by the nonlinearity $g(x) = \exp\{-\gamma x\} - 1$, where $\gamma > 0$ is a constant. Theorems 1–3 remain valid in this case as well. All the proofs need to be adjusted accordingly due to the limits in (14) now being $+\infty$. Therefore, for the DEPCA of the form

$$x'(t) = a(t)\left[\exp\{-\gamma x([t])\} - 1\right], \quad \gamma > 0, \; a(t) \geq 0, \tag{22}$$

the following holds

Proposition 9 *Suppose that coefficient $a(t)$ of Eq. (22) satisfies $0 \leq a(t) \leq 2$ and $\int_0^\infty a(t)\,dt = \infty$. Then its constant solution $x(t) \equiv 0$ is globally asymptotically stable: $\lim_{t\to\infty} x(t) = 0$ for any solution x with arbitrary initial condition $x(0) = x_0 \in R$.*

5 Discussion

The unimodal case of the family of maps F_λ subject to condition (14) considered in this paper is just one special case possible for the general Eq. (1). The approaches and techniques developed in Sect. 4 can be generalized and applied to many other cases. For example, it is a straightforward adaptation of the developed techniques and reasoning for the case when the family F_λ consists of monotone maps only (either increasing, or decreasing, or a combination of both). In fact most of the considerations of this paper would become simpler for the monotone cases. Thus, Theorems 1–3 are valid when the family F_λ consists of all monotone maps.

Another easy carry-over is the case of a unimodal family when the limits in (14) are $+\infty$. Example 3 of Sect. 4.4 demonstrates exactly this situation. One other distinct possibility is when the common domain of immediate attraction for the family F_λ, $\lambda \in \Lambda_*$, is a finite closed interval $D_* = [c, d]$, not the entire real axis **R**. The considerations and reasoning in this subcase may be even simpler in some instances than those in the current paper; however, they need to be considered and worked out in detail, which requires additional efforts.

A more complex case is when the family F_λ is not unimodal, but such that its maps have more than one critical points. Example 2 deals with such a situation, where the family F_λ of maps consists of representatives of two types: monotone maps, and maps with exactly two critical points. The simplifying factor there is that each map of the family is symmetric with respect to the bisector $y = x$, which allows us to essentially use the same techniques and reasoning as in the case of unimodal maps.

We believe that the unimodality of the family F_λ is not the crucial hypothesis that makes the main results of this paper valid. However, the techniques we use in the proofs rely heavily on many essential properties of unimodal maps. We think Theorems 1–3 are valid in a more general case, without the unimodality or monotonicity assumptions. However, we do not currently have clear ideas about possible path how to prove this.

Acknowledgements This work was initiated during A. Ivanov's visit and research stay at the Universidad Nacional de Educación a Distancia, Madrid, in November 2016. He is thankful to Professors D. Franco and J. Perán for their invitation, support and hospitality, as well as for the fruitful discussions that helped shape this paper. The author was also partially supported by the Alexander von Humboldt Stiftung, Germany, during his sabbatical leave from the Pennsylvania

State University in the fall 2016, when a substantial part of this research work was accomplished. Principal results of this paper were reported at the International Conference on Difference Equations and Applications held at Technische Universität Dresden, Germany, during May 21–25, 2018.

References

1. Franco, D., Ivanov, A., Perán, J.: Global asymptotic stability in differential equations with piecewise constant argument (2018), 24 pp
2. Cooke, K.L., Wiener, J.: A survey of differential equations with piecewise continuous arguments. Delay Differential Equations and Dynamical Systems (Claremont, CA, 1990). Lecture Notes in Mathematics, vol. 1475, pp. 1–15. Springer, Berlin (1991)
3. Wiener, J.: Generalized Solutions of Functional Differential Equations. World Scientific Publishing Co., Inc., River Edge (1993), xiv+410 pp
4. de Melo, W., van Strien, S.: One-dimensional dynamics. Ergebnisse der Mathematik und ihrer Grenzgebiete 3 [Results in Mathematics and Related Areas 3], vol. 25. Springer, Berlin (1993), 605 pp
5. Sharkovsky, A.N., Kolyada, S.F., Sivak, A.G., Fedorenko, V.V.: Dynamics of One-Dimensional Maps. Mathematics and Its Application, vol. 407. Kluwer Academic Publishers, Dordrecht (1997), 261 pp
6. Cooke, K.L., Győri, I.: Numerical approximation of the solutions of delay differential equations on an infinite interval using piecewise constant arguments. Comput. Math. Appl. **28**(1–3), 81–92 (1994)
7. Győri, I., Hartung, F.: On numerical approximation using differential equation with piecewise-constant arguments. Period. Math. Hung. **56**(1), 55–69 (2008)
8. Ivanov, A.F., Trofimchuk, S.I.: On global dynamics in a periodic differential equation with deviating argument. Appl. Math. Comput. **252**, 446–456 (2015)
9. Matsunaga, H., Hara, T., Sakata, S.: Global attractivity for a logistic equation with piecewise constant argument. Nonlinear Differ. Equ. Appl. (NoDEA) **8**, 45–52 (2001)
10. Wright, E.M.: A non-linear differential-difference equation. J. Reine Angew. Math. **194**, 66–87 (1955)
11. Singer, D.: Stable orbits and bifurcation of maps of the interval. SIAM J. Appl. Math. **35**(2), 260–267 (1978)

Sharp Conditions for Oscillation and Nonoscillation of Neutral Difference Equations

Başak Karpuz

Abstract In this paper, we give sufficient conditions for oscillation and nonoscillation of solutions to the neutral difference equation

$$\Delta[x(n) - r(n)x(n-\kappa)] + p(n)x(n-\tau) = 0 \quad \text{for } n = 0, 1, \ldots,$$

where Δ is the forward difference operator, $\kappa, \tau \in \mathbb{N}$, $\{r(n)\}_{n=0}^{\infty} \subset [0, 1]$ and $\{p(n)\}_{n=0}^{\infty} \subset (0, \infty)$. Our new oscillation and nonoscillation tests include product type conditions. We also show with an example that the new conditions are sharp.

Keywords Neutral difference equations · Oscillation · Nonoscillation

1 Introduction

We consider the difference equation

$$\Delta[x(n) - r(n)x(n-\kappa)] + p(n)x(n-\tau) = 0 \quad \text{for } n = 0, 1, \ldots, \tag{1}$$

where Δ is the forward difference operator, $\kappa, \tau \in \mathbb{N}$, $\{r(n)\}_{n=0}^{\infty} \subset [0, 1]$ and $\{p(n)\}_{n=0}^{\infty} \subset (0, \infty)$.

If $r(n) \equiv 0$, then (1) reduces to the classical form

$$\Delta x(n) + p(n)x(n-\tau) = 0 \quad \text{for } n = 0, 1, \ldots. \tag{2}$$

Let us quote some important results in the development of the oscillation and nonoscillation of (2).

B. Karpuz (✉)
Department of Mathematics, Faculty of Science, Tınaztepe Campus, Dokuz Eylül University, 35160 Buca, İzmir, Turkey
e-mail: bkarpuz@gmail.com
URL: http://kisi.deu.edu.tr/basak.karpuz

M. Bohner et al. (eds.), *Difference Equations and Discrete Dynamical Systems with Applications*, Springer Proceedings in Mathematics & Statistics 312,
https://doi.org/10.1007/978-3-030-35502-9_11

In [3], Erbe and Zhang obtained the first results on the oscillation of (2), and gave the following conditions.

Theorem A ([3, Theorems 2.2 and 2.3])

(i) Assume

$$\liminf_{n\to\infty} p(n) > \frac{\tau^{\tau}}{(\tau+1)^{\tau+1}}.$$

Then, every solution of (2) *is oscillatory.*

(ii) Assume

$$p(n) \leq \frac{\tau^{\tau}}{(\tau+1)^{\tau+1}} \quad \textit{for all large } n.$$

Then, (2) *has a nonoscillatory solution.*

By Ladas, Philos and Sficas, Theorem A is improved as follows.

Theorem B ([10, Theorem 1]) *Assume*

$$\liminf_{n\to\infty} \sum_{j=n-\tau}^{n-1} p(j) > \left(\frac{\tau}{\tau+1}\right)^{\tau+1}.$$

Then, every solution of (2) *is oscillatory.*

Note that Theorem B comments only on the oscillation of (2). In [9], Ladas conjectured that if the condition

$$\sum_{j=n-\tau}^{n-1} p(j) \leq \left(\frac{\tau}{\tau+1}\right)^{\tau+1} \quad \text{for all large } n$$

implies nonoscillation of (2). A negative answer to this conjecture is given by Yu, Zang and Wang in [17, Example 1] by means of the following theorem.

Theorem C ([17, Theorems 1, 4])

(i) Assume

$$\liminf_{n\to\infty} \inf_{\lambda\in E} \left\{ \frac{1}{\lambda \prod_{j=n-\tau}^{n-1}[1-\lambda p(j)]} \right\} > 1,$$

where $E := \{\lambda > 0 : 1 - \lambda p(n) > 0$ *for all large* $n\}$*. Then, every solution of* (2) *is oscillatory.*

(ii) Assume that there exists $\lambda_0 \in E$ *such that*

$$\frac{1}{\lambda_0 \prod_{j=n-\tau}^{n-1}[1-\lambda_0 p(j)]} \leq 1 \quad \textit{for all large } n.$$

Then, (2) *has a nonoscillatory solution.*

It is shown in [17, Remark 1] that Theorem C (i) implies Theorem B.

Next, we quote a recent result from [8] by the author.

Theorem D ([8, Theorems 1, 2])

(i) Assume

$$\liminf_{n\to\infty} \inf_{\lambda\geq 1} \left\{ \frac{1}{\lambda} \prod_{j=n-\tau}^{n} [1+\lambda p(j)] \right\} > 1.$$

Then, every solution of (2) *is oscillatory.*

(ii) Assume that there exists $\lambda_0 \geq 1$ *such that*

$$\frac{1}{\lambda_0} \prod_{j=n-\tau}^{n} [1+\lambda_0 p(j)] \leq 1 \quad \textit{for all large } n.$$

Then, (2) *has a nonoscillatory solution.*

In [8, Corollary 3], the author corrected Ladas' conjecture, and showed that

$$\sum_{j=n-\tau}^{n} p(j) \leq \left(\frac{\tau}{\tau+1}\right)^{\tau} \quad \text{for all large } n$$

implies nonoscillation of (2).

In the literature, researchers mostly consider

$$\Delta[x(n) - rx(n-\kappa)] + p(n)x(n-\tau) = 0 \quad \text{for } n = 0, 1, \ldots,$$

where $\kappa, \tau \in \mathbb{N}$, $r \in \mathbb{R}$ and $\{p(n)\} \subset (0, \infty)$, as an neutral extension of (2), i.e., when the neutral coefficient is constant (see [1, 4–7, 11–14]). There are less studies focusing on (1) compared to those on (2) (see, for example, [2, 16, 18]). In this paper, we will extend the results of [8] to the neutral difference equation (1). We will also show that the extended results are sharp.

The paper is organized as follows. In Sect. 2, we present our oscillation and nonoscillation tests. In Sect. 3, we prove some auxiliary results required in the proofs of the main results. In Sect. 4, we will give proofs of the main results presented in Sect. 2. Section 5 includes some numerical examples illustrating the significance of our new results. Finally, in Sect. 6, we make our final comments to conclude the paper.

2 Main Results

In this section, we present oscillation and nonoscillation tests without proofs. The proofs will be given after the necessary background is formed. We first state our oscillation test.

Theorem 1 *Assume*

$$\limsup_{n\to\infty} r(n) < 1 \tag{3}$$

and

$$\limsup_{n\to\infty} \max_{l-\tau+1\leq n+1\leq l} \left\{\left(\sum_{j=l-\tau}^{n} p(j)\right)\left(\sum_{j=n+1}^{l} p(j)\right)\right\} > 0. \tag{4}$$

Assume further

$$\liminf_{n\to\infty} \inf_{\lambda\geq 1} \left\{r(n-\tau) \prod_{j=n-\kappa+1}^{n} [1+\lambda p(j)] + \frac{1}{\lambda} \prod_{j=n-\tau}^{n} [1+\lambda p(j)]\right\} > 1. \tag{5}$$

Then, every solution of (1) *is oscillatory.*

Our next result can be regarded as the complement of the previous one.

Theorem 2 *Assume that there exists* $\lambda_0 \geq 1$ *such that*

$$r(n-\tau) \prod_{j=n-\kappa+1}^{n} [1+\lambda_0 p(j)] + \frac{1}{\lambda_0} \prod_{j=n-\tau}^{n} [1+\lambda_0 p(j)] \leq 1 \quad \text{for all large } n. \tag{6}$$

Then, (1) *has a nonoscillatory solution.*

3 Auxiliary Results

In this section, give several auxiliary results, which are required to prove the main results. For simplicity, we let

$$\mu := \begin{cases} \tau, & r(n) = 0 \text{ for all large } n \\ \max\{\kappa, \tau\}, & \text{otherwise.} \end{cases}$$

We start this section with a simple lemma.

Lemma 1 *If* (4) *holds, then*

$$\limsup_{n\to\infty} p(n) > 0. \tag{7}$$

Proof We will prove the contrapositive statement, i.e.,

$$\lim_{n\to\infty} p(n) = 0 \quad \Longrightarrow \quad \lim_{n\to\infty} q(n) = 0,$$

where

$$q(n) := \max_{l-\tau+1\leq n+1\leq l}\left\{\left(\sum_{j=l-\tau}^{n} p(j)\right)\left(\sum_{j=n+1}^{l} p(j)\right)\right\} \quad \text{for } n = \tau-1, \tau, \ldots.$$

Note that

$$\begin{aligned} q(n) := \max\Bigg\{&\left(\sum_{j=n-\tau+1}^{n} p(j)\right)p(n+1), \left(\sum_{j=n-\tau+2}^{n} p(j)\right)\left(\sum_{j=n+1}^{n+2} p(j)\right), \\ &\left(\sum_{j=n-\tau+3}^{n} p(j)\right)\left(\sum_{j=n+1}^{n+3} p(j)\right), \ldots, p(n)\left(\sum_{j=n+1}^{n+\tau} p(j)\right)\Bigg\} \end{aligned} \tag{8}$$

for $n = \tau - 1, \tau, \ldots$. Pick $\varepsilon \in (0, 1)$, then there exists n_0 sufficiently larger than $(\tau - 1)$ such that $p(n) < \frac{\varepsilon}{\tau}$ for $n = n_0, n_0 + 1, \ldots$, which by (8) yields

$$q(n) \leq \left(\tau\frac{\varepsilon}{\tau}\right)\left(\tau\frac{\varepsilon}{\tau}\right) = \varepsilon^2 < \varepsilon \quad \text{for } n_1, n_1 + 1, \ldots,$$

where $n_1 := n_0 + \tau - 1$. Thus, $\lim_{n\to\infty} q(n) = 0$, and the proof is completed.

The following result is extracted from [15].

Lemma 2 (See [15, Theorem 2.3]) *Assume* (3) *and*

$$\sum_{j}^{\infty} p(j) = \infty. \tag{9}$$

Then, every nonoscillatory solution of (1) *tends to zero asymptotically.*

Now, we give another lemma.

Lemma 3 *Assume* (4). *If* z *is an eventually positive solution of*

$$\Delta z(n) - \frac{r(n-\tau)p(n)}{p(n-\kappa)}\Delta z(n-\kappa) + p(n)z(n-\tau) = 0 \quad \textit{for } n = 0, 1, \ldots, \tag{10}$$

which is eventually decreasing, then

$$\liminf_{n\to\infty} \frac{z(n-\tau)}{z(n+1)} < \infty.$$

Proof By (4), there exist a constant $\varepsilon \in (0, 1)$ and two increasing sequences $\{n_k\}_{k=0}^{\infty}, \{l_k\}_{k=0}^{\infty} \subset \mathbb{N}$ such that $n_{k+1} \geq n_k + 2\tau$, $l_k - \tau + 1 \leq n_k + 1 \leq l_k$ and $\big(\sum_{j=l_k-\tau}^{n_k} p(j)\big)\big(\sum_{j=n_k+1}^{l_k} p(j)\big) \geq \varepsilon$ for $k = 0, 1, \ldots$. We may suppose without loss of generality that $z(n) > 0$, $z(n-\tau) > 0$, $\Delta z(n) < 0$ and $\Delta z(n-\kappa) < 0$ for $n = n_0, n_0 + 1, \ldots$. It follows from (10) that

$$\Delta z(n) + p(n)z(n-\tau) \leq 0 \quad \text{for } n = n_0, n_0 + 1, \ldots. \tag{11}$$

From (11), for $k = 1, 2, \ldots$, we have

$$\begin{aligned} z(n_k + 1) &> z(n_k + 1) - z(l_k + 1) = -\sum_{j=n_k+1}^{l_k} \Delta z(j) \\ &\geq \sum_{j=n_k+1}^{l_k} p(j)z(j-\tau) \geq \Bigg(\sum_{j=n_k+1}^{l_k} p(j)\Bigg) z(l_k - \tau) \end{aligned} \tag{12}$$

and

$$\begin{aligned} z(l_k - \tau) &> z(l_k - \tau) - z(n_k + 1) = -\sum_{j=l_k-\tau}^{n_k} \Delta z(j) \\ &\geq \sum_{j=l_k-\tau}^{n_k} p(j)z(j-\tau) \geq \Bigg(\sum_{j=l-\tau}^{n_k} p(j)\Bigg) z(n_k - \tau). \end{aligned} \tag{13}$$

Combining (12) and (13), we get

$$1 \leq \frac{z(n_k - \tau)}{z(n_k + 1)} < \frac{1}{\varepsilon} \quad \text{for } k = 1, 2, \ldots. \tag{14}$$

This completes the proof.

Finally, we give an equivalence result.

Lemma 4 *The following statements are equivalent.*

(i) Equation (1) *has an eventually positive solution.*
(ii) Equation (10) *has an eventually positive solution, which is eventually decreasing.*

Proof (i) $\Longrightarrow$ (ii) Let $\{x(n)\}$ be an eventually positive solution of (1). Set

$$z(n) := x(n) - r(n)x(n-\kappa) \quad \text{for } n = n_0, n_0 + 1, \ldots. \tag{15}$$

From (1), we have

$$\Delta z(n) + p(n)x(n-\tau) = 0 \quad \text{for } n = n_0, n_0 + 1, \ldots, \tag{16}$$

which shows that $\{z(n)\}$ is eventually decreasing. From (15) and (16), we have

$$\begin{aligned}0 &= \Delta z(n) + p(n)[r(n-\tau)x(n-\tau-\kappa) + z(n-\tau)] \\ &= \Delta z(n) + r(n-\tau)p(n)x(n-\tau-\kappa) + p(n)z(n-\tau) \\ &= \Delta z(n) - \frac{r(n-\tau)p(n)}{p(n-\kappa)}\Delta z(n-\kappa) + p(n)z(n-\tau)\end{aligned}$$

for $n = n_1, n_1+1, \ldots$, where $n_1 := n_0 + \mu$. This proves that $\{z(n)\}$ satisfies (10). Next, to show that $\{z(n)\}$ is eventually positive, we assume the contrary that it is eventually negative. Suppose that $z(n) < 0$ for $n = n_2, n_2+1, \ldots$, where $n_2 \in \mathbb{N}$ is sufficiently larger than n_1. Here, we consider the following two possible cases.

Case 1. $\{x(n)\}$ is unbounded. Then, there exists n_3 sufficiently larger than n_2 such that $x(n_3) = \max\{x(n) : n = n_2, n_2+1, \ldots, n_3\}$. Then, we get

$$0 > z(n_2) \geq z(n_3) = x(n_3) - r(n_3)x(n_3-\kappa) \geq [1 - r(n_3)]x(n_3) \geq 0,$$

which is a contradiction.

Case 2. $\{x(n)\}$ is bounded. Then, we have

$$\limsup_{n\to\infty} x(n) = \lim_{n\to\infty} z(n) + \limsup_{n\to\infty} x(n-\kappa) < \limsup_{n\to\infty} x(n),$$

which is a contradiction.

Thus, $\{z(n)\}$ is eventually positive.

(ii) $\Longrightarrow$ (i) Let $\{z(n)\}$ be an eventually positive solution of (10), which is eventually decreasing. Then, setting $x(n) := -\frac{\Delta z(n+\tau)}{p(n+\tau)}$, we can show that $\{x(n)\}$ is an eventually positive solution of (1).

This completes the proof.

4 Proofs of Main Results

Now, we are at a position to prove our main results given in Sect. 2. We keep the order, and start with the oscillation test.

Proof (*Proof of Theorem* 1) Assume the contrary that $\{x(n)\}$ is an eventually positive solution of (1). Define $\{z(n)\}$ as in (15). Then, by Lemma 4, $\{z(n)\}$ is an eventually positive solution of (10), which is eventually decreasing. By Lemma (1), we have (7), which yields (9). Thus, by Lemma 2, we have $\lim_{n\to\infty} x(n) = 0$. Now, by (5), we may assume for $n = n_0, n_0+1, \ldots$ that

$$\inf_{\lambda\geq 1}\left\{r(n-\tau)\prod_{j=n-\kappa+1}^{n}[1+\lambda p(j)]+\frac{1}{\lambda}\prod_{j=n-\tau}^{n}[1+\lambda p(j)]\right\}\geq\alpha,$$

where $\alpha > 1$ and $n_0 \in \mathbb{N}$ is sufficiently large. Define

$$\Lambda := \{\lambda \geq 1 : \ \Delta z(n) + \lambda p(n)z(n+1) \leq 0 \quad \text{for all large } n\}.$$

It follows from (11) that $1 \in \Lambda$, i.e., $\Lambda \neq \emptyset$. If $\lambda \in \Lambda$, then

$$\begin{aligned}0 &= \Delta z(n) + p(n)x(n-\tau)\\ &= \Delta z(n) + \lambda p(n)z(n+1) + p(n)[x(n-\tau) - \lambda z(n+1)]\end{aligned}$$

for $n = n_0, n_0+1, \ldots$. Thus, $x(n-\tau) \geq \lambda z(n+1)$ for $n = n_0, n_0+1, \ldots$. Let $\lambda \in \Lambda$ and set

$$\sigma(n) := z(n)\varphi(n) \quad \text{for } n = n_0, n_0+1, \ldots,$$

where $\varphi(n) := \prod_{j=0}^{n-1}[1+\lambda p(j)] > 0$ for $n = n_0, n_0+1, \ldots$. Then, for $n = n_0$, $n_0+1, \ldots$, $\Delta\varphi(n) = \lambda p(n)\varphi(n)$ and

$$\begin{aligned}\Delta\sigma(n) &= [\Delta z(n)]\varphi(n) + z(n+1)[\Delta\varphi(n)]\\ &= [\Delta z(n) + \lambda p(n)z(n+1)]\varphi(n) \leq 0,\end{aligned}$$

which shows that $\{\sigma(n)\}$ is nonincreasing. Then, for $n = n_0, n_0+1, \ldots$, we estimate

$$\begin{aligned}\Delta z(n) =& - p(n)x(n-\tau) = -p(n)\big[r(n-\tau)x(n-\kappa-\tau) + z(n-\tau)\big]\\ \leq& - p(n)\big[\lambda r(n-\tau)z(n-\kappa+1) + z(n-\tau)\big]\\ =& - \lambda p(n)\left[r(n-\tau)\frac{\sigma(n-\kappa+1)}{\varphi(n-\kappa+1)} + \frac{1}{\lambda}\frac{\sigma(n-\tau)}{\varphi(n-\tau)}\right]\\ =& - \lambda p(n)\left[r(n-\tau)\frac{\varphi(n+1)}{\varphi(n-\kappa+1)}\frac{\sigma(n-\kappa+1)}{\varphi(n+1)} + \frac{1}{\lambda}\frac{\varphi(n+1)}{\varphi(n-\tau)}\frac{\sigma(n-\tau)}{\varphi(n+1)}\right]\\ \leq& - \lambda p(n)\left[r(n-\tau)\prod_{j=n-\kappa+1}^{n}[1+\lambda p(j)] + \frac{1}{\lambda}\prod_{j=n-\tau}^{n}[1+\lambda p(j)]\right]\frac{\sigma(n+1)}{\varphi(n+1)}\\ =& - \lambda p(n)\left[r(n-\tau)\prod_{j=n-\kappa+1}^{n}[1+\lambda p(j)] + \frac{1}{\lambda}\prod_{j=n-\tau}^{n}[1+\lambda p(j)]\right]z(n+1)\\ \leq& - \alpha\lambda p(n)z(n+1),\end{aligned}$$

which shows that $\alpha\lambda \in \Lambda$. This yields $\sup \Lambda = \infty$. On the other hand, by Lemma 3, there exist a constant $\beta \in \mathbb{R}$ and an increasing divergent sequence $\{n_k\} \subset \mathbb{N}$ such that $\frac{z(n_k-\tau)}{z(n_k+1)} \leq \beta$ for $k = 1, 2, \ldots$. As $\lim_{n\to\infty} x(n) = 0$, there exists a subsequence $\{n_{k_\ell}\}$ of $\{n_k\}$ such that $x(n_{k_\ell} - \tau) = \min\{x(j) : \ j = n_0, n_0+1, \ldots, n_{k_\ell} - \tau\}$. Then, for

$\ell = 1, 2, \ldots$, we estimate

$$\begin{aligned} z(n_{k_\ell}+1) - z(n_{k_\ell}-\tau) &= -\sum_{j=n_{k_\ell}-\tau}^{n_{k_\ell}} p(j)x(j-\tau) \\ &\leq -x(n_{k_\ell}-\tau)\sum_{j=n_{k_\ell}-\tau}^{n_{k_\ell}} p(j) \\ &< -\varepsilon x(n_{k_\ell}-\tau), \end{aligned}$$

which shows that $\varepsilon x(n_{k_\ell}-\tau) \leq z(n_{k_\ell}-\tau)$ for $\ell = 1, 2, \ldots$. Thus, for $\ell = 1, 2, \ldots$, we obtain

$$\begin{aligned} 0 &= \Delta z(n_{k_\ell}) + p(n_{k_\ell})x(n_{k_\ell}-\tau) \\ &< \Delta z(n_{k_\ell}) + \frac{1}{\varepsilon}p(n_{k_\ell})z(n_{k_\ell}-\tau) \\ &\leq \Delta z(n_{k_\ell}) + \frac{\beta}{\varepsilon}p(n_{k_\ell})z(n_{k_\ell}+1), \end{aligned}$$

which shows that $\sup \Lambda < \frac{\beta}{\varepsilon}$. This contradiction completes the proof.

Next, we prove our nonoscillation test.

Proof (*Proof of Theorem* 2) Assume that (6) holds for $n = n_0, n_0+1, \ldots$, where $n_0 \in \mathbb{N}$ is sufficiently large. By (6), we obtain

$$\begin{aligned} &1 - \lambda_0 p(n)\left[r(n-\tau)\prod_{j=n-\kappa+1}^{n-1}[1+\lambda_0 p(j)] + \frac{1}{\lambda_0}\prod_{j=n-\tau}^{n-1}[1+\lambda_0 p(j)]\right] \\ &= 1 - \frac{\lambda_0 p(n)}{1+\lambda_0 p(n)}\left[r(n-\tau)\prod_{j=n-\kappa+1}^{n}[1+\lambda_0 p(j)] + \frac{1}{\lambda_0}\prod_{j=n-\tau}^{n}[1+\lambda_0 p(j)]\right] \\ &\geq 1 - \frac{\lambda_0 p(n)}{1+\lambda_0 p(n)} = \frac{1}{1+\lambda_0 p(n)} > 0 \end{aligned} \tag{17}$$

for $n = n_0, n_0+1, \ldots$. Further, from (6), we have $\frac{1}{\lambda_0}[1+\lambda_0 p(n)] \leq 1$ for $n = n_0, n_0+1, \ldots$, which yields $1 - p(n) > \frac{1}{\lambda_0}$ for $n = n_0, n_0+1, \ldots$, i.e., $1 > p(n) > 0$ for $n = n_0, n_0+1, \ldots$. Define the sequences $\{y(n)\}$ and $\{w(n)\}$ by

$$y(n) := \begin{cases} 1, & n = n_0-\mu, \ldots, n_0-1 \\ \dfrac{w(n)}{1-\lambda_0 p(n)w(n)}, & n = n_0, n_0+1, \ldots \end{cases} \tag{18}$$

and

$$w(n) := r(n-\tau)\bigg(\prod_{j=n-\kappa+1}^{n-1}[1+\lambda(j)y(j)p(j)]\bigg)y(n-\kappa+1)$$
$$+\frac{1}{\lambda_0}\prod_{j=n-\tau}^{n-1}[1+\lambda(j)y(j)p(j)] \quad \text{for } n = n_0, n_0+1, \ldots. \tag{19}$$

First, we claim that $y(n) > 0$ for $n = n_0, n_0 + 1, \ldots$. Assume the contrary that $y(l) \leq 0$ for some integer $l \geq n_0$. Without loss of generality, we may assume that $y(n) > 0$ for $n = n_0 - \mu, n_0 - \mu + 1, \ldots, l - 1$. Then, we have

$$y(l) \geq \frac{\frac{1}{\lambda_0}\prod_{j=l-\tau}^{l-1}[1+\lambda(j)y(j)p(j)]}{1-\lambda_0 p(l)\Big[\frac{1}{\lambda_0}\prod_{j=l-\tau}^{l-1}[1+\lambda(j)y(j)p(j)]\Big]}$$
$$\geq \frac{\frac{1}{\lambda_0}\prod_{j=l-\tau}^{l-1}[1+\lambda(j)y(j)p(j)]}{1-p(l)} > 0,$$

which is a contradiction. Thus, $y(n) > 0$ for $n = n_0 - \mu, n_0 - \mu + 1, \ldots$. Next, we claim that $y(n) \leq 1$ for $n = n_0, n_0 + 1, \ldots$. Assume the contrary that $y(l) > 1$ for some integer $l \geq n_0$. Without loss of generality, we may assume that $y(n) \leq 1$ for $n = n_0 - \mu, n_0 - \mu + 1, \ldots, l - 1$. From (17), (18) and (19), we have $1 - \lambda_0 p(l)$ $w(l) \geq \frac{1}{1+\lambda_0 p(l)}$ or equivalently $w(l) \leq \frac{1}{1+\lambda_0 p(l)}$. Thus,[1] (18) yields

$$y(l) \leq \frac{w(l)}{1-\lambda_0 p(l)w(l)} \leq \frac{\frac{1}{1+\lambda_0 p(l)}}{1-\frac{\lambda_0 p(l)}{1+\lambda_0 p(l)}} = 1,$$

which is a contradiction. Thus, $y(n) \leq 1$ for $n = n_0 - \mu, n_0 - \mu + 1, \ldots$. From (18), we have

$$y(n) = [1 + \lambda_0 p(n)y(n)]w(n) \quad \text{for } n = n_0, n_0 + 1, \ldots$$

or equivalently

$$y(n) = r(n-\tau)\bigg(\prod_{j=n-\kappa+1}^{n}[1+\lambda_0 y(j)p(j)]\bigg)y(n-\kappa+1)$$
$$+\frac{1}{\lambda_0}\prod_{j=n-\tau}^{n}[1+\lambda_0 y(j)p(j)] \quad \text{for } n = n_0, n_0+1, \ldots. \tag{20}$$

Now, let

[1] $\varphi(x) := \frac{x}{1-hx}$ is increasing for $x \in (0, \frac{1}{h})$, where $h > 0$.

$$z(n) := \prod_{j=n_0-\mu}^{n-1} \frac{1}{1+\lambda_0 y(j)p(j)} \quad \text{for } n = n_0-\mu, n_0-\mu+1, \ldots. \tag{21}$$

Clearly, $\{z(n)\} \subset (0, 1]$ is decreasing. From (20) and (21), for $n = n_0, n_0+1, \ldots$, we get

$$0 = \frac{\Delta z(n)}{p(n)z(n+1)} - r(n-\tau)\frac{z(n-\kappa+1)}{z(n+1)}\frac{\Delta z(n-\kappa)}{p(n-\kappa)z(n-\kappa+1)} + \frac{z(n-\tau)}{z(n+1)}$$

or equivalently

$$\Delta z(n) - r(n-\tau)\frac{p(n)}{p(n-\kappa)}\Delta z(n-\kappa) + z(n-\tau) = 0,$$

which shows that $\{z(n)\}$ satisfies (10). By Lemma 4, we see that (1) also has an eventually positive solution. This completes the proof.

5 An Example

In this section, we give an example to represent the sharpness of the results.

Example 1 Consider the difference equation

$$\Delta[x(n) - r_0 x(n-\kappa)] + p(n)x(n-\tau) = 0 \quad \text{for } n = 0, 1, \ldots, \tag{22}$$

where $r_0 \in [0, 1)$, $\{p(n)\}$ is both $(\tau+1)$ and κ periodic such that $p(a)p(b) > 0$ for $a, b \in \{0, 1, \ldots, \tau\}$ with $a \neq b$. Then, (22) has a nonoscillatory solution if and only if

$$\min_{\lambda \geq 1}\left\{r_0 \prod_{j=\mu-\kappa+1}^{\mu}[1+\lambda p(j)] + \frac{1}{\lambda}\prod_{j=\mu-\tau}^{\mu}[1+\lambda p(j)]\right\} \leq 1 \tag{23}$$

since

$$\begin{aligned} r_0 \prod_{j=\mu-\kappa+1}^{\mu}[1+\lambda p(j)] &+ \frac{1}{\lambda}\prod_{j=\mu-\tau}^{\mu}[1+\lambda p(j)] \\ &\geq \frac{1}{\lambda}[1+\lambda p(a)][1+\lambda p(b)] \to \infty \text{ as } \lambda \to \infty. \end{aligned}$$

If $p(n) \equiv p_0$, then (23) reduces to

$$\min_{\lambda \geq 1}\left\{r_0(1+\lambda p_0)^{\kappa} + \frac{1}{\lambda}(1+\lambda p_0)^{\tau+1}\right\} \leq 1. \tag{24}$$

If $r_0 = 0$, then (24) is equivalent to

$$p_0 \leq \frac{\tau^\tau}{(\tau+1)^{\tau+1}}.$$

If $r_0 \in (0, 1)$ and $\kappa = \tau$, then (24) is equivalent to

$$p(p + r_0 + 1)\tau \leq 1 + p_0$$

and

$$\frac{1}{2}\Big(2r_0 + p_0(\tau+1) + \sqrt{p_0}\sqrt{p_0(\tau+1)^2 + 4r_0\tau}\Big) \times \left(1 + \frac{-p_0(\tau-1) + \sqrt{p_0}\sqrt{p_0(\tau+1)^2 + 4r_0\tau}}{2(p_0+r_0)\tau}\right)^\tau \leq 1.$$

If $r_0 \in (0, 1)$ and $\kappa = \tau + 1$, then (24) is equivalent to

$$p_0\big(r_0(\tau+1) + \tau\big) \leq 1$$

and

$$\frac{1}{2}\Big(2r_0 + p_0\tau + \sqrt{p_0}\sqrt{p_0\tau^2 + 4r_0(\tau+1)}\Big) \times \left(1 + \frac{-p_0\tau + \sqrt{p_0}\sqrt{p_0\tau^2 + 4r_0(\tau+1)}}{2r_0(\tau+1)}\right)^{\tau+1} \leq 1.$$

6 Final Discussion

In this section, we will make our final remarks to finalize the paper. More precisely, we will remove the condition (3) by replacing (4) with a stronger one to give new oscillation conditions.

The proof of the following lemma is similar to that of Lemma 3, and hence we omit it here.

Lemma 5 *Assume*

$$\liminf_{n\to\infty} \max_{l-\tau+1 \leq n+1 \leq l} \left\{\left(\sum_{j=l-\tau}^{n} p(j)\right)\left(\sum_{j=n+1}^{l} p(j)\right)\right\} > 0. \tag{25}$$

If z is an eventually positive solution of (10), *which is eventually decreasing, then*

$$\lim_{n\to\infty} \frac{z(n-\tau)}{z(n+1)} < \infty.$$

Remark 1 Clearly, (25) implies (4). Thus, by summing up (16), one can show that $\liminf_{n\to\infty} x(n) = 0$ for any nonoscillatory solution $\{x(n)\}$ of (1).

Theorem 3 *Assume* (5) *and* (25). *Then, every solution of* (1) *is oscillatory.*

Proof The proof can be given similar to that of Theorem 1. Thus, we omit here.

Corollary 1 *Assume* (25) *and*

$$\liminf_{n\to\infty} \prod_{j=n-\min\{\kappa-1,\tau\}}^{n} [1 + p(j)] > 1 \quad or \quad \liminf_{n\to\infty} \prod_{j=n-\max\{\kappa-1,\tau\}}^{n-\min\{\kappa-1,\tau\}-1} [1 + p(j)] > 1.$$

Then, every solution of

$$\Delta[x(n) - x(n-\kappa)] + p(n)x(n-\tau) = 0 \quad for\ n = 0, 1, \ldots \tag{26}$$

is oscillatory.

Corollary 2 *Assume* (25) *and*

$$\liminf_{n\to\infty} \sum_{j=n-\kappa+1}^{n} p(j) > 0 \quad or \quad \liminf_{n\to\infty} \sum_{j=n-\tau}^{n} p(j) > 0.$$

Then, every solution of (26) *is oscillatory.*

Remark 2 If $\liminf_{n\to\infty} p(n) > 0$, then the conditions of Corollaries 1 and 2 hold.

Remark 3 If

$$\lim_{n\to\infty} r(n) = 0, \tag{27}$$

then

$$\liminf_{n\to\infty} \inf_{\lambda\geq 1} \left\{ r(n-\tau) \prod_{j=n-\kappa+1}^{n} [1 + \lambda p(j)] + \frac{1}{\lambda} \prod_{j=n-\tau}^{n} [1 + \lambda p(j)] \right\}$$
$$= \liminf_{n\to\infty} \inf_{\lambda\geq 1} \left\{ \frac{1}{\lambda} \prod_{j=n-\tau}^{n} [1 + \lambda p(j)] \right\}.$$

By virtue of [8, Lemma 1], the condition

$$\liminf_{n\to\infty} \inf_{\lambda\geq 1} \left\{ \frac{1}{\lambda} \prod_{j=n-\tau}^{n} [1 + \lambda p(j)] \right\} > 1 \tag{28}$$

implies (4) or

$$\limsup_{n\to\infty} \sum_{j=n-\tau}^{n} p(j) > 1.$$

Hence, we can give the following oscillation test for (1).

Theorem 4 *Assume* (27) *and* (28). *Then, every solution of* (1) *is oscillatory.*

Proof The proof can be given similar to that of Theorem 1. Thus, we omit here.

References

1. Chatzarakis, G.E., Miliaras, G.N., Stavroulakis, I.P., Thandapani, E.: Asymptotic behavior of non-oscillatory solutions of first-order neutral difference equations. Panam. Math. J. **23**(1), 111–129 (2013)
2. Chen, M.P., Lalli, B.S., Yu, J.S.: Oscillation in neutral delay difference equations with variable coefficients. Comput. Math. Appl. **29**(3), 5–11 (1995). https://doi.org/10.1016/0898-1221(94)00223-8
3. Erbe, L.H., Zhang, B.G.: Oscillation of discrete analogues of delay equations. Differ. Integr. Equ. **2**(3), 300–309 (1989)
4. Georgiou, D.A., Grove, E.A., Ladas, G.: Oscillations of neutral difference equations. Appl. Anal. **33**(3–4), 243–253 (1989). https://doi.org/10.1080/00036818908839876
5. Gong, X.H., Zhong, X.Z., Jia, J.Q., Ouyang, R., Han, H.Q.: Oscillation of first-order neutral difference equations. Mod. Appl. Sci. **3**(8), 90–94 (2009). https://doi.org/10.5539/mas.v3n8p90
6. Guo, S.B., Shen, Y.J.: Necessary and sufficient conditions for oscillation of first order neutral difference equations. Acta Math. Appl. Sin. **36**(5), 840–850 (2013). http://123.57.41.99/jweb_yysxxb/EN/article/downloadArticleFile.do?attachType=PDF&id=13647
7. Jia, J.Q., Zhong, X.Z., Gong, X.H., Ouyang, R., Han, H.Q.: Nonoscillation of first-order neutral difference equation. Mod. Appl. Sci. **3**(11), 30–33 (2009). https://doi.org/10.5539/mas.v3n11p30
8. Karpuz, B.: Sharp oscillation and nonoscillation tests for linear difference equations. J. Differ. Equ. Appl. **23**(12), 1929–1942 (2017). https://doi.org/10.1080/10236198.2017.1379515
9. Ladas, G.: Recent developments in the oscillation of delay difference equations. In: Differential equations (Colorado Springs, CO, 1989), Lecture Notes in Pure and Applied Mathematics, vol. 127, pp. 321–332. Dekker, New York (1991)
10. Ladas, G., Philos, C.G., Sficas, Y.G.: Sharp conditions for the oscillation of delay difference equations. J. Appl. Math. Simul. **2**(2), 101–111 (1989). https://doi.org/10.1155/S1048953389000080
11. Lalli, B.S.: Oscillation theorems for neutral difference equations. Comput. Math. Appl. **28**(1–3), 191–202 (1994). https://doi.org/10.1016/0898-1221(94)00107-3
12. Lalli, B.S., Zhang, B.G.: On existence of positive solutions and bounded oscillations for neutral difference equations. J. Math. Anal. Appl. **166**(1), 272–287 (1992). https://doi.org/10.1016/0022-247X(92)90342-B
13. Lalli, B.S., Zhang, B.G.: Oscillation and comparison theorems for certain neutral difference equations. J. Austral. Math. Soc. Ser. B **34**(2), 245–256 (1992). https://doi.org/10.1017/S0334270000008754
14. Lalli, B.S., Zhang, B.G., Li, J.Z.: On the oscillation of solutions and existence of positive solutions of neutral difference equations. J. Math. Anal. Appl. **158**(1), 213–233 (1991). https://doi.org/10.1016/0022-247X(91)90278-8

15. Parhi, N., Tripathy, A.K.: Oscillation of forced nonlinear neutral delay difference equations of first order. Czechoslovak Math. J. **53**(1), 83–101 (2003). https://doi.org/10.1023/A:1022975525370
16. Parhi, N., Tripathy, A.K.: Oscillation of a class of neutral difference equations of first order. J. Differ. Equ. Appl. **9**(10), 933–946 (2003). https://doi.org/10.1080/1023619021000047680
17. Yu, J.S., Zhang, B.G., Wang, Z.C.: Oscillation of delay difference equations. Appl. Anal. **53**(1–2), 117–124 (1994). https://doi.org/10.1080/00036819408840248
18. Zhang, G., Cheng, S.S.: Oscillation criteria for a neutral difference equation with delay. Appl. Math. Lett. **8**(3), 13–17 (1995). https://doi.org/10.1016/0893-9659(95)00022-I

On the Boundedness Character of a Rational System of Difference Equations with Non-constant Coefficients

Yevgeniy Kostrov, Zachary Kudlak and Patrick Vernon

Abstract We investigate the boundedness character of nonnegative solutions of the nonautonomous rational system

$$\begin{cases} x_{n+1} = \dfrac{\alpha_n + \gamma_n x_n}{\beta_n x_n + y_n} & \\ & \text{for } n = 0, 1, \ldots \\ y_{n+1} = g(x_n, \ldots, x_{n-k+1}, y_n, \ldots, y_{n-k+1}, n) & \end{cases}$$

where the coefficient sequences $\{\alpha_n\}$, $\{\beta_n\}$ and $\{\gamma_n\}$ are bounded both above and below by positive numbers, the initial conditions $(x_0, y_0), (x_{-1}, y_{-1}), \ldots, (x_{-k+1}, y_{-k+1})$ are positive, and g takes on only positive values for positive values of $x_n, \ldots, x_{n-k+1}, y_n, y_{n-k+1}$ and nonnegative integers n. Special cases of this system, such as with periodic coefficients, are also investigated.

Keywords Boundedness · Periodicity · Nonautonomous rational difference equations · Rational systems

1 Introduction

We will investigate the boundedness character of solutions of the system

Y. Kostrov
Manhattanville College, Purchase, NY 10577, USA
e-mail: yevgeniy.kostrov@mville.edu

Z. Kudlak (✉)
US Coast Guard Academy, New London, CT 06320, USA
e-mail: zachary.a.kudlak@uscga.edu

P. Vernon
Xavier University of Louisiana, New Orleans, LA 70125, USA
e-mail: rvernon@xula.edu

M. Bohner et al. (eds.), *Difference Equations and Discrete Dynamical Systems with Applications*, Springer Proceedings in Mathematics & Statistics 312,
https://doi.org/10.1007/978-3-030-35502-9_12

$$\begin{cases} x_{n+1} = \dfrac{\alpha_n + \gamma_n x_n}{\beta_n x_n + y_n} \\ \\ y_{n+1} = g(x_n, \ldots, x_{n-k+1}, y_n, \ldots, y_{n-k+1}, n) \end{cases} \quad \text{for } n = 0, 1, \ldots \tag{1}$$

where the coefficient sequences $\{\alpha_n\}$, $\{\beta_n\}$ and $\{\gamma_n\}$ are bounded both above and below by positive constants, the initial conditions (x_0, y_0), $(x_{-1}, y_{-1}), \ldots,$ (x_{-k+1}, y_{-k+1}) are positive, and $g\colon (0, \infty)^{2k} \times \mathbb{N}_0 \to (0, \infty)$ is a function, $g(u_1, \ldots,$ $u_k, v_1, \ldots, v_k, t)$ which takes on only positive values for positive values of $u_1, \ldots, u_k,$ $v_1, \ldots, v_k$ and nonnegative integers t. In some situations, we are interested in cases when the coefficients form periodic sequences, say of period p, for some positive integer $p \geq 2$.

One of our goals in this paper is to establish general results about System (1) without specifying the function g. After exploring these general properties, we will use the following first order rational system as an example to illustrate the usefulness of these results.

$$\begin{cases} x_{n+1} = \dfrac{\alpha_n + \gamma_n x_n}{\beta_n x_n + y_n} \\ \\ y_{n+1} = \dfrac{a_n + b_n x_n + c_n y_n}{A_n + B_n x_n + C_n y_n} \end{cases} \quad \text{for } n = 0, 1, \ldots \tag{2}$$

The boundedness character of autonomous cases similar to system (2) have been investigated in several papers (for reference see [2, 5–7, 14]). We were motivated by the work in [11]. For examples and reference purposes, we will use the same numbering system as defined in [9]. Please see Appendix A for a table of the special cases contained in System (2) with numbers for reference. We are particularly interested in discovering cases where "periodicity may destroy boundedness," see [10], which is when the autonomous analog of the system has all solutions bounded, but when periodic coefficients are introduced, unbounded solutions can appear.

Throughout the paper we will use the following notation. We will denote a sequence $\{x_1, x_2, \ldots\}$ by $\{x_n\}$, and use x_n to denote the nth element of $\{x_n\}$. If there exists M such that $x_n < M$ for all n, we say that $\{x_n\}$ is *bounded above*. Otherwise we say $\{x_n\}$ is *unbounded*. Furthermore, all sequences are nonnegative, and therefore are bounded below by zero.

We will say that a system contained in (1) has *boundedness character* (B, B) if every solution $\{(x_n, y_n)\}_{n=0}^{\infty}$ is bounded above in both the first and second variables. A system has *boundedness character* (U, U) if there exists some initial conditions, x_0 and y_0 and some sequences of parameters such that $\{x_n\}_{n=0}^{\infty}$ is unbounded and there exist possibly different initial conditions and sequences of parameters such that $\{y_n\}_{n=0}^{\infty}$ is unbounded. A system has *boundedness character* (B, U) if every solution $\{(x_n, y_n)\}_{n=0}^{\infty}$ is bounded above in the first variable and there exist initial conditions and sequences of parameters such that $\{y_n\}_{n=0}^{\infty}$ is unbounded. We similarly define *boundedness character* (U, B).

In Sect. 2 we will state and prove results about the boundedness of System (1), then in Sect. 3 we will demonstrate how to apply these results to settle the boundedness character of several special cases of System (2). In Sect. 4 we look at the few remaining special cases of System (2) for which we used alternate methods to provide the boundedness character, giving the complete boundedness character of (2). Appendix A contains a table of the 49 special cases of (2) and their boundedness character established in this paper.

2 General Results

The following inequality will be quite useful. For reference, see [15].

$$\min\left\{\frac{\alpha_1}{B_1},\frac{\alpha_2}{B_2},\dots,\frac{\alpha_n}{B_n}\right\} \le \frac{\sum_{i=1}^{n}\alpha_i}{\sum_{i=1}^{n}B_i} \le \max\left\{\frac{\alpha_1}{B_1},\frac{\alpha_2}{B_2},\dots,\frac{\alpha_n}{B_n}\right\}, \tag{3}$$

where $\alpha_1,\dots,\alpha_n$ are nonnegative real numbers and $B_1, B_2,\dots,B_n$ are positive real numbers.

The following theorem will show that if x_{n+1} is "big," then the term x_n and y_n must have been "small," and if the term x_n is "small," then the term y_{n-1} and the ratio $\frac{y_{n-1}}{x_{n-1}}$ are both "big." In the sequel, these ideas can be used to establish boundedness in the x_n variable for specific special cases of System (2). We formalize this below in Theorem 1. See Sect. 3 for several examples which use Theorem 1.

Theorem 1 *Let $\{(x_n, y_n)\}$ be a solution of System* (1) *where $\{\alpha_n\}$, $\{\beta_n\}$, and $\{\gamma_n\}$ are bounded above and below by positive numbers. Let $M > 0$ be defined such that for all n,*

$$\frac{\gamma_{n-1}}{\alpha_n}(M\beta_n-\gamma_n)-\beta_{n-1}>0\,. \tag{4}$$

Then, the following two statements are true:

1. *if $x_{n+1} > M$, then $x_n < \frac{\alpha_n}{M\beta_n-\gamma_n}$ and $y_n < \frac{\alpha_n}{M}$;*
2. *if $x_n < \frac{\alpha_n}{M\beta_n-\gamma_n}$, then*
   $y_{n-1} > \frac{\alpha_{n-1}}{\alpha_n}(M\beta_n-\gamma_n)$ and $\frac{y_{n-1}}{x_{n-1}} > \frac{\gamma_{n-1}}{\alpha_n}(M\beta_n-\gamma_n)-\beta_{n-1}$.

Proof First of all, from (4), we have that $M\beta_n-\gamma_n>0$ for all n. To prove the first part, suppose that $x_{n+1} > M$, then by iterating for x_{n+1} we obtain

$$\frac{\alpha_n+\gamma_n x_n}{\beta_n x_n+y_n} > M\,, \quad \text{which means} \quad \alpha_n > (M\beta_n-\gamma_n)x_n + My_n\,.$$

Then $\alpha_n > (M\beta_n-\gamma_n)x_n$ and $\alpha_n > My_n$; thus $x_n < \frac{\alpha_n}{M\beta_n-\gamma_n}$ and $y_n < \frac{\alpha_n}{M}$.

For the second part, suppose that $x_n < \frac{\alpha_n}{M\beta_n-\gamma_n}$, then by iterating for x_n we obtain

$$x_n = \frac{\alpha_{n-1} + \gamma_{n-1}x_{n-1}}{\beta_{n-1}x_{n-1} + y_{n-1}} < \frac{\alpha_n}{M\beta_n - \gamma_n}$$
$$(\alpha_{n-1} + \gamma_{n-1}x_{n-1})(M\beta_n - \gamma_n) < \alpha_n(\beta_{n-1}x_{n-1} + y_{n-1})$$
$$\frac{\alpha_{n-1}(M\beta_n - \gamma_n)}{\alpha_n} + x_{n-1}\left(\frac{\gamma_{n-1}}{\alpha_n}(M\beta_n - \gamma_n) - \beta_{n-1}\right) < y_{n-1}\,.$$

Then $\frac{\alpha_{n-1}}{\alpha_n}(M\beta_n - \gamma_n) < y_{n-1}$ and $x_{n-1}\left(\frac{\gamma_{n-1}}{\alpha_n}(M\beta_n - \gamma_n) - \beta_{n-1}\right) < y_{n-1}$, thus $\frac{\gamma_{n-1}}{\alpha_n}(M\beta_n - \gamma_n) - \beta_{n-1} < \frac{y_{n-1}}{x_{n-1}}$. □

Theorem 1 implies that if $\{x_n\}$ is not bounded above, then $\{x_n\}$ and $\{y_n\}$ are not bounded below by positive numbers, and that if $\{x_n\}$ is not bounded below by a positive number, then $\{y_n\}$ and $\{\frac{y_n}{x_n}\}$ are not bounded above. We will break the contrapositives of these statements into several corollaries.

Corollary 1 *Let $\{(x_n, y_n)\}$ be a solution of System* (1) *where $\{\alpha_n\}$, $\{\beta_n\}$, and $\{\gamma_n\}$ are bounded above and below by positive numbers. If $\{y_n\}$ is bounded below by a positive number, then $\{x_n\}$ is bounded above.*

Proof We will prove the contrapositive; namely, if $\{x_n\}$ is not bounded above, then $\{y_n\}$ is not bounded below by a positive number. Suppose $\{x_n\}$ is not bounded above. Let M be as in Theorem 1, and choose n such that $x_{n+1} > M$. Then by Theorem 1 we have $y_n < \frac{\alpha_n}{M}$. Then for any sufficiently large M there exists n such that $y_n < \frac{\alpha_n}{M}$, so $\{y_n\}$ is not bounded below by a positive number. □

Corollary 2 *Let $\{(x_n, y_n)\}$ be a solution of System* (1) *where $\{\alpha_n\}$, $\{\beta_n\}$, and $\{\gamma_n\}$ are bounded above and below by positive numbers. If $\{y_n\}$ is bounded above, then $\{x_n\}$ is bounded below by a positive number.*

Proof We will prove the contrapositive; namely, if $\{x_n\}$ is not bounded below by a positive number, then $\{y_n\}$ is not bounded above. Suppose $\{x_n\}$ is not bounded below by a positive number. Let M be as in Theorem 1, and choose n such that $x_n < \frac{\alpha_n}{M\beta_n - \gamma_n}$. Then by Theorem 1 we have $y_{n-1} > \frac{\alpha_{n-1}}{\alpha_n}(M\beta_n - \gamma_n)$. Then for any sufficiently large M there exists n such that $y_{n-1} > \frac{\alpha_{n-1}}{\alpha_n}(M\beta_n - \gamma_n)$, so $\{y_n\}$ is not bounded above. □

Remark 1 Inequality (3) allows Corollaries 1 and 2 to be extended to the more general kth order rational difference equation given below,

$$x_{n+1} = \frac{\alpha_n + \gamma_n f(x_n, \ldots, x_{n-k+1}, y_n, \ldots, y_{n-k+1})}{\beta_n f(x_n, \ldots, x_{n-k+1}, y_n, \ldots, y_{n-k+1}) + y_n},$$

where $f\colon (0, \infty)^{2k} \to (0, \infty)$.

Corollary 3 *Let $\{(x_n, y_n)\}$ be a solution of System* (1) *where $\{\alpha_n\}$, $\{\beta_n\}$, and $\{\gamma_n\}$ are bounded above and below by positive numbers. If $\{x_n\}$ is bounded below by a positive number, then $\{x_n\}$ is bounded above.*

Proof We will prove the contrapositive; namely, if $\{x_n\}$ is not bounded above, then $\{x_n\}$ is not bounded below by a positive number. Suppose $\{x_n\}$ is not bounded above. Let M be as in Theorem 1, and choose n such that $x_{n+1} > M$. Then by Theorem 1 we have $x_n < \frac{\alpha_n}{M\beta_n-\gamma_n}$. Then for any sufficiently large M there exists n such that $x_n < \frac{\alpha_n}{M\beta_n-\gamma_n}$, so $\{x_n\}$ is not bounded below by a positive constant. □

Remark 2 Inequality (3) allows Corollary 3 to be extended to the more general kth order rational difference equation given below,

$$x_{n+1} = \frac{\alpha_n + \gamma_n x_n + k_n f(\mathbf{x_n}, \mathbf{y_n})}{\beta_n x_n + l_n f(\mathbf{x_n}, \mathbf{y_n}) + d_n h(\mathbf{x_n}, \mathbf{y_n})},$$

where $\mathbf{x_n} = (x_n, \ldots, x_{n-k+1})$, $\mathbf{y_n} = (y_n, \ldots, y_{n-k+1})$, $f\colon (0,\infty)^{2k} \to (0,\infty)$, $h\colon (0,\infty)^{2k} \to (0,\infty)$, $\{d_n\}$ is bounded above and below by positive numbers, and either $\{k_n\}$ and $\{l_n\}$ are both always zero or both bounded above and below by positive numbers.

Corollary 4 *Let $\{(x_n, y_n)\}$ be a solution of System* (1) *where $\{\alpha_n\}$, $\{\beta_n\}$, and $\{\gamma_n\}$ are bounded above and below by positive numbers. If $\left\{\frac{y_n}{x_n}\right\}$ is bounded above, then $\{x_n\}$ and $\{y_n\}$ are bounded above.*

Proof We will begin by proving a contrapositive; namely, if $\{x_n\}$ is not bounded above, then $\{\frac{y_n}{x_n}\}$ is not bounded above. Suppose $\{x_n\}$ is not bounded above. Let M be as in Theorem 1, and choose n such that $x_{n+1} > M$. Then $x_n < \frac{\alpha_n}{M\beta_n-\gamma_n}$, so $\frac{y_{n-1}}{x_{n-1}} > \frac{\gamma_{n-1}}{\alpha_n}(M\beta_n - \gamma_n) - \beta_{n-1}$. Then for any sufficiently large M there exists n such that $\frac{y_{n-1}}{x_{n-1}} > \frac{\gamma_{n-1}}{\alpha_n}(M\beta_n - \gamma_n) - \beta_{n-1}$, so $\{\frac{y_n}{x_n}\}$ is not bounded above.

Now suppose there exists $M_r > 0$ such that $\frac{y_n}{x_n} < M_r$ for all n. Then by the previous paragraph we have that $\{x_n\}$ is bounded above, so there exists $M_x > 0$ such that $x_n < M_x$ for all n. Likewise, $y_{n+1} < M_r x_n < M_r M_x$ for all n, so $\{y_n\}$ is also bounded above. □

Example 1 Consider the system numbered (30, 33), namely,

$$\begin{cases} x_{n+1} = \frac{\alpha_n+\gamma_n x_n}{\beta_n x_n+y_n} \\ y_{n+1} = \frac{a_n+b_n x_n}{B_n x_n+y_n} \end{cases} \quad \text{for } n = 0, 1, 2 \ldots . \tag{5}$$

In the next theorem we verify a result from [4].

Theorem 2 *The boundedness character of* (5) *is* (B, B).

Proof Note the following:

$$\frac{y_{n+1}}{x_{n+1}} = \left(\frac{a_n + b_n x_n}{B_n x_n + y_n}\right)\left(\frac{\beta_n x_n + y_n}{\alpha_n + \gamma_n x_n}\right) \le \max\left\{\frac{a_n}{\alpha_n}, \frac{b_n}{\gamma_n}\right\} \cdot \max\left\{\frac{\beta_n}{B_n}, 1\right\}.$$

Then $\frac{y_{n+1}}{x_{n+1}}$ is bounded above, so by Corollary 4 we have that $\{x_n\}$ and $\{y_n\}$ are bounded. □

Lemma 1 *Let $\{(x_n, y_n)\}$ be a solution of System* (1) *where $\{\alpha_n\}$, $\{\beta_n\}$, and $\{\gamma_n\}$ are bounded above and below by positive numbers. If $\{x_n\}$ is bounded below by a positive number, then $\{y_n\}$ is bounded above.*

Proof Choose $m_x > 0$ such that $x_n > m_x$ for all n. By Corollary 3, we know there is some M_x such that $x_n < M_x$ for all n. Then

$$m_x < x_{n+1} = \frac{\alpha_n + \beta_n x_n}{\gamma_n x_n + y_n} < \frac{\alpha_n + \beta_n M_x}{y_n},$$

which implies

$$y_n < \frac{\alpha_n + \beta_n M_x}{m_x}.$$

Since $\{\alpha_n\}$, $\{\beta_n\}$ are bounded sequences, then $\{y_n\}$ is bounded. □

Theorem 3 *Let $\{(x_n, y_n)\}$ be a solution of System* (1) *where $\{\alpha_n\}$, $\{\beta_n\}$, and $\{\gamma_n\}$ are bounded above and below by positive numbers. The sequence $\{x_n\}$ is bounded above and below by positive numbers if and only if $\{y_n\}$ is bounded above.*

Proof Corollary 2 states that if $\{y_n\}$ is bounded above, then $\{x_n\}$ is bounded below by a positive number, and Corollary 3 states that if $\{x_n\}$ is bounded below by a positive number, then $\{x_n\}$ is bounded above. Lemma 1 states that if $\{x_n\}$ is bounded below by a positive number, then $\{y_n\}$ is bounded above. □

Theorem 3 establishes that System (1) can never have boundedness character (U, B). Furthermore, in order for a solution to be unbounded in the first variable, it will also have to be unbounded in the second, so a system with boundedness character (U, U) can be unbounded in both variables in the same solution.

Remark 3 One direction of Theorem 3 will hold for a more general kth order rational difference equation given below,

$$x_{n+1} = \frac{\alpha_n + \gamma_n x_n + k_n f(x_n, \ldots, x_{n-k+1}, y_n, \ldots, y_{n-k+1})}{\beta_n x_n + y_n + l_n f(x_n, \ldots, x_{n-k+1}, y_n, \ldots, y_{n-k+1})},$$

where $f\colon (0, \infty)^{2k} \to (0, \infty)$, and $\{k_n\}$ and $\{l_n\}$ are either both always zero or both bounded above and below by positive numbers. Namely, if $\{y_n\}$ is bounded above, then $\{x_n\}$ is bounded above and below by positive numbers.

We will now explore situations in which $\{y_n\}$ is unbounded.

Theorem 4 *Let $\{(x_n, y_n)\}$ be a solution of System* (1) *where $\{\alpha_n\}$, $\{\beta_n\}$, and $\{\gamma_n\}$ are bounded above and below by positive numbers. If* $\lim y_n = \infty$, *then* $\lim x_n = 0$.

Proof Since $\lim y_n = \infty$ we know that $\{y_n\}$ is bounded below by a positive number. By Corollary 1 we know that $\{x_n\}$ is bounded above by some M_x. Then

$$x_{n+1} = \frac{\alpha_n + \beta_n x_n}{\gamma_n x_n + y_n} < \frac{\alpha_n + \beta_n M_x}{y_n}.$$

Since the numerator of the above fraction is bounded above and the denominator goes to infinity, we see that $\lim x_n = 0$. □

The results given above show that if $\lim y_n = 0$, $\lim y_n = \infty$ or $\{y_n\}$ is bounded above or below by positive numbers, then $\{x_n\}$ is bounded above. Thus, to find an example in which $\{x_n\}$ is unbounded, we need a sequence $\{y_n\}$ such that $\liminf y_n = 0$ and $\limsup y_n = \infty$. In the following, we will study examples in which $\lim y_{2n} = 0$ and $\lim y_{2n+1} = \infty$. The following corollary applies the results of Theorem 1 to the even and odd subsequences of $\{x_n\}$ and $\{y_n\}$.

Corollary 5 *Let $\{(x_n, y_n)\}$ be a solution of System* (1) *where $\{\alpha_n\}$, $\{\beta_n\}$, and $\{\gamma_n\}$ are bounded above and below by positive numbers. If $\{y_{2n+1}\}$ is bounded below by a positive number, then $\{x_{2n}\}$ is bounded above. If $\{y_{2n}\}$ is bounded above, then $\{x_{2n+1}\}$ is bounded below by a positive number.*

Proof We will prove the contrapositive; namely, if $\{x_{2n}\}$ is not bounded above, then $\{y_{2n+1}\}$ is not bounded below by a positive number, and if $\{x_{2n+1}\}$ is not bounded below by a positive number, then $\{y_{2n}\}$ is not bounded above.

Suppose $\{x_{2n}\}$ is not bounded above. Let M be as in Theorem 1 and choose n such that $x_{2n} > M$. Then by Theorem 1 we have $y_{2n-1} < \frac{\alpha_n}{M}$. Then for any sufficiently large M there exists n such that $y_{2n-1} < \frac{\alpha_n}{M}$, so $\{y_{2n+1}\}$ is not bounded below by a positive number.

Now suppose $\{x_{2n+1}\}$ is not bounded below by a positive number. Let M be as in Theorem 1 and choose n such that $x_{2n+1} < \frac{\alpha_{2n+1}}{M\beta_{2n+1} - \gamma_{2n+1}}$. Then by Theorem 1 we have $y_{2n} > \frac{\alpha_{2n}}{\alpha_{2n+1}}(M\beta_{2n+1} - \gamma_{2n+1})$. Then for any sufficiently large M there exists n such that $y_{2n} > \frac{\alpha_{2n}}{\alpha_{2n+1}}(M\beta_{2n+1} - \gamma_{2n+1})$, so $\{y_{2n}\}$ is not bounded above. □

Theorem 5 *Let $\{(x_n, y_n)\}$ be a solution of System* (1) *where $\{\alpha_n\}$, $\{\beta_n\}$, and $\{\gamma_n\}$ are bounded above and below by positive numbers. If $\lim y_{2n} = 0$ and $\lim y_{2n+1} = \infty$, then $\{x_{2n}\}$ is not bounded below by a positive constant and $\{x_{2n+1}\}$ is not bounded above.*

Proof Suppose for the sake of contradiction that $\{x_{2n+1}\}$ is bounded above by M_x. Then

$$x_{2n+2} = \frac{\alpha_{2n+1} + \beta_{2n+1} x_{2n+1}}{\gamma_{2n+1} x_{2n+1} + y_{2n+1}} < \frac{\alpha_{2n+1} + \beta_{2n+1} M_x}{y_{2n+1}}.$$

Since the numerator of the above fraction is bounded above and the denominator goes to infinity, we see that $\lim x_{2n} = 0$. However, this would imply that $\lim x_{2n+1} = \infty$, contradicting our initial supposition. Thus, $\{x_{2n+1}\}$ is not bounded above.

Next, suppose $\{x_{2n}\}$ is bounded below by $m_x > 0$. Since $\{y_{2n+1}\} \to \infty$ we know that $\{y_{2n+1}\}$ is bounded below by a positive number. Then by Corollary 5 we know that there exists M'_x such that $x_{2n} < M'_x$ for all n. Then

$$x_{2n+1} = \frac{\alpha_{2n} + \beta_{2n}x_{2n}}{\gamma_{2n}x_{2n} + y_{2n}} < \frac{\alpha_{2n} + \beta_{2n}M'_x}{\gamma_{2n}m_x}.$$

Hence, $\{x_{2n+1}\}$ is bounded above, contradicting the previous result in this proof. Therefore, we conclude that $\{x_{2n}\}$ is not bounded below. □

We give an example to show how Theorem 5 can be used.

Example 2 Consider the system numbered (30, 2), which is given as follows:

$$\begin{cases} x_{n+1} = \frac{\alpha_n + \gamma_n x_n}{\beta_n x_n + y_n} \\ y_{n+1} = \frac{a_n}{y_n} \end{cases} \text{for } n = 0, 1, \ldots. \tag{6}$$

Theorem 6 *Let $\{(x_n, y_n)\}$ be a solution of (6), and suppose that $\{\alpha_n\}$, $\{\beta_n\}$, and $\{\gamma_n\}$ are bounded above and below by positive constants. Suppose $\{a_n\}$ is periodic with even period $2k$, and that $a_0 \cdots a_{2k-2} \neq a_1 \cdots a_{2k-1}$. Then both $\{x_n\}$ and $\{y_n\}$ are unbounded.*

Proof Consider that for any positive integers r and n,

$$y_{2(kn+r)} = y_{2kn+2r} = y_{2k(n-1)+2r}\left(\frac{a_1 \cdots a_{2k-1}}{a_0 \cdots a_{2k-2}}\right) = y_{2r}\left(\frac{a_1 \cdots a_{2k-1}}{a_0 \cdots a_{2k-2}}\right)^n$$

Since $a_0 \cdots a_{2k-2} \neq a_1 \cdots a_{2k-1}$, we know that $\left(\frac{a_1 \cdots a_{2k-1}}{a_0 \cdots a_{2k-2}}\right)^n$ converges to either 0 or ∞. Consider the case where it converges to 0. Also note that any nonnegative integer t can be written as $kn + r$ for some integer $r \geq 0$ and some integer $n \geq 0$. Since $\{y_{2(kn+r)}\}$ converges to 0 for any r, we have $\{y_{2n}\}$ converges to 0. Next, note that $y_{2n+1} = \frac{a_{2n}}{y_{2n}}$, so $\{y_{2n+1}\}$ converges to infinity. Then by Theorem 5 we know that $\{x_n\}$ is not bounded above.

The proof is similar in the case that $\left(\frac{a_1 \cdots a_{2k-1}}{a_0 \cdots a_{2k-2}}\right)^n$ converges to ∞ and will be omitted. □

Theorem 6 shows us that (6) is an example of a system in which "periodicity may destroy boundedness," since when the coefficients are constants, all solutions will be bounded, and when periodic coefficients are introduced unbounded solutions can emerge.

Further, it is interesting to note that when $\{a_n\} = \{a_0, a_1, a_2, \ldots, a_{2k}, \ldots\}$, a period-$(2k+1)$ sequence, for $k = 0, 1, \ldots$, then it can be shown that $y_{2k+1} = \frac{a_0 \cdots a_{2k}}{(a_1 \cdots a_{2k-1})y_0}$ and likewise $y_{4k+2} = y_0$. Hence, $\{y_n\}$ is a periodic sequence, which means that $\{y_n\}$ in this case is bounded above and below by positive constants, and by Theorem 3 this implies that all such solutions of (6) will be bounded in both x and y.

The next two examples will show why it is difficult to loosen the restrictions or strengthen the results of Theorem 5.

Example 3 Consider the solution to the special case of System (1), namely,

$$\begin{cases} x_{n+1} = \frac{1+x_n}{x_n+y_n} \\ \\ y_{n+1} = g(n) \end{cases} \quad n = 0, 1, \dots ,$$

where $g : \mathbb{N}_0 \to (0, \infty)$ is defined by

$$g(n) = \begin{cases} t \text{ if } n = 3t \text{ for some integer } t \geq 1 \\ \frac{1}{t} \text{ if } n = 3t + 2 \text{ for some integer } t \geq 1 \\ 1 \text{ otherwise} \end{cases},$$

and with initial condition $(x_0, y_0) = (1, 1)$. Then $\{y_{2n}\}$ is not bounded below by a positive constant, and $\{y_{2n+1}\}$ is not bounded above, however $\lim y_{2n} \neq 0$ and $\lim y_{2n+1} \neq \infty$. For $n > 1$, note that $x_{3n+2} = 1$, so $x_{3n+3} = \frac{1+1}{1+\frac{1}{n}}$, which implies that $1 < x_{3n+3} < 2$. Then $x_{3n+4} < \frac{1+2}{1+(n+1)} < 3$, and therefore $\{x_{3n+1}\}$ is bounded above. So $\{x_n\}$ is bounded above in this example. Note, however, that $\{x_n\}$ is not bounded below. This example shows why the limits on the even and odd subsequences of $\{y_n\}$ in the statement of Theorem 5 are necessary.

Example 4 Consider the solution to the special case of System (1), namely,

$$\begin{cases} x_{n+1} = \frac{1+x_n}{x_n+y_n} \\ \\ y_{n+1} = g(n) \end{cases} \quad n = 0, 1, \dots ,$$

where $g : \mathbb{N}_0 \to (0, \infty)$ is defined by

$$g(n) = \begin{cases} \frac{1}{t} \text{ if } n = 4t \text{ or } n = 4t + 2 \text{ for some integer } t \geq 1 \\ t \text{ if } n = 4t + 1 \text{ for some integer } t \geq 1 \\ t^2 \text{ if } n = 4t + 3 \text{ for some integer } t \geq 1 \\ 1 \text{ otherwise} \end{cases},$$

and with initial condition $(x_0, y_0) = (1, 1)$. Then, obviously, $\lim y_{2n} = 0$ and $\lim y_{2n+1} = \infty$. Also note that $y_{2n} < 1$ and $y_{2n+1} > 1$ for all $n > 4$. Then $x_{2n+2} < 1$ and $x_{2n+1} > 1$. So we have on one hand

$$x_{4n+4} = \frac{x_{4n+2} + y_{4n+2} + 1 + x_{4n+2}}{1 + x_{4n+2} + x_{4n+2}y_{4n+3} + y_{4n+2}y_{4n+3}} \leq \frac{1+1+1+1}{y_{4n+2}y_{4n+3}} = \frac{4}{n},$$

and on the other hand,

$$x_{4n+5} = \frac{1 + x_{4n+4}}{x_{4n+4} + y_{4n+4}} > \frac{1}{\frac{4}{n} + \frac{1}{n+1}} > \frac{1}{\frac{4}{n} + \frac{1}{n}} = \frac{n}{5}.$$

Note also that

$$x_{4n+5} = \frac{1 + x_{4n+4}}{x_{4n+4} + y_{4n+4}} < \max\left\{\frac{x_{4n+4}}{x_{4n+4}}, \frac{1}{y_{4n+4}}\right\} = \max\{1, n+1\} = n+1\,.$$

So, for $n > 4$,

$$x_{4n+6} = \frac{1 + x_{4n+5}}{x_{4n+5} + y_{4n+5}} > \frac{\frac{n}{5}}{n+1+n+1} > \frac{\frac{n}{5}}{4n} = \frac{1}{20}\,.$$

Furthermore,

$$x_{4n+7} = \frac{1 + x_{4n+6}}{x_{4n+6} + y_{4n+6}} < \frac{1 + x_{4n+6}}{x_{4n+6}} = \frac{1}{x_{4n+6}} + 1 < 21\,.$$

Then $\lim x_{4n+4} = 0$ and $\lim x_{4n+5} = \infty$. This shows that $\{x_{2n}\}$ is not bounded below by a positive number and $\{x_{2n+1}\}$ is not bounded above; however, we see that $\lim x_{2n} \neq 0$ since $x_{4n+6} > \frac{1}{20}$ for all n and $\lim x_{2n+1} \neq \infty$ since $x_{4n+7} < 21$ for all n, and hence the conclusion of Theorem 5 cannot be strengthened beyond what is stated.

We can however improve the result of Theorem 5 if we know about the product $y_{2n}y_{2n+1}$ or $x_{2n}y_{2n+1}$.

Theorem 7 *Let $\{(x_n, y_n)\}$ be a solution of System (1) where $\{\alpha_n\}$, $\{\beta_n\}$, and $\{\gamma_n\}$ are bounded above and below by positive numbers. Suppose $\lim y_{2n} = 0$ and $\lim y_{2n+1}=\infty$. If $\lim y_{2n}y_{2n+1} = \infty$ or $\lim x_{2n}y_{2n+1} = \infty$, then $\lim x_{2n} = 0$ and $\lim x_{2n+1} = \infty$.*

Proof By Corollary 5 we know that $\{x_{2n}\}$ is bounded above. Then

$$\begin{aligned} x_{2n+2} &= \frac{\alpha_{2n+1} + \beta_{2n+1}x_{2n+1}}{\gamma_{2n+1}x_{2n+1} + y_{2n+1}} \\ &= \frac{\alpha_{2n+1}\gamma_{2n}x_{2n} + \alpha_{2n+1}y_{2n} + \beta_{2n+1}\alpha_{2n} + \beta_{2n}\beta_{2n+1}x_{2n}}{\gamma_{2n+1}\alpha_{2n} + \gamma_{2n+1}\beta_{2n}x_{2n} + \gamma_{2n}x_{2n}y_{2n+1} + y_{2n}y_{2n+1}}\,. \end{aligned}$$

The numerator is bounded and the denominator goes to ∞, thus we have $\lim x_{2n+2}=0$ so $\lim x_{2n} = 0$. Moreover,

$$x_{2n+1} = \frac{\alpha_{2n} + \beta_{2n}x_{2n}}{\gamma_{2n}x_{2n} + y_{2n}}\,.$$

The denominator goes to 0 and the numerator is bounded below by a positive constant, so we have $\lim x_{2n+1} = \infty$. □

We give an example below to illustrate how Theorem 7 can be used.

Example 5 Consider the solution to the special case of System (2), namely,

$$\begin{cases} x_{n+1} = \frac{1+x_n}{x_n+y_n} \\ y_{n+1} = g(n) \end{cases} n = 0, 1, \dots,$$

where $g : \mathbb{N}_0 \to (0, \infty)$ is defined by

$$g(n) = \begin{cases} 2^{-\left(\frac{n+1}{2}\right)} & \text{if } n \text{ is odd} \\ 3^{\frac{n}{2}} & \text{if } n \text{ is even} \end{cases},$$

with initial condition $(x_0, y_0) = (1, 1)$. Then, for each n, we have $y_{2n} = \frac{1}{2^n}$ and $y_{2n+1} = 3^n$. Then $\lim x_{2n} = 0$ and $\lim x_{2n+1} = \infty$, since $y_{2n}y_{2n+1} = \left(\frac{3}{2}\right)^n$.

3 Examples Establishing Boundedness Character of Special Cases of System (2)

In this section, we will illustrate the effectiveness of the results given in Sect. 2 by establishing the boundedness character of several special cases of System (2). For convenience, Appendix A contains a list of these 49 systems with the numbering scheme from [9]. For completeness, in Sect. 4 we will investigate the remaining special cases of System (2) which require a more specific approach. This will resolve the boundedness character of all 49 systems contained in (2).

3.1 When g Is a Rational Function with Similar Numerator and Denominator

We see that System (2) is made up of rational functions. We will now investigate solutions of System (1) when g can be written as a ratio of linear combinations of functions. We are motivated by the work of Lugo and Pallidino, see [12, 13, 16].

Let N be a positive integer. For the following theorems, suppose that $\{\nu_n^{(j)}\}$ and $\{\mu_n^{(j)}\}$ are sequences of nonnegative real numbers for each $j = 1, \dots, N$, and there exist n and j such that $\mu_n^{(j)} \neq 0$. Further, for any particular j, if $\nu_n^{(j)} \neq 0$ (or $\mu_n^{(j)} \neq 0$) for some n then the sequence $\{\nu_n^{(j)}\}$ (or $\{\mu_n^{(j)}\}$) is bounded above and below by positive numbers. This means that either the sequence is bounded above and below by positive values, or it is a sequence of all zeros.

Let $f_j\colon (0, \infty)^{2k} \to (0, \infty)$ be a function $f_j(u_1, \dots, u_k, v_1, \dots, v_k)$ for each $j = 1, \dots, N$.

Let g be the function defined in System (1), and suppose that g can be written in the following form:

$$g(u_1, \ldots, u_k, v_1, \ldots, v_k, n) = \frac{\sum\limits_{j=1}^{N} \nu_n^{(j)} f_j(u_1, \ldots, u_k, v_1, \ldots, v_k)}{\sum\limits_{j=1}^{N} \mu_n^{(j)} f_j(u_1, \ldots, u_k, v_1, \ldots, y_k)}. \tag{7}$$

What this means is that for any given value of n, the function g can be written as a fraction where the numerator and denominator is a linear combination of terms made up of the functions $\{f_1, \ldots, f_N\}$. Further note that f_j does not depend on the index n of the solution to (1).

Theorem 8 *Let $\{(x_n, y_n)\}$ be a solution of System* (1) *where $\{\alpha_n\}$, $\{\beta_n\}$, and $\{\gamma_n\}$ are bounded above and below by positive numbers. Suppose that g is in the form of* (7), *with the property that $\nu_n^{(j)} > 0$ if and only if $\mu_n^{(j)} > 0$ for each $j = 1, \ldots, N$ and each $n = 1, 2 \ldots$. Then $\{y_n\}$ is bounded above and below by positive numbers and $\{x_n\}$ is bounded above and below by positive numbers.*

Proof Define the set

$$\mathcal{S} = \{j \in \{1, \ldots, N\} \, : \, \nu_n^{(j)} > 0\}.$$

Choose positive constants $m_{\nu^{(j)}}$, $M_{\nu^{(j)}}$, $m_{\mu^{(j)}}$, $M_{\mu^{(j)}}$ such that $m_{\nu^{(j)}} < \nu_n^{(j)} < M_{\nu^{(j)}}$ and $m_{\mu^{(j)}} < \mu_n^{(j)} < M_{\mu^{(j)}}$ for $j \in \mathcal{S}$ and all n. This is valid since we are saying that $\{w_n\} \in \left\{\left\{\mu_n^{(j)}\right\}, \left\{\nu_n^{(j)}\right\}\right\}$ is positive if and only if it is bounded above and below by positive constants. Consider the expression for y_{n+1}, using inequality (3):

$$\begin{aligned} y_{n+1} &= g(x_n, \ldots, x_{n-k}, y_n, \ldots, y_{n-k}, n) \\ &= \frac{\sum\limits_{j \in \mathcal{S}} \nu_n^{(j)} f_j(x_n, \ldots, x_{n-k+1}, y_n, \ldots, y_{n-k+1})}{\sum\limits_{j \in \mathcal{S}} \mu_n^{(j)} f_j(x_n, \ldots, x_{n-k+1}, y_n, \ldots, y_{n-k+1})} \\ &\le \max_{j \in \mathcal{S}} \left\{ \frac{\nu_n^{(j)}}{\mu_n^{(j)}} \right\} \le \max_{j \in \mathcal{S}} \left\{ \frac{M_{\nu^{(j)}}}{m_{\mu^{(j)}}} \right\}. \end{aligned}$$

Likewise, we find a lower bound on y_{n+1},

$$y_{n+1} \ge \min_{j \in \mathcal{S}} \left\{ \frac{m_{\nu^{(j)}}}{M_{\mu^{(j)}}} \right\} > 0\,.$$

Since $\{y_n\}$ is bounded above, Theorem 3 tells us that $\{x_n\}$ is bounded above and below by positive constants. □

Remark 4 It follows from Theorem 8 that the following special cases of System (2) have boundedness character (B, B).

(30, 1), (30, 5), (30, 9), (30, 28), (30, 32), (30, 36), (30, 49)

We can generalize Theorem 8 slightly by considering rational equations g as in (7) in which the denominator has additional terms compared to the numerator. The proof will be omitted since it follows similarly to the proof of Theorem 8.

Theorem 9 *Let $\{(x_n, y_n)\}$ be a solution of System* (1) *where $\{\alpha_n\}$, $\{\beta_n\}$, and $\{\gamma_n\}$ are bounded above and below by positive numbers. Suppose that g is in the form of* (7)*, with the property that if $\nu_n^{(j)} > 0$ then $\mu_n^{(j)} > 0$ for each $j = 1, \ldots, N$ and each $n = 1, 2 \ldots$. Then $\{y_n\}$ is bounded from above and $\{x_n\}$ is bounded from above and from below by positive constants.*

Remark 5 It follows from Theorem 9 that the following special cases of System (2) have boundedness character (B, B).

$$(30, 10), (30, 11), (30, 13), (30, 15), (30, 17), (30, 18),$$

$$(30, 37), (30, 38), (30, 39), (30, 43), (30, 44), (30, 45)$$

The more interesting case occurs when g has extra terms in the numerator that are not present in the denominator. It turns out that there is more than one possible boundedness character in this case. We now give a result for which solutions are bounded in both $\{x_n\}$ and $\{y_n\}$.

Theorem 10 *Let $\{(x_n, y_n)\}$ be a solution of System* (1) *where $\{\alpha_n\}$, $\{\beta_n\}$, and $\{\gamma_n\}$ are bounded above and below by positive numbers. Suppose that g is in the form of* (7)*, with the property that if $\mu_n^{(j)} > 0$ then $\nu_n^{(j)} > 0$ for each $j = 1, \ldots, N$ and each $n = 1, 2 \ldots$. Suppose that there exists a function g^* with the following properties:*

1. *g^* is also in the form of* (7)*, with*

$$g^*(u_1, \ldots, u_k, v_1, \ldots, v_k, n) = \frac{\sum_{j=1}^{N} \hat{\nu}_n^{(j)} f_j(u_1, \ldots, u_k, v_1, \ldots, v_k)}{\sum_{j=1}^{N} \hat{\mu}_n^{(j)} f_j(u_1, \ldots, u_k, v_1, \ldots, y_k)}.$$

   with the property that if $\hat{\mu}_n^{(j)} > 0$ then $\hat{\nu}_n^{(j)} > 0$ for each $j = 1, \ldots, N$ and each $n = 1, 2 \ldots$;
2. *g^* is non-decreasing in $u_1, \ldots, u_k$ and non-increasing in $v_1, \ldots, v_k$ for each fixed n.*

Additionally, suppose that for each fixed n,

$$g(u_1, \ldots, u_k, v_1, \ldots, v_k, n) \le g^*(u_1, \ldots, u_k, v_1, \ldots, v_k, n).$$

Then the sequences $\{x_n\}$ and $\{y_n\}$ are bounded above and below by positive numbers.

Proof Let S be the following set

$$\mathcal{S} = \{j \in \{1, \ldots, N\} : \mu_n^{(j)} > 0\}.$$

There exist positive constants $m_{\nu^{(j)}}, M_{\nu^{(j)}}, m_{\mu^{(j)}}, M_{\mu^{(j)}}$ such that $m_{\nu^{(j)}} < \nu_n^{(j)} < M_{\nu^{(j)}}$ and $m_{\mu^{(j)}} < \mu_n^{(j)} < M_{\mu^{(j)}}$ for $j \in \mathcal{S}$ and for each n. First note that

$$\begin{aligned} y_{n+1} &\geq \frac{\sum_{j\in\mathcal{S}} \nu_n^{(j)} f_j(x_n, \ldots, x_{n-k+1}, y_n, \ldots, y_{n-k+1})}{\sum_{j\in\mathcal{S}} \mu_n^{(j)} f_j(x_n, \ldots, x_{n-k+1}, y_n, \ldots, y_{n-k+1})} \\ &\geq \min_{j\in\mathcal{S}} \left\{ \frac{\nu_n^{(j)}}{\mu_n^{(j)}} \right\} \geq \min_{j\in\mathcal{S}} \left\{ \frac{m_{\nu^{(j)}}}{M_{\mu^{(j)}}} \right\}. \end{aligned}$$

Let m_y be a positive lower bound of $\{y_n\}$. By Corollary 1 there exists M_x, an upper bound of $\{x_n\}$. Suppose that there exists g^* with the given properties in the statement of Theorem 10. Let $M_{\hat{\nu}^{(j)}}$ be an upper bound of $\{\hat{\nu}_n^{(j)}\}$ for each $j = 1, \ldots, N$. Let $\hat{\mathcal{S}}$ be the following set

$$\hat{\mathcal{S}} = \{j \in \{1, \ldots, N\} : \hat{\mu}_n^{(j)} > 0\}.$$

Let $m_{\hat{\mu}^{(j)}}$ be a positive lower bound of $\{\hat{\mu}_n^{(j)}\}$ for each $j \in \hat{\mathcal{S}}$. Then, for all n,

$$y_{n+1} \leq g^*(M_x, \ldots, M_x, m_y, \ldots, m_y, n) \leq \frac{\sum_{j=1}^{N} M_{\hat{\nu}^{(j)}} f_j(M_x, \ldots, M_x, m_y, \ldots, m_y)}{\sum_{j\in\hat{\mathcal{S}}} m_{\hat{\mu}^{(j)}} f_j(M_x, \ldots, M_x, m_y, \ldots, m_y)}$$

which is an upper bound on $\{y_n\}$. Then by Theorem 3, $\{x_n\}$ is bounded from above and below by positive numbers. □

Example 6 Consider the system numbered (30, 48), given by

$$\begin{cases} x_{n+1} = \frac{\alpha_n + \gamma_n x_n}{\beta_n x_n + y_n} \\ y_{n+1} = \frac{a_n + b_n x_n + c_n y_n}{B_n x_n + y_n} \end{cases} \quad n = 0, 1, \ldots. \tag{8}$$

We can use Theorem 10 to establish that (8) has boundedness character (B, B). Notice that

$$g(u, v, n) = \frac{a_n + b_n u + c_n v}{B_n u + v} \leq \frac{a_n + b_n u + c_n v}{v} = g^*(u, v, n),$$

where $g^*(u, v, n)$ is in the form of (7) and has the monotonicity required by Theorem 10.

Remark 6 It follows from Theorem 10, and by following steps similar to Example 6, that systems

$$(30, 20), (30, 22), (30, 26), (30, 41), (30, 46), (30, 48)$$

have boundedness character (B, B).

The following lemma will establish that the sequence $\{x_n\}$ will be bounded in the case when g has the property that all the terms in the denominator are in the numerator.

Lemma 2 *Let $\{(x_n, y_n)\}$ be a solution of System* (1) *where $\{\alpha_n\}$, $\{\beta_n\}$, and $\{\gamma_n\}$ are bounded above and below by positive numbers. Let g be the function given in* (1) *and suppose that there exists a function g^* with the following properties:*

1. *for each fixed n,*

$$g^*(u_1, \ldots, u_k, v_1, \ldots, v_k, n) \leq g(u_1, \ldots, u_k, v_1, \ldots, v_k, n);$$

2. *g^* is in the form of* (7), *with the property that if $\mu_n^{(j)} > 0$ then $\nu_n^{(j)} > 0$ for each $j = 1, \ldots, N$ and each $n = 1, 2 \ldots$.*

Then $\{y_n\}$ is bounded below by a positive constant, and $\{x_n\}$ is bounded above.

Proof Let $\mathcal{S}$ be the following set

$$\mathcal{S} = \{j \in \{1, \ldots, N\} : \mu_n^{(j)} > 0\}.$$

There exist positive constants $m_{\nu^{(j)}}, M_{\nu^{(j)}}, m_{\mu^{(j)}}, M_{\mu^{(j)}}$ such that $m_{\nu^{(j)}} < \nu_n^{(j)} < M_{\nu^{(j)}}$ and $m_{\mu^{(j)}} < \mu_n^{(j)} < M_{\mu^{(j)}}$ for $j \in \mathcal{S}$ and for each n.

Now, consider y_{n+1} and use inequality (3) in a way that is similar to the way it is used in the proof of Theorem 8:

$$\begin{aligned} y_{n+1} &= g(x_n, \ldots, x_{n-k+1}, y_n, \ldots, y_{n-k+1}, n) \\ &\geq g^*(x_n, \ldots, x_{n-k+1}, y_n, \ldots, y_{n-k+1}, n) \\ &\geq \frac{\sum\limits_{j \in \mathcal{S}} \nu_n^{(j)} f_j(x_n, \ldots, x_{n-k+1}, y_n, \ldots, y_{n-k+1})}{\sum\limits_{j \in \mathcal{S}} \mu_n^{(j)} f_j(x_n, \ldots, x_{n-k+1}, y_n, \ldots, y_{n-k+1})} \\ &\geq \min_{j \in \mathcal{S}} \left\{ \frac{\nu_n^{(j)}}{\mu_n^{(j)}} \right\} \geq \min_{j \in \mathcal{S}} \left\{ \frac{m_{\nu^{(j)}}}{M_{\mu^{(j)}}} \right\} > 0 . \end{aligned}$$

Hence, we have shown that $\{y_n\}$ is bounded below by a positive number $m_y = \min\limits_{j \in \mathcal{S}} \left\{ \frac{m_{\nu^{(j)}}}{M_{\mu^{(j)}}} \right\}$. If we choose M_x such that $M_x > \max \left\{ \frac{\gamma_n}{\beta_n}, \frac{\alpha_n}{m_y} \right\}$ for all n, then M_x is an upper bound for $\{x_n\}$. This shows if we increase any of the $\nu_n^{(j)}$, then M_x and

m_y as previously defined will still be bounds (see the examples on the existence of unbounded solutions in the sequel). □

This next theorem will establish the existence of solutions of (1) for which $\{y_n\}$ is unbounded.

Theorem 11 *Let $\{(x_n, y_n)\}$ be a solution of System* (1) *where $\{\alpha_n\}$, $\{\beta_n\}$, and $\{\gamma_n\}$ are bounded above and below by positive numbers. Let g be the function given in* (1) *and suppose that there exists a function g^* with the following properties:*

1. *for each fixed n,*

$$g^*(u_1, \ldots, u_k, v_1, \ldots, v_k, n) \leq g(u_1, \ldots, u_k, v_1, \ldots, v_k, n);$$

2. *g^* is in the form of* (7), *with the property that if $\mu_n^{(j)} > 0$ then $\nu_n^{(j)} > 0$ for each $j = 1, \ldots, N$ and each $n = 1, 2 \ldots$;*
3. *g^* is non-increasing in $u_1, \ldots, u_k$, non-decreasing in $v_1, \ldots, v_k$;*
4. *there is at least one term in the numerator of g^* which is a term containing v_l, $l \in \{1, 2, \ldots, k\}$, and which is not present in the denominator.*

Then $\{x_n\}$ will be bounded for all solutions of (1) and there exist solutions of (1) for which $\{y_n\}$ is unbounded.

Proof Assume $g \geq g^*$ and g^* has a term of v_l which is in the numerator of g^* and not in its denominator.

By Lemma 2 we know that there is some $M_x > 0$ and $m_y > 0$ such that $x_n < M_x$ and $y_n > m_y$ for all n. We will use the monotonicity of g^* and substitute in these bounds, everywhere except for the v_l term.

$$\begin{aligned} y_{n+1} &\geq g^*(M_x, \ldots, M_x, m_y, \ldots, y_{n-l+1}, \ldots, m_y, n) \\ &\geq \frac{\sum_{j \in \mathcal{S}} \nu_n^{(j)} f_j(M_x, \ldots, M_x, m_y, \ldots, m_y)}{\sum_{j \in \mathcal{S}} \mu_n^{(j)} f_j(M_x, \ldots, M_x, m_y, \ldots, m_y)} + k_n y_{n-l+1} \geq k_n y_{n-l+1}\,. \end{aligned}$$

Where

$$k_n = \frac{\nu_n^{(l)}}{\sum_{j \in \mathcal{S}} \mu_n^{(j)} f_j(M_x, \ldots, M_x, m_y, \ldots, m_y)}\,.$$

Let $\epsilon > 0$ be fixed. Choose $\{\nu_n^{(j)}\}$ and $\{\mu_n^{(j)}\}$ such that $k_n > 1 + \epsilon$ for all n. Note that we can do this without changing M_x and m_y, as shown in Lemma 2. Then this shows that $\{y_n\}$ can be unbounded. □

Example 7 Consider the system numbered (30, 47), given by

$$\begin{cases} x_{n+1} = \frac{\alpha_n+\gamma_n x_n}{\beta_n x_n+y_n} \\ y_{n+1} = \frac{a_n+b_n x_n+c_n y_n}{A_n+x_n} \end{cases} \quad n = 0, 1, \dots . \tag{9}$$

We can use Theorem 11 to establish that (9) has boundedness character (B, U). It follows from Lemma 2 that $\{x_n\}$ is bounded above by some constant M_x. Then

$$y_{n+1} = \frac{a_n + b_n x_n + c_n y_n}{A_n + x_n} = \frac{a_n + b_n x_n}{A_n + x_n} + \frac{c_n}{A_n + x_n} y_n \geq \frac{c_n}{A_n + M_x} y_n . \tag{10}$$

By choosing coefficients such that $\frac{c_n}{A_n+M_x} > 1 + \epsilon$ for all n and for some $\epsilon > 0$, we have shown that $\lim y_n = \infty$.

Remark 7 It follows from Theorem 11, and by following steps similar to Example 7, that the following special cases of System (2) have boundedness character (B, U).

$$(30, 19), (30, 24), (30, 27), (30, 40), (30, 42), (30, 47)$$

For system (30, 24), one may iterate for x_n to get a function g which has the desired monotonicity required for Theorem 11.

3.2 Establishing Boundedness Character (B, U)

In this subsection we look at several examples which use Theorem 1. The main idea is to use a proof by contradiction to show that $\{x_n\}$ is bounded above, then find appropriate $\{c_n\}$ that would make $\{y_n\}$ unbounded. We will consider several examples. In Example 8, we find the contradiction on the size of some y_i. In Example 9, we find the contradiction by considering $\frac{y_i}{x_i}$ for some i. And in Example 10, we find there is a contradiction on either y_i or $\frac{y_i}{x_i}$ for some i.

Example 8 Consider the system numbered (30, 4), which is given as follows:

$$\begin{cases} x_{n+1} = \frac{\alpha_n+\gamma_n x_n}{\beta_n x_n+y_n} \\ y_{n+1} = c_n y_n \end{cases} \quad \text{for } n = 0, 1 \dots . \tag{11}$$

Theorem 12 *The boundedness character of* (11) *is* (B, U).

Proof Let M be defined as in (4) such that for all n, $\frac{\alpha_{n-1}}{\alpha_n}(M\beta_n - \gamma_n) > \frac{\alpha_n}{c_{n-1}M}$. Suppose $\{x_n\}$ is unbounded above, and choose n such that $x_{n+1} > M$. Then by Theorem 1, $y_{n-1} > \frac{\alpha_{n-1}}{\alpha_n}(M\beta_n - \gamma_n)$ and $y_n < \frac{\alpha_n}{M}$. But then $y_{n-1} < \frac{\alpha_n}{c_{n-1}M}$, contradicting our choice of M. So $\{x_n\}$ is bounded above by M. Note also that if there exists $\epsilon > 0$, such that $c_n > 1 + \epsilon$ for all n, then $\{y_n\}$ will be unbounded. □

Remark 8 The proof that the system numbered (30, 25) has boundedness character (B, U) follows similarly.

Example 9 Consider the system numbered (30, 29), which is given as follows:

$$\begin{cases} x_{n+1} = \frac{\alpha_n+\gamma_n x_n}{\beta_n x_n+y_n} \\ y_{n+1} = \frac{a_n+c_n y_n}{A_n+x_n} \end{cases} \text{for } n = 0, 1, 2 \ldots. \tag{12}$$

Theorem 13 *The boundedness character of* (12) *is* (B, U).

Proof Let M be defined such that for all n,

1. $\frac{a_{n-1}}{A_{n-1}} > \frac{\alpha_n}{M}$;
2. $c_{n-1}\left(\frac{\gamma_{n-1}}{\alpha_n}(M\beta_n - \gamma_n) - \beta_{n-1}\right) > \frac{\alpha_n}{M}$.

Suppose $\{x_n\}$ is unbounded above, and choose n such that $x_{n+1} > M$. Then by Theorem 1, $\frac{y_{n-1}}{x_{n-1}} > \frac{\gamma_{n-1}}{\alpha_n}(M\beta_n - \gamma_n) - \beta_{n-1}$ and

$$\min\left\{\frac{a_{n-1}}{A_{n-1}}, \frac{c_{n-1}y_{n-1}}{x_{n-1}}\right\} \leq \frac{a_{n-1} + c_{n-1}y_{n-1}}{A_{n-1} + x_{n-1}} = y_n < \frac{\alpha_n}{M}.$$

But $\frac{a_{n-1}}{A_{n-1}} > \frac{\alpha_n}{M}$ and $\frac{c_{n-1}y_{n-1}}{x_{n-1}} > c_{n-1}\left(\frac{\gamma_{n-1}}{\alpha_n}(M\beta_n - \gamma_n) - \beta_{n-1}\right) > \frac{\alpha_n}{M}$, contradicting our choice of M. So $\{x_n\}$ is bounded above by M. Then $y_{n+1} > \frac{c_n y_n}{A_n+M}$, so if we choose coefficients such that there exists $\epsilon > 0$ such that $\frac{c_n}{A_n+M} > 1 + \epsilon$ for all n, then $\{y_n\}$ will be unbounded. Note that replacing each c_n with a larger number to satisfy this inequality will not invalidate any of the previous inequalities in our definition of M, so M is still an upper bound for $\{x_n\}$. □

Remark 9 The proofs that the systems numbered (30, 6) and (30, 21) have boundedness character (B, U) follow similarly.

Example 10 Consider the system numbered (30, 14), which is given as follows:

$$\begin{cases} x_{n+1} = \frac{\alpha_n+\gamma_n x_n}{\beta_n x_n+y_n} \\ y_{n+1} = \frac{c_n y_n}{A_n+x_n} \end{cases} \text{for } n = 0, 1, \ldots. \tag{13}$$

Theorem 14 *The boundedness character of* (13) *is* (B, U).

Proof Let M be defined such that

1. $\frac{c_{n-1}}{2A_{n-1}}\frac{\alpha_{n-1}}{\alpha_n}(M\beta_n - \gamma_n) > \frac{\alpha_n}{M}$;
2. $\frac{c_{n-1}}{2}(\frac{\gamma_{n-1}}{\alpha_n}(M\beta_n - \gamma_n) - \beta_{n-1}) > \frac{\alpha_n}{M}$.

Suppose $\{x_n\}$ is unbounded above, and choose n such that $x_{n+1} > M$. Then by Theorem 1, $y_{n-1} > \frac{\alpha_{n-1}}{\alpha_n}(M\beta_n - \gamma_n)$, $\frac{y_{n-1}}{x_{n-1}} > \frac{\gamma_{n-1}}{\alpha_n}(M\beta_n - \gamma_n) - \beta_{n-1}$, and

$$\min\left\{\frac{c_{n-1}y_{n-1}}{2A_{n-1}}, \frac{c_{n-1}y_{n-1}}{2x_{n-1}}\right\} \leq \frac{c_{n-1}y_{n-1}}{A_{n-1} + x_{n-1}} = y_n < \frac{\alpha_n}{M}.$$

But

$$\frac{c_{n-1}y_{n-1}}{2A_{n-1}} > \frac{c_{n-1}}{2A_{n-1}}\frac{\alpha_{n-1}}{\alpha_n}(M\beta_n - \gamma_n) > \frac{\alpha_n}{M},$$

and

$$\frac{c_{n-1}y_{n-1}}{2x_{n-1}} > \frac{c_{n-1}}{2}\left(\frac{\gamma_{n-1}}{\alpha_n}(M\beta_n - \gamma_n) - \beta_{n-1}\right) > \frac{\alpha_n}{M},$$

contradicting our choice of M. Then $\{x_n\}$ is bounded above by M. Then $y_{n+1} > \frac{c_n y_n}{A_n+M}$, so if we choose coefficients such that there exists $\epsilon > 0$ such that $\frac{c_n}{A_n+M} > 1+\epsilon$ for all n, then $\{y_n\}$ will be unbounded. As in the proof of Theorem 13, we can increase c_n without increasing M, so such an example is possible. □

4 Remaining Special Cases of System (2)

In this section, we provide conditions and examples for certain systems contained in (2) to be (U, U).

4.1 Finding an Unbounded Solution Using Three Iterations

In this subsection, we focus on examples in which we use three iterations on x and y to find solutions that are unbounded in both variables. In the first example, nonconstant coefficients are necessary for unbounded solutions.

Example 11 Consider the system numbered (30, 8), namely

$$\begin{cases} x_{n+1} = \frac{\alpha_n+\gamma_n x_n}{\beta_n x_n + y_n} \\ y_{n+1} = \frac{b_n x_n}{y_n} \end{cases} \quad \text{for } n = 0, 1, \ldots . \tag{14}$$

It was shown in [6] that with constant coefficients, every solution of (14) is bounded in both variables. However, as we will show in the following theorem, this is an example of when "periodicity may destroy boundedness," since when the coefficients are periodic, unbounded solutions will exist.

Theorem 15 *The boundedness character of* (14) *is* (U, U).

Proof We will show that with a particular set of period-three coefficients, the solution will be unbounded. Let us define $\{b_n\}$, where $b_{3t+1} = 0.1$ for integers t and $b_n = 1$ otherwise, and take all other coefficients to be 1.

Choose positive constants K, L such that the following hold

1. $1 + \frac{7}{L} < K < 2 < L$;
2. $\frac{10}{8K^2} > 1$.

Pick initial conditions (x_0, y_0) such that $y_0 > L$ and $1 < x_0 < K$. Suppose for some n that $y_{3n} > L$ and $1 < x_{3n} < K$.

Then

$$x_{3n+1} = \frac{1 + x_{3n}}{x_{3n} + y_{3n}} < \frac{1 + K}{1 + L} \leq \max\left\{1, \frac{K}{L}\right\} = 1\,,$$

and

$$y_{3n+1} = \frac{x_{3n}}{y_{3n}} < \frac{K}{L}\,.$$

Now, by iteration we consider x_{3n+2}

$$x_{3n+2} = \frac{(x_{3n} + y_{3n} + 1 + x_{3n})y_{3n}}{y_{3n}(1 + x_{3n}) + x_{3n}(x_{3n} + y_{3n})}\,.$$

Similarly, by iterating in the expression for y_{3n+2} we find

$$\begin{aligned} y_{3n+2} &= \frac{0.1x_{3n+1}}{y_{3n+1}} = 0.1\frac{1 + x_{3n}}{x_{3n} + y_{3n}}\frac{y_{3n}}{x_{3n}} \\ &= 0.1\frac{1 + x_{3n}}{x_{3n}}\frac{y_{3n}}{x_{3n} + y_{3n}} < 0.1\frac{1 + K}{1} = 0.1(1 + 2) < 1\,. \end{aligned}$$

Now, we get a lower bound for x_{3n+3}. Namely,

$$x_{3n+3} = \frac{1 + x_{3n+2}}{x_{3n+2} + y_{3n+2}} > \frac{1 + x_{3n+2}}{x_{3n+2} + 1} = 1\,.$$

And, next, an upper bound for x_{3n+3},

$$\begin{aligned} x_{3n+3} &< \frac{1 + x_{3n+2}}{x_{3n+2}} = \frac{1}{x_{3n+2}} + 1 = 1 + \frac{y_{3n}(1 + x_{3n}) + x_{3n}(x_{3n} + y_{3n})}{(x_{3n} + y_{3n} + 1 + x_{3n})y_{3n}} \\ &= 1 + \frac{(1 + x_{3n})}{(x_{3n} + y_{3n} + 1 + x_{3n})} + \frac{x_{3n}(x_{3n} + y_{3n})}{(x_{3n} + y_{3n} + 1 + x_{3n})y_{3n}} \\ &< 1 + \frac{(1 + K)}{y_{3n}} + \frac{K(K + y_{3n})}{(y_{3n})y_{3n}} \\ &< 1 + \frac{3}{y_{3n}} + \frac{2(2y_{3n})}{(y_{3n})y_{3n}} = 1 + \frac{3}{y_{3n}} + \frac{4}{y_{3n}} = 1 + \frac{7}{y_{3n}} < 1 + \frac{7}{L} < K\,. \end{aligned}$$

And here, we will show y_{3n+3} has a lower bound. Let $p = \frac{10}{8K^2}$. Then

$$\begin{aligned} y_{3n+3} &= \frac{x_{3n+2}}{y_{3n+2}} = \frac{10(x_{3n} + y_{3n})x_{3n}}{(1 + x_{3n})y_{3n}}\frac{(x_{3n} + y_{3n} + 1 + x_{3n})y_{3n}}{y_{3n}(1 + x_{3n}) + x_{3n}(x_{3n} + y_{3n})} \\ &> \frac{10y_{3n}}{(1 + K)y_{3n}}\frac{y_{3n}^2}{y_{3n}(1 + K) + K(K + y_{3n})} > \frac{10}{(2K)}\frac{y_{3n}^2}{y_{3n}(2K) + K(2y_{3n})} \end{aligned}$$

$$= \frac{10}{(2K)} \frac{y_{3n}}{2K + 2K} = \left(\frac{10}{8K^2}\right) y_{3n} = p y_{3n} > pL > L\,.$$

Then $y_{3n+3k+3} > Lp^{k+1}$ for any k. Thus, $\lim y_{3n} = \infty$.
We can see that

$$x_{3n+3k+4} = \frac{1 + x_{3n+3k+3}}{x_{3n+3k+3} + y_{3n+3k+3}} < \frac{1+K}{1+Lp^{k+1}}\,,$$

and hence, $\lim x_{3n+3k+4} = 0$. While, similarly,

$$y_{3n+3k+4} = \frac{x_{3n+3k+3}}{y_{3n+3k+3}} < \frac{K}{Lp^{k+1}}\,,$$

which shows that $\lim y_{3n+3k+4} = 0$ also. But if this is the case, then

$$\lim x_{3n+3k+5} \geq \lim \frac{1}{x_{3n+3k+4} + y_{3n+3k+4}} = \infty\,.$$

Hence, we have shown that $\{x_n\}$ and $\{y_n\}$ are unbounded in this case. □

Question 1 Theorem 15 shows that (14) can be unbounded in $\{x_n\}$ and $\{y_n\}$ with period-three coefficients. For what prime periods k does system (14) have unbounded solutions with period k coefficients? In particular, do unbounded solutions of (14) exist when coefficients are period-two?

Example 12 Consider system (30, 16), which is given as follows:

$$\begin{cases} x_{n+1} = \frac{\alpha_n + \gamma_n x_n}{\beta_n x_n + y_n} \\ y_{n+1} = \frac{b_n x_n}{A_n + y_n} \end{cases} \quad \text{for } n = 0, 1, 2 \ldots \tag{15}$$

We give an alternative proof to what is presented in [6].

Theorem 16 *System* (15) *has boundedness character* (U, U).

Proof It suffices to find a solution of (15) with constant coefficients which is unbounded in both variables, namely $\alpha_n = \alpha$, $\gamma_n = \gamma$, etc. Choose coefficients and let $m > 0$ be a constant such that the following inequalities hold:

1. $\dfrac{(m(\beta+1)+\gamma)(\beta+1)(A^2+bm)}{\beta\alpha A} < 1;$
2. $\dfrac{(\beta\alpha+\gamma(m(\beta+1)+\gamma))(\beta+1)(A^2+bm)}{b\beta\alpha A} < 1.$

For example, let $m = 0.25$, $A = 1$, $b = 4$, $\beta = 0.5$, $\gamma = 1$, $\alpha = 10$.
Choose initial conditions $x_0 < m$ and $y_0 < m$. Suppose $x_{3n} < m$ and $y_{3n} < m$ for some n.

$$x_{3n+1} = \frac{\alpha + \gamma x_{3n}}{\beta x_{3n} + y_{3n}} > \frac{\alpha}{\beta m + m} = \frac{\alpha}{m(\beta+1)}$$

$$y_{3n+1} = \frac{bx_{3n}}{A + y_{3n}} < \frac{bm}{A}$$

$$\frac{y_{3n+1}}{x_{3n+1}} < \frac{bm}{A}\frac{(\beta+1)m}{\alpha}$$

$$y_{3n+2} = \frac{bx_{3n+1}}{A + y_{3n+1}} > \frac{b\frac{\alpha}{m(\beta+1)}}{A + \frac{bm}{A}} = \frac{b\alpha A}{m(\beta+1)(A^2+bm)}$$

$$x_{3n+2} = \frac{\alpha + \gamma x_{3n+1}}{\beta x_{3n+1} + y_{3n+1}} = \frac{\frac{\alpha}{x_{3n+1}} + \gamma}{\beta + \frac{y_{3n+1}}{x_{3n+1}}} > \frac{\gamma}{\beta + \frac{bm}{A}\frac{(\beta+1)m}{\alpha}}$$

$$x_{3n+2} = \frac{\frac{\alpha}{x_{3n+1}} + \gamma}{\beta + \frac{y_{3n+1}}{x_{3n+1}}} < \frac{m(\beta+1)+\gamma}{\beta}$$

$$\begin{aligned} y_{3n+3} &= \frac{bx_{3n+2}}{A + y_{3n+2}} < \frac{b\frac{m(\beta+1)+\gamma}{\beta}}{A + \frac{b\alpha A}{m(\beta+1)(A^2+bm)}} < \frac{b\frac{m(\beta+1)+\gamma}{\beta}}{\frac{b\alpha A}{m(\beta+1)(A^2+bm)}} \\ &= \frac{(m(\beta+1)+\gamma)(\beta+1)(A^2+bm)}{\beta\alpha A}m \end{aligned}$$

$$\begin{aligned} x_{3n+3} &= \frac{\alpha + \gamma x_{3n+2}}{\beta x_{3n+2} + y_{3n+2}} < \frac{\alpha + \gamma x_{3n+2}}{y_{3n+2}} < \frac{\alpha + \gamma\frac{m(\beta+1)+\gamma}{\beta}}{\frac{b\alpha A}{m(\beta+1)(A^2+bm)}} \\ &= \frac{(\beta\alpha + \gamma(m(\beta+1)+\gamma))(\beta+1)(A^2+bm)}{b\beta\alpha A}m\,. \end{aligned}$$

So, it follows that $\lim x_{3n} = 0$ and $\lim y_{3n} = 0$. Therefore the subsequence $\{x_{3n+1}\}$ is unbounded and by Theorem 3, $\{y_n\}$ must also be unbounded. □

Example 13 Consider system (30, 31), which is given as follows:

$$\begin{cases} x_{n+1} = \frac{\alpha_n + \gamma_n x_n}{\beta_n x_n + y_n} \\ y_{n+1} = \frac{a_n + b_n x_n}{A_n + y_n} \end{cases} \quad \text{for } n = 0, 1, 2 \ldots . \tag{16}$$

We give an alternative proof to what is presented in [6].

Theorem 17 *System* (16) *has boundedness character* (U, U).

Proof It suffices to find a solution of (15) with constant coefficients which is unbounded in both variables, namely $\alpha_n = \alpha$, $\gamma_n = \gamma$, etc. Choose coefficients and let $m_y, M_x > 0$ be constants such that the following inequalities hold:

1. $\dfrac{b\alpha A m_y}{(\beta(\alpha+\gamma M_x) + a + bM_x)(A^2 m_y + am_y + b(\alpha+\gamma M_x))} > 1;$
2. $\dfrac{\beta(\alpha+\gamma M_x) + a + bM_x}{\beta m_y} + \dfrac{\gamma}{\beta} < M_x.$

For example, let $m_y = 1000$, $M_x = A = a = 0.1$, $b = 10$, $\beta = \alpha = 1$, $\gamma = 0.01$. Choose initial conditions $x_0 < M_x$ and $y_0 > m_y$. Suppose $x_{3n} < M_x$ and $y_{3n} > m_y$ for some n. Then

$$
\begin{aligned}
x_{3n+1} &= \frac{\alpha + \gamma x_{3n}}{\beta x_{3n} + y_{3n}} < \frac{\alpha + \gamma x_{3n}}{y_{3n}} < \frac{\alpha + \gamma M_x}{y_{3n}} \\
y_{3n+1} &= \frac{a + b x_{3n}}{A + y_{3n}} < \frac{a + b M_x}{y_{3n}} \\
x_{3n+2} &= \frac{\alpha + \gamma x_{3n+1}}{\beta x_{3n+1} + y_{3n+1}} > \frac{\alpha y_{3n}}{\beta(\alpha + \gamma M_x) + a + b M_x} \\
\frac{a y_{3n}}{A y_{3n} + a + b M_x} < y_{3n+2} &= \frac{a + b x_{3n+1}}{A + y_{3n+1}} < \frac{a}{A} + \frac{b}{A} \frac{\alpha + \gamma M_x}{y_{3n}} \\
x_{3n+3} &= \frac{\alpha + \gamma x_{3n+2}}{\beta x_{3n+2} + y_{3n+2}} < \frac{\alpha}{\beta x_{3n+2}} + \frac{\gamma}{\beta} \\
&< \frac{\beta(\alpha + \gamma M_x) + a + b M_x}{\beta m_y} + \frac{\gamma}{\beta} < M_x \\
y_{3n+3} &= \frac{a + b x_{3n+2}}{A + y_{3n+2}} > \frac{b x_{3n+2}}{A + y_{3n+2}} > \frac{b \frac{\alpha y_{3n}}{\beta(\alpha+\gamma M_x)+a+b M_x}}{A + \frac{a}{A} + \frac{b}{A}\frac{\alpha+\gamma M_x}{y_{3n}}} \\
&> \frac{b \alpha A m_y}{(\beta(\alpha + \gamma M_x) + a + b M_x)(A^2 m_y + a m_y + b(\alpha + \gamma M_x))} y_{3n} .
\end{aligned}
$$

So, it follows that $\lim y_{3n} = \infty$, so $\lim x_{3n-1} = \infty$ as well. □

4.2 A Solution with Nonperiodic Coefficients

In all of the previous examples, we were able to reduce cases with bounded coefficients to those with periodic coefficients. In the following example, we use nonperiodic coefficients to produce a solution which is unbounded in both variables.

Example 14 Consider the system numbered (30, 3), namely

$$
\begin{cases} x_{n+1} = \frac{\alpha_n + \gamma_n x_n}{\beta_n x_n + y_n} \\ y_{n+1} = \frac{a_n}{x_n} \end{cases} \quad \text{for } n = 0, 1, \ldots . \tag{17}
$$

It was shown in Sect. 5 of [1] that with constant coefficients, (17) would have boundedness character (B, U). We will show in this example that with non-constant coefficients, this system has boundedness character (U, U). We begin with several technical lemmas.

Lemma 3 *Let $\{(x_n, y_n)\}$ be a solution of (17) where $\alpha_{2n} = \frac{1}{2}$, $a_{2n+1} = 2$, and $\alpha_{2n+1} = a_{2n} = \beta_n = \gamma_n = 1$ for all n. If $1 < x_0 < \frac{5}{4}$ and $1 < y_0 < \frac{5}{4}$, then $1 < x_{2m} < \frac{5}{4}$ and $\left(\frac{8}{7}\right)^m < y_{2m} < 4^m\left(\frac{5}{4}\right)$ for all $m \geq 0$.*

Proof Suppose, for some integer $k \geq 0$, that $1 < x_{2k} < \frac{5}{4}$ and $(\frac{8}{7})^k < y_{2k} < 4^k\left(\frac{5}{4}\right)$. Then $\frac{4}{5} < y_{2k+1} < 1$.

$$y_{2k+2} = \frac{2}{x_{2k+1}} = 2\frac{x_{2k} + y_{2k}}{\frac{1}{2} + x_{2k}} > 2\frac{0 + y_{2k}}{\frac{1}{2} + \frac{5}{4}} > \left(\frac{8}{7}\right)^{k+1}$$

$$y_{2k+2} \leq 2\max\{1, 2y_{2k}\} = 4y_{2k} < 4^{k+1}\left(\frac{5}{4}\right)$$

$$x_{2k+2} = \frac{1 + x_{2k+1}}{x_{2k+1} + y_{2k+1}} < \frac{1 + x_{2k+1}}{x_{2k+1} + \frac{4}{5}} \leq \frac{5}{4}$$

$$x_{2k+2} > \frac{1 + x_{2k+1}}{x_{2k+1} + 1} = 1\,.$$

Then $1 < x_{2(k+1)} < \frac{5}{4}$ and $\left(\frac{8}{7}\right)^{k+1} < y_{2(k+1)} < 4^{k+1}\left(\frac{5}{4}\right)$. By induction on k, if $1 < x_0 < \frac{5}{4}$ and $1 < y_0 < \frac{5}{4}$ then $1 < x_{2m} < \frac{5}{4}$ and $\left(\frac{8}{7}\right)^m < y_{2m} < 4^m\left(\frac{5}{4}\right)$ for all $m \geq 0$. □

Lemma 4 *Let $\{(x_n, y_n)\}$ be a solution of (17) where $\alpha_{2n+1} = a_{2n+1} = 2$ and $\alpha_{2n} = a_{2n} = \gamma_n = \beta_n = 1$ for all n. Let m be a positive integer. If $1 < x_0 < 2$ and $1.5(6^{6m^2}) < y_0$, then $\left(\frac{3}{2}\right)^m < x_{2m} < 2(6^m)$ and $1 < y_{2m} < 2^m y_0$.*

Proof Suppose for some integer $0 \leq k < m$, that $\left(\frac{3}{2}\right)^k < x_{2k} < 2(6^k)$ and $\frac{y_0}{6^{k^2}2^k} \leq y_{2k} \leq 2^k y_0$. Then by iterating on y we get

$$y_{2k+2} = 2\frac{x_{2k} + y_{2k}}{1 + x_{2k}} < 2\frac{x_{2k}y_{2k} + y_{2k}}{1 + x_{2k}} = 2y_{2k} < 2^{k+1}y_0\,.$$

$$y_{2k+2} > 2\frac{0 + y_{2k}}{x_{2k} + x_{2k}} = \frac{y_{2k}}{x_{2k}} > \frac{\frac{y_0}{6^{k^2}2^k}}{2(6^k)} > \frac{y_0}{6^{(k+1)^2}2^{k+1}}\,.$$

$$\begin{aligned}
x_{2k+2} &= x_{2k}\frac{(3x_{2k} + 2y_{2k} + 1)}{2x_{2k} + x_{2k}^2 + y_{2k}} > x_{2k}\frac{(\frac{1}{2}y_{2k} + \frac{3}{2}y_{2k})}{3x_{2k}^2 + y_{2k}} \\
&\geq x_{2k}\min\left\{\frac{y_{2k}}{6x_{2k}^2}, \frac{3}{2}\right\} \geq x_{2k}\min\left\{\frac{\frac{y_0}{6^{k^2}2^k}}{6(2(6^k))^2}, \frac{3}{2}\right\} \\
&\geq x_{2k}\min\left\{\frac{y_0}{6^{k^2+3k+2}}, \frac{3}{2}\right\} \geq x_{2k}\min\left\{\frac{1.5(6^{6m^2})}{6^{k^2+3k+2}}, \frac{3}{2}\right\} \\
&= x_{2k}\min\left\{1.5(6^{6m^2-k^2-3k-2}), \frac{3}{2}\right\}
\end{aligned}$$

$$\geq x_{2k} \min\left\{1.5(6^{6(k+1)^2-k^2-3k-2}), \frac{3}{2}\right\}$$

$$\geq x_{2k} \min\left\{1.5(6^{5k^2+9k+4}), \frac{3}{2}\right\} > \left(\frac{3}{2}\right)^{k+1}.$$

$$y_{2k} \geq \frac{y_0}{6^{k^2}2^k} > \frac{1.5(6^{6m^2})}{6^{k^2}2^k} > 2(6^k) > x_{2k}.$$

$$x_{2k+2} = \frac{x_{2k}(3x_{2k} + 2y_{2k} + 1)}{2x_{2k} + x_{2k}^2 + y_{2k}}$$

$$< \frac{x_{2k}(3y_{2k} + 2y_{2k} + y_{2k})}{0 + 0 + y_{2k}} < 2(6^{k+1}).$$

So $(\frac{3}{2})^{k+1} < x_{2(k+1)} < 2(6^{k+1})$ and $\frac{y_0}{6^{(k+1)^2}2^{k+1}} < y_{2(k+1)} < 2^{k+1}y_0$. Then by induction on k, if $1 < x_0 < 2$ and $1.5(6^{6m^2}) < y_0$, then $(\frac{3}{2})^m < x_{2m} < 2(6^m)$ and $1 < \frac{1.5(6^{6m^2})}{6^{m^2}2^m} < \frac{y_0}{6^{m^2}2^m} < y_{2m} < 2^m y_0$. □

Lemma 5 *Let $\{(x_n, y_n)\}$ be a solution of (17) where $\alpha_n = \beta_n = \gamma_n = a_n = 1$ for all n. Let m_1 and m_2 be positive integers. If $1 < x_0 < 1 + (\frac{4}{3})^{m_1}$ and $1 < y_0 < 1 + 2^{m_2}$, then $1 < x_{2(m_1+m_2)+14} < \frac{5}{4}$ and $1 < y_{2(m_1+m_2)+14} < \frac{5}{4}$.*

Proof For all n, let $c_n = x_n - 1$ and $d_n = y_n - 1$. Suppose, for some integer $k \geq 0$, that $1 < x_{2k} < 1 + (\frac{4}{3})^{m_1}$ and $1 < y_{2k} < 1 + 2^{m_2-k}$. Then $y_{2k+1} < 1$. By iterating on y, we get

$$y_{2k+2} = \frac{1}{x_{2k+1}} = 1 + \frac{d_{2k}}{2 + c_{2k}} > 1.$$

$$y_{2k+2} < 1 + \frac{d_{2k}}{2} < 1 + 2^{m_2-(k+1)}.$$

$$x_{2k+2} = \frac{1 + x_{2k+1}}{x_{2k+1} + y_{2k+1}} > 1.$$

$$x_{2k+2} = \frac{(2x_{2k} + y_{2k} + 1)x_{2k}}{2x_{2k} + y_{2k} + x_{2k}^2} < \frac{(2x_{2k} + y_{2k} + 1)x_{2k}}{2x_{2k} + y_{2k} + 1} = x_{2k}.$$

Then $1 < x_{2(k+1)} < 1 + (\frac{4}{3})^{m_1}$ and $1 < y_{2(k+1)} < 1 + 2^{m_2-(k+1)}$. By induction on k, if $1 < x_0 < 1 + (\frac{4}{3})^{m_1}$ and $1 < y_0 < 1 + 2^{m_2}$ then $1 < x_{2(m_2+2)} < 1 + (\frac{4}{3})^{m_1}$ and $1 < y_{2(m_2+2)} < 1 + 2^{-2} = \frac{5}{4}$.

Next, suppose for some integer $k \geq 0$, that $1 < x_{2k} < 1 + (\frac{4}{3})^{m_1-k}$ and $1 < y_{2k} < \frac{5}{4}$. (Then $y_{2k} < 2$.) Then

$$x_{2k+2} = \frac{(2x_{2k} + y_{2k} + 1)x_{2k}}{2x_{2k} + y_{2k} + x_{2k}^2}$$
$$= 1 + c_{2k}\frac{2 + c_{2k} + d_{2k}}{4 + 4c_{2k} + d_{2k} + c_{2k}^2} > 1$$
$$x_{2k+2} < 1 + c_{2k}\frac{2 + c_{2k} + 1}{4 + 4c_{2k} + 0 + 0} < 1 + \frac{3}{4}c_{2k} < 1 + \left(\frac{4}{3}\right)^{m_1-(k+1)}$$

$$1 < \frac{x_{2k} + y_{2k}}{1 + x_{2k}} = y_{2k+2} \leq \max\{y_{2k}, 1\} < \frac{5}{4}.$$

Then $1 < x_{2k+2} < 1 + \left(\frac{4}{3}\right)^{m_1-(k+1)}$ and $1 < y_{2k+2} < \frac{5}{4}$. By induction on k, if $1 < x_{2(m_2+2)} < 1 + (\frac{4}{3})^{m_1}$ and $1 < y_{2(m_2+2)} < \frac{5}{4}$ then $1 < x_{2(m_2+2+m_1+5)} < 1 + (\frac{4}{3})^{-5} < \frac{5}{4}$ and $1 < y_{2(m_2+2+m_1+5)} < \frac{5}{4}$. □

Lemmas 3, 4, and 5 together show us that we can, for any $M > 0$ and x_0 and y_0 each between 1 and $\frac{5}{4}$, construct a finite sequence which is not bounded above by M and whose last terms are between 1 and $\frac{5}{4}$. Repeating this process for an unbounded sequence $\{M_m\} = \left\{\left(\frac{3}{2}\right)^m\right\}$ and chaining the resulting sequences of x and y together (using the last terms of one finite sequence as the initial conditions of the next finite sequence) will give us a solution to system (17) that is unbounded in both variables. We formalize this idea in the following theorem.

Theorem 18 *There exists a solution to system* (17) *with bounded coefficients in which* $\{x_n\}$ *and* $\{y_n\}$ *are unbounded.*

Proof In this theorem we will create a sequence of indices $\{k_n\}$ such that $x_{k_{3m+2}} > m$ for each positive integer m, and thus a system in which $\{x_n\}$ and $\{y_n\}$ are unbounded above. Let $k_0 = 0$ and choose x_0 and y_0 each between 1 and $\frac{5}{4}$. We will define the other values of $\{k_n\}$ using induction on m.

Let m be a nonnegative integer, and suppose $1 < x_{k_{3m}} < \frac{5}{4}$ and $1 < y_{k_{3m}} < \frac{5}{4}$. Define $k_{3m+1} = k_{3m} + 2(84m^2 + 4)$. Note that we choose this value for k_i in preparation to use Lemma 3. For all n such that $k_{3m} \leq n < k_{3m+1}$, define coefficients as in Lemma 3; that is, let $\alpha_n = \frac{1}{2}$ if $n = 2t$ for some integer t, let $a_n = 2$ if $n = 2t + 1$ for some integer t, and let all other coefficients equal 1. Then by Lemma 3, we have

$$1 < x_{k_{3m+1}} < \frac{5}{4} < 2$$
$$y_{k_{3m+1}} > \left(\frac{8}{7}\right)^{84m^2+4} = \left(\frac{8}{7}\right)^4\left(\left(\frac{8}{7}\right)^{14}\right)^{6m^2} > \frac{3}{2}(6)^{6m^2}$$
$$y_{k_{3m+1}} < 4^{84m^2+4}\left(\frac{5}{4}\right).$$

Next, define $k_{3m+2} = k_{3m+1} + 2m$. Note that we choose this value for k_i in preparation to use in Lemma 4. For all n such that $k_{3m+1} \leq n < k_{3m+2}$, define coefficients as in Lemma 4; that is, let $\alpha_n = a_n = 2$ if $n = 2t + 1$ for some integer t, and let all other coefficients equal 1. Then by Lemma 4, we have

$$1 \le \left(\frac{3}{2}\right)^m < x_{k_{3m+2}} < 2(6^m) < \left(\frac{4}{3}\right)^3 \left(\left(\frac{4}{3}\right)^9\right)^m < 1 + \left(\frac{4}{3}\right)^{9m+3}$$

$$1 < y_{k_{3m+2}} < 2^m \left(4^{84m^2+4}\right)\left(\frac{5}{4}\right)$$

$$< 2^m (2^2)^{84m^2+4}(2) < 1 + 2^{168m^2+m+9}.$$

Finally, define $k_{3m+3} = k_{3m+2} + 2(168m^2 + 10m + 19)$. Note that we choose this value for k_i in preparation to use in Lemma 5. For all n such that $k_{3m+2} \le n < k_{3m+3}$, define coefficients as in Lemma 5; that is, let all coefficients equal 1. Then

$$\begin{aligned} x_{k_{3m+3}} &= x_{k_{3m+2}+2(168m^2+10m+19)} \\ &= x_{k_{3m+2}+2((168m^2+m+9)+(9m+3))+14} \\ y_{k_{3m+3}} &= y_{k_{3m+2}+2((168m^2+m+9)+(9m+3))+14} \end{aligned}$$

Then by Lemma 5, we have $1 < x_{k_{3m+3}} < \frac{5}{4}$ and $1 < y_{k_{3m+3}} < \frac{5}{4}$.

By induction on m, we have $x_{k_{3m}}$ and $y_{k_{3m}}$ are each between 1 and $\frac{5}{4}$ for all m. Note also that $\{x_n\}$ and $\{y_n\}$ are unbounded since $x_{k_{3m+2}} > \left(\frac{3}{2}\right)^m$ and $y_{3m+1} > \left(\frac{3}{2}\right)(6)^{6m^2}$ for each m. □

It is interesting to note that this example uses coefficients from the finite set $\left\{\frac{1}{2}, 1, 2\right\}$ to show that solutions of (17) are unbounded, but the coefficient sequences are not periodic.

Question 2 Are there solutions of (17) which have periodic coefficients such that $\{x_n\}$ is unbounded?

5 Conclusion

We have established several results concerning system (1) for which the function defining y_{n+1} is left general. Using these results, we were able to determine the boundedness character of many of the special cases contained in (2). The boundedness character of all 49 special cases of (2) has been established by us for coefficients that form sequences which are bounded above and below by positive constants, and the results are summarized in Appendix A. The boundedness character of several of the systems contained in (2) with bounded coefficient sequences follows directly from proofs that were given in the autonomous case, and these have been noted with footnotes in Appendix A.

In many of the systems contained in (2), the boundedness character for periodic coefficients matches that for coefficients that form bounded sequences. There is one system where the boundedness character for periodic coefficients is still open.

A Boundedness Character of System (2)

The second equation of every system contained within (2) is listed below. In each, $x_{n+1} = \frac{\alpha_n+\gamma_n x_n}{\beta_n x_n+y_n}$. The numbering from [9] is given. Coefficients are assumed to be from sequences bounded above and below by positive constants. Note that the systems that differ in boundedness from the autonomous case are (30, 2), (30, 3), and (30, 8).

1	$y_{n+1} = a_n$	**(B,B)**	2	$y_{n+1} = \frac{a_n}{y_n}$	**(U,U)**
3	$y_{n+1} = \frac{a_n}{x_n}$	**(U,U)**	4	$y_{n+1} = c_n y_n$	**(B,U)**
5	$y_{n+1} = b_n$	**(B,B)**	6	$y_{n+1} = \frac{c_n y_n}{x_n}$	**(B,U)**
7	$y_{n+1} = b_n x_n$	**(B,B)**[a]	8	$y_{n+1} = \frac{b_n x_n}{y_n}$	**(U,U)**
9	$y_{n+1} = c_n$	**(B,B)**	10	$y_{n+1} = \frac{a_n}{A_n+y_n}$	**(B,B)**
11	$y_{n+1} = \frac{a_n}{A_n+x_n}$	**(B,B)**	12	$y_{n+1} = \frac{a_n}{B_n x_n+y_n}$	**(B,B)**[b]
13	$y_{n+1} = \frac{y_n}{A_n+y_n}$	**(B,B)**	14	$y_{n+1} = \frac{c_n y_n}{A_n+x_n}$	**(B,U)**
15	$y_{n+1} = \frac{c_n y_n}{B_n x_n+y_n}$	**(B,B)**	16	$y_{n+1} = \frac{b_n x_n}{A_n+y_n}$	**(U,U)**
17	$y_{n+1} = \frac{x_n}{A_n+x_n}$	**(B,B)**	18	$y_{n+1} = \frac{b_n x_n}{B_n x_n+y_n}$	**(B,B)**
19	$y_{n+1} = a_n + c_n y_n$	**(B,U)**	20	$y_{n+1} = \frac{a_n+c_n y_n}{y_n}$	**(B,B)**
21	$y_{n+1} = \frac{a_n+c_n y_n}{x_n}$	**(B,U)**	22	$y_{n+1} = a_n + b_n x_n$	**(B,B)**
23	$y_{n+1} = \frac{a_n+b_n x_n}{y_n}$	**(B,B)**[c]	24	$y_{n+1} = \frac{a_n+x_n}{x_n}$	**(B,U)**
25	$y_{n+1} = b_n x_n + c_n y_n$	**(B,U)**	26	$y_{n+1} = \frac{b_n x_n+c_n y_n}{y_n}$	**(B,B)**
27	$y_{n+1} = \frac{b_n x_n+c_n y_n}{x_n}$	**(B,U)**	28	$y_{n+1} = \frac{a_n+c_n y_n}{A_n+y_n}$	**(B,B)**
29	$y_{n+1} = \frac{a_n+c_n y_n}{A_n+x_n}$	**(B,U)**	30	$y_{n+1} = \frac{a_n+c_n y_n}{B_n x_n+y_n}$	**(B,B)**[d]
31	$y_{n+1} = \frac{a_n+b_n x_n}{A_n+y_n}$	**(U,U)**	32	$y_{n+1} = \frac{a_n+b_n x_n}{A_n+x_n}$	**(B,B)**
33	$y_{n+1} = \frac{a_n+b_n x_n}{B_n x_n+y_n}$	**(B,B)**	34	$y_{n+1} = \frac{b_n x_n+c_n y_n}{A_n+y_n}$	**(B,B)**[e]
35	$y_{n+1} = \frac{b_n x_n+c_n y_n}{A_n+x_n}$	**(B,U)**[f]	36	$y_{n+1} = \frac{b_n x_n+c_n y_n}{B_n x_n+y_n}$	**(B,B)**
37	$y_{n+1} = \frac{a_n}{A_n+B_n x_n+y_n}$	**(B,B)**	38	$y_{n+1} = \frac{c_n y_n}{A_n+B_n x_n+y_n}$	**(B,B)**
39	$y_{n+1} = \frac{b_n x_n}{A_n+B_n x_n+y_n}$	**(B,B)**	40	$y_{n+1} = a_n + b_n x_n + c_n y_n$	**(B,U)**
41	$y_{n+1} = \frac{a_n+b_n x_n+c_n y_n}{y_n}$	**(B,B)**	42	$y_{n+1} = \frac{a_n+b_n x_n+c_n y_n}{x_n}$	**(B,U)**
43	$y_{n+1} = \frac{a_n+c_n y_n}{A_n+B_n x_n+y_n}$	**(B,B)**	44	$y_{n+1} = \frac{a_n+b_n x_n}{A_n+B_n x_n+y_n}$	**(B,B)**
45	$y_{n+1} = \frac{b_n x_n+c_n y_n}{A_n+B_n x_n+y_n}$	**(B,B)**	46	$y_{n+1} = \frac{a_n+b_n x_n+c_n y_n}{A_n+y_n}$	**(B,B)**
47	$y_{n+1} = \frac{a_n+b_n x_n+c_n y_n}{A_n+x_n}$	**(B,U)**	48	$y_{n+1} = \frac{a_n+b_n x_n+c_n y_n}{B_n x_n+y_n}$	**(B,B)**
49	$y_{n+1} = \frac{a_n+b_n x_n+c_n y_n}{A_n+B_n x_n+C_n y_n}$	**(B,B)**			

[a]The proof of Theorem 4 from [10] can be generalized to coefficients bounded above and below by positive numbers.
[b]See Theorem 3.7 in [11] for a proof.

[c]Theorem 2.3 in [8] can be generalized to coefficients bounded above and below by positive numbers.
[d]Theorem 6 in [14] can be generalized to coefficients bounded above and below by positive numbers.
[e]System (30, 34) in Theorem 8.1 in [6] can be generalized to coefficients bounded above and below by positive numbers.
[f]Theorem 7.1 in [3] can be generalized to coefficients bounded above and below by positive numbers.

References

1. Amleh, A.M., Camouzis, E., Ladas, G.: On the dynamics of a rational difference equation. II. Int. J. Difference Equ. **3**(2), 195–225 (2008)
2. Amleh, A.M., Camouzis, E., Ladas, G., Radin, M.A.: Patterns of boundedness of a rational system in the plane. J. Difference Equ. Appl. **16**(10), 1197–1236 (2010). https://doi.org/10.1080/10236190903325144
3. Brett, A.M., Camouzis, E., Ladas, G., Lynd, C.D.: On the boundedness character of a rational system. J. Numer. Math. Stoch. **1**(1), 1–10 (2009)
4. Camouzis, E.: Boundedness character of rational systems in the plane. In: Difference equations and applications, pp. 163–170. Uğur-Bahçeşehir Univ. Publ. Co., Istanbul (2009)
5. Camouzis, E., Drymonis, E., Ladas, G.: Patterns of boundedness of the rational system $x_{n+1} = \frac{\alpha_1+\beta_1 x_n}{A_1+C_1 y_n}$ and $y_{n+1} = \frac{\alpha_2+\beta_2 x_n+\gamma_2 y_n}{A_2+B_2 x_n+C_2 y_n}$. Fasc. Math. **44**, 9–18 (2010)
6. Camouzis, E., Drymonis, E., Ladas, G.: Patterns of boundedness of the rational system $x_{n+1} = \frac{\alpha_1+\beta_1 x_n}{A_1+B_1 x_n+C_1 y_n}$ and $y_{n+1} = \frac{\alpha_2+\beta_2 x_n+\gamma_2 y_n}{A_2+B_2 x_n+C_2 y_n}$. Commun. Appl. Nonlinear Anal. **18**(1), 1–23 (2011)
7. Camouzis, E., Drymonis, E., Ladas, G., Tikjha, W.: Patterns of boundedness of the rational system $x_{n+1} = \frac{\alpha_1}{A_1+B_1 x_n+C_1 y_n}$ and $y_{n+1} = \frac{\alpha_2+\beta_2 x_n+\gamma_2 y_n}{A_2+B_2 x_n+C_2 y_n}$. J. Difference Equ. Appl. **18**(1), 89–110 (2012). https://doi.org/10.1080/10236198.2010.515591
8. Camouzis, E., Kent, C.M., Ladas, G., Lynd, C.D.: On the global character of solutions of the system: $x_{n+1} = \frac{\alpha_1+y_n}{x_n}$ and $y_{n+1} = \frac{\alpha_2+\beta_2 x_n+\gamma_2 y_n}{A_2+B_2 x_n+C_2 y_n}$. J. Difference Equ. Appl. **18**(7), 1205–1252 (2012). https://doi.org/10.1080/10236198.2011.555406
9. Camouzis, E., Kulenović, M.R.S., Ladas, G., Merino, O.: Rational systems in the plane. J. Difference Equ. Appl. **15**(3), 303–323 (2009). https://doi.org/10.1080/10236190802125264
10. Camouzis, E., Ladas, G.: When does periodicity destroy boundedness in rational equations? J. Difference Equ. Appl. **12**(9), 961–979 (2006). https://doi.org/10.1080/10236190600822369
11. Kostrov, Y., Kudlak, Z.: On a first order rational system of difference equations with non-constant coefficients. Commun. Appl. Nonlinear Anal. **22**(3), 1–24 (2015)
12. Lugo, G., Palladino, F.J.: Unboundedness results for systems. Cent. Eur. J. Math. **7**(4), 741–756 (2009). https://doi.org/10.2478/s11533-009-0051-2
13. Lugo, G., Palladino, F.J.: Some boundedness results for systems of two rational difference equations. Cent. Eur. J. Math. **8**(6), 1058–1090 (2010). https://doi.org/10.2478/s11533-010-0063-y
14. Lugo, G., Palladino, F.J.: On the boundedness character of rational systems in the plane. J. Difference Equ. Appl. **17**(12), 1801–1811 (2011). https://doi.org/10.1080/10236198.2010.491513
15. Mitrinović, D.: Elementary Inequalities. P. Noordhoff LTD (1964)
16. Palladino, F.J.: Difference inequalities, comparison tests, and some consequences. Involve **1**(1), 91–100 (2008). https://doi.org/10.2140/involve.2008.1.91

Stability Investigation of Biosensor Model Based on Finite Lattice Difference Equations

Vasyl Martsenyuk, Aleksandra Klos-Witkowska and Andriy Sverstiuk

Abstract We consider the delayed antibody-antigen competition model for two-dimensional array of biopixels

$$
\begin{aligned}
x_{i,j}(n+1) &= x_{i,j}(n)\exp\left\{\beta - \gamma y_{i,j}(n-r) - \delta_x x_{i,j}(n-r)\right\} + \hat{S}\left\{x_{i,j}(n)\right\}, \\
y_{i,j}(n+1) &= y_{i,j}(n)\exp\left\{ - \mu_y + \eta\gamma x_{i,j}(n-r) - \delta_y y_{i,j}(n)\right\}, i, j = \overline{1, N},
\end{aligned}
$$

$n, r \in \mathbb{N}$. Here $x_{i,j}(t)$ is the concentration of antigens, $y_{i,j}(t)$ is the concentration of antibodies in biopixel (i, j), $i, j = \overline{1, N}$. $\hat{S}\{x_{i,j}(n)\} = (D/\Delta^2)\{x_{i-1,j}(n) + x_{i+1,j}(n) + x_{i,j-1}(n) + x_{i,j+1}(n) - 4x_{i,j}(n)\}$ is spatial diffusion-like operator. Permanence of the system is investigated. Stability research uses approach of Lyapunov functions. Numerical simulations are used in order to investigate qualitative behavior when changing the value of time delay $r \in \mathbb{N}$ and diffusion D/Δ^2. It was shown that when increasing the value of time delay r, we transit from steady state through Hopf bifurcation, increasing period and finally to chaotic behavior. The increase of diffusion causes an appearance of chaotic solutions also.

Keywords Finite lattice difference equations · Permanence · Global attractivity · Hopf bifurcation · Chaos

V. Martsenyuk (✉) · A. Klos-Witkowska
Department of Computer Science and Automatics, University of Bielsko-Biala, 2 Willowa, 43-309 Bielsko-Biala, Poland
e-mail: vmartsenyuk@ath.bielsko.pl

A. Klos-Witkowska
e-mail: awitkowska@ath.bielsko.pl

A. Sverstiuk
Department of Medical Informatics, Ternopil State Medical University, 1 Voli, Ternopil 46001, Ukraine
e-mail: sverstyuk@tdmu.edu.ua

M. Bohner et al. (eds.), *Difference Equations and Discrete Dynamical Systems with Applications*, Springer Proceedings in Mathematics & Statistics 312,
https://doi.org/10.1007/978-3-030-35502-9_13

1 Introduction

One of the most important current research is related with design of sensor devices. They are considered as cornerstones of Industry 4.0 [2, 20] relying primarily on cyber-physical systems, which include smart sensors with supervision and communication purposes. A lot of modern applications in biology, medicine, ecology, food industry are dealt with biosensors. They are kind of sensors aimed for measurements of biological substances. The design of biosensor devices includes estimation of parameters enabling us their stable functioning.

The most of models describing biosensors uses partial differential equations. However it does not take into account discrete nature of spatial coordinates of the biosensor, which is based on two- or three-dimensional biopixels array. That is why in series of works (see [15, 19]) finite lattice differential equations were offered to describe biosensor devices.

Differential equations which are used in the biosensors modelling are based on population dynamics for describing different biological species interaction. For example, in case of immunosensors, which are kind of biosensors, we use antigens and antibodies, which play roles of preys and predators respectively. In such a way we result in well-known predator-prey differential equations in the biosensor modelling.

A lot of results for the predator-prey models with discrete or distributed time delays has been obtained. As it was mentioned in [13], the problem is that the most of these results are related with the continuous-time systems but discrete-time modeling is more appropriate in cases "when populations have a short life expectancy, nonoverlapping generations in real world".

Let $x_{i,j}(t)$ be concentration of antigens, $y_{i,j}(t)$ be concentration of antibodies in biopixel (i, j), $i, j = \overline{1, N}$.

The model is based on biological assumption for arbitrary biopixel (i, j), which are described in [15]. It includes the following constant parameters: birthrate for antigen population, $\beta > 0$; death rate of antigens, δ_x; probability rate of neutralization of antigens by antibodies, $\gamma > 0$; birthrate of antibodies, μ_y; death rate of antibodies, δ_y; probability rate of immune response with respect to antibodies, η. Immune response appears with some constant time delay $\tau > 0$. For the sake of simplicity we assume the same delay τ for the death rate of antigens. We have some diffusion of antibodies from four neighboring pixels $(i-1, j)$, $(i+1, j)$, $(i, j-1)$, $(i, j+1)$. The complete biological reasoning and description of the model is presented in [15].

So we start from considering a very simple delayed antibody-antigen competition model for biopixels two-dimensional array which was offered and investigated in [15]

$$
\begin{aligned}
\frac{dx_{i,j}(t)}{dt} &= (\beta - \gamma y_{i,j}(t-\tau) - \delta_x x_{i,j}(t-\tau))x_{i,j}(t) + \hat{S}\{x_{i,j}\}, \\
\frac{dy_{i,j}(t)}{dt} &= \Big(- \mu_y + \eta\gamma x_{i,j}(t-\tau) - \delta_y y_{i,j}(t) \Big) y_{i,j}(t)
\end{aligned}
\tag{1}
$$

with given initial functions

$$x_{i,j}(t) = x^0_{i,j}(t) \geq 0, \quad y_{i,j}(t) = y^0_{i,j}(t) \geq 0, \quad t \in [-\tau, 0),$$
$$x_{i,j}(0), y_{i,j}(0) > 0. \tag{2}$$

For a square $N \times N$ array of traps, we use the following discrete diffusion form of the spatial operator, which was already applied for modelling biosensors (see [19])

$$\hat{S}\{x_{i,j}\} = D\Delta^{-2}\left[x_{i-1,j} + x_{i+1,j} + x_{i,j-1} + x_{i,j+1} - 4x_{i,j}\right] \quad i, j = \overline{1, N} \tag{3}$$

Each colony is affected by the antigen produced in four neighboring colonies, two in each dimension of the array, separated by the equal distance Δ.

We use the boundary condition $x_{i,j} = 0$ for the edges of the array $i, j = 0, N + 1$.

The techniques, which are used in the work for discretization, permanence and stability investigation are primarily based on the approach developed in [13] for predator-prey system. Here they were extended in case of finite lattice model with diffusion.

The paper is structured in the following way. In the Sect. 2 we employ the discretization technique to derive the discrete version of system (1). The conditions for quasi-permanence are investigated in Sect. 3. Global stability research is presented in Sect. 4. Results of modeling (1) are displayed in Sect. 5. It can be seen that qualitative behavior of the system is determined mostly by the time of immune response τ (or time delay) and diffusion rate D.

Within this paper we use the following notation:

- the symbol $i = \overline{m, n}$ for some integer i, m, n , $m < n$ means $i = m, m + 1, ..., n$;
- $\lfloor x \rfloor$ denotes the greatest integer less than or equal to the real value x;
- $a^u = \sup_{n \in \mathbb{N}} a(n)$ and $a^l = \inf_{n \in \mathbb{N}} a(n)$ for any bounded sequence $\{a(n)\}$;
- $\mathbb{R}^+$ denotes the set of nonnegative real numbers;
- $\mathbb{N}$ be the sets of nonnegative integers;
- $\otimes$ denotes the direct product of matrices.

2 Deriving the Difference Equations Model

The system (1) without diffusion is approximated by the following differential equations with piecewise constant arguments

$$\frac{dx_{i,j}(t)}{dt} = \left(\beta - \gamma y_{i,j}(\lfloor t/h \rfloor h - \lfloor \tau/h \rfloor h) - \delta_x x_{i,j}(\lfloor t/h \rfloor h - \lfloor \tau/h \rfloor h)\right) x_{i,j}(t),$$
$$\frac{dy_{i,j}(t)}{dt} = \left(-\mu_y + \eta\gamma x_{i,j}(\lfloor t/h \rfloor h - \lfloor \tau/h \rfloor h) - \delta_y y_{i,j}(\lfloor t/h \rfloor h)\right) y_{i,j}(t) \tag{4}$$

for $t \in [nh, (n + 1)h), n \in \mathbb{N}$.

Noting that $\lfloor t/h \rfloor = n$, $\lfloor \tau/h \rfloor = r \in \mathbb{N}$, we integrate (4) over $[nh, t)$, where $t < (n+1)h$, then (4) can be reformulated as

$$\frac{dx_{i,j}(t)}{dt} = \Big(\beta - \gamma y_{i,j}(nh - rh) - \delta_x x_{i,j}(nh - rh)\Big)x_{i,j}(t),$$
$$\frac{dy_{i,j}(t)}{dt} = \Big(-\mu_y + \eta\gamma x_{i,j}(nh - rh) - \delta_y y_{i,j}(nh)\Big)y_{i,j}(t).$$

Denoting $x_{i,j}(n) = x_{i,j}(nh)$, $y_{i,j}(n) = y_{i,j}(nh)$, then we have

$$\begin{aligned} x_{i,j}(t) &= x_{i,j}(n)\exp\Big\{\beta - \gamma y_{i,j}(n-r) - \delta_x x_{i,j}(n-r)\Big\}, \\ y_{i,j}(t) &= y_{i,j}(n)\exp\Big\{-\mu_y + \eta\gamma x_{i,j}(n-r) - \delta_y y_{i,j}(n)\Big\}. \end{aligned} \tag{5}$$

Setting $t \to (n+1)h$ in (5) and simplifying, adding diffusion to the first equation,[1] we get a discrete analogue of continuous time system (1) with the form

$$\begin{aligned} x_{i,j}(n+1) &= x_{i,j}(n)\exp\Big\{\beta - \gamma y_{i,j}(n-r) - \delta_x x_{i,j}(n-r)\Big\} + \hat{S}\Big\{x_{i,j}(n)\Big\}, \\ y_{i,j}(n+1) &= y_{i,j}(n)\exp\Big\{-\mu_y + \eta\gamma x_{i,j}(n-r) - \delta_y y_{i,j}(n)\Big\}, \end{aligned} \tag{6}$$

We pay attention that behavior of the system (6) may not be the same as for the differential one (1). The equivalence of differential versus difference equations trajectories, which are obtained with help of forward Euler, backward Euler or central difference schemes, may be for "sufficiently small discretization time steps" only [12]. The problems of equivalence of difference and differential Lotka–Volterra equations were firstly studied in [18]. The derivation of the discrete model (6) is the standard connection between Lotka–Volterra continuous time model and discrete time Nicholson–Bailey model [18]. The Nicholson–Bailey model was derived in order to have similar dynamical properties as Lotka–Volterra ones.

In [16] nonstandard Mickens scheme of time-discretization was described, which was further used in order to get dynamical consistency between discrete-time and continuous-time models in a lot of research and applications [17].

With respect to time-discretization of lattice reaction-diffusion equations we have a wide range of research—in the one-dimensional lattice [5, 6], higher dimensional lattices [9], papers dealing with the exact role of the time discretization on the existence of travelling waves [10] or validity of maximum and comparison principles [23] which is the main reason which led authors to the alternative discretization. There are also numerous contributions in the case of delayed lattice differential equations, for example [14, 25].

[1] Diffusion term is considered be additive in order to get clear permanence and stability results. Actually the diffusion on discrete space may be represented by a matrix multiplication also [4].

In Sect. 5 we investigate the problem of qualitative consistency between (6) and (1) numerically.

We introduce the following definitions for finite lattice difference equations (6).

Definition 1 It is said that system (6) is **quasi-permanent** if there exist positive constants m_x, M_x, $m_{y,i,j}$, $M_{y,i,j}$, $i, j = \overline{1, N}$ that every positive solution $\left\{\left(x_{i,j}(n), y_{i,j}(n)\right)\right\}$, $i, j = \overline{1, N}$ of system (6) satisfies

$$m_x \leq \liminf_{n\to\infty} \sum_{i,j=1}^{N} x_{i,j}(n) \leq \limsup_{n\to\infty} \sum_{i,j=1}^{N} x_{i,j}(n) \leq M_x,$$

$$m_{y,i,j} \leq \liminf_{n\to\infty} y_{i,j}(n) \leq \limsup_{n\to\infty} y_{i,j}(n) \leq M_{y,i,j}.$$

Definition 2 A positive solution $\left\{\left(x^{\star}_{i,j}(n), y^{\star}_{i,j}(n)\right)\right\}$, $i, j = \overline{1, N}$ of system (6) is **globally attractive** if each other positive solution $\left\{\left(x_{i,j}(n), y_{i,j}(n)\right)\right\}$, $i, j = \overline{1, N}$ of system (6) satisfies

$$\lim_{n\to\infty} |x_{i,j}(n) - x^{\star}_{i,j}(n)| = 0, \quad \lim_{n\to\infty} |y_{i,j}(n) - y^{\star}_{i,j}(n)| = 0, \quad i, j = \overline{1, N}.$$

3 Permanence

In spite of the fact that a series of results were obtained when considering permanence of Nicholson models without diffusion, e.g. recent ones are [7, 13, 26], much more less permanence results were established for discrete reaction-diffusion models. Here we mention work [1] for on one- and two-dimensional lattices. These results were approved numerically in [3].

Here we have already introduced the notion of quasi-permanence of the system (6) which is "weaker" as compared with traditional permanence. The reason is the taking into account diffusion of $x_{i,j}$ within the finite lattice. In turn, the system is permanent with respect to $y_{i,j}(n)$ in a traditional sense.

In order to prove quasi-permanence of the system (6), we need the following auxiliary results from [26].

Lemma 1 *It holds*

$$\max_{x\in\mathbb{R}} x \exp\left(\beta(1-x)\right) = \frac{\exp\left(\beta - 1\right)}{\beta} \tag{7}$$

for $\beta > 0$.

Lemma 2 *Assume that* $x(n)$ *satisfies* $x(n) > 0$ *and*

$$x(n+1) \leq x(n)\exp\{s(n)(1-ax(n))\} \tag{8}$$

for $n \in [n_1, \infty)$, *where* a *is a positive constant. Then*

$$\lim_{n\to\infty} \sup x(n) \leq \frac{1}{as^u}\exp\left\{s^u - 1\right\}. \tag{9}$$

Lemma 3 *Assume that* $\{x(n)\}$ *satisfies*

$$x(n+1) \geq x(n)\exp\{s(n)(1-ax(n))\}, \quad n \geq N_0, \tag{10}$$

$\lim_{n\to\infty} \sup x(n) \leq x^u$ *and* $x(N_0) > 0$, *where* a *is a constant such that* $ax^u > 1$ *and* $N_0 \in \mathbb{N}$. *Then*

$$\lim_{n\to\infty} \inf x(n) \geq \frac{1}{a}\exp\left\{s^u(1-ax^u)\right\}. \tag{11}$$

The next auxiliary result is related to necessary condition to the positive invariance of the positive orthant

$$\Omega = \left\{\left(x_{i,j}(n), y_{i,j}(n)\right) : i, j = \overline{1, N}, x_{i,j}(n) > 0, y_{i,j}(n) > 0\right\}$$

Lemma 4 *Assume that the positive orthant* Ω *is positive invariant for the system (6), i.e.* $x_{i,j}(0) > 0$, $y_{i,j}(0) > 0$ *implies* $x_{i,j}(n) > 0$, $y_{i,j}(n) > 0$, $n \in \mathbb{N}$, $i, j = \overline{1, N}$.
Then

$$e^{\beta} > \frac{4D}{\Delta^2} \tag{12}$$

holds.

Proof We assume for purposes of contradiction that $e^{\beta} \leq \frac{4D}{\Delta^2}$. Consider a counterexample, if $N = 1$. Then the first equation (6) together with the boundary conditions $x_{0,1} = x_{2,1} = x_{1,0} = x_{1,2} = 0 = y_{0,1} = y_{2,1} = y_{1,0} = y_{1,2}$ introduced above, yields

$$\begin{aligned} x_{1,1}(n+1) &= x_{1,1}(n)\exp\left\{\beta - \gamma y_{1,1}(n-r) - \delta_x x_{1,1}(n-r)\right\} - (4D/\Delta^2)x_{1,1}(n) \\ &\leq x_{1,1}(n)(e^{\beta} - 4D/\Delta^2) \leq 0 \end{aligned}$$

for any $x_{1,1}(0) > 0$, $y_{1,1}(0) > 0$, contradicting the original supposition.

The next result introduces a sufficient condition for the underlying grid size ensuring that the solution of (6) is non-vanishing.

Lemma 5 *Let for the system (6) the positive orthant* Ω *be positive invariant. Besides that, let* N *be such that* $f_{extnc}(N) < 1$ *holds, where*

$$f_{extnc}(N) = \max_{k,l=\overline{1,N}} \left| e^{\beta} - \frac{4D}{\Delta^2}\left(1 + \cos\frac{\pi(k+l)}{2(N+1)}\cos\frac{\pi(k-l)}{2(N+1)}\right)\right|. \tag{13}$$

Then $\lim_{n\to\infty} x_{i,j}(n) = 0,\ i,\ j = \overline{1, N}$.

Proof It requires a comparison principle for difference equations (see [24], Theorem 2.1, p. 241).

The following inequalities hold for $x_{i,j}(n+1)$

$$x_{i,j}(n+1) < x_{i,j}(n)e^{\beta} + \hat{S}\left\{x_{i,j}(n)\right\}.$$

Consider N^2-vector of the form

$$X(n) = \Big(x_{1,1}(n), x_{1,2}(n), \ldots, x_{1,N}(n), x_{2,1}(n), \ldots, x_{2,N}(n), \ldots, x_{N,1}(n), \ldots, x_{N,N}(n)\Big)^{\top}.$$

We compare $X(n+1) \leq CX(n)$, where $C = I_N \otimes A + B \otimes I_N$,

$$A = \begin{bmatrix} e^{\beta} - \frac{4D}{\Delta^2} & \frac{D}{\Delta^2} & & & & \\ \frac{D}{\Delta^2} & e^{\beta} - \frac{4D}{\Delta^2} & \frac{D}{\Delta^2} & & & \\ & \frac{D}{\Delta^2} & \ddots & & & \\ & & & \ddots & & \\ & & & & e^{\beta} - \frac{4D}{\Delta^2} & \frac{D}{\Delta^2} \\ & & & & \frac{D}{\Delta^2} & e^{\beta} - \frac{4D}{\Delta^2} \end{bmatrix} \in \mathbb{R}^{N\times N},$$

$$B = \begin{bmatrix} 0 & \frac{D}{\Delta^2} & & & & \\ \frac{D}{\Delta^2} & 0 & \frac{D}{\Delta^2} & & & \\ & \frac{D}{\Delta^2} & \ddots & & & \\ & & & \ddots & & \\ & & & & 0 & \frac{D}{\Delta^2} \\ & & & & \frac{D}{\Delta^2} & 0 \end{bmatrix} \in \mathbb{R}^{N\times N},$$

I_N is $N \times N$ identity matrix. The N^2 eigenvalues of C are of the form (see [11], Theorem 8.3.1) $\lambda_{k,l}(C) = \lambda_k(A) + \lambda_l(B),\ k, l = \overline{1, N}$, where the eigenvalues of A

$$\lambda_k(A) = e^{\beta} - \frac{4D}{\Delta^2} - \frac{2D}{\Delta^2}\cos\left(\pi k/(N+1)\right), \quad k = \overline{1, N},$$

the eigenvalues of B

$$\lambda_l(B) = -\frac{2D}{\Delta^2}\cos\left(\pi l/(N+1)\right), \quad l = \overline{1, N}.$$

The comparison system $Z(n+1) = CZ(n)$ tends asymptotically to zero if $\left|\lambda_{k,l}\right| < 1$. That is,

$$\max_{k,l=\overline{1,N}} \left| e^{\beta} - \frac{4D}{\Delta^2} - \frac{2D}{\Delta^2}\left(\cos\frac{\pi k}{N+1} + \cos\frac{\pi l}{N+1} \right) \right| < 1.$$

Theorem 1 *Consider the system (6) satisfying the positive invariance of the set* Ω.[2] *Let*

$$\begin{aligned} \alpha_{1,i,j} &= -\mu_y + \eta\gamma M_{x,i,j}\frac{\exp(\beta-1)}{\delta_x}, \\ \alpha_{2,i,j} &= \beta - \gamma\frac{\exp(\alpha_{1,i,j}-1)}{\delta_y}, \\ \alpha_{3,i,j} &= -\mu_y + \frac{\eta\gamma}{\delta_x}\exp(\alpha_{2,i,j}(1-\delta_x\zeta_{i,j})), \end{aligned} \tag{14}$$

where $\zeta_{i,j}$, $i, j = \overline{1, N}$ *are some constants.*

If there exist $M_{x,i,j}(r) > 1$, $i, j = \overline{1, N}$ *such that for any* $\zeta_{i,j} > 0$, $i, j = \overline{1, N}$ *conditions*

$$\min\left\{\alpha_{k,i,j}, k = \overline{1,3}, i, j = \overline{1,N}\right\} > 0 \tag{15}$$

hold, then system (6) is quasi-permanent.

Proof Since Ω is a positive invariant set of (6), we assume that $\left\{\left(x_{i,j}(n), y_{i,j}(n)\right),\right.$ $\left. i, j = \overline{1, N}\right\}$ is its arbitrary positive solution.

Firstly we prove that $\left\{\left(x_{i,j}(n), y_{i,j}(n)\right), i, j = \overline{1, N}\right\}$ is uniformly upper bounded. Consider the first equation of (6). We get

$$\begin{aligned} x_{i,j}(n+1) &\le x_{i,j}(n)\exp\left\{\beta - \delta_x x_{i,j}(n-r)\right\} + \hat{S}\left\{x_{i,j}(n)\right\} \\ &= x_{i,j}(n)\exp\left\{\beta\left(1 - \frac{\delta_x}{\beta}x_{i,j}(n-r)\right)\right\} + \hat{S}\left\{x_{i,j}(n)\right\}. \end{aligned} \tag{16}$$

Let $n_{1,i,j}(r) \in \mathbb{N}$ be such that

$$\frac{x_{i,j}(n)}{x_{i,j}(n-r)} < M_{x,i,j}, \quad n > n_1.$$

Applying Lemma 1 it implies for $n > n_{1,i,j}(r)$

[2]Lemma 4 offers necessary condition (12).

$$
\begin{aligned}
x_{i,j}(n+1) &\le M_{x,i,j}(r)x_{i,j}(n-r)\exp\left\{\beta\left(1-\frac{\delta_x}{\beta}x_{i,j}(n-r)\right)\right\}+\hat{S}\left\{x_{i,j}(n)\right\}\\
&\le M_{x,i,j}(r)\frac{\beta}{\delta_x}\frac{\exp(\beta-1)}{\beta}+\hat{S}\left\{x_{i,j}(n)\right\}\\
&= M_{x,i,j}(r)\frac{\exp(\beta-1)}{\delta_x}+\hat{S}\left\{x_{i,j}(n)\right\}.
\end{aligned}
$$

Hence, for $n > n_1(r) = \max_{i,j=\overline{1,N}} n_{1,i,j}(r)$ we have

$$
\limsup_{n\to\infty}\sum_{i,j=1}^{N} x_{i,j}(n) \le \frac{\exp(\beta-1)}{\delta_x}\sum_{i,j=1}^{N} M_{x,i,j}(r) =: M_x(r).
$$

Moreover, there exists a sufficiently large $n_2(\epsilon)\in\mathbb{N}$, that for any constant $\epsilon > 0$ it holds[3]

$$
x_{i,j}(n) \le M_{x,i,j}(r)\frac{\exp(\beta-1)}{\delta_x}+\epsilon, \quad n \ge n_2(\epsilon). \tag{17}
$$

Consider the second equation of (6). Hence we get

$$
\begin{aligned}
y_{i,j}(n+1) \le& y_{i,j}(n)\exp\left\{-\mu_y+\eta\gamma[M_{x,i,j}(r)\frac{\exp(\beta-1)}{\delta_x}+\epsilon]-\delta_y y_{i,j}(n)\right\}\\
=& y_{i,j}(n)\exp\left\{\alpha^{\epsilon}_{x,i,j}\left(1-\frac{\delta_y}{\alpha^{\epsilon}_{x,i,j}}y_{i,j}(n)\right)\right\}.
\end{aligned}
$$

Here

$$
\alpha^{\epsilon}_{1,i,j} := -\mu_y+\eta\gamma\left[M_{x,i,j}(r)\frac{\exp(\beta-1)}{\delta_x}+\epsilon\right]
$$

Applying Lemma 2 and letting $\epsilon \to 0$, we have

$$
\limsup_{n\to\infty} y_{i,j}(n) \le \frac{\exp(\alpha^{0}_{1,i,j}-1)}{\delta_y} =: F^{u}_{i,j}. \tag{18}
$$

---

[3]In order to substantiate it, we assume the contrary, namely, there are $\epsilon_1 > 0$ and $i^\star, j^\star \in \overline{1,N}$ such that $x_{i^\star,j^\star}(n) > M_{x,i^\star,j^\star}(r)\frac{\exp(\beta-1)}{\delta_x}+\epsilon_1$, for all $n > 0$. Then

$$
\begin{aligned}
\limsup_{n\to\infty}\sum_{i,j=1}^{N} x_{i,j}(n) &\le \frac{\exp(\beta-1)}{\delta_x}\sum_{i,j=1}^{N} M_{x,i,j}(r) < \frac{\exp(\beta-1)}{\delta_x}\sum_{i,j=1,i\ne i^\star,j\ne j^\star}^{N} M_{x,i,j}(r)\\
&+x_{i^\star,j^\star}-\epsilon_1 \le \frac{\exp(\beta-1)}{\delta_x}\sum_{i,j=1}^{N} M_{x,i,j}(r)+\hat{S}\left\{x_{i^\star,j^\star}(n-1)\right\}-\epsilon_1,
\end{aligned}
$$

which is a contradiction at $n \to \infty$.

Hence it follows that

$$\lim_{n\to\infty}\sup\sum_{i,j=1}^{N} y_{i,j}(n) \le \sum_{i,j=1}^{N} \frac{\exp(\alpha_{1,i,j}^{0}-1)}{\delta_y} =: M_y.$$

Further we prove that $\left\{(x_{i,j}, y_{i,j}), i, j = \overline{1, N}\right\}$ is uniformly ultimately lower bounded.

According to (18), there exists an $n_3(\epsilon) > n_2(\epsilon)$ such that

$$y_{i,j}(n) \le \frac{\exp(\alpha_{1,i,j}^{0}-1)}{\delta_y} + \epsilon$$

for $n > n_3$ and constant ϵ determined above.

Due to the first equation of (6) we have

$$x_{i,j}(n+1) \ge x_{i,j}(n)\exp\left\{\beta - \gamma\left(\frac{\exp(\alpha_{1,i,j}^{0}-1)}{\delta_y} + \epsilon\right) - \delta_x x_{i,j}(n-r)\right\}.$$

We have to differ two cases. *Case 1*. There exists $n_4(\epsilon) > n_3(\epsilon)$ such that $x_{i,j}(n) < x_{i,j}(n-r), n > n_4$. It implies

$$x_{i,j}(n+1) \ge x_{i,j}(n)\exp\left\{\alpha_{2,i,j}^{\epsilon}(1 - \delta_x x_{i,j}(n)\right\}.$$

Here

$$\alpha_{2,i,j}^{\epsilon} := \beta - \gamma\left(\frac{\exp(\alpha_{1,i,j}^{0}-1)}{\delta_y} + \epsilon\right).$$

Due to Lemma 3 we get

$$x_{i,j}(n) \ge \frac{1}{\delta_x}\exp\left\{\alpha_{2,i,j}^{\epsilon}(1 - \delta_x\left(M_{x,i,j}(r)\frac{\exp(\beta-1)}{\delta_x} + \epsilon\right)\right\}$$

Let $\epsilon \to 0$. It implies

$$\lim_{n\to\infty}\inf x_{i,j}(n) \ge \frac{1}{\delta_x}\exp\left\{\alpha_{2,i,j}^{0}\left(1 - \delta_x M_{x,i,j}(r)\frac{\exp(\beta-1)}{\delta_x}\right)\right\}$$

Case 2. For all $n > n_2(\epsilon)$ we have $x_{i,j}(n) \ge x_{i,j}(n-r)$. It follows that $\lim_{n\to\infty} x_{i,j}(n) = x_{i,j}^{u}$ exists, where $x_{i,j}^{u} := x_{i,j}(0)\exp\beta$. On the other hand $x_{i,j}^{u} \ge \frac{1}{\delta_x}$. It follows that

$$\lim_{n\to\infty}\inf x_{i,j}(n) \ge \frac{1}{\delta_x}\exp\left\{\alpha_{2,i,j}(1 - \delta_x x_{i,j}^{u})\right\}$$

Considering the second equation of (6), we get

$$y_{i,j}(n+1) \geq y_{i,j}(n) \exp\left\{\kappa_{i,j}\left(1 - \frac{\delta_y}{\kappa_{i,j}} y_{i,j}(n)\right)\right\}.$$

Here

$$\kappa_{i,j} := -\mu_y + \frac{\eta\gamma}{\delta_x} \exp\left\{\alpha^0_{2,i,j}(1 - \delta_x x^u_{i,j})\right\}.$$

Further we use the following inequality[4]

$$\frac{\delta_y}{\kappa_{i,j}} y^u_{i,j} = \frac{\exp(\alpha^0_{1,i,j}) - 1}{\kappa_{i,j}} \geq \frac{\exp(\alpha^0_{1,i,j}) - 1}{\alpha^0_{1,i,j}} > 1.$$

Applying Lemma 3, we have

$$\liminf_{n\to\infty} y_{i,j}(n) \geq \frac{\kappa_{i,j}}{\delta_y} \exp\left\{\kappa_{i,j}\left(1 - \frac{\delta_y}{\kappa_{i,j}} y^u_{i,j}\right)\right\}.$$

4 Stability Investigation for the Finite Lattice Difference Model of Immunosensor

4.1 Steady States

The complex topology of set of endemic steady states for lattice Nagumo reaction-diffusion dynamical systems were studied in [6]. Moreover, when considering Nagumo equation on graphs, in [22] they observed that for sufficiently strong reactions (or sufficiently weak diffusion) there are exponentional growth of the number of endemic steady states.

In general case steady state $\mathcal{E}_{i,j} \equiv \left(x_{i,j}, y_{i,j}\right)$, $i, j = \overline{1, N}$ for difference system (6) can be found as a result of solution of the algebraic system:

$$\begin{aligned} x_{i,j} &= x_{i,j} \exp\left\{\beta - \gamma y_{i,j} - \delta_x x_{i,j}\right\} + \hat{S}\left\{x_{i,j}\right\}, \\ y_{i,j} &= y_{i,j} \exp\left\{-\mu_y + \eta\gamma x_{i,j} - \delta_y y_{i,j}\right\}, \end{aligned} \tag{19}$$

with respect to $(x_{i,j}, y_{i,j})$, $i, j = \overline{1, N}$. We have to distinguish the following cases.

[4]Here we use that $\frac{1}{x}\exp(x-1) > 1$ for $x > 0$.

Antigen and antibody-free steady state $\mathcal{E}^{0,0}_{i,j} \equiv \mathcal{E}^{0,0} = \left(0, 0\right), i, j = \overline{1, N}$.

Antibody-free endemic[5] **steady state**. The system (6) has so-called antibody-free steady state, namely

$$\mathcal{E}^{*,0}_{i,j} \equiv \mathcal{E}^{*,0} = \left(\frac{\beta}{\delta_x}, 0\right), \quad i, j = \overline{1, N}$$

Identical endemic steady state. In case if $x_{i,j} \equiv x > 0$, $i, j = \overline{1, N}$ (it yields $\hat{S}\left\{x_{i,j}\right\} \equiv 0$) we get steady state $\mathcal{E}_{i,j} \equiv \mathcal{E}^{idnt} = \left(x^{idnt}, y^{idnt}\right)$, where

$$x^{idnt} = \frac{\beta\delta_y + \gamma\mu_y}{\eta\gamma^2 + \delta_x\delta_y}, \quad y^{idnt} = \frac{-\mu_y\delta_x + \eta\gamma\beta}{\eta\gamma^2 + \delta_x\delta_y}.$$

We see that if $-\mu_y\delta_x + \eta\gamma\beta > 0$, then $\mathcal{E}^{idnt}$ is endemic.

Nonidentical endemic steady state. In general case we need to solve the algebraic system (19) to find endemic steady state, which we call here as nonidentical steady state $\mathcal{E}^{nonidnt} = \left(x^{nonidnt}_{i,j}, y^{nonidnt}_{i,j}\right)$, $i, j = \overline{1, N}$. In case if all $(x^{nonidnt}_{i,j}, y^{nonidnt}_{i,j}) > 0$, then $\mathcal{E}^{nonidnt}$ is endemic. We note that the values of x^{idnt} and y^{idnt} can be used as initial approximations for numerical methods to solve nonlinear algebraic system (19).

4.2 Global Attractivity

In [21, 28] there were studied the global attractivity of the positive equilibrium of the discrete Nicholsons model. Global attractivity of Nicholson's differential equation with continuous diffusion was investigated in [27] with help of maximum principle. In case of continuous-time reaction-diffusion model from $\mathbb{R}^n$ on graphs, it was shown that at some parameters there are 3^n stationary solutions, out of which 2^n are asymptotically stable [22].

It is natural to expect a possible global attractivity of large number of positive solutions for difference model (6). The next result offers the sufficient conditions of global attractivity, which were obtained with help of Lyapunov functions.

Theorem 2 *Assume that conditions of the Theorem 1 hold and there exists a positive constant ξ such that*

[5]Here we use epidemiological term "endemic" meaning the state when the "infection" (in this context, antigen) is constantly maintained at a baseline level in an area without external inputs.

$$\begin{aligned} &\exp\left\{\gamma m_y + \delta_x m_x - \beta\right\} - \delta_x - \frac{1}{m_x} - \eta\gamma \geq \xi, \\ &\min\left\{\delta_y, \frac{2}{M_y} - \delta_y\right\} - \gamma \geq \xi. \end{aligned} \tag{20}$$

Then any positive solution $\left\{\left(x^{\star}_{i,j}(n), y^{\star}_{i,j}(n)\right), i, j = \overline{1, N}\right\}$ *of system (6) is globally attractive.*

Proof Consider $\left\{\left(x_{i,j}(n), y_{i,j}(n)\right), i, j = \overline{1, N}\right\}$ is arbitrary positive solution of system (6). Let

$$V_{1,1,i,j}(n) = |\ln\left(x_{i,j}(n) - \hat{S}\left\{x_{i,j}(n-1)\right\}\right) - \ln\left(x^{\star}_{i,j}(n) - \hat{S}\left\{x^{\star}_{i,j}(n-1)\right\}\right)|$$

Then it follows from the first equation of (6) that

$$\begin{aligned} V_{1,1,,i,j} \leq & |\ln x_{i,j}(n) - \ln x^{\star}_{i,j}(n)| + \gamma|y_{i,j}(n-r) - y^{\star}_{i,j}(n-r)| \\ & + \delta_v|x_{i,j}(n-r) - x^{\star}_{i,j}(n-r)|. \end{aligned} \tag{21}$$

By the Mean Value theorem, we get

$$\ln x_{i,j}(n) - \ln x^{\star}_{i,j}(n) = \frac{1}{\theta_1(n)}(x_{i,j}(n) - x^{\star}_{i,j}(n)),$$

where $\theta_1(n)$ lies between $x_{i,j}(n)$ and $x^{\star}_{i,j}(n)$,

$$\begin{aligned} &\ln(x_{i,j}(n) - \hat{S}\left\{x_{i,j}(n-1)\right\}) - \ln(x^{\star}_{i,j}(n) - \hat{S}\left\{x^{\star}_{i,j}(n-1)\right\}) \\ = &\frac{1}{\theta_2(n)}((x_{i,j}(n) - x^{\star}_{i,j}(n)) - (\hat{S}\left\{x_{i,j}(n-1)\right\} - \hat{S}\left\{x^{\star}_{i,j}(n-1)\right\})), \end{aligned}$$

where $\theta_2(n)$ lies between $x_{i,j}(n) - \hat{S}\left\{x_{i,j}(n-1)\right\}$ and $x^{\star}_{i,j}(n) - \hat{S}\left\{x^{\star}_{i,j}(n-1)\right\}$.

We consider

$$\begin{aligned} &|\ln x_{i,j}(n) - \ln x^{\star}_{i,j}(n)| \\ &= |\ln(x_{i,j}(n) - \hat{S}\left\{x_{i,j}(n-1)\right\}) - \ln(x^{\star}_{i,j}(n) - \hat{S}\left\{x^{\star}_{i,j}(n-1)\right\})| \\ &- |\ln(x_{i,j}(n) - \hat{S}\left\{x_{i,j}(n-1)\right\}) - \ln(x^{\star}_{i,j}(n) - \hat{S}\left\{x^{\star}_{i,j}(n-1)\right\})| \\ &+ |\ln x_{i,j}(n) - \ln x^{\star}_{i,j}(n)| \\ &\geq V_{1,1,i,j}(n) - \left(\frac{1}{\theta_2(n)} - \frac{1}{\theta_1(n)}\right)|x_{i,j}(n) - x^{\star}_{i,j}(n)| \\ &- \frac{1}{\theta_2(n)}\left(\hat{S}\left\{x_{i,j}(n-1)\right\} - \hat{S}\left\{x^{\star}_{i,j}(n-1)\right\}\right). \end{aligned} \tag{22}$$

Combining (21) and (22), we have

$$\begin{aligned}\Delta V_{1,1,i,j}(n) &= V_{1,1,i,j}(n+1) - V_{1,1,i,j}(n)\\ &\le -\left(\frac{1}{\theta_2(n)} - \frac{1}{\theta_1(n)}\right)|x_{i,j}(n) - x^{\star}_{i,j}(n)|\\ &+ \gamma|y_{i,j}(n-r) - y^{\star}_{i,j}(n-r)|\\ &+ \delta_x|x_{i,j}(n-r) - x^{\star}_{i,j}(n-r)|\\ &- \frac{1}{\theta_2(n)}\left(\hat{S}\left\{x_{i,j}(n-1)\right\} - \hat{S}\left\{x^{\star}_{i,j}(n-1)\right\}\right).\end{aligned} \tag{23}$$

Next, we let

$$V_{1,2,i,j}(n) = \sum_{s=n-r}^{n-1} \delta_v|x_{i,j}(s) - x^{\star}_{i,j}(s)| + \sum_{s=n-r}^{n-1} \gamma|y_{i,j}(s) - y^{\star}_{i,j}(s)|$$

Then we have

$$\begin{aligned}\Delta V_{1,2,i,j}(n) &= V_{1,2,i,j}(n+1) - V_{1,2,i,j}(n)\\ &= \sum_{s=n+1-r}^{n} \delta_x|x_{i,j}(s) - x^{\star}_{i,j}(s)| + \sum_{s=n+1-r}^{n} \gamma|y_{i,j}(s) - y^{\star}_{i,j}(s)|\\ &- \sum_{s=n-r}^{n-1} \delta_x|x_{i,j}(s) - x^{\star}_{i,j}(s)| - \sum_{s=n-r}^{n-1} \gamma|y_{i,j}(s) - y^{\star}_{i,j}(s)|\\ &= \delta_x|x_{i,j}(n) - x^{\star}_{i,j}(n)| - \delta_x|x_{i,j}(n-r) - x^{\star}_{i,j}(n-r)|\\ &+ \gamma|y_{i,j}(n) - y^{\star}_{i,j}(n)| - \gamma|y_{i,j}(n-r) - y^{\star}_{i,j}(n-r)|.\end{aligned} \tag{24}$$

We let $W_{1,i,j} = V_{1,1,i,j}(n) + V_{1,2,i,j}(n)$. Then it follows from (23) and (24) that

$$\begin{aligned}\Delta V_{1,i,j}(n) &= \Delta V_{1,1,i,j}(n) + \Delta V_{1,2,i,j}(n)\\ &\le \left(\delta_x - \frac{1}{\theta_2(n)} + \frac{1}{\theta_1(n)}\right)|x_{i,j}(n) - x^{\star}_{i,j}(n)|\\ &+ \gamma|y_{i,j}(n) - y^{\star}_{i,j}(n)|\\ &- \frac{1}{\theta_2(n)}\left(\hat{S}\left\{x_{i,j}(n-1)\right\} - \hat{S}\left\{x^{\star}_{i,j}(n-1)\right\}\right)\end{aligned} \tag{25}$$

In a similar way we define for the second equation of (6)

$$V_{2,i,j}(n) = V_{2,1,i,j}(n) + V_{2,2,i,j}(n),$$

where

$$V_{2,1,i,j}(n) = |\ln y_{i,j}(n) - \ln y^{\star}_{i,j}(n)|,$$
$$V_{2,2,i,j}(n) = \sum_{s=n-r}^{n-1} \eta\gamma |x_{i,j}(s) - x^{\star}_{i,j}(s)|.$$

Then we have

$$\begin{aligned} \Delta V_{2,1,i,j}(n) &= V_{2,1,i,j}(n+1) - V_{2,1,i,j}(n) \\ &\le -\left(\frac{1}{\theta_3(n)} - |\frac{1}{\theta_3} - \delta_y|\right)|y_{i,j}(n) - y^{\star}_{i,j}| \\ &+ \eta\gamma|x_{i,j}(n-r) - x^{\star}_{i,j}(n-r)|, \end{aligned}$$

where θ_3 is between $y_{i,j}(n)$ and $y^{\star}_{i,j}(n)$,

$$\Delta V_{2,2,i,j}(n) = \eta\gamma|x_{i,j}(n) - x^{\star}_{i,j}(n)| - \eta\gamma|x_{i,j}(n-r) - x^{\star}_{i,j}(n-r)|.$$

Hence

$$\begin{aligned} \Delta V_{2,i,j}(n) &= \Delta V_{2,1,i,j}(n) + \Delta V_{2,2,i,j}(n) \\ &\le -\left(\frac{1}{\theta_3(n)} - |\frac{1}{\theta_3} - \delta_y|\right)|y_{i,j}(n) - y^{\star}_{i,j}(n)| \\ &+ \eta\gamma|x_{i,j}(n) - x^{\star}_{i,j}(n)|. \end{aligned} \tag{26}$$

Now, we introduce for any pixel (i, j) Lyapunov function

$$V_{i,j}(n) = V_{1,i,j}(n) + V_{2,i,j}(n). \tag{27}$$

According to (25)–(27) we have

$$\begin{aligned} \Delta V_{i,j}(n) &= \Delta V_{1,i,j}(n) + \Delta V_{2,i,j}(n) \\ &\le \left(\delta_x - \frac{1}{\theta_2(n)} + \frac{1}{\theta_1(n)} + \eta\gamma\right)|x_{i,j}(n) - x^{\star}_{i,j}(n)| \\ &+ \left(\gamma - \frac{1}{\theta_3(n)} - |\frac{1}{\theta_3} - \delta_y|\right)|y_{i,j}(n) - y^{\star}_{i,j}(n)| \\ &- \frac{1}{\theta_2(n)}\left(\hat{S}\left\{x_{i,j}(n-1)\right\} - \hat{S}\left\{x^{\star}_{i,j}(n-1)\right\}\right). \end{aligned}$$

We let $V(n) = \sum_{i,j=1}^{N} V_{i,j}(n)$. When summing $\Delta V_{i,j}(n)$ through $i, j = \overline{1, N}$, and taking into account the diffusion properties of spatial operator, we get

$$\Delta V(n) \leq \left(\delta_x - \frac{1}{\theta_2(n)} + \frac{1}{\theta_1(n)} + \eta\gamma\right) \sum_{i,j=1}^{N} |x_{i,j}(n) - x^\star_{i,j}(n)|$$

$$+ \left(\gamma - \frac{1}{\theta_3(n)} - |\frac{1}{\theta_3} - \delta_y|\right) \sum_{i,j=1}^{N} |y_{i,j}(n) - y^\star_{i,j}(n)|$$

$$\leq \left(\delta_x - \exp\left\{\gamma m_y + \delta_x m_x - \beta\right\} + \frac{1}{m_x} + \eta\gamma\right) \sum_{i,j=1}^{N} |x_{i,j}(n) - x^\star_{i,j}(n)|$$

$$+ \left(\gamma - \min\left\{\delta_y, \frac{2}{M_y} - \delta_y\right\}\right) \sum_{i,j=1}^{N} |y_{i,j}(n) - y^\star_{i,j}(n)|$$

$$\leq -\xi\left(\sum_{i,j=1}^{N} |x_{i,j}(n) - x^\star_{i,j}(n)| + \sum_{i,j=1}^{N} |y_{i,j}(n) - y^\star_{i,j}(n)|\right).$$

It completes the proof.

5 Numerical Investigation

In work [15] we investigated numerically the continuous-time model of immunosensor (1) at parameters values:

$\beta = 2\,\text{min}^{-1}$, $\gamma = 2\,\frac{\text{mL}}{\text{min}\cdot\mu\text{g}}$, $\mu_y = 1\,\text{min}^{-1}$, $\eta = 0.8/\gamma$, $\delta_x = 0.5\,\frac{\text{mL}}{\text{min}\cdot\mu\text{g}}$, $\delta_y = 0.5\,\frac{\text{mL}}{\text{min}\cdot\mu\text{g}}$, $D/\Delta^2 = 2.22\,\text{min}^{-1}$.

Here we analyze its discrete analogue, which we obtain with help of scaling some of the corresponding parameters due to discretization step $h = 0.01$ and choosing the others experimentally[6,7]:

$\beta = 2h$, $\gamma = 2\,h$, $\mu_y = h$, $\eta = 0.01184/\gamma$, $\delta_x = 0.5\,h$, $\delta_y = 0.5\,h$, $D/\Delta^2 = 2.22\sqrt{h}$.

We see that the scaling of the parameters should be studied deeper. But the exact numerical consistency with the continuous-time system is not the objective of this work. We leave it for our future research.

We start from investigating of positivity of the solutions. Firstly, we see that the necessary condition (12) of positive invariance of the set Ω holds. Then, in order to check that $\lim_{n\to\infty} x_{i,j}(n) \neq 0$, $i, j = \overline{1, N}$, we analyze the function $f_{\text{extnc}}(N)$ (Fig. 1). We see that the value $N = 14$ is the threshold below which $f_{\text{extnc}}(N) < 1$, that is $\lim_{n\to\infty} x_{i,j}(n) \neq 0$. Moreover, since $f_{\text{extnc}}(16) > 1$, we conclude that $\lim_{n\to\infty} x_{i,j}(n) \neq 0$.

---

[6]Hereinafter we omit units of dimensions of parameters.

[7]After scaling of γ the value $\eta = 0.8/\gamma$ may not be applicable for Nicholson-type difference system (it causes number overflow). So, we have decreased it to 0.01184 experimentally.

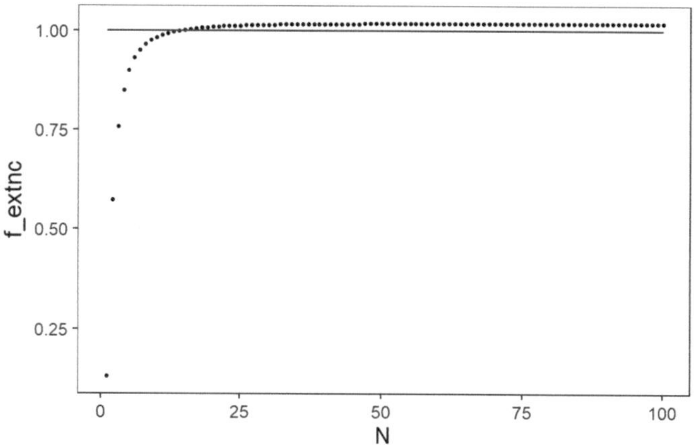

Fig. 1 The values of the function $f_{\text{extnc}}(N)$ for $N \in \{1, 2, \ldots, 100\}$ (black points) as compared with one (blue line)

Table 1 Numerical simulation of the system (6) at $r = 8$

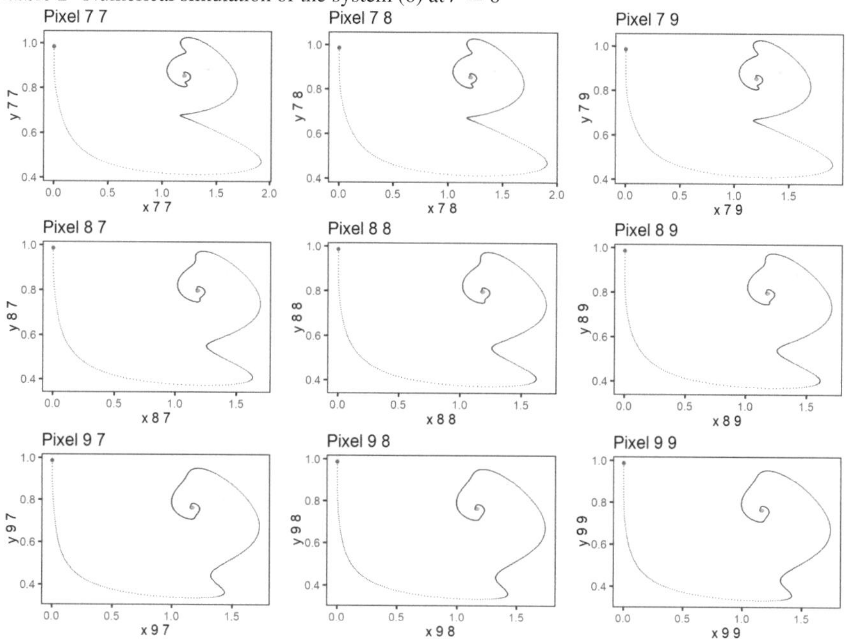

After calculating the steady states for pixels (identical and nonidentical ones), we can apply the global stability conditions (20).

Similarly to differential equations in the discrete-time model we can see that when changing the value of time delay r we have changes of qualitative behavior of pixels

Table 2 The phase planes of the system (6) for antibody populations $y_{i,j}$ versus antigen populations $x_{i,j}$, $i, j = \overline{7, 9}$. Numerical simulation of the system (6) at $r = 12$. Here • indicates initial state, • indicates identical steady state, • indicates nonidentical steady state. The solution converges to a stable limit cycle

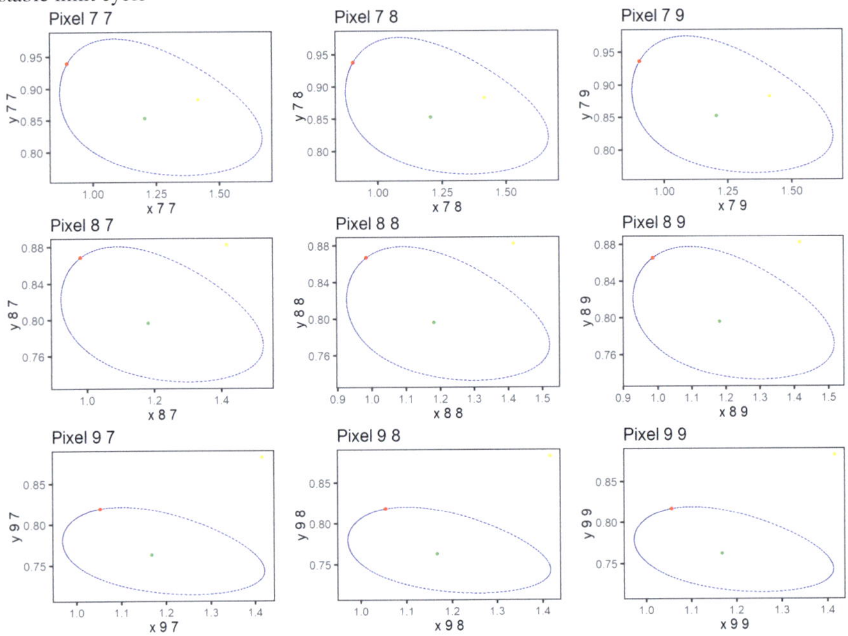

and entire model. We considered the parameter value set given above and computed the long-time behavior of the system (6) describing two-dimensional 16×16-pixels array ($N = 16$) for $r = 8$, 12 and 15. The phase diagrams of the antibody vs. antigen populations for the pixel (8, 8) and its neighborhood for these values of r are shown in Tables 1, 2, 3.

For example, at $r \leq 10$ we can see trajectories corresponding to stable node for all pixels (see Tables 1). At values $r = 10$ Hopf bifurcation occurs and further trajectories correspond to stable limit cycles of ellipsoidal form for all pixels (see Table 2). We note that in order that the numerical solutions regarding Hopf bifurcation were in agreement with the theoretical results, we should apply a Hopf bifurcation theorem from the work [8] which proves appearance of small invariant attracting cycles of radius $O(\sqrt{h})$.

For $r = 12$, the phase diagrams in Table 2 show that the solution is a limit cycle with two local extrema (one local maximum and one local minimum) per cycle. Then for $r = 14, 15$ the solution is a limit cycle with twelve local extrema per cycle (see Table 3). Finally, for $r = 16$, the behavior looks like chaotic one. Similarly as in continuous-time model [15], we have regarded behavior as chaotic if no periodic behavior could be found in the long-time behavior of the solutions.

Table 3 The limit cycles on the phase plane plots of the system (6) for antibody populations $y_{i,j}$ versus antigen populations $x_{i,j}, i, j = \overline{7, 9}$. Numerical simulation of the system (6) at $r = 15$. Here • indicates identical steady state, • indicates nonidentical steady state. Limit cycles are obtained as trajectories for $t \in [4000, 5000]$. The solution converges to a stable limit cycle with twelve local extrema per cycle

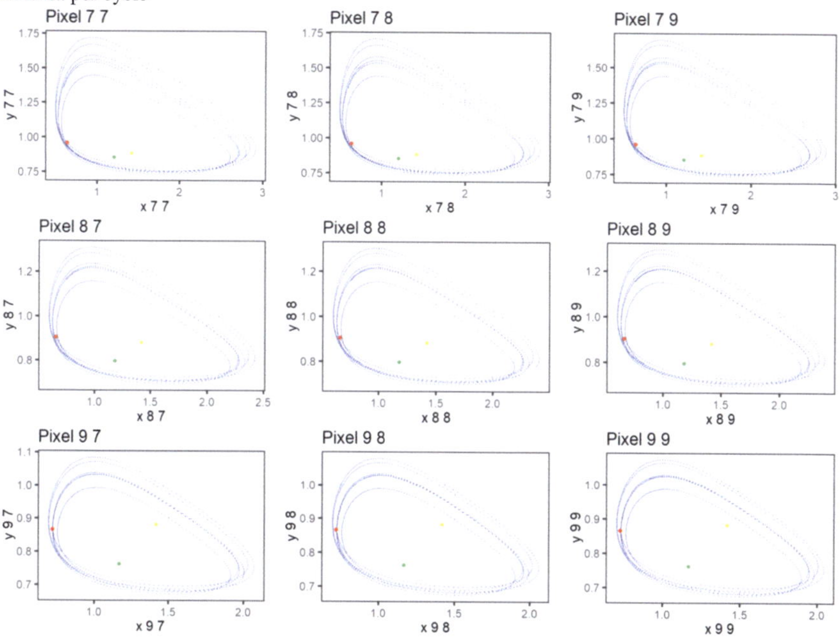

At $D = 0$ (i.e., without diffusion) a numerical bifurcation diagram showing the maximum and minimum points for the limit cycles for the antigen population $x_{1,1}$ as a function of time delay is given in Fig. 2. The Hopf bifurcation from the stable equilibrium point to a simple limit cycle can be clearly seen at $r = 18$. Further all dynamical behavior is characterized as limit cycles, which can be evidenced numerically (see Fig. 3).

At $D/\Delta^2 = 0.02$ a numerical bifurcation diagram showing the maximum and minimum points for the limit cycles for the antigen population $V_{1,1}$ as a function of time delay is given in Fig. 4. The Hopf bifurcation from the stable equilibrium point to a simple limit cycle and the sharp transitions at critical values of the time delay between limit cycles with increasing numbers of maximum and minimum points per cycle can be clearly seen.

As a check that the solution is chaotic for $r \geq 16$, we perturbed the initial conditions to test the sensitivity of the system. Figures 5, 6, 7 show a comparison of the solutions for the antigen population $x_{1,1}$ with initial conditions $x_{1,1}(n) = 1$ and $x_{1,1}(n) = 1.001$, $n \in [-r, 0]$, and identical all the rest ones. In Figs. 6, 7 near the initial time the two solutions appear to be the same, but as time increases there is a marked difference between the solutions supporting the conclusion that the system behavior is chaotic at $r \geq 16$.

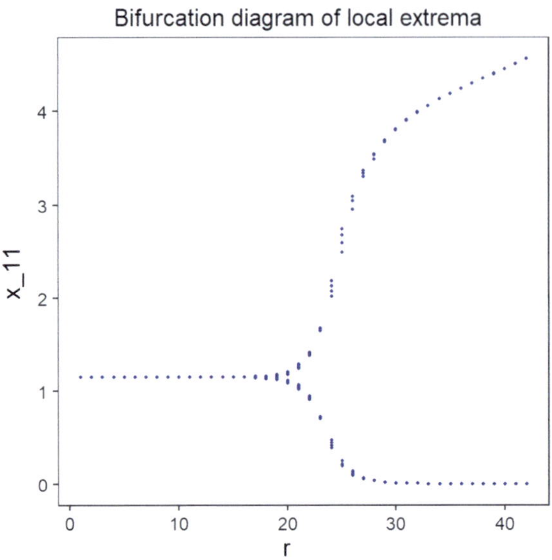

Fig. 2 A numerical bifurcation diagram at $D = 0$. The points show the local extreme points for the $V_{1,1}$ population at $n \in [3300, 5000]$. Hopf-type bifurcation appears at $r = 18$

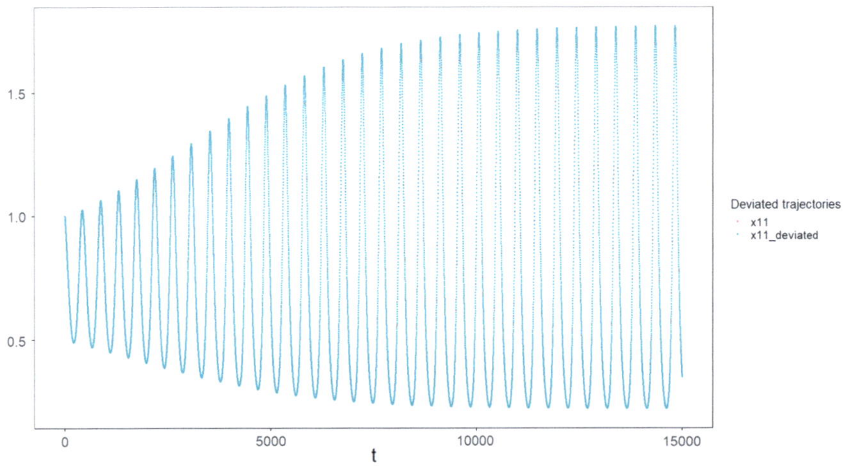

Fig. 3 The time series of the solutions to the system (6) for the antigen population $x_{1,1}$ from $n = 0$ to 5000 with $D/\Delta^2 = 0.0$ and $r = 24$ for initial conditions $x_{1,1}(n) = 1$ and $x_{1,1}(t) = 1.001$ (deviated), $n \in [-r, 0]$, and identical all the rest ones. The two solutions appear to be the same, supporting the conclusion that the system behavior is not chaotic

When analyzing an influence of diffusion on qualitative behavior of the model we pay attention on one more the way to chaos presented in Fig. 8. We see that increasing the values of D/Δ^2 we transit from steady state to limit cycles and finally to chaotic behavior at values $D/\Delta^2 \approx 0.025$.

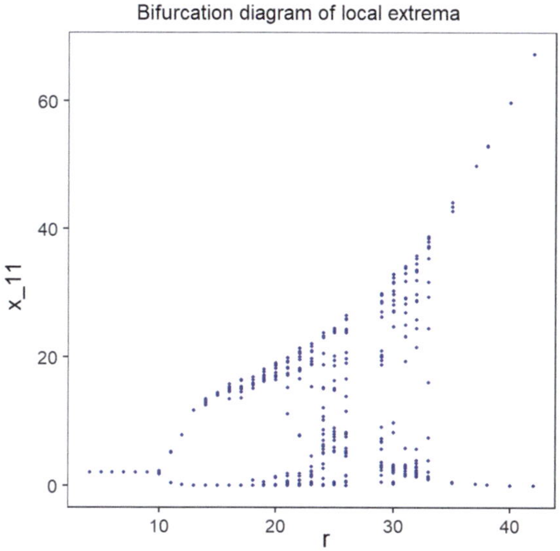

Fig. 4 A numerical bifurcation diagram at $D/\Delta^2 = 0.02$ showing the "bifurcation path to chaos" as the time delay r is increased. The points show the local extreme points per cycle for the $x_{1,1}$ population. Chaotic-type solutions occur at $r = 16$. Note that at $r = 27, 28$ we have unbounded solutions

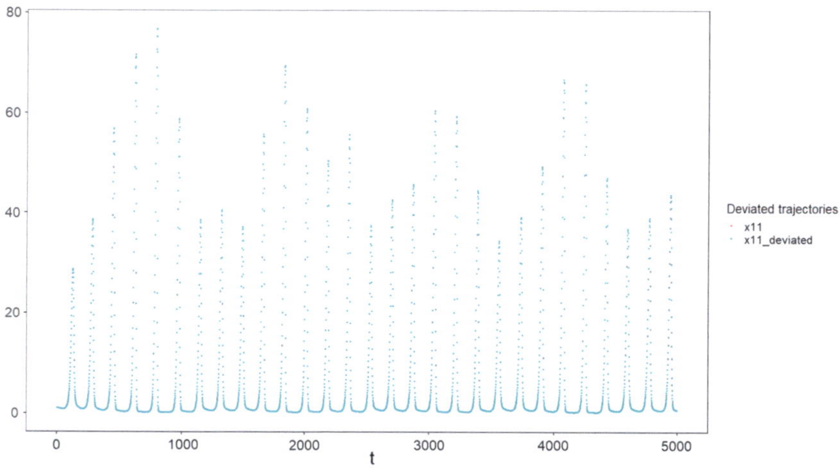

Fig. 5 The time series of the solutions to the system (6) for the antigen population $x_{1,1}$ from $n = 0$ to 5000 with $D/\Delta^2 = 0.02$ and $r = 15$ for initial conditions $x_{1,1}(n) = 1$ and $x_{1,1}(t) = 1.001$ (deviated), $n \in [-r, 0]$, and identical all the rest ones. The two solutions appear to be the same, supporting the conclusion that the system behavior is not chaotic

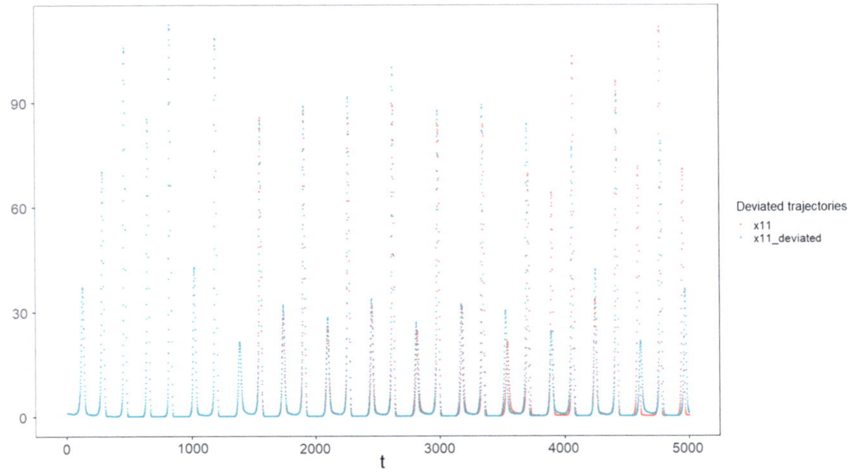

Fig. 6 The time series of the solutions to the system (6) for the antigen population $x_{1,1}$ from $n = 0$ to 5000 with $D/\Delta^2 = 0.02$ and $r = 16$ for initial conditions $x_{1,1}(n) = 1$ and $x_{1,1}(t) = 1.001$ (deviated), $n \in [-r, 0]$, and identical all the rest ones. At the beginning the two solutions appear to be the same, but as time increases there is a marked difference between the solutions supporting the conclusion that the system behavior is chaotic

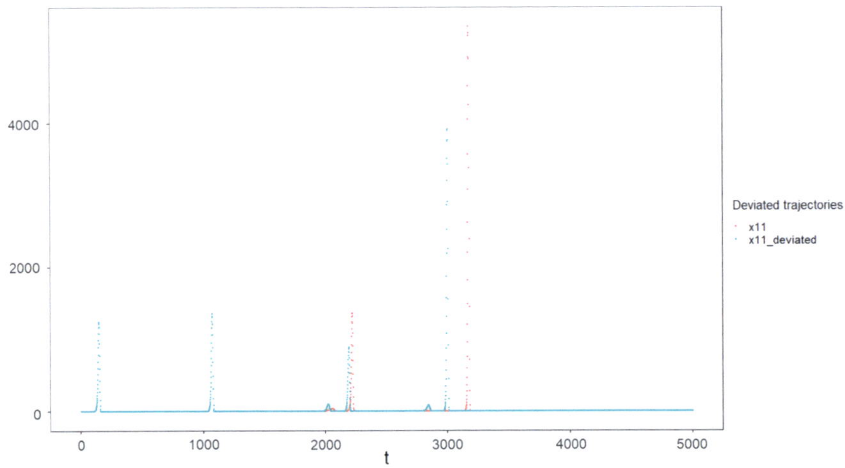

Fig. 7 The time series of the solutions to the system (6) for the antigen population $x_{1,1}$ from $n = 0$ to 5000 with $D/\Delta^2 = 0.02$ and $r = 26$ for initial conditions $x_{1,1}(n) = 1$ and $x_{1,1}(t) = 1.001$ (deviated), $n \in [-r, 0]$, and identical all the rest ones. At the beginning the two solutions appear to be the same, but as time increases there is a marked difference between the solutions supporting the conclusion that the system behavior is chaotic

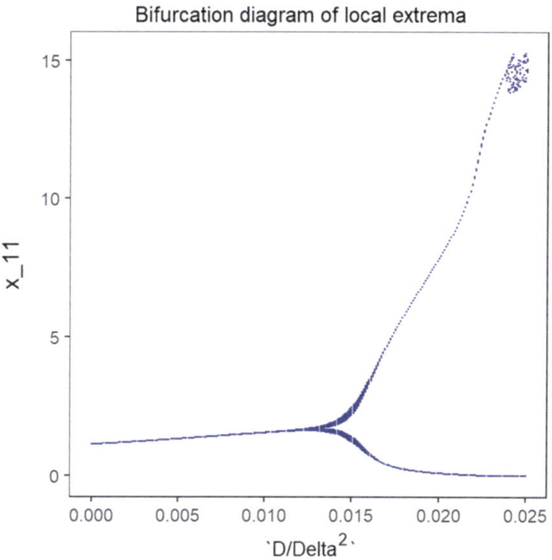

Fig. 8 A numerical bifurcation diagram at $r = 12$ showing the "bifurcation path to chaos" as the diffusion D/Δ^2 is increased. The points show the local extreme points per cycle for the $x_{1,1}$ population. Chaotic-type solutions occur at $\tau \approx 0.025$ with value 0 for the number of extreme points

6 Conclusions

In the work we offered model of immunosensor which is based on the reaction-diffusion system of the finite lattice difference equations with delay. The main results of the work are conditions of permanence and global asymptotic stability for endemic state. Unfortunately, we are not able to say about permanence of the solution in usual sense. So, here we introduced some "weaker" notion of quasi-permanence allowing us to get conditions in a clear form.

It was shown that the dimension of pixels array N should be large enough (sufficient condition) and the diffusion D/Δ^2 should be small enough (necessary condition) to guarantee the positive invariance.

For the purpose of stability investigation we have used method of Lyapunov functions (it is more correct to say "functionals" here). It combines general approach for construction of Lyapunov functions of predator-prey models with finite lattice differential equations.

Numerical examples showed us influence on stability of different parameters. Increasing time delay we transmit from stable node to limit cycles and finally to chaotic behavior. Such behavior is dynamically consistent with the behavior of continuous-time model, which was studied in [15].

We note that difference with continuous-time model is the way to chaos. Namely, in case of differential equations it was period-doubling, which was characterized by the doubling of the number of local extrema. In case of difference equations we have "chaos" as a result of increasing the number of local extrema (not doubling). It is caused by the discrete nature of the delay r in contrary to the continuous-time system.

Another parameter causing changes in dynamic behavior of the system (6) is the diffusion. Namely, it was numerically shown that when increasing D, we transit from periodic solutions to chaotic ones also.

References

1. Allen, L.J.S.: Persistence, extinction, and critical patch number for island populations. J. Math. Biol. **24**(6), 617–625 (1987)
2. Berger, C., Hees, A., Braunreuther, S., Reinhart, G.: Characterization of cyber-physical sensor systems. Proc. CIRP **41**, 638–643 (2016)
3. Bevers, M., Flather, C.H.: Numerically exploring habitat fragmentation effects on populations using cell-based coupled map lattices. Theor. Popul. Biol. **55**(1), 61–76 (1999)
4. Caswell, H.: Matrix Population Models: Construction, Analysis, and Interpretation. Sinauer Associates, Massachusetts (1989)
5. Chow, S.-N., Mallet-Paret, J., Shen, W.: Traveling waves in lattice dynamical systems. J. Differ. Equ. **149**(2), 248–291 (1998)
6. Chow, S.-N., Shen, W.: Dynamics in a discrete nagumo equation: spatial topological chaos. SIAM J. Appl. Math. **55**(6), 1764–1781 (1995)
7. Faria, T., Röst, G.: Persistece, permaece ad global stability for a n-dimesional nicholson system. J. Dyn. Differ. Equ. **26**(3), 723–744 (2014)
8. Hofbauer, J., Iooss, G.: A Hopf bifurcation theorem for difference equations approximating a differential equation. Monatshefte für Mathematik **98**(2), 99–113 (1984)
9. Hupkes, H.J., Van Vleck, E.S.: Negative diffusion and traveling waves in high dimensional lattice systems. SIAM J. Math. Anal. **45**(3), 1068–1135 (2013)
10. Hupkes, H.J., Van Vleck, E.S.: Travelling waves for complete discretizations of reaction diffusion systems. J. Dyn. Differ. Equ. **28**(3–4), 955–1006 (2016)
11. Lancaster, P., Tismenetsky, M.: The Theory of Matrices: with Applications. Elsevier, Amsterdam (1985)
12. Letellier, C., Elaydi, S., Aguirre, L.A., Alaoui, A.: Difference equations versus differential equations, a possible equivalence for the Rössler system? Phys. D: Nonlinear Phenom. **195**(1–2), 29–49 (2004)
13. Liu, L., Liu, Z.: Asymptotic behaviors of a delayed nonautonomous predator-prey system governed by difference equations. Discret. Dyn. Nat. Soc. **1–15**, 2011 (2011)
14. Ma, S., Weng, P., Zou, X.: Asymptotic speed of propagation and traveling wavefronts in a non-local delayed lattice differential equation. Nonlinear Anal.: Theory, Methods Appl. **65**(10), 1858–1890 (2006)
15. Martsenyuk, V., Kłos-Witkowska, A., Sverstiuk, A.: Stability, bifurcation and transition to chaos in a model of immunosensor based on lattice differential equations with delay. Electron. J. Qual. Theory Differ. Equ. **27**, 1–31 (2018)
16. Mickens, R.E.: Nonstandard Finite Difference Models of Differential Equations. World Scientific, Singapore (1994)
17. Mickens, R.E.: Advances in the Applications of Nonstandard Finite Difference Schemes. World Scientific, Singapore (2005)
18. Nicholson, A.J., Bailey, V.A.: The balance of animal populations. part i. In: Proceedings of the Zoological Society of London, vol. 105, pp. 551–598. Wiley Online Library (1935)
19. Prindle, A., Samayoa, P., Razinkov, I., Danino, T., Tsimring, L.S., Hasty, J.: A sensing array of radically coupled genetic 'biopixels'. Nature **481**(7379), 39–44 (2012)
20. Schütze, A., Helwig, N., Schneider, T.: Sensors 4.0 – smart sensors and measurement technology enable industry 4.0. J. Sens. Sens. Syst. **7**(1), 359–371 (2018)
21. So, J.O.S.E.P.H.W.-H., Yu, J.S.: On the stability and uniform persistence of a discrete model of nicholson's blowflies. J. Math. Anal. Appl. **193**(1), 233–244 (1995)

22. Stehlík, P.: Exponential number of stationary solutions for nagumo equations on graphs. J. Math. Anal. Appl. **455**(2), 1749–1764 (2017)
23. Stehlík, P., Volek, J.: Maximum principles for discrete and semidiscrete reaction-diffusion equation. Discret. Dyn. Nat. Soc. **2015**, (2015)
24. Wang, L., Wang, M.Q.: Ordinary Difference Equation. Xinjiang University Press, Xinjiang, China (1989)
25. Wang, Z.-C., Li, W.-T., Jianhong, W.: Entire solutions in delayed lattice differential equations with monostable nonlinearity. SIAM J. Math. Anal. **40**(6), 2392–2420 (2009)
26. Yang, X.: Uniform persistence and periodic solutions for a discrete predator-prey system with delays. J. Math. Anal. Appl. **316**(1), 161–177 (2006)
27. Yi, T., Zou, X.: Global attractivity of the diffusive nicholson blowflies equation with neumann boundary condition: a non-monotone case. J. Differ. Equ. **245**(11), 3376–3388 (2008)
28. Zhang, B.G., Xu, H.X.: A note on the global attractivity of a discrete model of nicholsonś blowflies. Discret. Dyn. Nat. Soc. **3**(1), 51–55 (1999)

Discrete Reaction-Dispersion Equation

Zdeněk Pospíšil

Abstract The paper introduces a discrete analogy of the reaction-diffusion partial differential equation. Both the time and the space are considered to be discrete, the space is represented by a simple graph. The equation is derived from "first principles". Basic qualitative properties, namely, existence and stability of equilibria are discussed. The results are demonstrated on a particular system that can be interpreted as a model of metapopulation on interconnected patches with a deadly boundary. A condition for size of habitat needed for population survival is established.

Keywords Diffusion · Random walk · Graph theory · Stability of equilibria

1 Motivation

An auto-catalytic reaction of a "substance" and its diffusion on one dimensional continuous space is described by the parabolic PDE

$$\frac{\partial u}{\partial t} = D\frac{\partial^2 u}{\partial x^2} + f(u)\ , \quad t > 0\ ,\ x \in (0, L)\ ; \tag{1}$$

the state variable u represents a substance density, the function f (non-linear, in general) describes the reaction, and D is the diffusivity (the coefficient appearing in the Fickian law). The same equation stands for a model of spatially distributed population—it can be found in any textbook on mathematical ecology or population dynamics, cf. e.g. [4, pp. 265–343]. Let us remind two results concerning the reaction-diffusion equation that are of importance in a context of theoretical ecology.

Let $f : [0, \infty) \to [0, \infty)$ be a continuous function satisfying the conditions $f(0) = 0$, $(u - K)f(u) < 0$ for some positive constant K and $0 < u \neq K$. Then

Z. Pospíšil (✉)
Masaryk University, Brno, Czech Republic
e-mail: pospisil@math.muni.cz
URL: https://www.math.muni.cz/~pospisil

M. Bohner et al. (eds.), *Difference Equations and Discrete Dynamical Systems with Applications*, Springer Proceedings in Mathematics & Statistics 312,
https://doi.org/10.1007/978-3-030-35502-9_14

the Eq. (1) possesses the spatially homogeneous equilibrium $u \equiv 0$ (the extinction equilibrium). The Eq. (1) with the homogeneous Neumann boundary condition

$$\frac{\partial u(0)}{\partial x} = 0 = \frac{\partial u(L)}{\partial x} \tag{2}$$

has the positive asymptotically stable equilibrium $u \equiv K$ and the extinction equilibrium is unstable. If f is differentiable in a right neighbourhood of 0 with $r = f'(0) > 0$ and $L > \pi\sqrt{D/r}$ then the trivial equilibrium $u \equiv 0$ of the Eq. (1) with the homogeneous Dirichlet boundary condition

$$u(0) = 0 = u(L) \tag{3}$$

is unstable and the boundary value problem (1), (3) possesses an asymptotically stable spatially non-homogeneous equilibrium $u(t, x) = u^*(x)$ such that

$$0 < \max_{0<x<L} u^*(x) = u\left(\tfrac{1}{2}L\right) < K\ .$$

The last case serves as a theoretical explanation of the fact that a spatially distributed population needs some minimal size of its habitat.

The aim of the contribution is to present discrete counterparts of the reminded results. The discrete diffusion (heat) equation is a matter of textbooks, cf. [3, p. 168–172]. Here, the discrete space is a grid and the equation is linear autonomous difference system with Toeplitz matrix. The reaction term can be simply added to such a system. Systems of this type represent accurate discretization of the mentioned PDEs and they were studied not only on grids [2] but, recently, also on general graphs [8]. This approach brings reasonable results. But, by my opinion, this is not the only possibility how to model phenomena consisting of a reaction and dispersal of a "substance" on a discrete space, namely, on networks or graphs.

First, the word "diffusion" takes different meanings in different contexts. Humanities mean by it a very complex and complicated process [7]. In metapopulation ecology, it denotes a form of dispersal, in which the flux of individuals between two patches is proportional to the difference in their densities [1, p. 65]. Such a process coincide with a random walk. A more precise determination distinguishes diffusion from random walk. Diffusion is described by the Laplacian matrix (discrete Laplacian operator), while random walk by a matrix derived from the Markovian one. An important difference between the two processes is that diffusion tends to a state with uniformly distributed values in all nodes, while random walk tends to a state where the values are proportional to node degrees. Hence, the two processes coincide for regular graphs [6, Sects. 6.13 and 6.14]. Because of these terminological reasons, I use the more general word "dispersion" instead of the specific "diffusion".

Second, the process of reaction can be separated (at least mentally) from the one of dispersion. This idea was utilized already in framework of linear models of (meta)population dynamics [5] and it forms a basis for the subsequent considerations.

In the next section, the proposed reaction-dispersion system is derived from mechanistic presuppositions ("first principles"). The dispersion process is represented by the random walk and the model presented is more general than necessary for the rest of the paper. This generality may show that a scope of applicability exceeds the one indicated in the third section where a spatially homogeneous reaction and simple dispersion is considered. Results that can be regarded as analogues to the mentioned results from continuous spatial population dynamics are shown.

The notation used is standard. The vectors are considered to be the column ones, $\mathbf{v} = (v_1, v_2, \ldots, v_k)^\mathsf{T}$; we will also denote the entries of the vector $\mathbf{v}$ by $v_i = (\mathbf{v})_i$. The symbol $\operatorname{diag} \mathbf{v}$ denotes the diagonal matrix M with diagonal entries $m_{ii} = v_i$, the symbols $\mathbf{o}$, $\mathbf{1}$, and I denote zero vector, vector with all of entries equal to 1, and identity matrix, respectively.

2 The Model

Of course, a reaction-dispersion equation should describe two processes—a reaction, i.e. constitution and dissociation of a substance or birth and death of individuals forming a population, and dispersion, i.e. random spreading of it to a space. Assume that the space is represented by a simple connected graph $\mathcal{G} = \{N, E\}$, where $N = \{1, 2, \ldots, k\}$ is a set of nodes and $E \subseteq \big\{\{i, j\} : i, j \in N, i \neq j\big\}$ is the set of edges. Let A be the adjacency matrix of the graph $\mathcal{G}$ and denote $\sigma_j = \Big|\big\{i : \{i, j\} \in E\big\}\Big| = \sum_{i=1}^{k} a_{ij} = \big(\mathbf{1}^\mathsf{T}\mathsf{A}\big)_j$ the degree of the node $j \in N$.

Let $x_i(t)$ denote an expected concentration of a substance in the node $i \in N$ at the time t; naturally, we suppose $x_i \geq 0$.

Let us suppose that the reaction take place entirely in the nodes and the dispersion consists in relocation of particles (individuals) along the edges. Let us start with a description of the two processes separately.

2.1 Reaction

Let $\tau_\mathrm{r} > 0$ be a "typical reaction time". We assume that the change of concentration in the node $i \in N$ depends on the concentration x_i only, i.e. it is given by the recurrence relation

$$x_i(t + \tau_\mathrm{r}) = f_i\big(x_i(t)\big) .$$

Here $f_i : [0, \infty) \to [0, \infty)$, $i = 1, 2, \ldots, k$ are functions satisfying

$$f_i(0) = 0 \tag{4}$$

that are non-linear in general. Denoting $\mathbf{f}(\mathbf{x}) = \big(f_1(x_1), f_2(x_2), \ldots, f_k(x_k)\big)^\mathsf{T}$, we can write the relation in the simple vector form

$$\mathbf{x}(t+\tau_\mathrm{r}) = \mathbf{f}\big(\mathbf{x}(t)\big)\ . \tag{5}$$

2.2 Dispersion

Let us suppose that a particle in a node can either remain in it or it can move to a randomly chosen neighbor node. We can call this process "elementary dispersion event". Let us denote by $\tau_\mathrm{d} > 0$ the time necessary for replacement of a particle to a neighbor node and by d, $0 < d \le 1$, the probability that a particle leaves its node. The adopted assumptions yield the recurrence relations

$$x_i(t+\tau_\mathrm{d}) = (1-d)x_i(t) + \sum_{\{i,j\}\in E} \frac{d}{\sigma_j} x_j(t)\ , \quad i \in N\ ,$$

for the expected concentrations x_i. Let us define the matrix K with the entries $\kappa_{ij} = a_{ij}/\sigma_j$, that is $\mathsf{K} = \mathsf{A}(\operatorname{diag}\mathsf{A}\mathbf{1})^{-1}$. Since

$$\sum_{\{i,j\}\in E} \frac{d}{\sigma_j} x_j(t) = d\sum_{j=1}^{k} \frac{a_{ij}}{\sigma_j} x_j(t) = d\big(\mathsf{K}\mathbf{x}(t)\big)_i\ ,$$

we can rewrite the derived recurrence relation in the vector form

$$\mathbf{x}(t+\tau_\mathrm{d}) = \big(\mathsf{I} - d(\mathsf{I}-\mathsf{K})\big)\mathbf{x}(t)\ . \tag{6}$$

The matrix on the right hand side characterizes the described dispersion process. Obviously,

$$\sum_{j=1}^{k} \kappa_{ij} = \sum_{j=1}^{k} \frac{a_{ij}}{\sigma_j} = 1.$$

that is, the matrix K is left (row) stochastic and, subsequently, the dispersion matrix

$$\mathsf{D} = \mathsf{I} - d(\mathsf{I}-\mathsf{K}) \tag{7}$$

is left stochastic as well. The matrix D is used to describe random walk on a graph, cf. comment in the first section and the reference therein.

2.3 Reaction and Dispersion

Let us suppose that the two processes are separated in time. First, the reaction occurs and then it is followed by the dispersion. Such an assumption may be satisfied in population dynamic—e.g., a plant grows, flowers and produces seeds and then the seeds are spread. If this is the case, we can define the time unit (length of projection interval) to be the sum of the "typical reaction and dispersion times", $1 = \tau_\mathrm{r} + \tau_\mathrm{d}$ and, combining (5), (6) and (7), we obtain the non-linear recurrence relation

$$\mathbf{x}(t+1) = \mathbf{x}\big((t+\tau_\mathrm{r}) + \tau_\mathrm{d}\big) = \mathsf{D}\mathbf{x}(t+\tau_\mathrm{r}) = \mathsf{D}\mathbf{f}\big(\mathbf{x}(t)\big)\ .$$

Let us note that the above form of the recurrence fixes the "observation times" to moments before "reaction event" and after "dispersion event". This is just an arbitrary choice. The other one leads to the relation $\mathbf{x}(t+1) = \mathbf{f}\big(\mathsf{D}\mathbf{x}(t)\big)$; we will not deal with this option in the present paper.

The situation need not to be simple alternation of reaction and "elementary dispersion event". We can consider the possibility that the reaction is followed by several "elementary dispersions". If the number of them is n then we define the time unit by $1 = \tau_\mathrm{r} + n\tau_\mathrm{d}$ and the process is described by the recurrence relation

$$\mathbf{x}(t+1) = \mathsf{D}^n\mathbf{f}\big(\mathbf{x}(t)\big)\ . \tag{8}$$

2.4 Homogeneous Boundary Conditions

The Neumann boundary condition (2) for the continuous equation (1) states that no amount of the "substance" goes out of the region where it disperses, that is, the process is spatially isolated. The discrete equation (8) is defined on the underlying graph $\mathcal{G}$ only. That is, no "outer space" is considered and hence no additional condition is needed.

The Dirichlet boundary condition (3) determines the concentration on the two boundary points to be zero. This property can be mimic for the discrete equation (8) as well. Let us determine some of the nodes to be the boundary ones and prescribe the zero value of the concentration on them for each time t. This is equivalent to left multiplication of the vector $\mathbf{x}(t)$ by the "boundary" matrix B defined as

$$b_{ij} = \begin{cases} 1,\ i = j \notin N_B, \\ 0,\ \text{otherwise}, \end{cases}$$

where $N_B \subseteq N$ is the set of boundary nodes. The matrix B is obviously idempotent. Of course, if $N_B = \emptyset$ then the "boundary" matrix $\mathsf{B} = \mathsf{I}$ expresses the "Neumann-type" condition.

Let us suppose first that each "elementary dispersion event" is followed by an immediate "disappearance" of the substance in the boundary nodes. E.g., considering metapopulations we can regard the boundary patches to be be deadly ones. In this case, the matrix describing dispersion equals BD. Due to the condition (4), the identities $\mathbf{x} = \mathsf{B}\mathbf{x}$, $\mathbf{f}(\mathbf{x}) = \mathsf{B}\mathbf{f}(\mathbf{x})$ hold true for any vector $\mathbf{x}$ satisfying the boundary condition $x_i = 0$ for $i \in N_B$. Hence, the discrete reaction-dispersion equation with homogeneous boundary condition can be written in one of the equivalent forms

$$\mathbf{x}(t+1) = (\mathsf{BD})^n \mathbf{f}\big(\mathbf{x}(t)\big)\,, \quad \text{or} \quad \mathbf{x}(t+1) = (\mathsf{BD})^n \mathsf{B}\mathbf{f}\big(\mathbf{x}(t)\big)\,. \tag{9}$$

But the process of "disappearance" of particles/individuals in the boundary nodes may be a matter of the reaction only. If this is the case, the model can be written in one of the equivalent forms

$$\mathbf{x}(t+1) = \mathsf{BD}^n \mathbf{f}\big(\mathbf{x}(t)\big)\,, \quad \text{or} \quad \mathbf{x}(t+1) = \mathsf{BD}^n \mathsf{B}\mathbf{f}\big(\mathbf{x}(t)\big)\,. \tag{10}$$

3 Equilibria and Their Stability

Now, let us consider the system (9) or (10) with $n = 1$, i.e. the system

$$\mathbf{x}(t+1) = \mathsf{BDB}\mathbf{f}\big(\mathbf{x}(t)\big)\,. \tag{11}$$

Let $\mathbf{x}^*$ be the equilibrium of the equation (11), i.e. $\mathsf{BDB}\,\mathbf{f}\,(\mathbf{x}^*) = \mathbf{x}^*$. If the vector function $\mathbf{f}$ is differentiable one can check the stability of the equilibrium using linearization of the system. Since

$$\frac{\partial}{\partial x_j}\big(\mathbf{f}(\mathbf{x}^*)\big)_i = \frac{\partial}{\partial x_j} f_i(x_i^*) = \begin{cases} f_j'(x_j^*), & i = j, \\ 0, & i \neq j, \end{cases}$$

the variation matrix of the system (11) equals $\mathsf{J}(\mathbf{x}^*) = \mathsf{BDB}\mathrm{diag}\,\mathbf{f}'(\mathbf{x}^*)$, where $\mathbf{f}'(\mathbf{x}^*)$ stands for $\big(f_1'(x_1^*), f_2'(x_2^*), \ldots, f_k'(x_k^*)\big)^\mathsf{T}$. Consequently, the stability of the equilibrium $\mathbf{x}^*$ is determined by the spectrum of the matrix $\mathsf{BDB}\mathrm{diag}\,\mathbf{f}'(\mathbf{x}^*)$.

Let us pay attention to eigenvalues of the matrices BDB and BKB. Obviously, if $N_B \neq \emptyset$ then the both matrices possesses zero eigenvalues with corresponding eigenvectors $\mathbf{w}$ satisfying $w_i = 0$ for all $i \notin N_B$, i.e. $\mathsf{B}\mathbf{w} = \mathbf{o}$. That is, the geometric multiplicity of zero eigenvalues is at least $|N_B|$. The following statement says something on the others eigenvalues.

Proposition 1 *Let vector $\mathbf{w}$ satisfy the boundary condition, i.e., $\mathbf{w} = \mathsf{B}\mathbf{w}$. Then $\mathbf{w}$ is the eigenvector of the matrix BDB if and only if it is the eigenvector of the matrix BKB. The corresponding eigenvalues $\tilde{\lambda}$ and λ of the matrices BDB and BKB, respectively, satisfy the equality $\tilde{\lambda} = 1 - d(1-\lambda)$.*

Proof Let $\mathbf{w} = \mathsf{B}\mathbf{w}$. Then by (7),

$$\mathsf{BDB}\mathbf{w} = \mathsf{B}\big(\mathsf{I} - d(\mathsf{I} - \mathsf{K})\big)\mathsf{B}\mathbf{w} = \mathbf{w} - d\mathbf{w} + d\mathsf{BKB}\mathbf{w} .$$

Now, if $\mathsf{BKB}\mathbf{w} = \lambda\mathbf{w}$ then $\mathsf{BDB}\mathbf{w} = \big(1 - d(1 - \lambda)\big)\mathbf{w}$, if $\mathsf{BDB}\mathbf{w} = \tilde{\lambda}\mathbf{w}$ then $\big((\tilde{\lambda} - 1 + d)/d\big)\mathbf{w} = \mathsf{BKB}\mathbf{w}$. □

The Proposition 1 allows to prove a statement on stability of particular equilibria of the system (11).

Proposition 2 *Let the reaction function be the same on all non-boundary nodes, i.e.* $\mathbf{f}(\mathbf{x}) = \big(f(x_1), f(x_2), \ldots, f(x_k)\big)^{\mathsf{T}}$, *where* $f : [0, \infty) \to [0, \infty)$ *is differentiable function, and let* $x^*\mathsf{B}\mathbf{1}$ *be an equilibrium of the system (11). If*

$$\big|f'(x^*)\big(1 - d(1 - \lambda)\big)\big| < 1 \tag{12}$$

for each eigenvalue λ *of the matrix* BKB *such that the corresponding eigenvector* $\mathbf{w}$ *satisfies the boundary condition* $\mathbf{w} = \mathsf{B}\mathbf{w}$, *then the equilibrium is stable.*

If there exists an eigenvalue λ *with corresponding eigenvector* $\mathbf{w}$ *of the matrix* BKB *satisfying the boundary condition and the inequality is reversed, then the equilibrium is unstable.*

Proof We have $\mathsf{B}\mathrm{diag}\,\mathbf{f}'(x^*\mathsf{B}\mathbf{1}) = f'(x^*)\mathsf{B}$. Consequently $\mathsf{J}(x^*\mathsf{B}\mathbf{1}) = f'(x^*)\mathsf{BDB}$. Hence, if $|f'(x^*)\tilde{\lambda}| < 1$ for all eigenvalues $\tilde{\lambda}$ of the matrix BDB, then the equilibrium is stable. The condition is fulfilled for $\tilde{\lambda} = 0$ such that the corresponding eigenvector $\tilde{\mathbf{w}}$ satisfies $\tilde{\mathbf{w}} \neq \mathsf{B}\tilde{\mathbf{w}} = \mathbf{o}$. Now, the statement follows from the Proposition 1. □

Let us consider the system (11) with symmetric matrix K and no boundary node. Since $\kappa_{ij} = a_{ij}/\sigma_j$, the assumption on symmetry is satisfied if and only if $\sigma_i = \sigma_j$ for all i, j, i.e., for regular graphs only. In this case, a random walk coincides with diffusion. Since the matrix K is left stochastic and symmetric, it is double stochastic and its spectral radius equals 1 and one eigenvalue equals 1. Hence, if $\mathsf{B} = \mathsf{I}$, then the condition (12) holds if and only if it holds for $\lambda = 1$. If x^* is a fixed point of the function f, then

$$\mathsf{D}(x^*\mathbf{1}) = x^*\big(\mathsf{I} - d(\mathsf{I} - \mathsf{K})\big)\mathbf{1} = x^*(\mathbf{1} - d\mathbf{1} + d\mathbf{1}) = x^*\mathbf{1}$$

and $x^*\mathbf{1}$ is an equilibrium of the system (11) with $\mathsf{B} = \mathsf{I}$ and $\mathsf{D} = \mathsf{D}^{\mathsf{T}}$.

Hence, we have obtained an analogy to the statement that, under some condition, the continuous boundary value problem (1), (2) possesses a spatially homogeneous stable equilibrium: Let $K > 0$ and differentiable function $f : [0, \infty) \to [0, \infty)$ satisfy $\big(f(x) - K\big)(x - K) < 0$ for $x \neq K$. Then the system

$$\mathbf{x}(t + 1) = \mathsf{D}\mathbf{f}\big(\mathbf{x}(t)\big) \tag{13}$$

with symmetric matrix D has the positive stable equilibrium $K\mathbf{1}$.

The Proposition 2 speaks on stability of spatially homogeneous equilibria of the system (11) with equal "reaction functions" on each node. Due to the property (4), the zero vector $\mathbf{o}$ is a spatially homogeneous equilibrium; if the system is interpreted as a model of metapopulation the vector $\mathbf{o}$ represents the extinction equilibrium. The system (13) without boundary nodes and symmetric matrix D may possess a non-zero spatially homogeneous equilibrium. An example of a system with a non-empty boundary and a non-zero spatially homogeneous equilibrium is presented in the following subsection.

3.1 Particular Case

Let us consider the complete graph with $k \geq 3$ nodes, one of nodes (the last one, say) be the boundary one. Let the reaction be the same on each non-boundary node and let it be of the form

$$f(x) = xg(x) ,$$

where the function $g : [0, \infty) \to [0, \infty)$ is differentiable with

$$g(0) = r > 1, \quad g'(x) < 0, \quad (x - K)\big(g(x) - 1\big) < 0 \text{ for } x \neq K > 0 .$$

Finally, let one reaction be followed by one dispersion event. Now, the system (11) might represent a model of metapopulation on k mutually interconnected patches, one of them is deadly and the others exhibit the same condition for the local populations.

We have $f'(x) = g(x) + xg'(x)$ and

$$\mathsf{B} = \begin{pmatrix} 1\,0 \dots 0\,0 \\ 0\,1 \dots 0\,0 \\ \vdots\;\vdots\;\ddots\;\vdots\;\vdots \\ 0\,0 \dots 1\,0 \\ 0\,0 \dots 0\,0 \end{pmatrix}, \; \mathsf{K} = \begin{pmatrix} 0 & \frac{1}{k-1} & \cdots & \frac{1}{k-1} \\ \frac{1}{k-1} & 0 & \cdots & \frac{1}{k-1} \\ \vdots & \vdots & \ddots & \vdots \\ \frac{1}{k-1} & \frac{1}{k-1} & \cdots & 0 \end{pmatrix}, \; \mathsf{BKB} = \begin{pmatrix} 0 & \frac{1}{k-1} & \cdots & \frac{1}{k-1} & 0 \\ \frac{1}{k-1} & 0 & \cdots & \frac{1}{k-1} & 0 \\ \vdots & \vdots & \ddots & \vdots & \vdots \\ \frac{1}{k-1} & \frac{1}{k-1} & \cdots & 0 & 0 \\ 0 & 0 & \cdots & 0 & 0 \end{pmatrix}.$$

One can easily verify that the matrix BKB has the following eigenvalues with corresponding eigenvectors

$$\lambda_1 = \frac{k-2}{k-1}, \; \mathbf{w_1} = \begin{pmatrix} 1 \\ 1 \\ \vdots \\ 1 \\ 0 \end{pmatrix}, \quad \lambda_2 = 0 \, , \; \mathbf{w_2} = \begin{pmatrix} 0 \\ 0 \\ \vdots \\ 0 \\ 1 \end{pmatrix},$$

$$\lambda_3 = \lambda_4 = \cdots = \lambda_k = \frac{1}{1-k}, \; \mathbf{w_3} = \begin{pmatrix} 1 \\ 0 \\ 0 \\ \vdots \\ 0 \\ -1 \\ 0 \end{pmatrix}, \; \mathbf{w_4} = \begin{pmatrix} 0 \\ 1 \\ 0 \\ \vdots \\ 0 \\ -1 \\ 0 \end{pmatrix}, \; \ldots, \; \mathbf{w_k} = \begin{pmatrix} 0 \\ 0 \\ 0 \\ \vdots \\ 1 \\ -1 \\ 0 \end{pmatrix}.$$

Thus, we have

$$1 - d(1-\lambda_1) = 1 - \frac{d}{k-1}, \quad 1 - d(1-\lambda_j) = 1 - \frac{dk}{k-1}, \; j = 3, 4, \ldots, k.$$

Obviously, $1 - dk/(k-1) < 1 - d/(k-1)$. The assumptions $d \leq 1$ and $k \geq 3$ yield

$$0 \leq 1 - \frac{d}{k-1} \quad \text{and} \quad \frac{dk}{k-1} - 1 \leq 1 - \frac{d}{k-1},$$

since $\frac{1}{2}d \leq \frac{1}{2} \leq (k-1)/(k+1)$. Consequently

$$\left|1 - \frac{dk}{k-1}\right| \leq 1 - \frac{d}{k-1}.$$

That is, the criterion of equilibrium stability from the Proposition 2 reads

$$\left|f'(x^*)\right| \left(1 - \frac{d}{k-1}\right) < 1.$$

The zero vector $\mathbf{o}$ is one equilibrium (the extinction equilibrium) of the system. The assumption $f'(0) = g(0) = r > 1$ now implies that if

$$r\left(1 - \frac{d}{k-1}\right) < 1, \quad \text{i.e.} \quad k < 1 + \frac{dr}{r-1},$$

then the extinction equilibrium is stable.

Now, let us search conditions for the existence of equilibrium of the form $x^*\mathsf{B}\mathbf{1}$ with $x^* > 0$, that is, the equilibrium spatially homogeneous on the non-boundary nodes.

Let us note that under the assumptions listed above, the system (11) becomes

$$x_i(t+1) = (1-d)x_i(t)g\big(x_i(t)\big) + \frac{d}{k-1} \sum_{\substack{\{i,j\} \in E \\ j \neq k}} x_j(t)g\big(x_j(t)\big), \; i = 1, 2, \ldots, k-1$$

and $x_k(t) \equiv 0$. Consequently

$$g(x^*)\left(1-d+(k-2)\frac{d}{k-1}\right)=1, \quad \text{or,} \quad g(x^*)=\frac{k-1}{k-1-d}\,.$$

Since the function g is strictly decreasing with $g(0)=r$, $g(K)=1$, the last equation possesses unique solution x^* satisfying $0<x^*<K$ if and only if

$$\frac{k-1}{k-1-d}<r\,, \quad \text{i.e.} \quad k>1+\frac{dr}{r-1}\,. \tag{14}$$

If this is the case, we have

$$f'(x^*)=g(x^*)+x^*g'(x^*)=\frac{k-1}{k-1-d}+x^*g'(x^*),$$

$$\left|f'(x^*)\right|\left(1-\frac{d}{k-1}\right)=\left|1+\frac{k-1-d}{k-1}x^*g'(x^*)\right|=\left|1+x^*\frac{g'(x^*)}{g(x^*)}\right|$$

and the criterion of stability becomes

$$-2<x^*\frac{g'(x^*)}{g(x^*)}<0\,.$$

The last inequality is satisfied due the assumption $g'<0$.

Now, we can conclude: If $k<1+dr/(1-r)$ then the extinction equilibrium $\mathbf{o}$ is stable, if $k>1+dr/(1-r)$ then the equilibrium $\mathbf{o}$ is unstable and there exists the interior equilibrium $x^*\mathsf{B}\mathbf{1}$ which is stable provided $-2<x^*g'(x^*)/g(x^*)$.

This statement (namely, the second inequality (14)) can be considered to be one of possible analogies to the one that the boundary value problem (1), (3) possesses non-trivial equilibrium just for L "large enough", that is for sufficient "size" k of the space.

Acknowledgements This research was partially supported by the grant No. GA16-03796S of the Czech Grant Agency.

I am grateful to the both of the anonymous referees for their comments that allow to clarify the matter and to substantially improve the text.

References

1. Caswell, H.: Matrix Population Models. Sinauer Associates, Sunderland (2001)
2. Chow, S.-N., Mallet-Paret, J., Shen, W.: Traveling waves in lattice dynamical systems. J. Differ. Equ. **149**, 248–291 (1998)
3. Elaydi, S.: An Introduction to Difference Equations. Springer, New York (2015)
4. Kot, M.: Elements of Mathematical Ecology. Cambridge University Press, Cambridge (2001)
5. Lebreton, J.-D., Gonzales-Davila, G.: An introduction to models of subdivided populations. J. Biol. Syst. **1**, 389–423 (1993)
6. Newman, M.: Networks. Oxford University Press, Oxford (2018)

7. Rogers, E.M.: Diffusion of Innovations. The Free Press, New York (1983)
8. Stehlík, P.: Exponential number of stationary solutions for Nagumo equations on graphs. J. Math. Anal. Appl. **455**, 1749–1764 (2017)

A Note on Transformations of Independent Variable in Second Order Dynamic Equations

Pavel Řehák

Abstract The main purpose of this paper is to show how a transformation of independent variable in dynamic equations combined with suitable statements on a general time scale can yield new results or new proofs to known results. It seems that this approach has not been extensively used in the literature devoted to dynamic equations. We present, in particular, two types of applications. In the first one, an original dynamic equation is transformed into a simpler equation. In the second one, a dynamic equation in a somehow critical setting is transformed into a noncritical case. These ideas will be demonstrated on problems from oscillation theory and asymptotic theory of second order linear and nonlinear dynamic equations.

Keywords Transformation · Chain rule · Dynamic equation · Time scale · Oscillation · Asymptotic formulae

1 Introduction

It is well known that the chain rule in the "pure" form $(f \circ g)^{\Delta}(t) = f'(g(t))g^{\Delta}(t)$ does not hold on a general time scale $\mathbb{T}$, even if the derivative f' and the delta derivative g^{Δ} exist, see, e.g., [3]. This is the reason why the transformation of independent variable, a useful tool in the theory of differential equations, is not fully at our disposal for dynamic equations. Naturally, the problems can occur also with using the substitution method in delta integrals. There exist variants of chain rule on time scales, such as $(f \circ g)^{\Delta}(t) = \left[\int_0^1 f'(g(t) + h\mu(t)g^{\Delta}(t))\,\mathrm{d}h\right] g^{\Delta}(t)$ [5, Theorem 1.90] involving the classical, say Riemann, integral, or $(f \circ g)^{\Delta}(t) = f'(g(\xi))g^{\Delta}(t)$ [5, Theorem 1.87] with an unspecified value ξ coming from the Lagrange mean value theorem; they are however unsuitable for the use in many situations. Another version of the chain rule, $(f \circ g)^{\Delta} = (f^{\widetilde{\Delta}} \circ g)g^{\Delta}$ [5, Theorem 1.93], involves two

P. Řehák (✉)
Institute of Mathematics, FME,
Brno University of Technology, Technická 2, 61669 Brno, Czech Republic
e-mail: rehak.pavel@fme.vutbr.cz

M. Bohner et al. (eds.), *Difference Equations and Discrete Dynamical Systems with Applications*, Springer Proceedings in Mathematics & Statistics 312,
https://doi.org/10.1007/978-3-030-35502-9_15

generally different time scales $\mathbb{T}, \widetilde{\mathbb{T}}$, which are related through the inner function g by $\widetilde{\mathbb{T}} = g(\mathbb{T})$. The problem with applications of the last variant, when being interested in the transformation of independent variable, is that a dynamic equation on a time scale is transformed into a dynamic equation on a different time scale. On the other hand, if we take into account that nowadays a lot of results for dynamic equations hold on a general time scale, we can successfully use this approach, for instance, in the below described way.

As it is well known, transformations of independent and dependent variables in differential equations are useful, among others, when it is possible to transform a given differential equation into a differential equation which is in some way simpler. In this paper, we deal with dynamic equations and use some existing results that are valid on a general time scale to obtain new results or new proofs of known statements. We consider, in particular, two types of applications. In one we transform certain second order dynamic equations with a general coefficient at the leading term into dynamic equations of a similar type, on a different time scale, but with the leading coefficient equaling to one. In the other type of application we transform a dynamic equation under somehow critical setting (this will be specified later) into a dynamic equation on a different time scale under certain non-critical setting which can be handled by existing results.

Some analysis of basic aspects related to transformations of difference and dynamic equations that are close to our topic has already occurred in the literature. Transformations in linear Hamiltonian systems on time scales are treated in [3], Sturm–Liouville expressions on Sturmian time scales (the time scales that contain only isolated or l-d/r-d points) are, from this point of view, studied in [4], and the transformations for even order difference operators are considered in [25]. On the other hand, it seems that the ideas from our paper, although being practically known in the differential equations case, have not been extensively applied in dynamic equations in the literature, in spite of availability of various results on a general time scale.

For an outline of the first mentioned type of applications, let us give one problem from oscillation theory. Let us say we have oscillation criteria (for definiteness, the so-called Hille–Nehari criteria, see Sect. 2) for the difference equation $\Delta^2 y_k + p_k y_{k+1} = 0$ at disposal, and we are interested in criteria for the more general equation $\Delta(r_k \Delta y_k) + p_k y_{k+1} = 0$. It is problematic, in contrast to the corresponding differential equations case, to transform the latter form of the equation into the former one. Thus one would say that we have to analyse the difference equation directly in the more general form. However, there is also other possibility which is characteristic for what we do in this paper. If we know the criteria for the dynamic equation of the form $y^{\Delta\Delta} + p(t)y^\sigma = 0$ on a general time scale, then a suitable transformation of independent variable (which preserves oscillation properties) in the equation $\Delta(r_k \Delta y_k) + p_k y_{k+1} = 0$ can transform it into the former dynamic equation, which is then examined by existing results. Note that historically, the Hille–Nehari type criteria for the former, and simpler, difference equation were obtained as first, see, e.g., [7, 10]. It is worthy of mention also the following interesting fact related to this discrete oscillation problem. The Hille–Nehari type criteria for the simpler equation

$\Delta^2 y_k + p_k y_{k+1} = 0$ involves the well known critical constant $1/4$. But passing to the more general case, to the criteria for the equation $\Delta(r_k \Delta y_k) + p_k y_{k+1} = 0$, we find out that the critical constant can have a value different from $1/4$, in contrast to the differential equations case, and depends on the coefficient r. In fact, as we will see later, for dynamic equations we reveal also its dependence on the graininess of time scale.

To outline the second type of applications, we consider the difference equation $\Delta(r_k \Delta y_k) = p_k y_{k+1}$, where $r, p > 0$. It is known [18] that under the condition (which in fact can be weakened) $\lim_{k\to\infty} k\Delta p_k/p_k = \delta$ we are able to establish quite precise asymptotic formulae via the discrete theory of regular variation provided $\delta \neq -1$. The critical case $\delta = -1$ leads to a somehow delicate setting; it turns out that a suitable transformation of independent variable can help here, and brings us to dynamic equations which are in a non-critical setting in the above sense.

The paper is organized as follows. In the next section we deal with the so-called Hille–Nehari criteria for half-linear dynamic equations; we give a new proof to existing results. New results are obtained in Sect. 3 where we derive oscillation criteria for nonlinear dynamic equations. In Sect. 4 we study two variants of Euler type equations for which we establish the values of their oscillation constant. In addition to transformation of independent variable, a transformation of dependent variable plays a role, too. In the last section, we present also a new result. We do an asymptotic analysis of linear dynamic equations and establish asymptotic formulae for solutions in a critical case which is missing in the existing literature.

Let, as usually, $\mathbb{T}$ denote a time scale, which is assumed to be unbounded from above in our paper. We use the standard time scale notation, see [5, 6]. In particular, the symbols σ, μ, f^σ, f^Δ, $\int_a^b f(s)\,\Delta s$, $[a,b]_\mathbb{T}$, and $C_{\rm rd}$ stand for forward jump operator, graininess, $f \circ \sigma$, delta derivative, delta integral, a time scale interval, and the class of rd-continuous functions, respectively. The symbols $\widetilde{\sigma}$, $\widetilde{\mu}$, $f^{\widetilde{\Delta}}$, $\int_a^b f(s)\,\widetilde{\Delta} s$ have an analogous meaning, with being associated to a time scale $\widetilde{\mathbb{T}}$. For functions defined on $\mathbb{T}$ we denote: $f(t) \sim g(t)$ as $t \to \infty$ if $\lim_{t\to\infty} f(t)/g(t) = 1$; $f(t) = o(g(t))$ as $t \to \infty$ if $\lim_{t\to\infty} f(t)/g(t) = 0$; $f(t) = O(g(t))$ as $t \to \infty$ if $\exists c \in (0,\infty)$ such that $|f(t)| \le c|g(t)|$ for large $t \in \mathbb{T}$.

2 Oscillation of Half-Linear Dynamic Equations

We start with a simple observation concerning Hille–Nehari type oscillation criteria for half-linear dynamic equations. For information about these and other criteria for linear and half-linear differential and difference equations see [1, 2, 8, 24]. The result presented in the next theorem is actually known, see [16]. However, as already indicated, our aim is to demonstrate how the ideas based on a transformation can serve to establish a new proof where the result for the original equation will be obtained from the result for a simpler equation. A similar approach can be used whenever criteria for half-linear dynamic equations (including functional ones) or some other types of second order equations with the leading coefficient 1 are at disposal.

Consider the half-linear dynamic equation

$$(r(t)\Phi(y^{\Delta}))^{\Delta} + p(t)\Phi(y^{\sigma}) = 0, \tag{1}$$

where $\Phi(u) = |u|^{\alpha-1}\mathrm{sgn} u$ with $\alpha > 1$, and assume $p \in C_{\mathrm{rd}}([a,\infty)_{\mathbb{T}}, \mathbb{R})$, $1/r \in C_{\mathrm{rd}}([a,\infty)_{\mathbb{T}}, (0,\infty))$, $\int_a^{\infty} r^{1-\beta}(s)\,\Delta s = \infty$, $1/\alpha + 1/\beta = 1$, and $\int_t^{\infty} p(s)\,\Delta s$ exists, is nonnegative and eventually nontrivial for large t. Thanks to the Sturm type separation result, one (nontrivial) solution of (1) is oscillatory (i.e., it is neither eventually positive nor eventually negative) if and only if all solutions are oscillatory. Hence, we can classify Eq. (1) as oscillatory (all its solutions are oscillatory) or nonoscillatory (all its solutions are nonoscillatory).

Note that the constants which appear on the right-hand sides of the criteria in Theorem 1 depend on the graininess of time scale and the coefficient r. Denote

$$R_{\alpha}(t) = \int_a^t r^{1-\beta}(s)\,\Delta s, \quad M_* = \liminf_{t\to\infty} \frac{\mu(t) r^{1-\beta}(t)}{R_{\alpha}(t)}, \quad M^* = \limsup_{t\to\infty} \frac{\mu(t) r^{1-\beta}(t)}{R_{\alpha}(t)},$$

and

$$\gamma_{\alpha}(x) = \lim_{t\to x} \left(\frac{(t+1)^{\frac{\alpha-1}{\alpha}} - 1}{t} \right)^{\alpha-1} \left(1 - \frac{1-(t+1)^{-\frac{(\alpha-1)^2}{\alpha}}}{1-(t+1)^{1-\alpha}} \right).$$

Examples of particular settings are presented in Remark 1 and Sect. 4. The form of the constant $\gamma_{\alpha}(x)$ is related to the roots of a certain algebraic equation which is associated to a generalized Riccati type dynamic equation [16]; here the generalized Riccati dynamic equation is the first order nonlinear equation arising from (1) through the substitution $w = r\Phi(y^{\Delta}/y)$ [1, 13, 16].

Theorem 1 *If*

$$\liminf_{t\to\infty} R_{\alpha}^{\alpha-1}(t) \int_t^{\infty} p(s)\,\Delta s > \gamma_{\alpha}(M_*), \tag{2}$$

then (1) *is oscillatory. If*

$$\limsup_{t\to\infty} R_{\alpha}^{\alpha-1}(t) \int_t^{\infty} p(s)\,\Delta s < \gamma_{\alpha}(M^*), \tag{3}$$

then (1) *is nonoscillatory.*

Proof Let y be a solution of (1). Set $u(s) = y(t)$, $s = \tau(t)$, where $\tau : \mathbb{T} \to \mathbb{R}$, being more precisely defined later, is strictly increasing. Denote $\widetilde{\mathbb{T}} = \{\tau(t) : t \in \mathbb{T}\}$. Here we assume that the delta derivatives which are involved in our computations exist; later we will see that it indeed holds under our particular setting. Moreover, our τ will always be at least in C_{rd}^1 (so, in particular, will be continuous) and our $\widetilde{\mathbb{T}}$ will be (an unbounded) time scale. In view of the chain rule [5, Theorem 1.93], we have $y^\Delta = (u^{\widetilde{\Delta}} \circ \tau)\tau^\Delta$. Using the chain rule again,

$$\begin{aligned}(r\Phi(y^\Delta))^\Delta &= \left(r\Phi(\tau^\Delta)\Phi(u^{\widetilde{\Delta}} \circ \tau)\right)^\Delta = \left[[(r\Phi(\tau^\Delta)) \circ \tau^{-1} \circ \tau]\,\Phi(u^{\widetilde{\Delta}} \circ \tau)\right]^\Delta \\ &= \left[[(r\Phi(\tau^\Delta)) \circ \tau^{-1}]\Phi(u^{\widetilde{\Delta}})\right]^{\widetilde{\Delta}} \circ \tau\,\tau^\Delta.\end{aligned} \tag{4}$$

Thanks to the properties of τ, we have $\tau \circ \sigma = \widetilde{\sigma} \circ \tau$, and so $(u \circ \tau)^\sigma = u^{\widetilde{\sigma}} \circ \tau$. Therefore, in view of (4), u satisfies the equation

$$\left(\widetilde{r}(s)\Phi(u^{\widetilde{\Delta}})\right)^{\widetilde{\Delta}} + \widetilde{p}(s)\Phi(u^{\widetilde{\sigma}}) = 0 \tag{5}$$

on $\widetilde{\mathbb{T}}$, where

$$\widetilde{r} = (r\Phi(\tau^\Delta)) \circ \tau^{-1} \text{ and } \widetilde{p} = \frac{p}{\tau^\Delta} \circ \tau^{-1}.$$

Now we set $\tau = R_\alpha$. Then $\widetilde{\mathbb{T}} = \tau(\mathbb{T})$ is an unbounded time scale and, in particular, the interval $[a, \infty)_\mathbb{T}$ is transformed into the interval of the form $[\widetilde{a}, \infty)_{\widetilde{\mathbb{T}}}$. Further, $\tau^\Delta = r^{1-\beta}$, thus $\widetilde{r} = (rr^{(\alpha-1)(1-\beta)}) \circ \tau^{-1} = (r/r) \circ \tau^{-1} = 1$. From [16] we know that (5) is oscillatory provided $\liminf_{s\to\infty} s^{\alpha-1} \int_s^\infty \widetilde{p}(\eta)\,\widetilde{\Delta}\eta > \gamma_\alpha(\widetilde{M}_*)$ and nonoscillatory provided $\limsup_{s\to\infty} s^{\alpha-1} \int_s^\infty \widetilde{p}(\eta)\,\widetilde{\Delta}\eta < \gamma_\alpha(\widetilde{M}^*)$, where $\widetilde{M}_* = \liminf_{s\to\infty} \widetilde{\mu}(s)/s$, $\widetilde{M}^* = \limsup_{s\to\infty} \widetilde{\mu}(s)/s$, and the integral $\int_s^\infty \widetilde{p}(\eta)\,\widetilde{\Delta}\eta$ is nonnegative and eventually nontrivial for large s. We have

$$\begin{aligned}\frac{\widetilde{\mu}(s)}{s} &= \frac{\widetilde{\sigma}(s) - s}{s} = \frac{(\widetilde{\sigma} \circ R_\alpha)(t) - R_\alpha(t)}{R_\alpha(t)} \\ &= \frac{R_\alpha^\sigma(t) - R_\alpha(t)}{R_\alpha(t)} = \frac{\mu(t)R_\alpha^\Delta(t)}{R_\alpha(t)} = \frac{\mu(t)r^{1-\beta}(t)}{R_\alpha(t)},\end{aligned}$$

and so $\widetilde{M}_* = M_*$ and $\widetilde{M}^* = M^*$. Further, applying the substitution method in delta integrals [5, Theorem 1.98], see also [6, Theorem 5.40], we obtain

$$\begin{aligned}s^{\alpha-1} \int_s^S \widetilde{p}(\eta)\,\widetilde{\Delta}\eta &= R_\alpha^{\alpha-1}(t) \int_{R_\alpha(t)}^{R_\alpha(T)} \left(\frac{p}{R^\Delta} \circ R^{-1}\right)(\eta)\,\widetilde{\Delta}\eta \\ &= R_\alpha^{\alpha-1}(t) \int_t^T \frac{p(\xi)}{R^\Delta(\xi)} R^\Delta(\xi)\,\Delta\xi = R_\alpha^{\alpha-1}(t) \int_t^T p(\xi)\,\Delta\xi,\end{aligned}$$

where $S = R_\alpha(T)$, $T \in [t, \infty)_\mathbb{T}$. Letting T to ∞, we get

$$s^{\alpha-1} \int_s^\infty \widetilde{p}(\eta)\,\widetilde{\Delta}\eta = R_\alpha^{\alpha-1}(t) \int_t^\infty p(\xi)\,\Delta\xi.$$

Since our transformation preserves (non)oscillation of the equation, the statement is now clear. □

Remark 1 (i) Let $M := M_* = M^*$. Then the constants on the right-hand sides of (2) and (3) are the same and we have

$$\gamma_\alpha(M) = \begin{cases} \frac{1}{\alpha}\left(\frac{\alpha-1}{\alpha}\right)^{\alpha-1} & \text{if } M = 0, \\ \left(\frac{(M+1)^{\frac{\alpha-1}{\alpha}}-1}{M}\right)^{\alpha-1}\left(1 - \frac{1-(M+1)^{-\frac{(\alpha-1)^2}{\alpha}}}{1-(M+1)^{1-\alpha}}\right) & \text{if } 0 < M < \infty, \\ 0 & \text{if } M = \infty. \end{cases}$$

For example, if $\mu(t) = 0$ or $r(t) = 1$ with $\mu(t) = o(t)$ as $t \to \infty$, then $M_* = M^* = 0$. If $\mu(t) = (q-1)t$, $q > 1$, (as in q-calculus), then $M_* = M^* = q - 1 > 0$. In the case corresponding to linear equations we have

$$\gamma_2(M) = \begin{cases} \frac{1}{4} & \text{if } M = 0, \\ \frac{1}{(\sqrt{M+1}+1)^2} & \text{if } 0 < M < \infty, \\ 0 & \text{if } M = \infty. \end{cases}$$

In particular, $\gamma_2(0) = 1/4$, which is the well known constant from oscillation theory of linear DEs, ee e.g. [24]. See also [7, 10] for the linear discrete case where $r(t) = 1$. We again emphasize that if $r(t) \not\equiv 1$, then even in the difference equation case, the value $1/4$ does not need to be maintained; taking, e.g., $r(t) = 2^{-t}$, we get $M = 2$, and so $\gamma_2(M) = (\sqrt{2}+1)^{-2}$. Further, the constant $\gamma_2(M)$ can differ from $1/4$ also when $r(t) = 1$. For instance, if $\mathbb{T} = q^{\mathbb{N}_0} := \{q^k : k \in \mathbb{N}_0\}$, $q > 1$, then $M = q - 1$, and so $\gamma_2(M) = (\sqrt{q}+1)^{-2}$.

(ii) In view of the previous remark, as very special cases of Theorem 1, we get the following criteria. Let $\int_t^\infty p(s)\,\Delta s \geq 0$ for large t. Assuming $\mathbb{T} = \mathbb{R}$, $\alpha = 2$, $r(t) = 1$, if $\liminf_{t\to\infty} t\int_t^\infty p(s)\,\mathrm{d}s > 1/4$ ($\limsup_{t\to\infty} t\int_t^\infty p(s)\,\mathrm{d}s < 1/4$), then $y'' + p(t)y = 0$ is oscillatory (nonoscillatory), cf. [24]. Assuming $\mathbb{T} = \mathbb{Z}$, $\alpha = 2$, $r(t) = 1$, if $\liminf_{t\to\infty} t\sum_{j=t}^\infty p(j) > 1/4$ ($\limsup_{t\to\infty} t\sum_{j=t}^\infty p(j) < 1/4$), then $\Delta^2 y(t) + p(t)y(t+1) = 0$ is oscillatory (nonoscillatory), cf. [7, 10].

(iii) In the previous proof we used the criteria from [16]. The proof of those results is based on the function sequence technique combined with Riccati type transformation. Note that another possibility how to prove those criteria could be, for example, to combine the information about the oscillation constant of a certain Euler type half-linear dynamic equation (provided we have it at disposal) with integral comparison theorem, see [14, Theorem 11] and [16, Sect. 7].

(iv) Looking at the conditions posed on the coefficients of (1) in Theorem 1, a natural problem arises out, namely to obtain analogous criteria when the integral $\int_a^\infty r^{1-\beta}(s)\,\Delta s$ converges. The trouble in this case is that the same transformation as in the previous proof transforms the range of definition into a bounded set. We do not aim to treat this problem in our paper. Note only that in the linear case, we can use the transformation of dependent variable $y = hu$, $h(t) = \int_t^\infty 1/r(s)\,\Delta s$

(similarly as below in Sect. 4), where the original equation is transformed into an equation of the same type, but with the divergent integral of the reciprocal of its leading term. This approach is not at our disposal in the half-linear case, since it requires the linearity of solution space. However, we can think—and we believe it could work—about replacement for half-linear equations in the sense that the corresponding transformation is made in terms of the associated generalized Riccati equations; the nonlinear term in the Riccati equation associated to the original half-linear equation is somehow quadrated in asymptotic sense. For some applications of this idea in the differential equations case see e.g. [9, 19]. Other possibility is to utilize the so-called reciprocity principle (see e.g. [1]); the original equation is transformed via the relation $u = r\Phi(y^{\Delta})$ into an equation of the same type, where the new equation satisfies the assumption of the divergence of the integral containing the leading term. This approach however requires to overcome some technical problems since the delta derivative and the jump operator do not commute on a general time scale; a possibility is to consider the transformed equation in an integral form or to work with systems of two first order equations. Finally note that Hille–Nehari type criteria under the condition $\int_a^{\infty} r^{1-\beta}(s)\,\Delta s < \infty$ are directly in this setting proved in [17] via the function sequence technique involving a weighted Riccati transformation. For further information related to oscillation and other qualitative properties of half-linear equations see [8] (differential equations case), [2, 12] (difference equations case), and [1, 13] (dynamic equations case).

3 Oscillation of Nonlinear Dynamic Equations

Here we prove sharp criteria, which generalize existing ones (see Remark 2), for nonlinear dynamic equations of the form

$$(r(t)y^{\Delta})^{\Delta} + p(t)f(y) = 0, \tag{6}$$

where $p \in C_{\rm rd}([a,\infty)_{\mathbb{T}}, \mathbb{R})$ $1/r \in C_{\rm rd}([a,\infty)_{\mathbb{T}}, (0,\infty))$, $\int_a^{\infty} 1/r(s)\,\Delta s = \infty$, and f is a continuous function on $\mathbb{R}$ satisfying $xf(x) > 0$ for $x \neq 0$. Denote $R(t) = \int_a^t 1/r(s)\,\Delta s$.

Theorem 2 *(a) If there exists $\lambda \in \mathbb{R}$ with $\lambda > 1/4$ such that*

$$R(t)R^{\sigma}(t)r(t)p(t)\frac{f(x)}{x} \geq \lambda \tag{7}$$

for $t \in [a,\infty)_{\mathbb{T}}$ large and $|x|$ large, then all nontrivial solutions of (6) are oscillatory.

(b) If

$$R(t)R^{\sigma}(t)r(t)p(t)\frac{f(x)}{x} \leq \frac{1}{4} \tag{8}$$

for $t \in [a, \infty)_{\mathbb{T}}$ *large and* $x > 0$ *or* $x < 0$ *with* $|x|$ *large, then* (6) *has a nonoscillatory solution.*

Proof (a) Let y be a solution of (6). Set $u(s) = y(t)$, $s = R(t)$. Then, using the arguments similar to those in the proof of Theorem 1, we get that u satisfies the equation

$$u^{\widetilde{\Delta}\widetilde{\Delta}} + \widetilde{p}(s) f(u) = 0 \tag{9}$$

on the time scale $\widetilde{\mathbb{T}} = \{R(t) : t \in \mathbb{T}\}$, where $\widetilde{p} = (p/R^{\Delta}) \circ R^{-1} = (pr) \circ R^{-1}$. The coefficient in the leading term of (9) is equal to 1 since $rR^{\Delta} = r/r = 1$. The result now follows from the transformation relations and [22, Theorem 5.1] applied to Eq. (9); that theorem says that all nontrivial solutions of (9) are oscillatory provided there is $\lambda > 1/4$ such that $s\widetilde{\sigma}(s)\widetilde{p}(s) f(x)/x \geq \lambda$ for large $s \in \widetilde{\mathbb{T}}$ and $|x|$.

(b) The proof is similar to that of part (a); here we apply [22, Theorem 5.2] to transformed equation (9). □

Remark 2 (i) If $r(t) = 1$ and $p(t) = 1/(r(t)R(t)R^{\sigma}(t))$ with $a = 0$, i.e., $p(t) = 1/(t\sigma(t))$, then Theorem 2-(a) reduces to [22, Theorem 1.1] and Theorem 2-(b) reduces to [22, Theorem 1.2]. If $p = 1/(rRR^{\sigma})$, then the left-hand sides of (7) and (8) read as $f(x)/x$ and depend only on f. If $p = 1/(rR^2)$, then (6) can be seen as a "more natural time scale discretization" (when compared with the setting $p = 1/(rRR^{\sigma})$) of the differential equation $(r(t)y')' + f(y)/(r(t)R^2(t)) = 0$ on $\mathbb{T} = \mathbb{R}$. Conditions (7) and (8) read as $f(x)/x \geq \lambda R(t)/R^{\sigma}(t)$ and $f(x)/x \leq R(t)/(4R^{\sigma}(t))$, respectively, and we see how a larger graininess is "more favorable" to oscillation in this case.

(ii) For more information on the criteria of the type presented in Theorem 2 see [23] (in differential equations case) and [26] (in difference equations case). See also the last paragraph of the next section.

4 Oscillation Constants for Euler Type Linear Dynamic Equations and Their Perturbations

In this section we establish the so-called oscillation constant for two variants of Euler type dynamic equation. By oscillation constant of the equation $y^{\Delta\Delta} + p(t; \lambda)y^{\sigma} = 0$ we mean the number λ_0 such that the equation is oscillatory for $\lambda > \lambda_0$ and nonoscillatory for $\lambda < \lambda_0$. For other equations we define this concept similarly. As indicated in several points in this paper, Euler type equations (or their perturbations) are important for comparison purposes, see also e.g. [2, 8, 16, 22–24, 26]. By (non)oscillation of the equation we mean (non)oscillation of all its nontrivial solutions.

It is worthy of note that while in Theorem 3, the oscillation constant has the fixed value $1/4$, the oscillation constant for the equation considered in Theorem 4 depends on the coefficient r and the graininess of a time scale.

Consider first the equation

$$(r(t)y^{\Delta})^{\Delta} + \lambda p(t)y = 0, \tag{10}$$

where $1/r \in C_{\mathrm{rd}}([a,\infty)_{\mathbb{T}},(0,\infty))$ and $p(t)$ will be specified in Theorem 3. An interesting fact is that for both equations considered in Theorem 3 we can write their general solutions, see Remark 3. The Sturmian theory (in particular, the separation result) does not hold in general for equations of the form (10), in contrast to equations of the form (11). This means that we have not guaranteed the implication: one solution is (non)oscillatory implies all solutions are (non)oscillatory. In spite of this fact, under our special setting, we state the "true" oscillation constant which is defined as above, i.e., via (non)oscillation of equation.

The next result is new, it is an improvement of [22, Proposition 2.3]. Actually, Theorem 3-(a) can be obtained as an immediate consequence of Theorem 2. However, we offer an alternative way of the proof—it is based on a suitable transformation.

Theorem 3 *(a) Let* $\int_a^\infty 1/r(s)\,\Delta s=\infty$. *Then Eq.* (10) *with*

$$p(t)=\frac{1}{r(t)R(t)R^\sigma(t)},\quad R(t)=\int_a^t\frac{1}{r(s)}\,\Delta s,$$

has the oscillation constant $\lambda=1/4$.

(b) Let $\int_a^\infty 1/r(s)\,\Delta s<\infty$. *Then Eq.* (10) *with*

$$\frac{1}{p(t)}=r(t)R_c(t)R_c^\sigma(t)\left(1-\frac{R_c(t)}{R_c(a)}\right)\left(1-\frac{R_c^\sigma(t)}{R_c(a)}\right),\quad R_c(t)=\int_t^\infty\frac{1}{r(s)}\,\Delta s,$$

has the oscillation constant $\lambda=1/4$.

Proof (a) Let y be a solution of (10). Set $u(s)=y(t)$ and $s=R(t)$. Then, similarly as in the proof of Theorem 1, we get that u satisfies the equation $(\widetilde{r}(s)u^{\widetilde{\Delta}})^{\widetilde{\Delta}}+\lambda\widetilde{p}(s)u=0$ on the time scale $\widetilde{\mathbb{T}}=\{R(t):t\in\mathbb{T}\}$, where $\widetilde{r}(s)=(rR^\Delta)\circ R^{-1}(s)=1$ and $\widetilde{p}(s)=(p/R^\Delta)\circ R^{-1}(s)=(1/(RR^\sigma))\circ R^{-1}(s)=1/(s\widetilde{\sigma}(s))$, i.e., the equation $u^{\widetilde{\Delta}\widetilde{\Delta}}+(\lambda/(s\widetilde{\sigma}(s)))u=0$. Applying [22, Proposition 2.3] and the ideas of its proof, we obtain that the oscillation constant of the latter equation is $\lambda=1/4$. From the transformation relations it is clear that $\lambda=1/4$ is the oscillation constant also for original Eq. (10).

(b) First we introduce the dynamic operators $\mathcal{L}[y]=(ry^\Delta)^\Delta+py$ (note that here r and p can be general) and $\widehat{\mathcal{L}}[y]=(\widehat{r}y^\Delta)^\Delta+\widehat{p}y$, where $\widehat{r}=rhh^\sigma$ and $\widehat{p}=phh^\sigma$. It can be shown that if $h=R_c$, but also if $h=R$, then $h^\sigma\mathcal{L}[hz]=\widehat{\mathcal{L}}[z]$ for a sufficiently smooth z. Consequently, if y is a solution of (10) and we set $y=hz$, where $h=R_c$, then z satisfies the equation $(\widehat{r}(t)z^\Delta)^\Delta+\widehat{p}(t)z=0$, where $\widehat{r}=rR_cR_c^\sigma$ and

$$\widehat{p}(t)=\frac{\lambda}{r(t)(1-R_c(t)/R_c(a))(1-R_c^\sigma(t)/R_c(a))}.$$

We have

$$\int_a^t\frac{1}{\widehat{r}(s)}\,\Delta s=\int_a^t\left(\frac{1}{R_c(s)}\right)^\Delta\Delta s=\frac{1}{R_c(t)}-\frac{1}{R_c(a)}\to\infty$$

as $t \to \infty$. Denote $\widehat{R}(t) = \int_a^t 1/\widehat{r}(s)\,\Delta s$. Then

$$\frac{\lambda}{\widehat{r}(t)\widehat{R}(t)\widehat{R}^\sigma(t)} = \frac{\lambda}{r(t)R_c(t)R_c^\sigma(t)(1/R_c(t) - 1/R_c(a))(1/R_c^\sigma(t) - 1/R_c(a))}$$
$$= \frac{\lambda}{r(t)(1 - R_c(t)/R_c(a))(1 - R_c^\sigma(t)/R_c(a))} = \widehat{p}(t).$$

Now we can apply part (a) to the equation $(\widehat{r}(t)z^\Delta)^\Delta + \widehat{p}(t)z = 0$ to obtain that its oscillation constant is $\lambda = 1/4$ and, in view of the transformation relations, it is also the oscillation constant of (10). □

Remark 3 (i) Note that for the coefficient $p(t)$ in part (b) of the previous theorem we have $p(t) \sim 1/(r(t)R_c(t)R_c^\sigma(t))$ as $t \to \infty$. This means that for typical (namely asymptotic) comparison purposes involving (10) the coefficient $p(t)$ in the setting of (b) can be replaced by $1/(r(t)R_c(t)R_c^\sigma(t))$.

(ii) It is worthy of note that for the equations considered in Theorem 3 (in contrast for those in Theorem 4), we can establish the exact form of a general solution. We omit details. Let us note just that the arguments for such a statement are based on the transformation of independent variable in case (a) and the transformation of independent and dependent variable in case (b) and the knowledge (see [11]) of the general solution for the equation $u^{\widetilde{\Delta}\widetilde{\Delta}} + (\lambda/(s\widetilde{\sigma}(s)))u = 0$.

In the next theorem we consider a different variant of Euler type equation, namely

$$(r(t)y^\Delta)^\Delta + \lambda p(t)y^\sigma = 0, \tag{11}$$

where $1/r \in C_{\mathrm{rd}}([a,\infty)_{\mathbb{T}}, (0,\infty))$ and $p(t)$ will be specified in Theorem 4. As we will see, the difference between (10) and (11) is only seemingly slight—notice how the form of the oscillation constant is affected. The next result is known, but here we offer an alternative approach to its proof.

Theorem 4 *(a) Let $\int_a^\infty 1/r(s)\,\Delta s = \infty$ and the limit*

$$\lim_{t\to\infty} \frac{\mu(t)}{r(t)\int_a^t 1/r(s)\,\Delta s} =: M \in [0,\infty) \cup \{\infty\}$$

exist. Then equation (11) *with*

$$p(t) = \frac{1}{r(t)R(t)R^\sigma(t)}, \quad R(t) = \int_a^t \frac{1}{r(s)}\,\Delta s,$$

has the oscillation constant $\lambda = (\sqrt{M+1} + 1)^{-2}$ provided $M < \infty$. If $M = \infty$, then this equation is oscillatory for all $\lambda > 0$ (thus it is strongly oscillatory) and nonoscillatory otherwise.

(b) Let $\int_a^\infty 1/r(s)\,\Delta s < \infty$ *and the limit*

$$\lim_{t\to\infty} \frac{\mu(t)}{r(t)\int_{\sigma(t)}^\infty 1/r(s)\,\Delta s} =: N \in [0,\infty) \cup \{\infty\}$$

exist. Then Eq. (11) *with*

$$p(t) = \frac{1}{r(t)(R_c^\sigma(t))^2}, \quad R_c(t) = \int_t^\infty \frac{1}{r(s)}\,\Delta s,$$

has the oscillation constant $\lambda = (\sqrt{N+1}+1)^{-2}$ *provided* $N < \infty$*. If* $N = \infty$*, then this equation is oscillatory for all* $\lambda > 0$ *(thus it is strongly oscillatory) and nonoscillatory otherwise.*

Proof (a) Let y be a solution of (11). Set $u(s) = y(t)$ and $s = R(t)$. Then by the arguments similar to those in the proof of Theorem 1, u can be shown to satisfy the equation $u^{\widetilde{\Delta}\widetilde{\Delta}} + (\lambda/(s\widetilde{\sigma}(s)))u^{\widetilde{\sigma}} = 0$ on the time scale $\widetilde{\mathbb{T}} = \{R(t) : t \in \mathbb{T}\}$. From [15] we know that the oscillation constant for this equation is $\lambda = (\sqrt{M_0+1}+1)^{-2}$, where $M_0 := \lim_{s\to\infty} \widetilde{\mu}(s)/s \in [0,\infty)$. Since (assuming here that $M < \infty$)

$$\frac{\widetilde{\mu}(s)}{s} = \frac{\widetilde{\sigma}(s)-s}{s} = \frac{R^\sigma(t)-R(t)}{R(t)} = \frac{\mu(t)R^\Delta(t)}{R(t)} = \frac{\mu(t)}{r(t)R(t)}.$$

the (finite) limit M_0 indeed exists and we have

$$M_0 = \lim_{s\to\infty} \frac{\widetilde{\mu}(s)}{s} = \lim_{t\to\infty} \frac{\mu(t)}{r(t)R(t)} = M.$$

Thus the oscillation constant for the original Eq. (11) is $\lambda = (\sqrt{M+1}+1)^{-2}$. Oscillation of (11) for all $\lambda > 0$ when $M = \infty$ follows from strong oscillation (see [15]) of the equation $u^{\widetilde{\Delta}\widetilde{\Delta}} + (\lambda/(s\widetilde{\sigma}(s)))u^{\widetilde{\sigma}} = 0$ when $M_0 = \infty$. Alternatively, the result can be proved by a direct application of Theorem 1.

(b) The statement can be proved via transforming the equation under consideration into an equation satisfying the setting of (a) in the following sense. If we denote $\mathcal{L}_s[y] = (ry^\Delta)^\Delta + py^\sigma$ and $\widehat{\mathcal{L}}_s[y] = (\widehat{r}y^\Delta)^\Delta + \widehat{p}y^\sigma$, where $\widehat{r} = rhh^\sigma$ and $\widehat{p} = h^\sigma\mathcal{L}_s[h]$, then $h^\sigma\mathcal{L}_s[hz] = \widehat{\mathcal{L}}_s[z]$. Therefore, being y a solution of $\mathcal{L}_s[y] = 0$ and setting $y = hz, h \neq 0$, we get that z is a solution of $\widehat{\mathcal{L}}_s[z] = 0$. Since $\int_a^\infty 1/\widehat{r}(s)\,\Delta s = \infty$ when $h = R_c$, we find ourselves in the setting of (a). The details are left to the reader. □

Remark 4 We emphasize that the fact whether we consider or not the jump operators in the coefficient $\lambda/(t\sigma(t))$ of the equations which appear in the proofs (or in their generalized versions presented in the theorems) plays an important role. Let us demonstrate it on the time scale $\mathbb{T} = q^{\mathbb{N}_0} := \{q^k : k \in \mathbb{N}_0\}$, $q > 1$, for the equation $y^{\Delta\Delta} + \frac{\lambda}{t\sigma(t)}y^\sigma = 0$ and its variants. Since $M = \lim_{t\to\infty}\mu(t)/t = q-1$, the oscillation constant for this equation is $\lambda = \frac{1}{(\sqrt{q}+1)^2}$ and it (linearly) decreases to zero as $q \to \infty$. The oscillation constant for the q-difference equation $y^{\Delta\Delta} + \frac{\lambda}{(\sigma(t))^2} y^\sigma = 0$ is $\lambda = \frac{q}{(\sqrt{q}+1)^2}$ and it increases to 1 as $q \to \infty$. Finally, the oscillation constant for the q-difference equation $y^{\Delta\Delta} + \frac{\lambda}{t^2} y^\sigma = 0$ is $\lambda = \frac{1}{q(\sqrt{q}+1)^2}$ and it (quadratically) decreases to 0 as $q \to \infty$. In all these cases the oscillation constant tends to $\frac{1}{4}$ as $q \to 1+$.

Perturbed Euler type equations In view of the previous considerations a natural problem arises out: to consider an Euler type equation where the parameter in the coefficient reaches its critical value and to study how perturbations of this term affect oscillatory properties of the equation; note that there can be revealed the relation of this critical setting with the double root case of the associated algebraic equation in some instances. Let us recall that a suitable combination of transformations of dependent and independent variable, precisely, $s = \ln t$ and $u(s) = t^{-1/2} y(t)$, can transform the perturbed Euler differential equation (the so-called Riemann–Weber equation)

$$y'' + \frac{1}{t^2}\left(\frac{1}{4} + \frac{\lambda}{\ln^2 t}\right) y = 0$$

into the equation

$$\frac{\mathrm{d}^2 u}{\mathrm{d}s^2} + \frac{\lambda}{s^2} u = 0.$$

This trick can be applied repeatedly, thus equations such as

$$y'' + \frac{1}{t^2}\left(\frac{1}{4} + \frac{1}{4}\sum_{k=1}^{n-1} \frac{1}{\mathrm{Ln}_k^2 t} + \frac{\lambda}{\mathrm{Ln}_n^2 t}\right) y = 0,$$

where $\mathrm{Ln}_k t = \prod_{j=1}^{k} \ln_j t$ and $\ln_{j+1} t = \ln(\ln_j t)$, can be treated. We believe that utilizing suitable transformations of dependent and independent variable in combination with existing theory on time scales, will enable us to examine, for example, the equation

$$y^{\Delta\Delta} + \frac{1}{t\sigma(t)}\left(\omega_0 + \frac{\lambda}{\ell(t)\ell^\sigma(t)}\right) y^\sigma = 0 \tag{12}$$

or equations of similar forms, where ω_0 is the oscillation constant of the equation $y^{\Delta\Delta} + (\omega/t\sigma(t)) y^\sigma = 0$ and ℓ is a function which is somehow related to a logarithmic function. For instance, we conjecture that under the assumption $\mu(t) = o(t)$, the substitutions $y(t) = (u(t)/2) \int_a^t s^{-1/2}\,\Delta s$, $z(s) = u(t)$, $s = \ell(t) = \int_a^t (1/s)\,\Delta s$ transforms (12) with $\omega_0 = 1/4$ into the equation

$$\left((1 + o(1)) z^{\widetilde{\Delta}}\right)^{\widetilde{\Delta}} + (1 + o(1)) \frac{\lambda}{s\widetilde{\sigma}(s)} z^{\widetilde{\sigma}} = 0$$

on the time scale $\widetilde{\mathbb{T}} = \ell(\mathbb{T})$, and this equation can be further transformed introducing new independent variable η (similarly as in the proof of Theorem 1) into the equation

$$w^{\overline{\Delta}\overline{\Delta}} + (1 + o(1))\frac{\lambda}{\eta\overline{\sigma}(\eta)} w^{\overline{\sigma}} = 0$$

on a certain (unbounded) time scale $\overline{\mathbb{T}}$. Having at disposal information about perturbed Euler type dynamic equations, we can apply them, in combination with some comparison principle, to obtain (non)oscillation criteria for related linear or nonlinear dynamic equations, see, e.g. [1, 2, 14, 16, 22]. Note that there are also other equations, which can be understood as a perturbation of the equation under a certain critical setting, for example,

$$(t y^{\Delta})^{\Delta} + \frac{L(t)}{t} y^{\sigma} = 0, \tag{13}$$

where L varies slowly in some way (see Remark 5); likewise this setting corresponds somehow with the above mentioned critical double root case. The applications are expected not only in oscillation theory, but also in asymptotic theory. Indeed, the setting which is considered in below given Theorem 5 (see also Remark 5-(ii), (vi)), where we deal with asymptotic formulae, includes Eq. (13) as a special case.

5 Asymptotic Formulae

Consider the equation

$$(r(t) y^{\Delta})^{\Delta} = p(t) y^{\sigma}, \tag{14}$$

where $p, 1/r \in C_{\mathrm{rd}}([a, \infty)_{\mathbb{T}}, (0, \infty))$. This equation is nonoscillatory (see e.g. [1]) and any its nontrivial solution is eventually monotone (see [21]). Thus the set of eventually positive solutions of (14) consists of eventually positive decreasing solutions and eventually positive increasing solutions. Both these classes are nonempty [21]. Next we prove a new result which can cover the missing case in [21, Theorem 4], see also Remark 5-(ii) below.

Theorem 5 *Let* $\tau(t) = \int_a^t 1/s \, \Delta s$ *and* $\Psi(t) = t\tau(t)p(t)/r(t)$. *Assume that* $\mu(t) = o(t\tau(t))$ *as* $t \to \infty$,

$$\exists q \in C_{\mathrm{rd}}^1 \text{ such that } q(t) \sim tp(t), \quad \frac{q^{\Delta}(t)}{q(t)} t\tau(t) \to \gamma \text{ as } t \to \infty, \tag{15}$$

and

$$\lim_{t \to \infty} (t\tau(t))^2 \frac{p(t)}{r(t)} = 0. \tag{16}$$

(a) Let $\int_a^\infty 1/r(s)\,\Delta s = \infty$ *and* $\gamma < -1$*. Then any eventually positive decreasing solution* y *of* (14) *satisfies* $t\tau(t)y^\Delta(t)/y(t) \to 0$, $r(t)y^\Delta(t) \to 0$ *as* $t \to \infty$*, and one has:*

(a1) If $\int_a^\infty \Psi(s)\,\Delta s = \infty$*, then*

$$y(t) = \exp\left\{\int_a^t (1+o(1))\frac{\Psi(s)}{\gamma+1}\,\Delta s\right\} \tag{17}$$

and $y(t) \to 0$ *as* $t \to \infty$*.*

(a2) If $\int_a^\infty \Psi(s)\,\Delta s < \infty$*, then*

$$y(t) = \ell_y \exp\left\{-\int_t^\infty (1+o(1))\frac{\Psi(s)}{\gamma+1}\,\Delta s\right\} \tag{18}$$

and $y(t) \to \ell_y \in (0,\infty)$ *as* $t \to \infty$*.*

(b) Let $\int_a^\infty 1/r(s)\,\Delta s < \infty$ *and* $\gamma > -1$*. Then any eventually positive increasing solution* y *of* (14) *satisfies* $t\tau(t)y^\Delta(t)/y(t) \to 0$, $r(t)y^\Delta(t) \to \infty$ *as* $t \to \infty$*, and one has:*

(b1) If $\int_a^\infty \Psi(s)\,\Delta s = \infty$*, then* (17) *holds and* $y(t) \to \infty$ *as* $t \to \infty$*.*
(b2) If $\int_a^\infty \Psi(s)\,\Delta s = \infty$*, then* (18) *holds and* $y(t) \to \ell_y \in (0,\infty)$ *as* $t \to \infty$*.*

Proof Let y be a solution of (14). Set $u(s) = y(t)$, $s = \tau(t)$, where τ can be, at this moment, a general strictly increasing function in $C^1_{\mathrm{rd}}(\mathbb{T})$ with $\widetilde{\mathbb{T}} = \tau(\mathbb{T})$ unbounded from above. Then, similarly as in the proof of Theorem 1, u satisfies the equation

$$(\widetilde{r}(s)u^{\widetilde{\Delta}})^{\widetilde{\Delta}} = \widetilde{p}(s)u^{\widetilde{\sigma}}, \tag{19}$$

where $\widetilde{r} = (r\tau^\Delta)\circ\tau^{-1}$ and $\widetilde{p} = (p/\tau^\Delta)\circ\tau^{-1}$. Applying the substitution method in delta integrals [5, Theorem 1.98], we have, with $S = \tau(T)$ and $\widetilde{a} = \tau(a)$,

$$\int_{\widetilde{a}}^S \frac{1}{\widetilde{r}(s)}\,\widetilde{\Delta}s = \int_{\tau(a)}^{\tau(T)} \left(\frac{1}{r\tau^\Delta}\circ\tau^{-1}\right)(s)\,\widetilde{\Delta}s = \int_a^T \frac{\tau^\Delta(t)}{r(t)\tau^\Delta(t)}\,\Delta t = \int_a^T \frac{1}{r(t)}\,\Delta t.$$

Hence, in particular, $\int_{\widetilde{a}}^\infty 1/\widetilde{r}(s)\,\widetilde{\Delta}s$ converges if and only if $\int_a^\infty 1/r(s)\,\Delta s$ converges. Further, with $s = \tau(t)$, using the chain rule [5, Theorem 1.93] and the formula for delta derivative of the inverse [5, Theorem 1.97], we obtain

$$\begin{aligned}\frac{s(q\circ\tau^{-1})^{\widetilde{\Delta}}(s)}{(q\circ\tau^{-1})(s)} &= \frac{s(q^\Delta\circ\tau^{-1})(s)(\tau^{-1})^{\widetilde{\Delta}}(s)}{(q\circ\tau^{-1})(s)}\\ &= \frac{s(q^\Delta\circ\tau^{-1})(s)}{(q\circ\tau^{-1})(s)(\tau^\Delta\circ\tau^{-1})(s)} = \frac{\tau(t)q^\Delta(t)}{\tau^\Delta(t)q(t)}.\end{aligned} \tag{20}$$

Set $\tau(t)=\int_a^t 1/s\,\Delta s$. Then $\lim_{t\to\infty}\tau(t)=\infty$ by [22, Lemma 2.1] and $\tau^\Delta(t)=1/t$. Hence, in view of (15) and (20),

$$\frac{s(q\circ\tau^{-1})^{\widetilde{\Delta}}(s)}{(q\circ\tau^{-1})(s)}=t\tau(t)\frac{q^\Delta(t)}{q(t)}\to\gamma \tag{21}$$

as $t\to\infty$ (i.e., as $s\to\infty$). By (15), we also have

$$\widetilde{p}(s)=\left(\frac{p}{\tau^\Delta}\circ\tau^{-1}\right)(s)=\tau^{-1}(s)p(\tau^{-1}(s))\sim(q\circ\tau^{-1})(s) \tag{22}$$

as $s\to\infty$. Relations (21) and (22) mean that

$$\widetilde{p}\in\mathcal{RV}_{\widetilde{\mathbb{T}}}(\gamma), \tag{23}$$

where $\mathcal{RV}_{\widetilde{\mathbb{T}}}(\gamma)$ denotes the class of regularly varying functions of index γ on the time scale $\widetilde{\mathbb{T}}$, see, e.g., [21] and also Remark 5-(i). Further,

$$\frac{s^2\widetilde{p}(s)}{\widetilde{r}(s)}=s^2\left(\frac{p}{\tau^\Delta r\tau^\Delta}\right)\circ\tau^{-1}(s)=\left(\frac{\tau(t)}{\tau^\Delta(t)}\right)^2\frac{p(t)}{r(t)}=(t\tau(t))^2\frac{p(t)}{r(t)}.$$

Hence, in view of (16),

$$\lim_{s\to\infty}\frac{s^2\widetilde{p}(s)}{\widetilde{r}(s)}=0. \tag{24}$$

Finally,

$$\frac{\widetilde{\mu}(s)}{s}=\frac{\widetilde{\sigma}(s)-s}{s}=\frac{\widetilde{\sigma}(\tau(t))-\tau(t)}{\tau(t)}=\frac{\tau^\sigma(t)-\tau(t)}{\tau(t)}=\frac{\mu(t)\tau^\Delta(t)}{\tau(t)}=\frac{\mu(t)}{t\tau(t)}\to 0 \tag{25}$$

as $t\to\infty$ (i.e., as $s\to\infty$). Conditions (23)–(25) guarantee that the assumptions of [21, Theorem 4] are fulfilled for Eq. (19), and thus we can apply that result in the next steps.

Consider now case (a), thus we assume that $\gamma<-1$ and $\int_a^\infty 1/r(s)\,\Delta s=\infty$. Take an eventually positive decreasing solution y of (14). Since $u\circ\tau=y$ and $y^\Delta=(u^{\widetilde{\Delta}}\circ\tau)\tau^\Delta$, u is eventually positive decreasing solution of (19). By [21, Theorem 4-(i)], $u\in\mathcal{NSV}_{\widetilde{\mathbb{T}}}$, i.e., u is normalized slowly varying on $\widetilde{\mathbb{T}}$, i.e., $su^{\widetilde{\Delta}}(s)/u(s)\to 0$ as $s\to\infty$. Consequently, $y\circ\tau^{-1}\in\mathcal{NSV}_{\widetilde{\mathbb{T}}}$. Therefore, with using the ideas similar to those in (20),

$$t\tau(t)\frac{y^\Delta(t)}{y(t)}=\frac{s(y\circ\tau^{-1})^{\widetilde{\Delta}}(s)}{(y\circ\tau^{-1})(s)}\to 0$$

as $s\to\infty$. Further, with $S=\tau(T)$ and $\widetilde{a}=\tau(a)$, applying the substitution method in delta integrals, we obtain

$$\int_{\tilde a}^{S} \frac{s\widetilde{p}(s)}{\widetilde{r}(s)}\,\widetilde{\Delta}s = \int_{\tau(a)}^{\tau(T)} s\left(\frac{p}{\tau^\Delta r\tau^\Delta}\right)\circ\tau^{-1}(s)\,\widetilde{\Delta}s$$
$$= \int_a^T \frac{\tau(t)p(t)}{(\tau^\Delta(t))^2 r(t)}\tau^\Delta(t)\,\Delta t = \int_a^T \frac{\tau(t)p(t)}{\tau^\Delta(t)r(t)}\,\Delta t = \int_a^T \Psi(t)\,\Delta t. \tag{26}$$

In particular, $\int_{\tilde a}^\infty s\widetilde{p}(s)/\widetilde{r}(s)\,\widetilde{\Delta}s$ converges if and only if $\int_a^\infty \Psi(t)\,\Delta t$ converges. Assume that $\int_a^\infty \Psi(t)\,\Delta t = \infty$. Then $\int_{\tilde a}^\infty s\widetilde{p}(s)/\widetilde{r}(s)\,\widetilde{\Delta}s = \infty$, and, by [21, Theorem 4-(i)],

$$u(s) = \exp\left\{\int_{\tilde a}^{s} (1+o(1))\frac{\eta\widetilde{p}(\eta)}{(\gamma+1)\widetilde{r}(\eta)}\,\widetilde{\Delta}\eta\right\} \tag{27}$$

and $u(s)\to 0$ as $s\to\infty$. From $y(t)=u(s)$, (26), and (27), we get (17). If $\int_a^\infty \Psi(t)\,\Delta t < \infty$, then $\int_{\tilde a}^\infty s\widetilde{p}(s)/\widetilde{r}(s)\,\widetilde{\Delta}s < \infty$, and, by [21, Theorem 4-(i)],

$$u(s) = \ell_u \exp\left\{-\int_{s}^{\infty} (1+o(1))\frac{\eta\widetilde{p}(\eta)}{(\gamma+1)\widetilde{r}(\eta)}\,\widetilde{\Delta}\eta\right\} \tag{28}$$

as $s\to\infty$ with $\ell_u := \lim_{s\to\infty} u(s) \in (0,\infty)$. Similarly as before, using now (26) and (28), we get (18). Moreover, $\ell_y = \lim_{t\to\infty} y(t) = \lim_{t\to\infty} u(\tau(t)) = \lim_{s\to\infty} u(s) = \ell_u$.

(b) This part can be proved similarly as part (a); we apply [21, Theorem 4-(ii)] to transformed equation (19). □

Remark 5 (i) A closer examination of the proof shows that condition (15) can (equivalently) be replaced by $\tau^{-1}\cdot(p\circ\tau^{-1}) \in \mathcal{RV}_{\widetilde{\mathbb{T}}}(\gamma)$. Here, $\mathcal{RV}_{\mathbb{T}}(\vartheta)$ denotes the class of regularly varying functions of index ϑ on time scale $\mathbb{T}$. An rd-continuous positive function f belongs to $\mathcal{RV}_{\mathbb{T}}(\vartheta)$ if and only if there is $g \in (C^1_{\mathrm{rd}}(\mathbb{T}), (0,\infty))$ such that $f(t)\sim g(t)$ and $tg^\Delta(t)/g(t)\to\vartheta$ as $t\to\infty$, see [21].

(ii) Let us consider Eq. (14) and assume that $p\in\mathcal{RV}_{\mathbb{T}}(\delta)$. Notice that [21, Theorem 4] (which was applied in the previous proof to the transformed equation) requires $\delta\neq -1$. A natural problem is therefore to consider the critical case $\delta=-1$. This setting is somehow delicate and the method of the proof of [21, Theorem 4] does not work. However, the previous theorem enables us to treat this case, as the following example shows. Let $\mu(t)=o(t)$ as $t\to\infty$, $p(t)=L(t)/t$, where $L(t)=(\ln t)^{-2}(\ln(\ln t))^{-\omega}$ with $\omega>0$, and $r(t)=t$. If $\tau(t)$ is as in the theorem, then $\tau(t)\sim\ln t$ as $t\to\infty$, see the next item (iii). We have $p\in\mathcal{RV}_{\mathbb{T}}(-1)$, $\int_a^\infty 1/r(s)\,\Delta s=\infty$, and

$$(t\tau(t))^2\frac{p(t)}{r(t)} \sim (t\ln t)^2\frac{p(t)}{r(t)} = \frac{1}{(\ln(\ln t))^\omega}\to 0$$

as $t\to\infty$. Further, as $s\to\infty$,

$$\tau^{-1}(s)p(\tau^{-1}(s)) = \frac{1}{(\ln\tau^{-1}(s))^2(\ln(\ln\tau^{-1}(s)))^\omega} \sim \frac{1}{s^2\ln^\omega s} \in \mathcal{RV}_{\widetilde{\mathbb{T}}}(-2).$$

In view of the previous item (i), condition (15) is fulfilled with $\gamma = -2 < -1$. Thus Theorem 5-(a) can be applied and we get that any eventually positive decreasing solution y of (14) satisfies $t \ln t\, y^{\Delta}(t)/y(t) \to 0$, $r(t)y^{\Delta}(t) \to 0$ as $t \to \infty$, and obeys one of the formulae (17) or (18), according to whether $\int_a^{\infty} \Psi(s)\,\Delta s = \infty$ or $\int_a^{\infty} \Psi(s)\,\Delta s < \infty$, respectively, where, because of $\tau(t) \sim \ln t$ as $t \to \infty$, the function Ψ can be taken as

$$\Psi(t) = \frac{1}{t \ln t\, (\ln \ln t)^{\omega}}.$$

(iii) The condition $\mu(t) = o(t\tau(t))$ as $t \to \infty$ in Theorem 5, where $\tau(t) = \int_a^t 1/s\,\Delta s$, allows us to cover any time scale with a bounded graininess, for example, $\mathbb{R}$, $\mathbb{Z}$, $h\mathbb{Z}$, or the harmonic numbers $\mathbb{H} := \{\sum_{j=1}^{k-1} 1/j : k \in \mathbb{N}\}$, but also some time scales with an unbounded graininess such as $\mathbb{N}^{\alpha} := \{n^{\alpha} : n \in \mathbb{N}\}$ with $\alpha > 1$ (here we have $\mu(t) = \alpha t^{(\alpha-1)/\alpha} + O(t^{-1/\alpha})$) or the quantum calculus case $q^{\mathbb{N}_0} := \{q^n : n \in \mathbb{N}_0\}$ with $q > 1$ (here we have $\mu(t) = (q-1)t$). If we strengthen the condition $\mu(t) = o(t\tau(t))$ to the condition $\mu(t) = o(t)$ as $t \to \infty$ (which is satisfied, e.g., for $\mathbb{R}$, $\mathbb{Z}$, $h\mathbb{Z}$, $\mathbb{H}$, $\mathbb{N}^{\alpha}$), then $\tau(t) \sim \ln t$ as $t \to \infty$. Indeed, by the time scale L'Hospital rule and the Lagrange mean value theorem, we have

$$\lim_{t\to\infty} \frac{\ln t}{\tau(t)} = \lim_{t\to\infty} \frac{(\ln t)^{\Delta}}{1/t} = \lim_{t\to\infty} \frac{t}{\xi(t)},$$

where $t \le \xi(t) \le \sigma(t)$. Further,

$$1 = \frac{t}{t} \le \frac{\xi(t)}{t} \le \frac{\sigma(t)}{t} = \frac{t + \mu(t)}{t} = 1 + \frac{\mu(t)}{t}.$$

Since $\mu(t) = o(t)$, we get $\lim_{t\to\infty} t/\xi(t) = 1$ and so $\lim_{t\to\infty} \ln t/\tau(t) = 1$ Consequently, the formulae in Theorem 5 can be rewritten as follows: Instead of $\tau(t)$ in $\Psi(t)$, (15), and (16), we can write directly $\ln t$. Finally note that if $\mathbb{T} = q^{\mathbb{N}_0}$, then $\lim_{t\to\infty} \ln t/\tau(t) = \lim_{t\to\infty}[(\ln q)/t]/[(q-1)/t] = \ln q/(q-1)$, and so $\tau(t) \sim (q-1)\ln t/\ln q$ as $t \to \infty$.

(iv) Consider the transformation involving $\tau(t) = \int_a^t 1/s\,\Delta s$, as in the proof of Theorem 5. Then q-difference equations (that is, equations defined on $\mathbb{T} = q^{\mathbb{N}_0}$) are actually transformed into difference equations since $\tau(t) = \int_1^t 1/s\,\Delta s = \sum_{s\in[1,t)_{\mathbb{T}}} (1/s)(q-1)s = \sum_{s\in[1,t)_{\mathbb{T}}} (q-1) = \sum_{j=0}^{k-1} (q-1) = (q-1)k$, where $t = q^k$. Further, difference equations are by the same form of the substitution transformed into dynamic equations on the harmonic numbers. Relations between difference and q-difference equations are utilized in [20, Theorems 7.1–7.4], where asymptotic properties of solutions to q-difference equations under the critical setting are studied.

(v) In the proof of Theorem 5 we have used the transformation involving mapping τ in the special form $\tau(t) = \int_a^t 1/s\,\Delta s$. This suggests an idea to consider a transformation in a general form which could lead to a generalization of the existing results and a refinement of the concept of regular variation on time scales.

(vi) We believe that a combination of suitable transformations (see also the last paragraph in Sect. 4) along with existing results will enable us to examine asymptotics of other equations in critical cases which are close, for instance, to Euler type or Riemann–Weber type equations (see (12)), such as the equation $y^{\Delta\Delta} + p(t)y^{\sigma} = 0$, with $\mu(t) = o(t)$, where $t^2 p(t) \to 1/4$ and $(1/4 - t^2 p(t)) \ln^2 t \to 0$ as $t \to \infty$, and $|1/4 - t^2 p(t)|$ belongs to a suitable subclass of slowly varying functions on $\mathbb{T}$.

Acknowledgements The research has been supported by Brno University of Technology, specific research plan no. FSI-S-17-4464 and by the grant 17-03224S of the Czech Science Foundation. The author would like to thank two anonymous reviewers for their helpful and constructive comments.

References

1. Agarwal, R.P., Bohner, M., Řehák, P.: Half-linear dynamic equations. Nonlinear Analysis and Applications: To V. Lakshmikantham on His 80th Birthday, pp. 1–56. Kluwer Academic Publishers, Dordrecht (2003)
2. Agarwal, R.P., Bohner, M., Grace, S.R., O'Regan, D.: Discrete Oscillation Theory. Hindawi, New York (2005)
3. Ahlbrandt, C.D., Bohner, M., Ridenhour, J.: Hamiltonian systems on time scales. J. Math. Anal. Appl. **250**, 561–578 (2000)
4. Ahlbrandt, C.D., Bohner, M., Voepel, T.: Variable change for Sturm-Liouville differential expressions on time scales. J. Differ. Equ. Appl. **9**, 93–107 (2003)
5. Bohner, M., Peterson, A.: Dynamic Equations on Time Scales. An Introduction with Applications. Birkhäuser, Boston (2001)
6. Bohner, M., Peterson, A.: Advances in Dynamic Equations on Time Scales. Birkhäuser, Boston (2003)
7. Cheng, S.S., Yan, T.C., Li, H.J.: Oscillation criteria for second order difference equation. Funkc. Ekvacioj **34**, 223–239 (1991)
8. Došlý, O., Řehák, P.: Half-Linear Differential Equations. Elsevier, North Holland, Amsterdam (2005)
9. Došlý, O., Ünal, M.: Half-linear differential equations: linearization technique and its application. J. Math. Anal. Appl. **335**, 450–460 (2007)
10. Hinton, D.B., Lewis, R.T.: Spectral analysis of second order difference equations. J. Math. Anal. Appl. **63**, 421–438 (1978)
11. Huff, S., Olumolode, G., Pennington, N., Peterson, A.: Oscillation of an Euler-Cauchy dynamic equation. Discret. Contin. Dyn. Syst. **2003**, 423–431 (2003)
12. Řehák, P.: Oscillatory properties of second order half-linear difference equations. Czechoslov. Math. J. **51**, 303–321 (2001)
13. Řehák, P.: Half-linear dynamic equations on time scales: IVP and oscillatory properties. J. Nonlinear Funct. Anal. Appl. **7**, 361–404 (2002)
14. Řehák, P.: Function sequence technique for half-linear dynamic equations on time scales. Panam. Math. J. **16**, 31–56 (2006)
15. Řehák, P.: How the constants in Hille-Nehari theorems depend on time scales. Adv. Differ. Equ. **2006**, Art. ID 64534 (2006)
16. Řehák, P.: A critical oscillation constant as a variable of time scales for half-linear dynamic equations. Math. Slovaca **60**, 237–256 (2010)
17. Řehák, P.: New results on critical oscillation constants depending on a graininess. Dyn. Syst. Appl. **19**, 271–287 (2010)
18. Řehák, P.: Asymptotic formulae for solutions of linear second-order difference equations. J. Differ. Equ. Appl. **22**, 107–139 (2016)

19. Řehák, P.: Asymptotic formulae for solutions of half-linear differential equations. Appl. Math. Comput. **292**, 165–177 (2017)
20. Řehák, P.: An asymptotic analysis of nonoscillatory solutions of q-difference equations via q-regular variation. J. Math. Anal. Appl. **454**, 829–882 (2017)
21. Řehák, P.: The Karamata integration theorem on time scales and its applications in dynamic and difference equations. Appl. Math. Comput. **338**, 487–506 (2018)
22. Řehák, P., Yamaoka, N.: Oscillation constants for second-order nonlinear dynamic equations of Euler type on time scales. J. Differ. Equ. Appl. **23**, 1884–1900 (2017)
23. Sugie, J., Kita, K.: Oscillation criteria for second order nonlinear differential equations of Euler type. J. Math. Anal. Appl. **253**, 414–439 (2001)
24. Swanson, C.A.: Comparison and Oscillation Theory of Linear Differential Equations. Academic, New York (1968)
25. Voepel, T.: Discrete variable transformations on symplectic systems and even order difference operators. J. Math. Anal. Appl. **220**, 146–163 (1998)
26. Yamaoka, N.: Oscillation criteria for second-order nonlinear difference equations of Euler type. Adv. Differ. Equ. **218** (2012), 14 pp

Stability and Instability Regions for a Three Term Difference Equation

Petr Tomášek

Abstract The paper discusses stability and instability properties of difference equation $y(n+1) + ay(n-\ell+1) + by(n-\ell) = 0$ with real parameters a, b. Beside known results about its asymptotic stability conditions a deeper analysis of instability properties is introduced. An instability degree of difference equation's solution is introduced in analogy with theory of differential equations. Instability regions of a fixed degree are introduced and described in the paper. It is shown that dislocation of instability regions of various degrees obeys some rules and qualitatively depends on parity of difference equation's order.

Keywords Stability · Instability degree · Linear difference equation

1 Introduction

We are going to consider a three term linear difference equation

$$y(n+1) + ay(n-\ell+1) + by(n-\ell) = 0, \qquad n = 0, 1, 2, \ldots, \tag{1}$$

where $a, b \in \mathbb{R}$ and $\ell \in \mathbb{N}$. Our aim is to analyze asymptotic stability and instability conditions for the equation with respect to parameters a, b and ℓ. We recall that a linear difference equation with constant parameters is asymptotically stable if any of its solution tends to zero while n tends to infinity. This property is ensured if and only if all zeros $\lambda_1, \lambda_2, \ldots, \lambda_{\ell+1}$ of the equation's characteristic polynomial

P. Tomášek is supported by the Czech Science Foundation under the grant GA17-03224S: Asymptotic theory of ordinary and fractional differential equations and their numerical discretizations.

P. Tomášek (✉)
Faculty of Mechanical Engineering, Brno University of Technology, Institute of Mathematics, Technická 2896/2, 616 69 Brno, Czech Republic
e-mail: tomasek@fme.vutbr.cz
URL: http://www.vutbr.cz/en/people/petr-tomasek-13984

M. Bohner et al. (eds.), *Difference Equations and Discrete Dynamical Systems with Applications*, Springer Proceedings in Mathematics & Statistics 312,
https://doi.org/10.1007/978-3-030-35502-9_16

$$P(\lambda) = \lambda^{\ell+1} + a\lambda + b \quad (2)$$

lie inside the unit disk in a complex plane.

While the stability conditions are widely analyzed, the instability counterpart analysis remains submarginal. Since we are going to make a more detailed insight to instability properties of (1), we introduce a degree of instability of difference equation (1) in analogy with theory of delay differential equations (see [9, 11]). The following definition is formulated for (1), but it can be analogously considered also for another difference equations.

Definition 1 A number of polynomial (2) roots λ_k counted with their multiplicities, which satisfy $|\lambda_k| > 1$, is called *instability degree* of Eq. (1).

The degree of instability of (1) splits the parameter's plain (a, b) to disjoint domains. The subject of investigation in this paper is a description and some properties of the instability regions of (1).

2 Stability and Instability Regions

We start with a notion of stability region:

Definition 2 Let $\ell \in \mathbb{N}$. The set S_ℓ of all pairs $(a, b) \in \mathbb{R}^2$ for which (1) is asymptotically stable is called *asymptotic stability region*.

Remark 1 A fundamental necessary restriction for asymptotic stability region location is the following: Let (1) be asymptotically stable. Then $|b| < 1$. The assertion is obvious with respect to the fact that $|b|$ is a modulus of product of all the roots of (2). If (1) is asymptotically stable, then all roots of (2) have modulus lower then 1 and hence for modulus of their product it holds $|b| < 1$. From a graphical point of view it means that S_ℓ must be dislocated within the stripe $-1 < b < 1$ in (a, b) plane.

Remark 2 We can also mention a sufficient condition of asymptotic stability, which constructs the so-called Cohn domain of asymptotic stability, which is in the case of (1) in a form $|a| + |b| < 1$. This condition defines an opened square in (a, b) plane with circumradius one and with vertices situated on axes a and b symmetrically.

In a connection with a description of the region S_ℓ, we can recall necessary and sufficient conditions for (1) to be asymptotically stable. In [5] such conditions had been introduced, but as it was later shown, they were incorrect. It was pointed out in [14] that there was some ambiguity in a proof in [4] for a more general case of trinomial difference equation, but similar one depreciates the result in [5]. Correct conditions were later obtained and can be found in various forms in [2, 3, 12, 13]. The last mentioned recent paper also introduces a generalization of difference equation stability notion called $r-$stability. We present the asymptotic stability conditions for (1) in a form which can be obtained as a conclusion of result introduced in [1].

Lemma 1 *Let $a, b \in \mathbb{R}$ and $\ell \in \mathbb{N}$. If ℓ is odd then (1) is asymptotically stable if and only if $|a| < 1 + b$ and either*

$$b - 1 < |a| \leq 1 - b$$

or

$$|a| > |1 - b|, \quad \ell < \frac{\arccos \frac{-a^2+b^2-1}{2|a|}}{\arccos \frac{-a^2-b^2+1}{2|ab|}}.$$

If ℓ is even then (1) is asymptotically stable if and only if $|b| < 1 + a$ and either

$$a - 1 < |b| \leq 1 - a$$

or

$$|b| > |1 - a|, \quad \ell < \frac{\arccos \frac{-a^2+b^2-1}{2|a|}}{\arccos \frac{-a^2-b^2+1}{2|ab|}}.$$

Definition 3 Let $\ell, k \in \mathbb{N}$. The set $I_{\ell,k}$ of all pairs $(a, b) \in \mathbb{R}^2$ for which (1) has degree of instability k is called *region of the kth degree of instability*.

For a detailed description of S_ℓ and $I_{\ell k}$, $k = 1, 2, \dots \ell + 1$ we employ the boundary locus technique. We consider $\lambda = e^{\omega i}$, $\omega \in \mathbb{R}$ as a root of polynomial (2), i.e.

$$e^{(\ell+1)\omega i} + ae^{\omega i} + b = 0.$$

Applying the Euler's rule and considering real and imaginary parts separately we get

$$\cos((\ell + 1)\omega) + a \cos(\omega) + b = 0, \tag{3}$$

$$\sin((\ell + 1)\omega) + a \sin(\omega) = 0, \tag{4}$$

respectively. It is enough to consider $\omega \geq 0$ since the left-hand sides of the above equations are even and odd, respectively. In the sequel we determine representation of curves in (a, b) plane of such pairs of parameter (a, b) for which $P(\lambda^*) = 0$. There are only three possible cases to consider with respect to value of ω:

Case 1. For $\omega = 0$ we have a straight line $b = -a - 1$.

Case 2. For $\omega = m\pi, m \in \mathbb{N}$ Eq. (4) is fulfilled trivially and (3) gives straight lines

$$\begin{aligned} b &= a - 1 \quad \text{for} \quad \ell \quad \text{odd}, \\ b &= a + 1 \quad \text{for} \quad \ell \quad \text{even}. \end{aligned}$$

Case 3. For $\omega \neq r\pi, r \in \mathbb{N}_0$. Equations (4) and (3) give

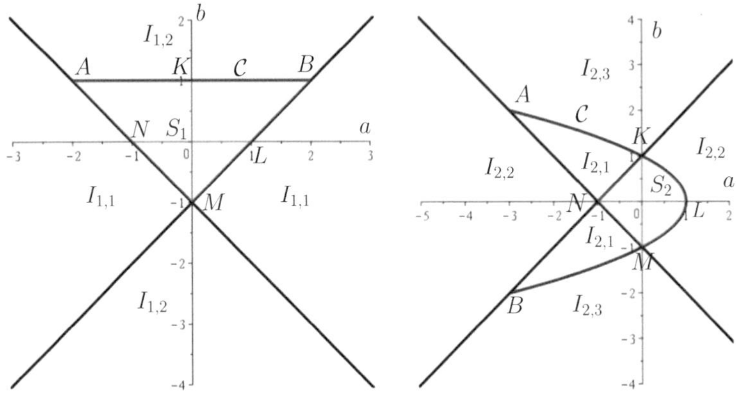

Fig. 1 Stability and instability regions for $\ell = 1$ (left) and $\ell = 2$ (right)

$$a = \frac{-\sin((\ell+1)\omega)}{\sin(\omega)} \tag{5}$$

$$b = \sin((\ell+1)\omega)\cot(\omega) - \cos((\ell+1)\omega) = \frac{\sin(\ell\omega)}{\sin(\omega)}. \tag{6}$$

Considering $\omega \in (0, \pi)$ the above introduced pair of a, b gives parametric expression of a curve $\mathcal{C}$ where some root of $P(\lambda)$ stands on the boundary of the unit circle. The straight lines from cases 1 and 2 together with the curve $\mathcal{C}$ represent the boundary between stability region and regions of various instability degrees.

It is enough to consider $\omega \in (0, \pi)$ to express the boundary curve, since for $\omega \neq r\pi$, $r \in \mathbb{N}_0$ the points of the same curve are obtained. Analyzing the limits of (a, b) for $\omega \to 0^+$ and $\omega \to \pi^-$ we obtain the boundary curve $\mathcal{C}$ endpoints A and B, respectively.

$$A = \lim_{\omega \to 0^+} (a, b) = (-\ell - 1, \ell)$$

and

$$B = \lim_{\omega \to \pi^-} (a, b) = (\ell + 1, \ell) \qquad \text{for } \ell \text{ odd},$$

$$B = \lim_{\omega \to \pi^-} (a, b) = (-\ell - 1, -\ell) \qquad \text{for } \ell \text{ even}.$$

In the following figures the stability and instability regions are introduced. They are separated by a bold curve $\mathcal{C}$ and bold straight lines corresponding to cases 1 and 2. The asymptotic stability region S_ℓ is highlighted by grey color (Figs. 1 and 2).

To have a better insight into the regions dislocation we introduce some of their properties.

Lemma 2 *The axis b is a part of the asymptotic stability region S_ℓ for $b \in (-1, 1)$ and of instability region of the highest degree $I_{\ell,\ell+1}$ for $b \notin [-1, 1]$.*

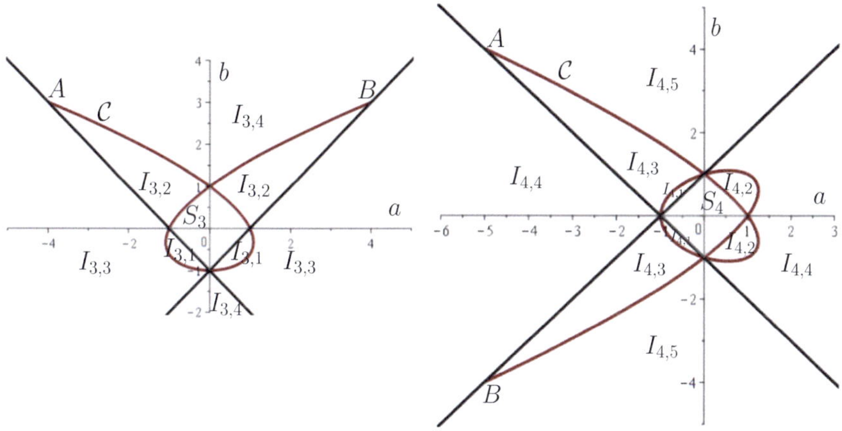

Fig. 2 Stability and instability regions for $\ell = 3$ (left) and $\ell = 4$ (right)

Proof In the case of $a = 0$ the characteristic polynomial $P(\lambda)$ of (1) has the form

$$P(\lambda) = \lambda^{\ell+1} + b$$

which all $\ell + 1$ roots have modulus $|b|^{1/(\ell+1)}$. □

A similar situation occurs for the axis a:

Lemma 3 *The axis a is a part of the asymptotic stability region S_ℓ for $a \in (-1, 1)$ and of instability region of the second highest degree $I_{\ell,\ell}$ for $a \notin [-1, 1]$.*

Proof In the case of $b = 0$ the characteristic polynomial $P(\lambda)$ of (1) has the form

$$P(\lambda) = \lambda^{\ell+1} + a\lambda,$$

which has one root $\lambda = 0$ and all the other ℓ roots have modulus $|a|^{1/\ell}$. □

Now we move our attention to another property, which can be observed from the above figures. The bold straight lines represent the pairs (a, b) for which a real root of modulus one occurs in (2). Thence there is a change just for 1 degree of instability between the neighboring regions on the segments of these lines, where there is no intersection with another curve of boundary. Similarly, on the curve segments of $\mathcal{C}$ free of any intersections points with another part of region boundary a switch for two degrees of instability is expected between the neighboring regions since two complex conjugate roots with unit modulus are present. There is sketched a situation for general odd and even ℓ at Figs. 3 and 4. It can be also observed that curve $\mathcal{C}$ is dislocated within two stripes of width $\sqrt{2}$ which long axes coincide with quadrants symmetry axes. In another words $\mathcal{C} \in T$, where $T = \{(a, b) \in \mathbb{R}^2 : (-a - 1 \le b \le -a + 1) \vee (a - 1 \le b \le a + 1)\}$. This observation can be formulated and proved as follows:

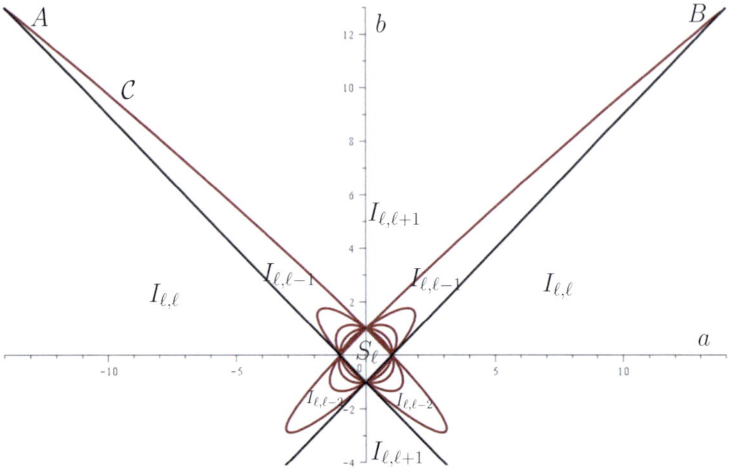

Fig. 3 Stability and instability regions for odd $\ell = 13$

Fig. 4 Stability and instability regions for even $\ell = 14$

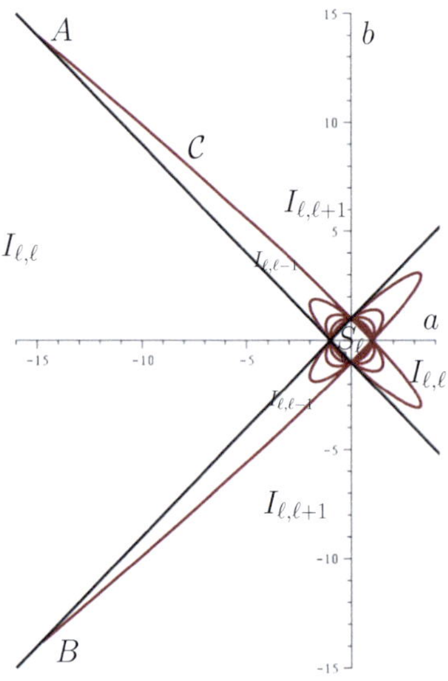

Theorem 1 *Let $\mathcal{C}$ be curve in plane (a, b) defined by (5), (6), where $\omega \in (0, \pi)$, $\ell \in \mathbb{N}$. Then $\mathcal{C} \subset T$, where $T = \{(a, b) \in \mathbb{R}^2, |b| \leq |a| + 1, |a| \leq |b| + 1\}$.*

Proof From (6) we get $|b| = |a\cos(\omega) + \cos((\ell + 1)\omega)| \leq |a\cos(\omega)| + |\cos((\ell + 1)\omega)|$ which gives $|b| \leq |a| + 1$.

To show $|a| \leq |b| + 1$ we consider (5) which implies $|a| = \left|\frac{\sin((\ell+1)\omega)}{\sin(\omega)}\right| = |b\cos(\omega) + \cos(\ell\omega)| \leq |b\cos(\omega)| + |\cos(\ell\omega)| \leq |b| + 1$. □

Corollary 1 *Let $U = \{(a, b) \in \mathbb{R}^2, |b| > |a| + 1\}$. Then $U \subset I_{\ell,\ell+1}$.*

Proof The assertion follows from Lemma 2 and Theorem 1 with respect to dislocation of instability regions boundaries. □

Corollary 2 *Let* $V = \{(a, b) \in \mathbb{R}^2, |a| > |b| + 1\}$. *Then* $V \subset I_{\ell,\ell}$.

Proof The assertion follows from Lemma 3 and Theorem 1 with respect to dislocation of instability regions boundaries. □

From the above assertions it follows that for instability regions up to instability degree $\ell - 1$ it holds $I_{\ell,k} \subset T$, $k = 1, 2, \ldots, \ell - 1$. Naturally, the same holds for the stability region: $S_\ell \subset T$. A general instability degree regions portrait can be assembled in the (a, b) plane with respect to the previous considerations.

Instability regions can be determined by their boundary, which consists of the curve $\mathcal{C}$ segments and eventually segments of straight lines $b = -a - 1, b = -a + 1$ and $b = a - 1$. From the parametric Eq. (5), (6) of the curve $\mathcal{C}$ it follows that crossing the curve $\mathcal{C}$ in the point (a, b) in the direction (a, b) (faraway from the origin of the plane) the instability degree rises for two in the points where $\mathcal{C}$ does not intersect itself. In such points the curve $\mathcal{C}$ has just two simple complex conjugate roots. The fact that the conjugate pair of the roots crosses the unit disk boundary outwards follows immediately from the parametric expression of the curve in (a, b). Indeed, consider roots of the characteristic polynomial $P(\lambda)$ on the circle with a general radius ρ, i.e. $\lambda = \rho \exp(i\omega)$. Then using the same steps as in the case of the curve $\mathcal{C}$ description we introduce curve $\mathcal{C}_\subset$, which is connecting pairs (a, b) where roots of $P(\lambda)$ with modulus ρ are presented. Parametric expression of $\mathcal{C}_\subset$ is then

$$a = -\rho^\ell \frac{\sin((\ell + 1)\omega)}{\sin(\omega)}$$
$$b = \rho^{\ell+1} \frac{\sin(\ell\omega)}{\sin(\omega)}.$$

where $\omega \in (0, \pi)$. Considering fixed parameters ω and ℓ a positive perturbation of modulus ρ in any pair $(a, b) \in \mathcal{C}$ moves this point away from the origin in a direction of $(\ell a, (\ell + 1)b)$ vector.

For description of instability regions we split the curve $\mathcal{C}$ to 2ℓ curve segments $\mathcal{C}_i$ given by (5), (6) with $\omega \in J_i = [(i - 1)\frac{\pi}{2\ell}, i\frac{\pi}{2\ell}]$, $i = 1, 2, \ldots, 2\ell$. Next we denote $K = (0, 1)$, $L = (1, 0)$, $M = (0, -1)$, $N = (-1, 0)$ points in (a, b) plane (see Fig. 1). These points are boundary points of $\mathcal{C}_i$, $i = 1, 2, \ldots, 2\ell$ curve segments. Notice that they are the only points where $\mathcal{C}$ can intersect itself. The sequence of these points along $\mathcal{C}$ by increasing $\omega \in [0, \pi]$ is $A, K, L, M, [: N, K, L, M :]_{\ell/2-1}, B$ for ℓ even and $A, [: K, L, M, N :]_{(\ell-1)/2}, K, B$ for ℓ odd, where subsequence between symbols $[: \ :]_s$ repeats s times.

Let us consider $\mathcal{C}$ as a function $b = f(a)$ in a suitable neighbourhood of points K, L, M, N. Local analysis of this function enables us to determine which segments $\mathcal{C}_i$ bound the considered instability regions. Particularly the slopes of consequential segments $\mathcal{C}_i, \mathcal{C}_{i+1}$ connected in these points give the sequence of boundaries we cross going along the axis of appropriate quadrant away from the origin. We illustrate the analysis in the point K. In this point there are connected neighbour segment pairs $(\mathcal{C}_{4k+1}, \mathcal{C}_{4k+2}), k = 0, 1, 2, \ldots, k_f$, where $k_f = (\ell - 1)/2$ for ℓ odd and $k_f = \ell/2 - 1$ for ℓ even. For these segment pairs the corresponding points K representations on $\mathcal{C}$ are in $\omega = \omega_k := \frac{(4k+1)\pi}{2\ell}, k = 0, 1, \ldots, k_f$, respectively. From (5), (6) we get

$$f'(a(\omega)) = \frac{\ell \cos(\ell\omega)\sin(\omega) - \sin(\ell\omega)\cos(\omega)}{\sin((\ell+1)\omega) - (\ell+1)\cos((\ell+1)\omega)}. \tag{7}$$

Particularly for $\omega = \omega_k$ we obtain the values representing the slopes of $\mathcal{C}$ in the point K as $f'(a(\omega_k)) = -\cos(\omega_k)/(1 + \ell\sin^2(\omega_k)), \quad k = 0, 1, 2, \ldots, k_f$. Now considering continuous function $\vartheta(u) = -\cos(u)/(1 + \ell\sin^2(u)), \ u \in (0, \pi)$, we have $\vartheta'(u) = (1 + \ell\cos^2(u) + \ell)\sin(u)/(\ell\cos^2(u) - 1 - \ell)^2 > 0$ for $u \in (0, \pi)$. Since $\vartheta(u)$ is increasing in $u \in (0, \pi)$, the studied sequence of the slopes is increasing too. On that account moving along the axis of the first quadrant from the origin we cross the segments $\mathcal{C}_{4k+2}, k = 0, 1, 2, \ldots, k_f$ in sequence. We recall that each crossing corresponds to the instability degree shift by 2. This gives (in restriction to the first quadrant) that $I_{\ell,2k+2}$ is bounded by $\mathcal{C}_{4k+2}$ and $\mathcal{C}_{4k+6}, k = 0, 1, 2, \ldots, k_f - 1$. In the case of odd ℓ the last segment $\mathcal{C}_{4k_f+6}$ connects K with B and therefore the region $I_{\ell,\ell-1}$ is bounded by $\mathcal{C}_{2\ell}, \mathcal{C}_{2\ell-4}$ and straight line segment BL. On the other hand, the increasing slope sequence in K gives in the second quadrant case conclusion that moving along the axis of the quadrant away from the origin we cross the segments $\mathcal{C}_{4k+1}, k = k_f, k_f - 1, \ldots, 2, 1, 0$ in sequence. Summarizing the previous analysis and considering analogous steps in other points L, M, N enables us to formulate the survey of stability and instability regions given by their boundary.

Theorem 2 *Consider Eq. (1), where $a, b \in \mathbb{R}$ and ℓ is odd integer. Then the stability region S_ℓ boundary and instability regions boundaries are given by the sets of curves*

S_ℓ	$\{\mathcal{C}_2; LM; MN; \mathcal{C}_{2\ell-1}\}$
$I_{\ell,1}$	$\{MN; \mathcal{C}_4\}, \{LM; \mathcal{C}_{2\ell-3}\}$
$I_{\ell,p}$, even $p = 2, 4, \ldots, \ell - 3$	$\{\mathcal{C}_{2p-2}; \mathcal{C}_{2p+2}\}, \{\mathcal{C}_{2\ell-2p+3}; \mathcal{C}_{2\ell-2p-1}\}$
$I_{\ell,m}$, odd $m = 3, 5, \ldots, \ell - 2$	$\{\mathcal{C}_{\ell-2m+6}; \mathcal{C}_{\ell-2m+2}\}, \{\mathcal{C}_{\ell+2m-5}; \mathcal{C}_{\ell+2m-1}\}$
$I_{\ell,\ell-1}$	$\{\mathcal{C}_1; \mathcal{C}_5; AN\}, \{\mathcal{C}_{2\ell}; \mathcal{C}_{2\ell-4}; BL\}$
$I_{\ell,\ell}$	$\{b = -a - 1, a \in (-\infty, -1]; \mathcal{C}_{2\ell-2}; b = a - 1, a \in (-\infty, 0]\}$ $\{b = -a - 1, a \in [0, \infty); \mathcal{C}_3; b = a - 1, a \in [1, \infty)\}$
$I_{\ell,\ell+1}$	$\{b = -a - 1, a \in (-\infty, -\ell - 1]; \mathcal{C}_1; \mathcal{C}_{2\ell}; b = a - 1, a \in [\ell + 1, \infty)\}$

Notice that the stability and instability regions dislocation is symmetric with respect to the axis b for ℓ odd.

Theorem 3 *Consider Eq. (1), where $a, b \in \mathbb{R}$ and ℓ is even integer. Then the stability region S_ℓ boundary and instability regions boundaries are given by the sets of curves*

S_ℓ	$\{C_2; C_{2\ell-1}; MN; NK\}$
$I_{\ell,1}$	$\{MN; C_4\}, \{NK; C_{2\ell-3}\}$
$I_{\ell,p}$, even $p = 2, 4, \dots, \ell - 2$	$\{C_{2p-2}; C_{2p+2}\}, \{C_{2\ell-2p+3}; C_{2\ell-2p-1}\}$
$I_{\ell,m}$, odd $m = 3, 5, \dots, \ell - 3$	$\{C_{2m-2}; C_{2m+2}\}, \{C_{2\ell-2m+3}; C_{2\ell-2m-1}\}$
$I_{\ell,\ell-1}$	$\{C_1; C_5; AN\}, \{C_{2\ell}; C_{2\ell-4}; NB\}$
$I_{\ell,\ell}$	$\{a = -\lvert b\rvert - 1\}$, $\{b = -a - 1, a \in [0, \infty); C_3; C_{2\ell-2}; b = a + 1, a \in [0, \infty)\}$
$I_{\ell,\ell+1}$	$\{b = -a - 1, a \in (-\infty, -\ell - 1]; C_1; b = a + 1, a \in [0, \infty)\}$ $\{b = -a - 1, a \in [0, \infty); C_{2\ell}; b = a + 1, a \in (-\infty, -\ell - 1]\}$

Notice that the stability and instability regions dislocation is symmetric with respect to the axis a for ℓ even.

3 Final Remarks

As it was remarked in the introduction, instability degree regions are not investigated in the literature as wide as the stability regions are, especially in the case of difference equations. As it was shown above, the dislocation of $I_{\ell,k}$ regions obey some rules.

Notice that with respect to the structure of linear difference equations solution there exists a periodic solution of (1) for any point $[a, b]$ from the curve C and from relevant straight line boundaries of instability regions, where a $P(\lambda)$ root with modulus one occurs. Deeper analysis considering this phenomena can be found in [6, 7], where a more general difference system was analyzed from the periodic solution existence point of view.

The introduced considerations have also an impact to the theory of polynomials: a dislocation of pairs (a, b) for which the polynomial (2) has a fixed number of roots inside the unit disk in complex plane, is introduced. There are several kinds of algebraic criteria to determine number of polynomial roots in specified area of complex plane. The description of such criteria including their proofs can be found in [8] or [10]. Most common are questions about location of all characteristic polynomial roots with respect to the left half-plane and unit circle in study of stability of differential equations and difference equations, respectively. But these criteria also enable us to determine a number of roots inside the specified area and outside of it. On that account we can numerically determine the instability degree of studied equation for the given parameters. Then we can develop an instability regions portrait in a computational way. But the above discussion presents another approach: it gives some rules of stability and instability regions dislocation of Eq. (1) in analytical way using boundary locus technique and supplementary considerations. The author believes that research of instability regions dislocation properties can be interesting and fruitful also in another cases of difference equations.

References

1. Čermák, J., Jánský, J., Kundrát, P.: On necessary and sufficient conditions for the asymptotic stability of higher order linear difference equations. J. Differ. Equ. Appl. **18**(11), 1781–1800 (2012)
2. Čermák, J., Jánský, J.: Explicit stability conditions for a linear trinomial delay difference equation. Appl. Math. Lett. **43**, 56–60 (2015)
3. Cheng, S.S., Huang, S.Y.: Alternate derivations of the stability region of a difference equation with two delays. Appl. Math. E-Notes **9**, 225–253 (2009)
4. Dannan, F.: The asymptotic stability of $x(n+k) + ax(n) + bx(n-l) = 0$. J. Differ. Equ. Appl. **10**(6), 589–599 (2004)
5. Dannan, F., Elaydi, S.: Asymptotic stability of linear difference equations of advanced type. J. Comput. Anal. Appl. **6**(2), 173–187 (2004)
6. Györi, I., Horváth, L.: Existence of periodic solutions in al linear higher order system of difference equations. Comput. Math. Appl. **66**, 2239–2250 (2013)
7. Györi, I., Horváth, L.: Utilization of the circulant matrix theory in periodic higher order autonomous difference equations. Int. J. Differ. Equ. **9**, 163–185 (2014)
8. Jury, E.I.: Inners and Stability of Dynamic Systems. Krieger, Malabar (1982)
9. Kolmanovskii, V., Myshkis, A.: Introduction to the Theory and Applications of Functional Differential Equations. Mathematics and Its Applications, vol. 463. Kluwer Academic Publishers, Dordrecht (1999)
10. Marden, M.: Geometry of Polynomials. Mathematical Surveys and Monographs, vol. 3. American Mathematical Society, Providence (1966)
11. Michiels, W., Niculescu, S.I.: Stability and Stabilization of Time-Delay Systems: An Eigenvalue-Based Approach. Advances in Design and Control. SIAM, Philadelphia (2007)
12. Kipnis, M.M., Nigmatulin, R.: Stability of the trinomial linear difference equations with two delays. Autom. Remote Control **65**(11), 1710–1723 (2004)
13. Kipnis, M.M., Nigmatulin, R.: D-decomposition method for stability checking for trinomial linear difference equation with two delays. Int. J. Pure Appl. Math. **111**(3), 479–489 (2016)
14. Ren, H.: Stability analysis of second order delay difference equations. Funkcial. Ekvac. **50**, 405–419 (2007)

MIX
Papier aus verantwortungsvollen Quellen
Paper from responsible sources
FSC® C105338

If you have any concerns about our products,
you can contact us on
ProductSafety@springernature.com

In case Publisher is established outside the EU,
the EU authorized representative is:
Springer Nature Customer Service Center GmbH
Europaplatz 3, 69115 Heidelberg, Germany

Printed by Libri Plureos GmbH
in Hamburg, Germany